P9-CEC-042

Mean square within

$$MSW = \frac{SSW}{dfw}$$

Mean square between

$$MSB = \frac{SSB}{dfb}$$

F ratio

$$F = \frac{MSB}{MSW}$$

CHAPTER 11

Chi square

$$\chi^2 \text{ (obtained)} = \sum \frac{(f_o - f_e)^2}{f_e}$$

CHAPTER 13

Phi

$$\phi = \sqrt{\frac{\chi^2}{N}}$$

Cramer's *V*

$$V = \sqrt{\frac{\chi^2}{(N)(\text{Minimum of } r - 1, c - 1)}}$$

Lambda

$$\lambda = \frac{E_1 - E_2}{E_1}$$

CHAPTER 14

Gamma

$$G = \frac{N_s - N_d}{N_s + N_d}$$

Spearman's rho

$$r_s = 1 - \frac{6 \sum D^2}{N(N^2 - 1)}$$

CHAPTER 15

Least-squares regression line

$$Y = a + bX$$

Slope

$$b = \frac{\Sigma(X - \overline{X})(Y - \overline{Y})}{\Sigma(X - \overline{X})^2}$$

Y intercept

$$a = \overline{Y} - b\overline{X}$$

Pearson's *r*

$$r = \frac{\Sigma(X - \overline{X})(Y - \overline{Y})}{\sqrt{[\Sigma(X - \overline{X})^2][\Sigma(Y - \overline{Y})^2]}}$$

CHAPTER 17

Partial correlation coefficient

$$r_{yx.z} = \frac{r_{yx} - (r_{yz})(r_{xz})}{\sqrt{1 - r_{yz}^2}\sqrt{1 - r_{xz}^2}}$$

Least-squares multiple regression line

$$Y = a + b_1X_1 + b_2X_2$$

Partial slope for X_1

$$b_1 = \left(\frac{s_y}{s_1}\right)\left(\frac{r_{y1} - r_{y2}r_{12}}{1 - r_{12}^2}\right)$$

Partial slope for X_2

$$b_2 = \left(\frac{s_y}{s_2}\right)\left(\frac{r_{y2} - r_{y1}r_{12}}{1 - r_{12}^2}\right)$$

Y intercept

$$a = Y - b_1X_1 - b_2X_2$$

Beta-weight for X_1

$$b_1^* = b_1\left(\frac{s_1}{s_y}\right)$$

Beta-weight for X_2

$$b_2^* = b_2\left(\frac{s_2}{s_y}\right)$$

Standardized least-squares regression line

$$Z_y = b_1^*Z_1 + b_2^*Z_2$$

Coefficient of multiple determination

$$R^2 = r_{y1}^2 + r_{y2.1}^2(1 - r_{y1}^2)$$

STATISTICS

A Tool for Social Research

STATISTICS

A Tool for Social Research

Eighth Edition

Joseph F. Healey

Christopher Newport University

WADSWORTH
CENGAGE Learning™

Australia • Brazil • Japan • Korea • Mexico
Singapore • Spain • United Kingdom
United States

WADSWORTH
CENGAGE Learning™

Statistics: A Tool for Social Research, Eighth Edition
Joseph F. Healey

Publisher: Michele Sordi

Acquisitions Editor: Chris Caldeira

Assistant Editor: Christina Beliso

Editorial Assistant: Erin Parkins

Marketing Manager: Michelle Williams

Marketing Communications Manager: Linda Yip

Project Manager, Editorial Production: Matt Ballantyne

Creative Director: Rob Hugel

Art Director: Caryl Gorska

Print Buyer: Karen Hunt

Permissions Editor: Tim Sisler

Production Service: Newgen

Copy Editor: Elliot Simon

Cover Designer: Yvo

Cover Image: GettyImages/Photodisc/Paul Burns; GettyImages/Photodisc/Paul Burns; GettyImages/Digital Vision; GettyImages/Stockbyte; GettyImages/Westend61/i2 Stock; GettyImages/Photodisc; GettyImages/Digital Vision; GettyImages/Lauren Nicole; GettyImages/Digital Vision RF; GettyImages/Photodisc/Connie Coleman; iStockPhoto/June and Aaron Photography

Compositor: Newgen

Printed in the United States of America
1 2 3 4 5 6 7 12 11 10 09 08

Library of Congress Control Number: 2007936259

ISBN-13: 978-0-495-09655-9

ISBN-10: 0-495-09655-5

Wadsworth Cengage Learning
10 Davis Drive
Belmont, CA 94002-3098
USA

Cengage Learning products are represented in Canada by Nelson Education, Ltd.

Brief Contents

Detailed Contents

Preface

Statistics are part of the everyday language of sociology and the other social sciences (including political science, social work, public administration, criminal justice, urban studies, and gerontology). These disciplines are research-based and routinely use statistics to express knowledge and to discuss theory and research. To join the conversations being conducted in these disciplines, you must be literate in the vocabulary of research, data analysis, and scientific thinking. Knowledge of statistics will enable you to understand the professional research literature, conduct quantitative research yourself, contribute to the growing body of social science knowledge, and reach your full potential as a social scientist.

Although essential, learning (and teaching) statistics can be a challenge. Students in social science statistics courses typically bring with them a wide range of mathematical backgrounds and an equally diverse set of career goals. They are often puzzled about the relevance of statistics for them, and, not infrequently, there is some math anxiety to deal with. This text introduces statistical analysis for the social sciences while addressing these realities.

The text makes minimal assumptions about mathematical background (the ability to read a simple formula is sufficient preparation for virtually all of the material in the text), and a variety of special features help students analyze data successfully. The text has been written especially for sociology and social work programs but is sufficiently flexible to be used in any program with a social science base.

The text is written at a level intermediate between a strictly mathematical approach and a mere "cookbook." I emphasize interpretation and understanding statistics in the context of social science research, but I have not sacrificed comprehensive coverage or statistical correctness. Mathematical explanations are kept at an elementary level, as is appropriate in a first exposure to social statistics. For example, I do not treat formal probability theory per se in the text.[1] Rather, the background necessary for an understanding of inferential statistics is introduced, informally and intuitively, in Chapters 5 and 6 while considering the concepts of the normal curve and the sampling distribution.

The text does not claim that statistics are "fun" or that the material can be mastered without considerable effort. At the same time, students are not overwhelmed with abstract proofs, formula derivations, or mathematical theory, which can needlessly frustrate the learning experience at this level.

GOAL OF THE TEXT AND CHANGES IN THIS EDITION

The goal of this text is to develop basic statistical literacy. The statistically literate person understands and appreciates the role of statistics in the research process, is competent to perform basic calculations, and can read and appreciate the professional research literature in his or her field as well as any research reports he or she may encounter in everyday life. This goal has not changed since

[1]A presentation of probability is available at the Web site for this text for those who are interested.

the first edition of this text. However, in recognition of the fact that "mere computation" has become less of a challenge in this high-tech age, this edition continues to increase the stress on interpretation and computer applications while deemphasizing computation. This will be apparent in several ways. For example, the feature called "Interpreting Statistics" has been continued and updated. These noncomputational sections are included in about half of the chapters and present detailed examples of "what to say after the statistics have been calculated." They use real data and real research situations to illustrate the process of developing meaning and understanding, and they exemplify how statistics can be used to answer important questions. The issues addressed in these sections include workplace surveillance, the gender gap in income, and the correlates of street crime.

Also, in recognition of the fact that modern technology has rendered long, tedious hand calculation obsolete, the end-of-chapter problems feature smaller, easier-to-handle data sets, although more challenging problems are also included in the chapters and in the Cumulative Exercises at the end of each part. Computer applications using SPSS are presented at the end of virtually every chapter. The data set used in the computer exercises has been updated to the 2006 General Social Survey.

This edition continues to focus on the development of basic statistical literacy, the three aspects of which provide a framework for discussing the additional features of this text.

1. An Appreciation of Statistics. A statistically literate person understands the relevance of statistics for social research, can analyze and interpret the meaning of a statistical test, and can select an appropriate statistic for a given purpose and a given set of data. This textbook develops these qualities, within the constraints imposed by the introductory nature of the course, in the following ways.

- *The relevance of statistics.* Chapter 1 includes a discussion of the role of statistics in social research and stresses their usefulness as ways of analyzing and manipulating data and answering research questions. Each example problem is framed in the context of a research situation. A question is posed and then, with the aid of a statistic, answered. The relevance of statistics for answering questions is thus stressed throughout the text. This central theme of usefulness is further reinforced by a series of boxes labeled "Application," each of which illustrates some specific way statistics can be used to answer questions.

  Almost all end-of-chapter problems are labeled by the social science discipline or subdiscipline from which they are drawn: [SOC] for sociology, [SW] for social work, [PS] for political science, [CJ] for criminal justice, [PA] for public administration, and [GER] for gerontology. By identifying problems with specific disciplines, students can more easily see the relevance of statistics to their own academic interests. (Not incidentally, they will also see that the disciplines have a large subject matter in common.)

- *Interpreting statistics.* For most students, interpretation—saying what statistics mean—is a big challenge. The ability to interpret statistics can be developed only by exposure and experience. To provide exposure, I have been careful, in the example problems, to express the meaning of the statistic in terms of the original research question. To provide experience, the end-of-chapter problems almost always call for an interpretation of the statistic calculated. To provide examples, many of the Answers to Odd-Numbered Computational Problems in the back of the text are expressed in words as well as numbers. The "Interpreting Statistics" sections provide additional, detailed examples of how to express the meaning of statistics.

- *Using statistics: Ideas for research projects.* Appendix E offers ideas for independent data-analysis projects for students. The projects require students to use SPSS to analyze a data set. They can be assigned at intervals throughout the semester or at the end of the course. Each project provides an opportunity for students to practice and apply their statistical skills and, above all, to exercise their ability to understand and interpret the meaning of the statistics they produce.

2. Computational Competence. Students should emerge from their first course in statistics with the ability to perform elementary forms of data analysis—to execute a series of calculations and arrive at the correct answer. While computers and calculators have made computation less of an issue today, computation is inseparable from statistics, so I have included a number of features to help students cope with these mathematical challenges.

- *"One Step at a Time" boxes* have been added to this edition. For every statistic, this important new feature breaks computation down into individual steps for maximum clarity and ease.
- *Extensive problem sets* are provided at the end of each chapter. For the most part, these problems use fictitious data and are designed for ease of computation.
- *Cumulative exercises* are included at the end of each part to provide practice in choosing, computing, and analyzing statistics. These exercises present only data sets and research questions. Students must choose appropriate statistics as part of the exercise.
- *Solutions* to odd-numbered computational problems are provided so that students may check their answers.
- *SPSS for Windows* continues as a feature (as in previous editions), to give students access to the computational power of the computer. This is explained in more detail later.

3. The Ability to Read the Professional Social Science Literature. The statistically literate person can comprehend and critically appreciate research reports written by others. The development of this skill is a particular problem at the introductory level since (1) the vocabulary of professional researchers is so much more concise than the language of the textbook, and (2) the statistics featured in the literature are more advanced than those covered at the introductory level. To help bridge this gap, I have included, beginning in Chapter 1, a series of boxes labeled "Reading Statistics." In each box, I briefly describe the reporting style typically used for the statistic in question and try to alert students about what to expect when they approach the professional literature. These inserts have been updated in this edition and include excerpts from the research literature and illustrate how statistics are actually applied and interpreted by social scientists.

Additional Features. A number of other features make the text more meaningful for students and more useful for instructors.

- *Readability and clarity.* The writing style is informal and accessible to students without ignoring the traditional vocabulary of statistics. Problems and examples have been written to maximize student interest and to focus on issues of concern and significance. For the more difficult material (such as hypothesis testing), students are first walked through an example problem before being confronted by formal terminology and concepts. Each chapter ends with a summary of major points and formulas and a glossary of important concepts. A list of frequently used formulas inside the front cover and a glossary of symbols inside the back cover can be used for quick reference.
- *Organization and coverage.* The text is divided into four parts, with most of the coverage devoted to univariate descriptive statistics, inferential statistics, and bivariate measures of association. The distinction between description and inference is introduced in the first chapter and maintained throughout the text. In selecting statistics for inclusion, I have tried to strike a balance between the essential concepts with which students must be familiar and the amount of material students can reasonably be expected to learn in their first (and perhaps only) statistics course, all the while bearing in mind that different instructors will naturally wish to stress different aspects of the subject. Thus, the text covers a full gamut of the usual statistics, with each chapter broken into subsections so that instructors may choose the particular statistics they wish to include. In this edition, the text has been shortened and streamlined by moving some infrequently used techniques and statistical procedures to the companion Web site.
- *Learning objectives.* Learning objectives are stated at the beginning of each chapter. These are intended to serve as "study guides" and to help students identify and focus on the most important material.
- *Review of mathematical skills.* A comprehensive review of all of the mathematical skills that will be used in this text is included. To increase visibility and usefulness, in this edition this section has been moved from an appendix to the Prologue. Students who are inexperienced or out of practice with mathematics can study this review early in the course and/or refer to it as needed. A self-test is included so that students can check their level of preparation for the course.
- *Statistical techniques and end-of-chapter problems are explicitly linked.* After a technique is introduced, students are directed to specific problems for practice and review. The "how-to-do-it" aspects of calculation are reinforced immediately and clearly.
- *End-of-chapter problems are organized progressively.* Simpler problems with small data sets are presented first. Often, explicit instructions or hints accompany the first several problems in a set. The problems gradually become more challenging and require more decision making by the student (e.g., choosing the most appropriate statistic for a certain situation). Thus, each problem set develops problem-solving abilities gradually and progressively.
- *Computer applications.* This text integrates SPSS, the leading social science statistics package, to help students take advantage of the power of the com-

puter. Appendix F provides an introduction to SPSS. The demonstrations at the end of each chapter explain how to use the statistical package to produce the statistics presented in the chapter. Student exercises analyzing data with SPSS are also included. The student version of SPSS is available as a supplement to this text.

- *Realistic, up-to-date data.* The database for computer applications in the text is a shortened version of the 2006 General Social Survey. This database will give students the opportunity to practice their statistical skills on "real-life" data. The database is described in Appendix G.
- *Companion Web Site.* The Web site for this text includes additional material and some less frequently used techniques, a table of random numbers, hypothesis tests for ordinal-level variables, "find-the-test" flowcharts, and a number of other features.
- *Instructor's Manual/Testbank.* The Instructor's Manual includes chapter summaries, a test item file of multiple-choice questions, answers to even-numbered computational problems, and step-by-step solutions to selected problems. In addition, the Instructor's Manual includes cumulative exercises (with answers) that can be used for testing purposes.
- *Study Guide.* The Study Guide, written by Professor Roy Barnes, contains additional examples to illuminate basic principles, review problems with detailed answers, SPSS projects, and multiple-choice questions and answers that complement but do not duplicate the test item file.

Summary of Key Changes in this Edition. The following are the most important changes in this edition.

- "One Step at a Time" boxes have been added.
- The sections of some chapters (e.g., Section 6.3 and Section 8.6) have been divided into subsections for ease of comprehension.
- Diagrams, tables, and graphs have been added at key points to present the material visually and to enhance understanding (for an example, see the beginning of Chapter 3).
- Sections on open-ended and unequal intervals in frequency distributions have been added to Chapter 2.
- The data set used in the text has been updated to the 2006 General Social Survey. Likewise, the data cited in the text and in end-of-chapter problems have been updated.
- Computational formulas have been deleted from Chapter 10 (ANOVA) and Chapter 15 (correlation and regression).
- A section on "dummy" variables has been added to Chapter 15.
- The coverage of direct, spurious, intervening, and interactive relationships in Chapter 17 has been expanded for those classes that do not cover Chapter 16.
- There is more emphasis on interpretation throughout the text.
- The Reading Statistics inserts have been updated.

The text has been thoroughly reviewed for clarity and readability. As with previous editions, my goal is to offer a comprehensive, flexible, and student-oriented book that will provide a challenging first exposure to social statistics.

ACKNOWLEDGMENTS

This text has been in development, in one form or another, for over 25 years. An enormous number of people have made contributions, both great and small, to this project and, at the risk of inadvertently omitting someone, I am bound to at least attempt to acknowledge my many debts.

This edition reflects the thoughtful guidance of both Bob Jucha and Chris Caldeira of Wadsworth, and I thank them for their contributions. Much of whatever integrity and quality this book has is a direct result of the very thorough (and often highly critical) reviews that have been conducted over the years. I am consistently impressed by the sensitivity of my colleagues to the needs of the students, and, for their assistance in preparing this edition, I would like to thank the following reviewers: Akbar Aghajanian, *Fayetteville State University;* Robert Bausch, *Cameron University;* Beata J. Breg, *CUNY, Queens College;* Gregory A. Guagnano, *George Mason University;* Matt G. Mutchler, *University of California, Dominguez Hills;* Lisa Pellerin, *Ball State University;* Anastasia Prokos, *University of Nevada, Las Vegas.* Whatever failings are contained in the text are, of course, my responsibility and are probably the result of my occasional decisions not to follow the advice of my colleagues.

I would like to thank the instructors who made statistics understandable to me (Professors Satoshi Ito, Noelie Herzog, and Ed Erikson) and all of my colleagues at Christopher Newport University for their support and encouragement (especially Professors F. Samuel Bauer, Cheryl Chambers, Robert Durel, Marcus Griffin, Mai Lan Gustafsson, Ruth Kernodle, Michael Lewis, Marion Manton, Lea Pellet, Eduardo Perez, and Linda Waldron). I owe a special debt of gratitude to Professor Roy Barnes of the University of Michigan—Flint for his help with the "Interpreting Statistics" feature, and I would be very remiss if I did not acknowledge the constant support and excellent assistance of Iris Price of Christopher Newport University. Also, I thank all of my students for their patience and thoughtful feedback, and I am grateful to the Literary Executor of the late Sir Ronald A. Fisher, F.R.S., to Dr. Frank Yates, F.R.S., and to Longman Group Ltd., London, for permission to reprint Appendixes B, C, and D from their book *Statistical Tables for Biological, Agricultural and Medical Research* (6th edition, 1974).

Finally, I want to acknowledge the support of my family and rededicate this work to them. I have the extreme good fortune to be a member of an extended family that is remarkable in many ways and that continues to increase in size. Although I cannot list everyone, I would especially like to thank the older generation (my mother, Alice T. Healey), the next generation (my sons, Kevin and Christopher, my daughters-in-law, Jennifer and Jessica), the new members (my wife, Patricia Healey, and Christopher, Katherine, and Jennifer Schroen), and the youngest generation (Benjamin and Caroline Healey).

Prologue: Basic Mathematics Review

You will probably be relieved to hear that this text, your first exposure to statistics for social science research, is not particularly mathematical and does not stress computation per se. While you will encounter many numbers to work with and numerous formulas to use, the major emphasis will be on understanding the role of statistics in research and the logic by which we attempt to answer research questions empirically. You will also find that, in this text, the example problems and many of the homework problems have been intentionally simplified so that the computations will not unduly distract you from the task of understanding the statistics themselves.

On the other hand, you may regret to learn that there is, inevitably, some arithmetic that you simply cannot avoid if you want to master this material. It is likely that some of you haven't had any math in a long time, others have convinced themselves that they just cannot do math under any circumstances, and still others are just rusty and out of practice. All of you will find that mathematical operations that might seem complex and intimidating at first glance can be broken down into simple steps. If you have forgotten how to cope with some of these steps or are unfamiliar with these operations, this prologue is designed to ease you into the skills you will need to do all of the computations in this textbook. Also, you can use this section for review whenever you feel uncomfortable with the mathematics in the chapters to come.

CALCULATORS AND COMPUTERS

A calculator is a virtual necessity for this text. Even the simplest, least expensive model will save you time and effort and is definitely worth the investment. However, I recommend that you consider investing in a more sophisticated calculator with memories and preprogrammed functions, especially the statistical models that can compute means and standard deviations automatically. Calculators with these capabilities are available for less than $20.00 and will almost certainly be worth the small effort it will take to learn to use them.

In the same vein, there are several computerized statistical packages (or **statpaks**) commonly available on college campuses that can further enhance your statistical and research capabilities. The most widely used of these is the Statistical Package for the Social Sciences (**SPSS**). This program comes in a student version that is available bundled with this text (for a small fee).[1] Statistical packages like SPSS are many times more powerful than even the most sophisticated handheld calculators, and it will be well worth your time to learn how to use them because they will eventually save you time and effort. SPSS is introduced in Appendix F of this text, and exercises at the end of almost every

[1]Another statpak, called MicroCase, is available free and can be downloaded from the Web site for this text. All of the computer exercises in this text are available in MicroCase format at the Web site.

chapter will show you how to use the program to generate and interpret the statistics covered in the chapter

There are many other programs to help you accomplish the goal of generating accurate statistical results with a minimum of effort and time. Even spreadsheet programs such as Microsoft Excel, which is included in many versions of Microsoft Office, have some statistical capabilities. You should be aware that all of these programs (other than the simplest calculators) will require some effort to learn, but the rewards will be worth the effort.

In summary, you should find a way at the beginning of this course—with a calculator, a statpak, or both—to minimize the tedium and hassle of mere computing. This will permit you to devote maximum effort to the truly important goal of increasing your understanding of the meaning of statistics in particular and social research in general.

VARIABLES AND SYMBOLS

Statistics are a set of techniques by which we can describe, analyze, and manipulate **variables.** A variable is a trait that can change value from case to case or from time to time. Examples of variables include height, weight, level of prejudice, and political party preference. The possible values or scores associated with a given variable might be numerous (for example, income) or relatively few (for example, gender). I will often use symbols, usually the letter X, to refer to variables in general or to a specific variable.

Sometimes we will need to refer to a specific value or set of values of a variable. This is usually done with the aid of subscripts. So the symbol X_1 (read "X-sub-one") would refer to the first score in a set of scores, X_2 ("X-sub-two") to the second score, and so forth. Also, we will use the subscript i to refer to all the scores in a set. Thus, the symbol X_i ("X-sub-eye") refers to all of the scores associated with a given variable (for example, the test grades of a particular class).

OPERATIONS

You are all familiar with the four basic mathematical operations of addition, subtraction, multiplication, and division and the standard symbols (+, −, ×, ÷) used to denote them. I should remind you that multiplication and division can be symbolized in a variety of ways. For example, the operation of multiplying some number *a* by some number *b* may be symbolized in (at least) six different ways:

$$a \times b$$
$$a \cdot b$$
$$a * b$$
$$ab$$
$$a(b)$$
$$(a)(b)$$

In this text, we will commonly use the "adjacent symbols" format (that is, *ab*), the conventional times sign (×), or adjacent parentheses to indicate multiplication. On most calculators and computers, the asterisk (*) is the symbol for multiplication.

The operation of division can also be expressed in several different ways. In this text, we will use either of these two methods:

$$a/b \quad \text{or} \quad \frac{a}{b}$$

Several of the formulas with which we will be working require us to find the square of a number. To do this, multiply the number by itself. This operation is symbolized as X^2 (read "X squared"), which is the same thing as $(X)(X)$. If X has a value of 4, then

$$X^2 = (X)(X) = (4)(4) = 16$$

or we could say that "4 squared is 16."

The square root of a number is the value that, when multiplied by itself, results in the original number. So the square root of 16 is 4 because (4)(4) is 16. The operation of finding the square root of a number is symbolized as

$$\sqrt{X}$$

A final operation with which you should be familiar is summation, or the addition of the scores associated with a particular variable. When a formula requires the addition of a series of scores, this operation is usually symbolized as ΣX_i. Σ is the uppercase Greek letter sigma and stands for "the summation of." So the combination of symbols ΣX_i means "the summation of all the scores" and directs us to add the value of all the scores for that variable. If four people had family sizes of 2, 4, 5, and 7, then the summation of these four scores for this variable could be symbolized as

$$\Sigma X_i = 2 + 4 + 5 + 7 = 18$$

The symbol Σ is an operator, just like the + and $\times$ signs. It directs us to add all of the scores on the variable indicated by the X symbol.

There are two other common uses of the summation sign, and, unfortunately, the symbols denoting these uses are not, at first glance, sharply different from each other or from the symbol used earlier. A little practice and some careful attention to these various meanings should minimize the confusion. The first set of symbols is ΣX_i^2, which means "the sum of the squared scores." This quantity is found by *first* squaring each of the scores and *then* adding the squared scores together. A second common set of symbols will be $(\Sigma X_i)^2$, which means "the sum of the scores, squared." This quantity is found by *first* summing the scores and *then* squaring the total.

These distinctions might be confusing at first, so let's see if an example helps to clarify the situation. Suppose we had a set of three scores: 10, 12, and 13. So

$$X_i = 10,\ 12,\ 13$$

The sum of these scores would be indicated as

$$\Sigma X_i = 10 + 12 + 13 = 35$$

The sum of the squared scores would be

$$(\Sigma X_i)^2 = (10)^2 + (12)^2 + (13)^2 = 100 + 144 + 169 = 413$$

Take careful note of the order of operations here. First the scores are squared one at a time and then the squared scores are added. This is a completely different operation from squaring the sum of the scores:

$$(\Sigma X_i)^2 = (10 + 12 + 13)^2 = (35)^2 = 1225$$

To find this quantity, first the scores are summed and then the total of all the scores is squared. The squared sum of the scores (1225) is *not* the same as the sum of the squared scores (413). In summary, the operations associated with each set of symbols can be summarized as follows:

Symbols	Operations
ΣX_i	Add the scores
ΣX_i^2	First square the scores and then add the squared scores
$(\Sigma X_i)^2$	First add the scores and then square the total

OPERATIONS WITH NEGATIVE NUMBERS

A number can be either positive (if it is preceded by a + sign or by no sign at all) or negative (if it is preceded by a − sign). Positive numbers are greater than zero, and negative numbers are less than zero. It is very important to keep track of signs because they will affect the outcome of virtually every mathematical operation. This section briefly summarizes the relevant rules for dealing with negative numbers.

First, adding a negative number is the same as subtraction. For example,

$$3 + (-1) = 3 - 1 = 2$$

Second, subtraction changes the sign of a negative number:

$$3 - (-1) = 3 + 1 = 4$$

Note the importance of keeping track of signs here. If you neglected to change the sign of the negative number in the second expression, you would get the wrong answer.

For multiplication and division, you need to be aware of various combinations of negative and positive numbers. Ignoring the case of all positive numbers, this leaves several possible combinations. A negative number times a positive number results in a negative value:

$$(-3)(4) = -12$$

or

$$(3)(-4) = -12$$

A negative number multiplied by a negative number is always positive:

$$(-3)(-4) = 12$$

Division follows the same patterns. If there is a negative number in the calculations, the answer will be negative. If both numbers are negative, the answer will be positive. So

$$\frac{-4}{2} = -2$$

and

$$\frac{4}{-2} = -2$$

but

$$\frac{-4}{-2} = 2$$

Negative numbers do not have square roots, since multiplying a number by itself cannot result in a negative value. Squaring a negative number always results in a positive value (see the multiplication rules earlier).

ACCURACY AND ROUNDING OFF

A possible source of confusion in computation involves the issues of accuracy and rounding off. People work at different levels of accuracy and precision and, for this reason alone, may arrive at different answers to problems. This is important because our answers can be at least slightly different if you work at one level of precision and I (or your instructor or your study partner) work at another. You may sometimes think you've gotten the wrong answer when all you've really done is round off at a different place in the calculations or in a different way.

There are two issues here: when to round off and how to round off. In this text, I have followed the convention of working in as much accuracy as my calculator or statistics package will allow and then rounding off to two places of accuracy (two places beyond, or to the right of, the decimal point) only at the very end. If a set of calculations is lengthy and requires the reporting of intermediate sums or subtotals, I will round the subtotals off to two places as I go.

In terms of how to round off, begin by looking at the digit immediately to the right of the last digit you want to retain. If you want to round off to 100ths (two places beyond the decimal point), look at the digit in the 1000ths place (three places beyond the decimal point). If that digit is 5 or more, round up. For example, 23.346 would round off to 23.35. If the digit to the right is less than 5, round down. So, 23.343 would become 23.34.

Let's look at some more examples of how to follow these rules of rounding. If you are calculating the mean value of a set of test scores and your calculator shows a final value of 83.459067 and you want to round off to two places, look at the digit three places beyond the decimal point. In this case the value is 9 (greater than 5), so we would round the second digit beyond the decimal point up and report the mean as 83.46. If the value had been 83.453067, we would have reported our final answer as 83.45.

FORMULAS, COMPLEX OPERATIONS, AND THE ORDER OF OPERATIONS

A mathematical formula is a set of directions, stated in general symbols, for calculating a particular statistic. To "solve a formula," you replace the symbols with the proper values and then perform a series of calculations. Even the most complex formula can be simplified by breaking the operations down into smaller steps. Working through these steps requires some knowledge of general procedure and the rules of precedence of mathematical operations. This is because the order in which you perform calculations may affect your final answer. Consider the following expression:

$$2 + 3(4)$$

If you add first, you will evaluate the expression as

$$5(4) = 20$$

but if you multiply first, the expression becomes

$$2 + 12 = 14$$

Obviously, it is crucial to complete the steps of a calculation in the correct order.

The basic rules of precedence are to find all squares and square roots first, then do all multiplication and division, and finally complete all addition and subtraction. So the expression

$$8 + 2 \times 2^2/2$$

would be evaluated as

$$8 + 2 \times \frac{4}{2} = 8 + \frac{8}{2} = 8 + 4 = 12$$

The rules of precedence may be overridden by parentheses. Solve all expressions within parentheses before applying the rules stated earlier. For most of the complex formulas in this text, the order of calculations will be controlled by the parentheses. Consider the following expression:

$$(8 + 2) - 4(3)^2/(8 - 6)$$

Resolving the parenthetical expressions first, we would have

$$(10) - 4 \times 9/(2) = 10 - 36/2 = 10 - 18 = -8$$

Without the parentheses, the same expression would be evaluated as

$$\begin{aligned} &8 + 2 - 4 \times 3^2/8 - 6 \\ &= 8 + 2 - 4 \times 9/8 - 6 \\ &= 8 + 2 - 36/8 - 6 \\ &= 8 + 2 - 4.5 - 6 \\ &= 10 - 10.5 \\ &= -0.5 \end{aligned}$$

A final operation you will encounter in some formulas in this text involves denominators of fractions that themselves contain fractions. In this situation, solve the fraction in the denominator first and then complete the division. For example,

$$\frac{15 - 9}{6/2}$$

would become

$$\frac{15 - 9}{6/2} = \frac{6}{3} = 2$$

When you are confronted with complex expressions such as these, don't be intimidated. If you're patient with yourself and work through them step by step, beginning with the parenthetical expression, even the most imposing formulas can be managed.

EXERCISES

You can use the following problems as a "self-test" on the material presented in this review. If you can handle these problems, you're ready to do all of the arithmetic in this text. If you have difficulty with any of these problems, please review the appropriate section of this prologue. You might also want to use this section as an opportunity to become more familiar with your calculator. The answers are given immediately following these exercises, along with commentary and some reminders.

1. Complete each of the following:
a. $17 \times 3 =$
b. $17(3) =$
c. $(17)(3) =$
d. $17/3 =$
e. $(42)^2 =$
f. $\sqrt{113} =$

2. For the set of scores (X_i) of 50, 55, 60, 65, and 70, evaluate each of the following expressions:
$\Sigma X_i =$
$\Sigma X_i^2 =$
$(\Sigma X_i)^2 =$

3. Complete each of the following:
a. $17 + (-3) + (4) + (-2) =$
b. $15 - 3 - (-5) + 2 =$
c. $(-27)(54) =$
d. $(113)(-2) =$
e. $(-14)(-100) =$
f. $-34/-2 =$
g. $322/-11 =$
h. $\sqrt{-2} =$
j. $(-17)^2 =$

4. Round off each of the following to two places beyond the decimal point:
a. 17.17532
b. 43.119
c. 1076.77337
d. 32.4641152301
e. 32.4751152301

5. Evaluate each of the following:
a. $(3 + 7)/10 =$
b. $3 + 7/10 =$
c. $\dfrac{(4 - 3) + (7 + 2)(3)}{(4 + 5)(10)} =$
d. $\dfrac{22 + 44}{15/3} =$

ANSWERS TO EXERCISES

1. a. 51 **b.** 51 **c.** 51

(The obvious purpose of these first three problems is to remind you that there are several different ways of expressing multiplication.)

d. 5.67 (Note the rounding off.) **e.** 1764 **f.** 10.63

2. The first expression translates to "the sum of the scores," so this operation would be

$$\Sigma X_i = 50 + 55 + 60 + 65 + 70 = 300$$

The second expression is the "sum of the squared scores." So

$$\Sigma X_i^2 = (50)^2 + (55)^2 + (60)^2 + (65)^2 + (70)^2$$
$$\Sigma X_i^2 = 2500 + 3025 + 3600 + 4225 + 4900$$
$$\Sigma X_i^2 = 18{,}250$$

The third expression is "the sum of the scores, squared":

$$(\Sigma X_i)^2 = (50 + 55 + 60 + 65 + 70)^2$$
$$(\Sigma X_i)^2 = (300)^2$$
$$(\Sigma X_i)^2 = 90{,}000$$

Remember that ΣX_i^2 and $(\Sigma X_i)^2$ are two completely different expressions with very different values.

3. a. 16 **b.** 19 (Remember to change the sign of −5.) **c.** −1458

d. −226 **e.** 1400 **f.** 17 **g.** −29.27

h. Your calculator probably gave you some sort of error message for this problem, since negative numbers do not have square roots.

i. 289

4. a. 17.17 **b.** 43.12 **c.** 1076.77

d. 32.46 **e.** 32.48

5. a. 1 **b.** 3.7 (Note again the importance of parentheses.) **c.** 0.31

d. 13.2

1 Introduction

LEARNING OBJECTIVES

By the end of this chapter, you will be able to

1. Describe the limited but crucial role of statistics in social research.
2. Distinguish between three applications of statistics (univariate descriptive, bivariate descriptive, and inferential) and identify situations in which each is appropriate.
3. Distinguish between discrete and continuous variables and cite examples of each.
4. Identify and describe three levels of measurement and cite examples of variables from each.

1.1 WHY STUDY STATISTICS?

Students sometimes approach their first course in statistics with questions about the value of the subject matter. What, after all, do numbers and statistics have to do with understanding people and society? In a sense, this entire book will attempt to answer this question, and the value of statistics will become clear as we move from chapter to chapter. For now, the importance of statistics can be demonstrated, in a preliminary way, by briefly reviewing the research process as it operates in the social sciences. These disciplines are scientific, in the sense that social scientists attempt to verify their ideas and theories through research. Broadly conceived, **research** is any process by which information is systematically and carefully gathered for the purpose of answering questions, examining ideas, or testing theories. Research is a disciplined inquiry that can take numerous forms. Statistical analysis is relevant only for those research projects where the information collected is represented by numbers. Numerical information is called **data**, and the sole purpose of statistics is to manipulate and analyze data. **Statistics**, then, are a set of mathematical techniques used by social scientists to organize and manipulate data for the purpose of answering questions and testing theories.

What is so important about learning how to manipulate data? On one hand, some of the most important and enlightening works in the social sciences do not utilize statistics at all. There is nothing magical about data and statistics. The mere presence of numbers guarantees nothing about the quality of a scientific inquiry. On the other hand, data can be the most trustworthy information available to the researcher and, consequently, deserve special attention. Data that have been carefully collected and thoughtfully analyzed are the strongest, most objective foundations for building theory and enhancing understanding. Without a firm base in data, the social sciences would lose the right to the name *science* and would be of far less value.

Thus, the social sciences rely heavily on data analysis for the advancement of knowledge. Let us be very clear about one point: It is never enough merely to gather data (or, for that matter, any kind of information). Even the most objective and carefully collected numerical information does not and cannot speak for itself. The researcher must be able to use statistics effectively to organize, evaluate, and analyze the data. Without a good understanding of the principles of statistical analysis, the researcher will be unable to make sense of the data. Without the appropriate application of statistical techniques, the data will remain mute and useless.

Statistics are an indispensable tool for the social sciences. They provide the scientist with some of the most useful techniques for evaluating ideas, testing theory, and discovering the truth. The next section describes the relationships between theory, research, and statistics in more detail.

1.2 THE ROLE OF STATISTICS IN SCIENTIFIC INQUIRY

Figure 1.1 graphically represents the role of statistics in the research process. The diagram is based on the thinking of Walter Wallace and illustrates how the knowledge base of any scientific enterprise grows and develops. One point the diagram makes is that scientific theory and research continually shape each other. Statistics are one of the most important means by which research and theory interact. Let's take a closer look at the wheel.

Since the figure is circular, it has no beginning or end, and we could start our discussion at any point. For the sake of convenience, let's begin at the top and follow the arrows around the circle. A **theory** is an explanation of the relationships between phenomena. People naturally (and endlessly) wonder about problems in society (like prejudice, poverty, child abuse, and serial murders), and, in their attempt to understand these phenomena, they develop explanations ("low levels of education cause prejudice"). This kind of informal "theorizing" about society is no doubt very familiar to you, but, in contrast to our informal, everyday explanations, scientific theory is subject to a rigorous testing process. Let's take the problem of racial prejudice as an example to illustrate how the research process works.

What causes racial prejudice? One possible answer to this question is provided by a theory called the *contact hypothesis*. This theory was stated over 50 years ago by the social psychologist Gordon Allport, and it has been tested on

FIGURE 1.1 THE WHEEL OF SCIENCE

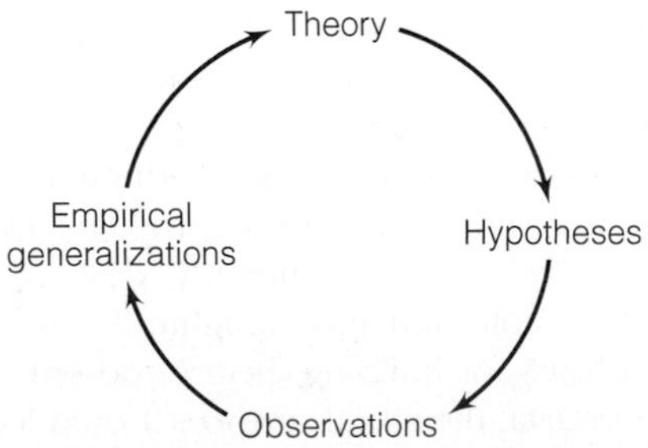

Source: Adapted from Walter Wallace, *The Logic of Science in Sociology* (Chicago: Aldine-Atherton, 1971).

a number of occasions since that time.[1] The theory links prejudice to the volume and nature of interaction between members of different racial groups. Specifically, the hypothesis states that contact situations in which different groups have equal status and are engaged in cooperative behavior will result in a reduction of prejudice on all sides. The more equal and cooperative the contact, the more likely people will see each other as individuals and not as representatives of a particular group. For example, the contact hypothesis predicts that members of a racially mixed athletic team that cooperate with each other to achieve victory would tend to experience a decline in prejudice. On the other hand, when different groups compete for jobs, housing, or other valuable resources, prejudice will tend to increase.

The contact hypothesis is not a complete explanation of prejudice, of course, but it will serve to illustrate a sociological theory. This theory offers an explanation for the relationship between two social phenomena: (1) prejudice and (2) equal-status, cooperative contact between members of different groups. People who have little equal-status contact will be more prejudiced, and those with more equal-status contact will be less prejudiced.

Before moving on, let's examine theory in a little more detail. The contact hypothesis, like most theories, is stated in terms of causal relationships between variables. A **variable** is any trait that can change values from case to case. Examples of variables would be gender, age, income, and political party affiliation. In any specific theory, some variables will be identified as causes and others will be identified as effects or results. In the language of science, the causes are called **independent variables** and the effects or result variables are called **dependent variables**. In our theory, equal-status contact would be the independent variable (or the cause) and prejudice would be the dependent variable (the result or effect). In other words, we are arguing that an individual's level of prejudice depends on (or is caused by) the extent to which he or she participates in equal-status, cooperative contacts with other groups.

A diagram such as the following can be a useful way of representing the relationships between variables.

Equal Status Contact → Prejudice
Independent variable → Dependent Variable
$X \rightarrow Y$

The arrow represents the direction of the causal relationship, and "X" and "Y" are general symbols for the independent and dependent variables, respectively.

So far, we have a theory of prejudice and independent and dependent variables. What we don't know yet is whether the theory is true or false. To find out, we need to compare our theory with the facts; we need to do some research. The next steps in the process would be to define our terms and ideas more specifically and exactly. One problem we often face in doing research is

[1]Allport, Gordon, 1954. *The Nature of Prejudice.* Reading, Massachusetts: Addison-Wesley. For recent attempts to test this theory, see: McLaren, Lauren. 2003. "Anti-Immigrant Prejudice in Europe: Contact, Threat Perception, and Preferences for the Exclusion of Migrants." *Social Forces,* 81:909–937; Pettigrew, Thomas. 1997. "Generalized Intergroup Contact Effects on Prejudice." *Personality and Social Psychology Bulletin.* 23:173–185, and Sigelman, Lee and Susan Welch. 1993. "The Contact Hypothesis Revisited: Black–White Interaction and Positive Racial Attitudes." *Social Forces.* 71:781–795.

that scientific theories are too complex and abstract to be fully tested in a single research project. To conduct research, one or more hypotheses must be derived from the theory. A **hypothesis** is a statement about the relationship between variables that, while logically derived from the theory, is much more specific and exact.

For example, if we wished to test the contact hypothesis, we would have to say exactly what we mean by prejudice and we would need to describe "equal-status, cooperative contact" in great detail. We would also review the research literature to help develop and clarify our definitions and our understanding of these concepts. As our definitions develop and the hypotheses take shape, we begin the next step of the research process, during which we will decide exactly how we will gather our data. We must decide how cases will be selected and tested, how exactly the variables will be measured, and a host of related matters. Ultimately, these plans will lead to the observation phase (the bottom of the wheel of science), where we actually measure social reality. Before we can do this, we must have a very clear idea of what we are looking for and a well-defined strategy for conducting the search.

To test the contact hypothesis, we would begin with people from different racial or ethnic groups. We might place some subjects in situations that required them to cooperate with members of other groups and other subjects in situations that feature intergroup competition. We would need to measure levels of prejudice before and after each type of contact. We might do this by administering a survey that asked subjects to agree or disagree with statements such as "Greater efforts must be made to racially integrate the public school system" and "Skin color is irrelevant and people are just people." Our goal would be to see if the people exposed to the cooperative contact situation actually become less prejudiced.

Now, finally, we come to statistics. As the observation phase of our research project comes to an end, we will be confronted with a large collection of numerical information or data. If our sample consisted of 100 people, we would have 200 completed surveys measuring prejudice: 100 completed before the contact situation and 100 filled out afterwards. Try to imagine dealing with 200 completed surveys. If we had asked each respondent just five questions to measure their prejudice, we would have a total of 1000 separate pieces of information to deal with. What do we do? We have to have some systematic way to organize and analyze this information; at this point, statistics will become very valuable. Statistics will supply us with many ideas about "what to do" with the data, and we will begin to look at some of the options in the next chapter. For now, let me stress two points about statistics.

First, statistics are crucial. Simply put, without statistics, quantitative research is impossible. Without quantitative research, the development of the social sciences would be severely impaired. Only by the application of statistical techniques can mere data help us shape and refine our theories and understand the social world better. Second, and somewhat paradoxically, the role of statistics is rather limited. As Figure 1.1 makes clear, scientific research proceeds through several mutually interdependent stages. Statistics become directly relevant only at the end of the observation stage. Before any statistical analysis can be legitimately applied, the preceding phases of the process must have been successfully completed. If the researcher has asked poorly conceived questions or has made serious errors of design or method, then even the most sophisticated statistical analysis is valueless. As useful as they can be, statistics cannot substitute for rigorous conceptualization, detailed and careful planning, or creative use of theory. Statistics cannot salvage a poorly conceived or designed research project. They cannot make sense out of garbage.

On the other hand, inappropriate statistical applications can limit the usefulness of an otherwise carefully done project. Only by successfully completing *all* phases of the process can a quantitative research project hope to contribute to understanding. A reasonable knowledge of the uses and limitations of statistics is as essential to the education of the social scientist as is training in theory and methodology.

As the statistical analysis comes to an end, we would begin to develop empirical generalizations. While we would be primarily focused on assessing our theory, we would also look for other trends in the data. Assuming that we found that equal-status, cooperative contact reduces prejudice in general, we might go on to ask if the pattern applies to males as well as to females, to the well educated as well as to the poorly educated, to older respondents as well as to the younger. As we probed the data, we might begin to develop some generalizations based on the empirical patterns we observe. For example, what if we found that contact reduced prejudice for younger respondents but not for older respondents? Could it be that younger people are less "set in their ways" and have attitudes and feelings that are more open to change? As we developed tentative explanations, we would begin to revise or elaborate our theory.

If we change the theory to take account of these findings, however, a new research project designed to test the revised theory is called for, and the wheel of science would begin to turn again. We (or perhaps some other researchers) would go through the entire process once again with this new—and, hopefully, improved—theory. This second project might result in further revisions and elaboration that would (you guessed it) require still more research projects, and the wheel of science would continue to turn as long as scientists were able to suggest additional revisions or develop new insights. Every time the wheel turned, our understandings of the phenomena under consideration would (hopefully) improve.

This description of the research process does not include white-coated, clipboard-carrying scientists who, in a blinding flash of inspiration, discover some fundamental truth about reality and shout, "Eureka!" The truth is that, in the normal course of science, it is a rare occasion when we can say with absolute certainty that a given theory or idea is definitely true or false. Rather, evidence for (or against) a theory will gradually accumulate over time, and ultimate judgments of truth will likely be the result of many years of hard work, research, and debate.

Let's briefly review our imaginary research project. We began with an idea or theory about intergroup contact and racial prejudice. We imagined some of the steps we would have to take to test the theory and took a quick look at the various stages of the research project. We wound up back at the level of theory, ready to begin a new project guided by a revised theory. We saw how theory can motivate a research project and how our observations might cause us to revise the theory and, thus, motivate a new research project. Wallace's wheel of science illustrates how theory stimulates research and how research shapes theory. This constant interaction between theory and research is the lifeblood of science and the key to enhancing our understandings of the social world.

The dialog between theory and research occurs at many levels and in multiple forms. Statistics are one of the most important links between these two realms. Statistics permit us to analyze data, to identify and probe trends and relationships, to develop generalizations, and to revise and improve our theories. As you will see throughout this text, statistics are limited in many ways. They are also an indispensable part of the research enterprise. Without statistics, the interaction between theory and research would become extremely difficult, and the progress of our disciplines would be severely retarded. *(For practice in describing the relationship between theory and research and the role of statistics in research, see problems 1.1 and 1.2.)*

1.3 THE GOALS OF THIS TEXT

In the preceding section, I argued that statistics are a crucial part of social science research and that every social scientist needs some training in statistical analysis. In this section, we address the questions of how much training is necessary and what the purposes of that training are. First, this textbook takes the point of view that statistics are tools. They can be a very useful means of increasing our knowledge of the social world, but they are not ends in themselves. Thus, we will not take a "mathematical" approach to the subject, although we will cover enough material so that you can develop a basic understanding of why statistics "do what they do." Instead, statistics will be presented as tools that can be used to answer important questions, and our focus will be on how these techniques are applied in the social sciences.

Second, all of you will soon become involved in advanced coursework in your major fields of study, and you will find that much of the literature used in these courses assumes at least basic statistical literacy. Furthermore, many of you, after graduation, will find yourselves in positions—either in a career or in graduate school—where some understanding of statistics will be very helpful or perhaps even required. Very few of you will become statisticians per se (and this text is not intended for the preprofessional statistician), but you must have a grasp of statistics in order to read and critically appreciate your own professional literature. As a student in the social sciences and in many careers related to the social sciences, you simply cannot realize your full potential without a background in statistics.

Within these constraints, this textbook is an introduction to statistics as they are utilized in the social sciences. The general goal of the text is to develop an appreciation—a "healthy respect"—for statistics and their place in the research process. You should emerge from this experience with the ability to use statistics intelligently and to know when other people have done so. You should be familiar with the advantages and limitations of the more commonly used statistical techniques, and you should know which techniques are appropriate for a given set of data and a given purpose. Lastly, you should develop sufficient statistical and computational skills and enough experience in the interpretation of statistics to be able to carry out some elementary forms of data analysis by yourself.

1.4 DESCRIPTIVE AND INFERENTIAL STATISTICS

As noted earlier, the general function of statistics is to manipulate data so that research question(s) can be answered. Two general classes of statistical techniques, depending on the research situation, are available to accomplish this task; both are introduced in this section.

Descriptive Statistics. The first class of techniques, called **descriptive statistics**, is relevant in several different situations:

1. When a researcher needs to summarize or describe the distribution of a single variable. These statistics are called *univariate* ("one variable") descriptive statistics.
2. When the researcher wishes to describe the relationship between two or more variables. These statistics are called *bivariate* ("two variable") or *multivariate* (more than two variable) descriptive statistics.

To describe a single variable, we would arrange the values or scores of that variable so that the relevant information can be quickly understood and appreciated. Many of the statistics that might be appropriate for this summarizing task are probably familiar to you. For example, percentages, graphs, and charts can all be used to describe single variables.

To illustrate the usefulness of univariate descriptive statistics, consider the following problem: Suppose you wanted to summarize the distribution of the variable "family income" for a community of 10,000 families. How would you do it? Obviously, you couldn't simply list all incomes in the community and let it go at that. Imagine trying to make sense of an unorganized list of 10,000 different incomes! You would want to develop some summary measures of the overall distribution of incomes—perhaps an arithmetic average or the proportions of incomes that fall in various ranges (such as low, middle, and high). Or perhaps a graph or a chart would be more useful. Whatever specific method you choose, its function is the same: to reduce these thousands of individual items of information into a few easily understood numbers. The process of allowing a few numbers to summarize many numbers, called **data reduction**, is the basic goal of univariate descriptive statistical procedures. Part I of this text is devoted to these statistics, the primary goal of which is simply to report, clearly and concisely, essential information about a variable.

The second type of descriptive statistics is designed to help the investigator understand the relationship between two or more variables. These statistics, called **measures of association**, allow the researcher to quantify the strength and direction of a relationship. These statistics are very useful because they enable us to investigate two matters of central theoretical and practical importance to any science: causation and prediction. These techniques help us disentangle and uncover the connections between variables. They help us trace the ways in which some variables might have causal influences on others; and, depending on the strength of the relationship, they enable us to predict scores on one variable from the scores on another. Note that measures of association cannot, by themselves, prove that two variables are causally related. However, these techniques can provide valuable clues about causation and are therefore extremely important for theory testing and theory construction.

For example, suppose you were interested in the relationship between "time spent studying statistics" and "final grade in statistics" and had gathered data on these two variables from a group of college students. By calculating the appropriate measure of association, you could determine the strength of the bivariate relationship and its direction. Suppose you found a strong, positive relationship. This would indicate that "study time" and "grade" were closely related (strength of the relationship) and that as one increased in value, the other also increased (direction of the relationship). You could make predictions from one variable to the other ("the longer the study time, the higher the grade").

As a result of finding this strong, positive relationship, you might be tempted to make causal inferences. That is, you might jump to such conclusions as "longer study time leads to (causes) higher grades." Such a conclusion might make a good deal of common sense and would certainly be supported by your statistical analysis. However, the causal nature of the relationship cannot be proven by the statistical analysis. Measures of association can be taken as important clues about causation, but the mere existence of a relationship can never be taken as conclusive proof of causation.

In fact, other variables might have an effect on the relationship. In our example, we probably would not find a perfect relationship between "study time" and "final grade." That is, we would probably find some individuals who spend a great deal of time studying but receive low grades and some individuals who fit the opposite pattern. We know intuitively that other variables besides study time affect grades (such as efficiency of study techniques, amount of background in mathematics, and even random chance). Fortunately, researchers can incorporate these other variables into the analysis and measure their effects. Part III of this text is devoted to bivariate (two-variable) and Part IV to multivariate (more than two variables) descriptive statistics.

Inferential Statistics. This second class of statistical techniques becomes relevant when we wish to generalize our findings from a **sample** to a **population**. A population is the total collection of all cases in which the researcher is interested and that he or she wishes to understand better. Examples of possible populations would be voters in the United States, all parliamentary democracies, unemployed Puerto Ricans in Atlanta, or sophomore college football players in the Midwest.

Populations can theoretically range from inconceivable in size ("all humanity") to quite small (all 35-year-old red-haired belly dancers currently residing in downtown Cleveland) but are usually fairly large. In fact, they are almost always too large to be measured. To put the problem another way, social scientists almost never have the resources or time to test every case in a population. Hence the need for **inferential statistics**, which involve using information from a sample (a carefully chosen subset of the population) to make inferences about a population. Since they have fewer cases, samples are much cheaper to assemble, and—if the proper techniques are followed—generalizations based on these samples can be very accurate representations of the population.

Many of the concepts and procedures involved in inferential statistics may be unfamiliar. However, most of us are experienced consumers of inferential statistics—most familiarly, perhaps, in the form of public-opinion polls and election projections. When a public-opinion poll reports that 42% of the American electorate plans to vote for a certain presidential candidate, it is essentially reporting a generalization to a population ("the American electorate"—which numbers about 100 million people) from a carefully drawn sample (usually about 1500 respondents). Matters of inferential statistics will occupy our attention in Part II of this book. *(For practice in describing different statistical applications, see problems 1.3 and 1.7.)*

1.5 DISCRETE AND CONTINUOUS VARIABLES

In the next chapter, you will begin to encounter some of the broad array of statistics available to the social scientist. One aspect of using statistics that can be puzzling is deciding when to use which statistic. You will learn specific guidelines as you go along, but I will introduce some basic and general guidelines at this point. The first of these concerns discrete and continuous variables; the second, covered in the next section, concerns level of measurement.

A variable is said to be **discrete** if it has a basic unit of measurement that cannot be subdivided. For example, number of people per household is a discrete variable. The basic unit is people, a variable that will always be measured in whole numbers; you'll never find 2.7 people living in a specific household. The scores of discrete variables will be 0, 1, 2, 3, or some other whole integer. Other examples of discrete variables include number of siblings, children, or cars. To measure these variables, we count the number of units (people, cars, siblings) for each case (household, person) and record results in whole numbers.

A variable is **continuous** if it has scores that can be subdivided infinitely (at least theoretically). One example of a continuous variable would be time, which can be measured in minutes, seconds, milliseconds (thousands of a second), nanoseconds (billionths of a second), or even smaller units. In a sense, when we measure a continuous variable, we are always approximating and rounding off the scores. We could report somebody's time in the 100-yard dash as 10.7 seconds or 10.732451 seconds, but, since time can be infinitely subdivided (if we have the technology to make the precise measurements), we will never be able to report the exact time elapsed. Since we cannot work with infinitely long numbers, we must report the scores on continuous variables as if they were discrete.

The scores of discrete and continuous variables may look the same even though we measure and process them differently. This distinction is one of the most basic in statistics and will be one of the criteria we use to choose among various statistics and graphic devices. *(For practice in distinguishing between discrete and continuous variables, see problems 1.4 through 1.8.)*

1.6 LEVEL OF MEASUREMENT

A second basic and very important guideline for the selection of statistics is the **level of measurement**, or the mathematical nature of the variables under consideration. Variables at the highest level of measurement have numerical scores and can be analyzed with a broad range of statistics. Variables at the lowest level of measurement have "scores" that are really just labels, not numbers at all. Statistics that require numerical variables are not appropriate and, often, completely meaningless when used with non-numerical variables. When selecting statistics, you must be sure that the level of measurement of the variable matches the mathematical operations required to compute the statistic.

For example, consider the variables age (measured in years) and income (measured in dollars). Both of these variables have numerical scores and could be summarized with a statistic such as the mean, or average (e.g., "The average income of this city is $43,000." "The average age of students on this campus is 19.7").

The mean, or average, would be meaningless as a way of describing gender or zip codes, variables with non-numerical scores. Your personal zip code might *look* like a number, but it is merely an arbitrary label that happens to be expressed in digits. These "numbers" cannot be added or divided, and statistics like the average cannot be applied to this variable: The average zip code of a group of people is a meaningless statistic.

Determining the level at which a variable has been measured is one of the first steps in any statistical analysis, and we will consider this matter at some length. I will make it a practice throughout this text to introduce level-of-measurement considerations for each statistical technique.

There are three levels of measurement. In order of increasing sophistication, they are nominal, ordinal, and interval-ratio. Each is discussed separately.

The Nominal Level of Measurement. Variables measured at the nominal level have "scores" or categories that are not numerical. Examples of variables at this level include gender, zip code, race, religious affiliation, and place of birth. At this lowest level of measurement, the only mathematical operation permitted is comparing the relative sizes of the categories (e.g., "There are more females than males in this dorm"). The categories or scores of nominal-level variables cannot be ranked with respect to each other and cannot be added, divided, or otherwise manipulated mathematically. Even when the scores or categories are expressed in digits (like zip codes and street addresses), all we can do is compare relative sizes of categories (e.g., "The most common zip code on this campus is 22033"). The categories themselves are not a mathematical scale; they are different from each other but not more or less or higher or lower than each other. Males and females differ in terms of gender, but neither category has more or less gender than the other. In the same way, a zip code of 54398 is different from but not "more than" a zip code of 13427.

Nominal variables are rudimentary, but there are criteria and procedures that we need to observe in order to measure them adequately. In fact, these criteria apply to variables measured at all levels, not just nominal variables. First, the categories of nominal-level variables must be mutually exclusive of each other so that no ambiguity exists concerning the category or score of any given case. Each case must have one and only one score or category. Second, the categories must be exhaustive. In other words, there must be a category—at least an "other" or miscellaneous category—for every possible score that might be found.

Third, the categories of nominal variables should be relatively homogeneous. That is, our categories should include cases that are truly comparable, or, to put it another way, we need to avoid categories that lump apples with oranges. There are no hard-and-fast guidelines for judging if a set of categories is appropriately homogeneous. The researcher must make that decision in terms of the specific purpose of the research, and categories that are too broad for some purposes may be perfectly adequate for others.

Table 1.1 demonstrates some errors in measuring religious preference. Scale A in the table violates the criterion of mutual exclusivity because of overlap between the categories Protestant and Episcopalian. Scale B is not exhaustive because it does not provide a category for people with no religious preference (None) or for people who belong to religions other than the three listed. Scale C uses a category (Non-Protestant, which would include Catholics, Jews, Buddhists, and so forth) that seems too broad for meaningful research. Scale D represents the way religious preference is often measured in North America, but these categories may be too general for some research projects and not comprehensive

TABLE 1.1 FOUR SCALES FOR MEASURING RELIGIOUS PREFERENCE

Scale A (not mutually exclusive)	Scale B (not exhaustive)	Scale C (not homogeneous)	Scale D (an adequate scale)
Protestant	Protestant	Protestant	Protestant
Episcopalian	Catholic	Non-Protestant	Catholic
Catholic	Jew		Jew
Jew			None
None			Other
Other			

enough for others. For example, an investigation of issues that have strong moral and religious content (assisted suicide, abortion, and capital punishment, for example) might need to distinguish between the various Protestant denominations, and an effort to document religious diversity would need to add categories for Buddhists, Muslims, and numerous other religious faiths.

As is the case with zip codes, numerical labels are sometimes used to identify the categories of a variable measured at the nominal level. This practice is especially common when the data are being prepared for computer analysis. For example, the various religions might be labeled with a 1 indicating Protestant, a 2 signifying Catholic, and so on. Remember that these numbers are merely labels or names and have no numerical quality to them. They cannot be added, subtracted, multiplied, or divided. The only mathematical operation permissible with nominal variables is counting and comparing the number of cases in each category of the variable.

The Ordinal Level of Measurement. Variables measured at the ordinal level are more sophisticated than nominal-level variables. They have scores or categories that can be ranked from high to low, so, in addition to classifying cases into categories, we can describe the categories in terms of "more or less" with respect to each other. Thus, with ordinal-level variables, not only can we say that one case is different from another; we can also say that one case is higher or lower, more or less than another.

For example, the variable socioeconomic status (SES) is usually measured at the ordinal level. The categories of the variable are often ordered according to the following scheme:

4. Upper class
3. Middle class
2. Working class
1. Lower class

Individual cases can be compared in terms of the categories into which they are classified. Thus, an individual classified as a 4 (upper class) would be ranked higher than an individual classified as a 2 (working class) and a lower-class person (1) would rank lower than a middle-class person (3). Other variables that are usually measured at the ordinal level include attitude and opinion scales, such as those that measure prejudice, alienation, or political conservatism.

The major limitation of the ordinal level of measurement is that a particular score represents only a position with respect to some other score. We can distinguish between high and low scores, but the distance between the scores cannot be described in precise terms. Although we know that a score of 4 is more than a score of 2, we do not know if it is twice as much as 2.

Since we don't know what the exact distances are from score to score on an ordinal scale, our options for statistical analysis are limited. For example, addition (and most other mathematical operations) assumes that the intervals between scores are exactly equal. If the distances from score to score are not equal, 2 + 2 might equal 3 or 5 or even 15. Thus, strictly speaking, statistics such as the average, or mean (which requires that the scores be added together and then divided by the number of scores), are not permitted with ordinal-level variables. The most sophisticated mathematical operation fully justified with an ordinal variable is the ranking of categories and cases (although, as we will see, it is not unusual for social scientists to take some liberties with this criterion).

The Interval-Ratio Level of Measurement.[2] The categories of nominal-level variables have no numerical quality to them. Ordinal-level variables have categories that can be arrayed along a scale from high to low, but the exact distances between categories or scores are undefined. Variables measured at the interval-ratio level not only permit classification and ranking but also allow the distance from category to category (or score to score) to be exactly defined.

Interval-ratio variables have two characteristics. First, they are measured in units that have equal intervals. For example, asking people how old they are will produce an interval-ratio-level variable (age) because the unit of measurement (years) has equal intervals (the distance from year to year is 365 days). Similarly, if we ask people how many siblings they have, we would produce a variable with equal intervals: Two siblings are one more than one and 13 is 1 more than 12.

The second characteristic of interval-ratio variables is that they have a true zero point. That is, the score of zero for these variables is not arbitrary; it indicates the absence or complete lack of whatever is being measured. For example, the variable "number of siblings" has a true zero point because it is possible to have no siblings at all. Similarly, it is possible to have zero years of education, no income at all, a score of zero on a multiple-choice test, and to be zero years old (although not for very long). Other examples of interval-ratio variables would be number of children, life expectancy, and years married. All mathematical operations are permitted for variables measured at the interval-ratio level.

Table 1.2 summarizes this discussion by presenting the basic characteristics of the three levels of measurement. Note that the number of permitted mathematical operations increases as we move from nominal to ordinal to interval-ratio levels of measurement. Ordinal-level variables are more sophisticated and flexible than nominal-level variables, and interval-ratio-level variables permit the broadest range of mathematical operations.

[2]Many statisticians distinguish between the interval level (equal intervals) and the ratio level (true zero point). I find the distinction unnecessarily cumbersome in an introductory text and will treat these two levels as one.

TABLE 1.2 BASIC CHARACTERISTICS OF THE THREE LEVELS OF MEASUREMENT

Levels	Examples	Measurement Procedures	Mathematical Operations Permitted
Nominal	Sex, race, religion, marital status	Classification into categories	Counting number in each category, comparing sizes of categories
Ordinal	Social class, attitude, and opinion scales	Classification into categories plus ranking of categories with respect to each other	All of the foregoing plus judgments of "greater than" and "less than"
Interval-ratio	Age, number of children, income	All of the foregoing plus description of distances between scores in terms of equal units	All of the foregoing plus all other mathematical operations (addition, subtraction, multiplication, division, square roots, etc.)

Level of Measurement: Final Points. Let us end this section by making four points. The first stresses the importance of level of measurement, and the next three discuss some common points of confusion in applying this concept.

First, knowing the level of measurement of a variable is crucial because it tells us which statistics are appropriate and useful. Not all statistics can be used with all variables. As displayed in Table 1.2, different statistics require different mathematical operations. For example, computing an average requires addition and division, and finding a median (or middle score) requires that the scores be ranked from high to low. Addition and division are appropriate only for interval-ratio-level variables, and ranking is possible only for variables that are at least ordinal in level of measurement. Your first step in dealing with a variable and selecting appropriate statistics is *always* to determine its level of measurement.

Second, the distinction made earlier between discrete and continuous variables is a concern only for interval-ratio-level variables. Nominal- and ordinal-level variables are almost always discrete. That is, researchers measure these variables by asking respondents to select the single category that best describes them, as illustrated in Table 1.3. The scores of these survey items, as they are stated here, are discrete because respondents must select only one category and because these scores cannot be subdivided.

TABLE 1.3 MEASURING A NOMINAL VARIABLE (MARITAL STATUS) AND AN ORDINAL VARIABLE (SUPPORT FOR CAPITAL PUNISHMENT) AS DISCRETE VARIABLES

What is your marital status? Are you presently:		*Do you support the death penalty for persons convicted of homicide?*	
Score	Category	Score	Category
1	Married	1	Strongly support
2	Divorced	2	Slightly support
3	Separated	3	Neither support or oppose
4	Widowed	4	Slightly oppose
5	Single	5	Strongly oppose

READING STATISTICS 1: Introduction

By this point in your education you have developed an impressive array of skills for reading words. Although you may sometimes struggle with a difficult idea or stumble over an obscure meaning, you can comprehend virtually any written work that you are likely to encounter.

As you continue your education in the social sciences, you must develop an analogous set of skills for reading numbers and statistics. To help you reach a reasonable level of literacy in statistics, I have included a series of boxed inserts in this text titled "Reading Statistics." These appear in most chapters and discuss how statistical results are typically presented in the professional literature. Each installment includes an extract or quotation from the professional literature so that we can analyze a realistic example.

As you will see, professional researchers use a reporting style that is quite different from the statistical language you will find in this text. Space in research journals and other media is expensive, and the typical research project requires the analysis of many variables. Thus, a large volume of information must be summarized in very few words. Researchers may express in a word or two a result or an interpretation that will take us a paragraph or more to state.

Because this is an introductory textbook, I have been careful to break down the computation and logic of each statistic and to identify, even to the point of redundancy, what we are doing when we use statistics. In this text we will never be concerned with more than a few variables at a time. We will have the luxury of analysis in detail and of being able to take pages or even entire chapters to develop a statistical idea or analyze a variable. Thus, a major theme of these boxed inserts will be to summarize how our comparatively long-winded (but more careful) vocabulary is translated into the concise language of the professional researcher.

When you have difficulty reading words, your tendency is (or at least should be) to consult reference books (especially dictionaries) to help you identify and analyze the elements (words) of the passage. When you have difficulty reading statistics, you should do exactly the same thing. I hope you will find this text a valuable reference book, but if you learn enough from this text to be able to use any source to help you read statistics, this text will have fulfilled one of its major goals.

Interval-ratio variables can be either discrete (number of times you've been divorced, which must be a whole integer) or continuous (age, which could be measured to the nanosecond or more). Remember that, since we cannot work with infinitely long numbers, continuous interval-ratio variables have to be rounded off at some level and reported as if they were discrete. The distinction relates more to our options for appropriate statistics or graphs than to the appearance of the variables.

Third, in determining level of measurement, always examine the way in which the scores of the variable are *actually stated.* This is particularly a problem with interval-ratio variables that have been measured at the ordinal level. To illustrate, consider income as a variable. If we asked respondents to list their exact income in dollars, we will generate scores that are interval-ratio in level of measurement. Measured in this way, the variable would have a true zero point (no income at all) and equal intervals from score to score (one dollar). It is more convenient for respondents, however, to ask them simply to check the appropriate category from a broad list, as in Table 1.4. The four scores or categories in Table 1.4 are ordinal in level of measurement because they are unequal in size. It is common for researchers to sacrifice precision (income in actual dollars) for the convenience of the respondents in this way. You should be careful

ONE STEP AT A TIME Determining the Level of Measurement of a Variable*

Step 1: Inspect the scores or values of the variable *as they are actually stated*, keeping in mind the basic definition of the three levels of measurement (see Table 1.2).

Step 2: Change the order of the scores. Do the scores still make sense? If the answer is *yes*, the variable is **nominal**. If the answer is *no*, proceed to step 3. To Illustrate: Gender is a nominal variable, and its scores can be stated in any order: (1) male, (2) female or (1) female, (2) male. Each statement of the scores is just as sensible as the other. On a nominal-level variable, no score is higher or lower than any other score and the order in which they are stated is arbitrary.

Step 3: Is the distance between the scores unequal or undefined? If the answer is *yes*, the variable is **ordinal**. If the answer is *no*, proceed to step 4. To illustrate, consider the scale measuring support for capital punishment presented in Table 1.3. People who "strongly" support the death penalty are more in favor than people who "slightly" support it, but the distance from one level of support to the next (from a score of 1 to a score of 2) is undefined. We do not have enough information to ascertain how much more or less one score is than another.

Step 4: If you answered *no* in steps 2 and 3, the variable is **interval-ratio**. Variables at this level have scores that are actual numbers: They have an order with respect to each other and are a defined, equal distance apart. They also have a true zero point. Examples of interval-ratio variables include age, income, and number of siblings.

*This system for determining level of measurement was suggested by Michael G. Bisciglia, Louisiana State University.

to look at the way in which the variable is measured before making a decision about its level of measurement.

Fourth, there is a mismatch between the variables that are usually of most interest to social scientists (race, sex, marital status, attitudes and opinions) and the most powerful and interesting statistics (such as the mean). The former are typically nominal or, at best, ordinal in level of measurement, but the more sophisticated statistics require measurement at the interval-ratio level. This mismatch creates some very real difficulties for social science researchers. On one hand, researchers will want to measure variables at the highest, most precise level of measurement. If income is measured in exact dollars, for example, researchers can make very precise descriptive statements about the differences between people. For example: "Ms. Smith earns $12,547 more than Mr. Jones." If the same variable is measured in broad, unequal categories, such as those in Table 1.4, comparisons between individuals would be less precise and provide less information: "Ms. Smith earns more than Mr. Jones."

On the other hand, given the nature of the disparity, researchers are more likely to treat variables as if they were higher in level of measurement than they

TABLE 1.4 MEASURING INCOME AT THE ORDINAL LEVEL

Score	Income Range
1	Less than $24,999
2	$25,000 to $49,999
3	$50,000 to $99,999
4	More than $100,000

actually are. In particular, variables measured at the ordinal level, especially when they have many possible categories or scores, are often treated as if they were interval-ratio and analyzed with the more powerful, flexible, and interesting statistics available at the higher level. This practice is common, but researchers should be cautious in assessing statistical results and developing interpretations when the level of measurement criterion has been violated

In conclusion, level of measurement is a very basic characteristic of a variable, and we will always consider it when presenting statistical procedures. Level of measurement is also a major organizing principle for the material that follows, and you should make sure that you are familiar with these guidelines. *(For practice in determining the level of measurement of a variable, see problems 1.4 through 1.8.)*

SUMMARY

1. Within the context of social research, the purpose of statistics is to organize, manipulate, and analyze data so that researchers can test their theories and answer their questions. Along with theory and methodology, statistics are a basic tool by which social scientists attempt to enhance their understanding of the social world.
2. There are two general classes of statistics. Descriptive statistics are used to summarize the distribution of a single variable and the relationships between two or more variables. Inferential statistics provide us with techniques by which we can generalize to populations from random samples.
3. Two basic guidelines for selecting statistical techniques were presented. Variables may be either discrete or continuous and may be measured at any of three different levels. At the nominal level, we can compare category sizes. At the ordinal level, categories and cases can be ranked with respect to each other. At the interval-ratio level, all mathematical operations are permitted. Interval-ratio-level variables can be either discrete or continuous. Variables at the nominal or ordinal level are almost always discrete.

GLOSSARY

Continuous variable. A variable with a unit of measurement that can be subdivided infinitely.

Data. Any information collected as part of a research project and expressed as numbers.

Data reduction. Summarizing many scores with a few statistics. A major goal of descriptive statistics.

Dependent variable. A variable that is identified as an effect, result, or outcome variable. The dependent variable is thought to be caused by the independent variable.

Descriptive statistics. The branch of statistics concerned with (1) summarizing the distribution of a single variable or (2) measuring the relationship between two or more variables.

Discrete variable. A variable with a basic unit of measurement that cannot be subdivided.

Hypothesis. A statement about the relationship between variables that is derived from a theory. Hypotheses are more specific than theories, and all terms and concepts are fully defined.

Independent variable. A variable that is identified as a causal variable. The independent variable is thought to cause the dependent variable.

Inferential statistics. The branch of statistics concerned with making generalizations from samples to populations.

Level of measurement. The mathematical characteristic of a variable and the major criterion for selecting statistical techniques. Variables can be measured at any of three levels, each permitting certain mathematical operations and statistical techniques. The characteristics of the three levels are summarized in Table 1.2.

Measures of association. Statistics that summarize the strength and direction of the relationship between variables.

Population. The total collection of all cases in which the researcher is interested.

Research. Any process of gathering information systematically and carefully to answer questions or test theories. Statistics are useful for research projects in which the information is represented in numerical form or as data.

Sample. A carefully chosen subset of a population. In inferential statistics, information is gathered from a sample and then generalized to a population.

Statistics. A set of mathematical techniques for organizing and analyzing data.

Theory. A generalized explanation of the relationship between two or more variables.

Variable. Any trait that can change values from case to case.

PROBLEMS

1.1 In your own words, describe the role of statistics in the research process. Using the "wheel of science" as a framework, explain how statistics link theory with research.

1.2 Find a research article in any social science journal. Choose an article on a subject of interest to you, and don't worry about being able to understand all of the statistics that are reported.

a. How much of the article is devoted to statistics per se (as distinct from theory, ideas, discussion, and so on)?

b. Is the research based on a sample from some population? How large is the sample? How were subjects or cases selected? Can the findings be generalized to some population?

c. What variables are used? Which are independent and which are dependent? For each variable, determine the level of measurement and whether the variable is discrete or continuous.

d. What statistical techniques are used? Try to follow the statistical analysis and see how much you can understand. Save the article and read it again after you finish this course and see if you do any better.

1.3 Distinguish between descriptive and inferential statistics. Describe a research situation that would use both types.

1.4 Following are some items from a public-opinion survey. For each item, indicate the level of measurement and whether the variable will be discrete or continuous. (*HINT: Remember that only interval-ratio-level variables can be continuous*).

a. What is your occupation? __________

b. How many years of school have you completed? __________

c. If you were asked to use one of these four names for your social class, which would you say you belonged in?

__________ Upper __________ Middle
__________ Working __________ Lower

d. What is your age? _____

e. In what country were you born? _____

f. What is your grade-point average? _____

g. What is your major? _____

h. The only way to deal with the drug problem is to legalize all drugs.

_____ Strongly agree
_____ Agree
_____ Undecided
_____ Disagree
_____ Strongly disagree

i. What is your astrological sign? _____

j. How many brothers and sisters do you have? _____

1.5 Following are brief descriptions of how researchers measured a variable. For each situation, determine the level of measurement of the variable and whether it is continuous or discrete.

a. Race. Respondents were asked to select a category from the following list:

_____ Black
_____ White
_____ Other

b. Honesty. Subjects were observed as they passed by a spot on campus where an apparently lost wallet was lying. The wallet contained money and complete identification. Subjects were classified into one of the following categories:

_____ Returned the wallet with money

_____ Returned the wallet but kept the money
_____ Did not return wallet.

c. Social class. Subjects were asked about their family situation when they were 16 years old. Was their family:

_____ Very well off compared to other families?
_____ About average?
_____ Not so well off?

d. Education. Subjects were asked how many years of schooling they and each parent had completed.

e. Racial integration on campus. Students were observed during lunchtime at the cafeteria for a month. The number of students sitting with students of other races was counted for each meal period.

f. Number of children. Subjects were asked: "How many children have you ever had? Please include any that may have passed away."

g. Student seating patterns in classrooms. On the first day of class, instructors noted where each student sat. Seating patterns were remeasured every two weeks until the end of the semester. Each student was classified as

____ same seat as last measurement;
____ adjacent seat;
____ different seat, not adjacent;
____ absent.

h. Physicians per capita. The number of practicing physicians was counted in each of 50 cities, and the researchers used population data to compute the number of physicians per capita.

i. Physical Attractiveness. A panel of 10 judges rated each of 50 photos of a mixed-race sample of males and females for physical attractiveness on a scale from 0 to 20, with 20 being the highest score.

j. Number of accidents. The number of traffic accidents for each of 20 busy intersections in a city was recorded. Also, each accident was rated as

_____ minor damage, no injuries;
_____ moderate damage, personal injury requiring hospitalization;
_____ severe damage and injury.

1.6 For each of the first 20 items in the General Social Survey (see Appendix G), indicate the level of measurement and whether the variable is continuous or discrete.

1.7 For each of the following research situations, identify the level of measurement of all variables and indicate whether they are discrete or continuous. Also, decide which statistical applications are used: descriptive statistics (single variable), descriptive statistics (two or more variables), or inferential statistics. Remember that it is quite common for a given situation to require more than one type of application.

a. The administration of your university is proposing a change in parking policy. You select a random sample of students and ask each one if he or she favors or opposes the change.

b. You ask everyone in your social research class to tell you the highest grade he or she ever received in a math course and the grade on a recent statistics test. You then compare the two sets of scores to see if there is any relationship.

c. Your aunt is running for mayor and hires you (for a huge fee, incidentally) to question a sample of voters about their concerns in local politics. Specifically, she wants a profile of the voters that will tell her what percentage belong to each political party, what percentage are male or female, and what percentage favor or oppose the widening of the main street in town.

d. Several years ago, a state reinstituted the death penalty for first-degree homicide. Supporters of capital punishment argued that this change would reduce the homicide rate. To investigate this claim, a researcher has gathered information on the number of homicides in the state for the two-year periods before and after the change.

e. A local automobile dealer is concerned about customer satisfaction. He wants to mail a survey form to all customers for the past year and ask them if they are satisfied, very satisfied, or not satisfied with their purchases.

1.8 For each of the following research situations, identify the independent and dependent variables. Classify each in terms of level of measurement and whether or not the variable is discrete or continuous.

a. A graduate student is studying sexual harassment on college campuses and asks 500 female students if they personally have experienced any such incidents. Each student is asked to estimate the frequency of these incidents as either "often, sometimes, rarely, or never." The researcher also gathers data on age and major to see if there is any connection between these variables and frequency of sexual harassment.

b. A supervisor in the Solid Waste Management Division of a city government is attempting to assess two different methods of trash collection. One area of the city is served by trucks with two-man crews who do "backyard" pickups, and the rest of the city is served by "hi-tech" single-person trucks with curbside pickup. The assessment measures include the number of complaints received from the two different areas over a six-month period, the amount of time per day required to service each area, and the cost per ton of trash collected.

c. The adult bookstore near campus has been raided and closed by the police. Your social research class has decided to poll the student body and get their reactions and opinions. The class decides to ask each student if he or she supports or opposes the closing of the store, how many times each one has visited the store, and if he or she agrees or disagrees that "pornography is a direct cause of sexual assaults on women." The class also collects information on the sex, age, religious and political philosophy, and major of each student to see if opinions are related to these characteristics.

d. For a research project in a political science course, a student has collected information about the quality of life and the degree of political democracy in 50 nations. Specifically, she used infant mortality rates to measure quality of life and the percentage of all adults who are permitted to vote in national elections as a measure of democratization. Her hypothesis is that quality of life is higher in more democratic nations.

e. A highway engineer wonders if a planned increase in the speed limit on a heavily traveled local avenue will result in any change in number of accidents. He plans to collect information on traffic volume, number of accidents, and number of fatalities for the six-month periods before and after the change.

f. Students are planning a program to promote "safe sex" and awareness of a variety of other health concerns for college students. To measure the effectiveness of the program, they plan to give a survey measuring knowledge about these matters to a random sample of the student body before and after the program.

g. Several states have drastically cut their budgets for mental health care. Will this increase the number of homeless people in these states? A researcher contacts a number of agencies serving the homeless in each state and develops an estimate of the size of the population before and after the cuts.

h. Does tolerance for diversity vary by race, ethnicity, or gender? Samples of white, black, Asian, Hispanic, and Native Americans have been given a survey that measures their interest in and appreciation of cultures and groups other than their own.

SPSS for Windows

Introduction to SPSS and the General Social Survey

The problems at the end of chapters in this text have been written so that they can be solved with just a simple hand calculator. I've purposely kept the number of cases involved unrealistically low so that the tedium of mere calculation would not interfere unduly with the learning process. To provide a more realistic experience in the analysis of social science data, we will analyze a shortened version of the 2006 General Social Survey (GSS). This database can be downloaded from our Web site (www.thomsonedu.com/sociology/healey).

The GSS is a public-opinion poll that has been conducted on nationally representative samples of citizens of the United States since 1972. The full survey includes hundreds of questions covering a broad range of social and political issues. The version supplied with this text has a limited number of variables and cases but is still actual, "real-life" data, so you have the opportunity to practice your statistical skills in a more realistic context.

One of the problems with reality, of course, is that it is often cumbersome and confusing. It's hard enough to do your homework with simplified problems, and you should be a little leery, in terms of your own time and effort, of promises of relevance and realism. This brings us to the second purpose of this section: computers and statistical packages. A statistical package is a set of computer programs for the analysis of data. The advantage of these packages is that, since the programs are already written, you can capitalize on the power of the computer with minimal computer literacy and virtually no programming experience.

This text utilizes a computerized statistical package called the Statistical Package for the Social Sciences (SPSS). In these sections at the ends of chapters, I explain how to use this package to manipulate and analyze the GSS data, and I illustrate and interpret the results. Be sure to read Appendix F before attempting any data analysis.

Part I Descriptive Statistics

Part I consists of four chapters, each devoted to a different application of univariate descriptive statistics. Chapter 2 covers "basic" descriptive statistics, including percentages, ratios, rates, frequency distributions, and graphs. It is a lengthy chapter, but the material is relatively elementary and at least vaguely familiar to most people. Although the statistics covered in this chapter are "basic," they are not necessarily simple or obvious, and the explanations and examples should be considered carefully before attempting the end-of-chapter problems or using them in actual research.

Chapter 3 and 4 cover measures of central tendency and dispersion, respectively. Measures of central tendency describe the typical case or average score (e.g., the mean), while measures of dispersion describe the amount of variety or diversity among the scores (e.g., the range, or the distance from the high score to the low score). These two types of statistics are presented in separate chapters to stress the point that centrality and dispersion are independent, separate characteristics of a variable. You should realize, however, that both measures are necessary and commonly reported together (along with some of the statistics presented in Chapter 2). To reinforce the idea that measures of centrality and dispersion are complementary descriptive statistics, many of the problems at the end of Chapter 4 require the computation of a measure of central tendency from Chapter 3.

Chapter 5 is a pivotal chapter in the flow of the text. It takes some of the statistics from Chapters 2 through 4 and applies them to the normal curve, a concept of great importance in statistics. The normal curve is a type of line chart or frequency polygon (see Chapter 2), which can be used to describe the position of scores using means (Chapter 3) and standard deviations (Chapter 4). Chapter 5 also uses proportions and percentages (Chapter 2).

In addition to its role in descriptive statistics, the normal curve is a central concept in inferential statistics, the topic of Part II of this text. Thus, Chapter 5 serves a dual purpose: It ends the presentation of univariate descriptive statistics and lays essential groundwork for the material to come.

2

Basic Descriptive Statistics
Percentages, Ratios and Rates, Tables, Charts, and Graphs

LEARNING OBJECTIVES

By the end of this chapter, you will be able to

1. Explain the purpose of descriptive statistics in making data comprehensible.
2. Compute and interpret percentages, proportions, ratios, rates, and percentage change.
3. Construct and analyze frequency distributions for variables at each of the three levels of measurement.
4. Construct and analyze bar and pie charts, histograms, and line graphs.

Research results do not speak for themselves. They must be organized and manipulated so that whatever meaning they have can be quickly and easily understood by the researcher and by his or her readers. Researchers use statistics to clarify their results and communicate effectively. In this chapter, we consider some commonly used techniques for presenting research results: percentages and proportions, ratios and rates, percentage change, tables, charts, and graphs. Mathematically speaking, these univariate descriptive statistics are not very complex (although they are not as simple as they might seem at first glance), but they are extremely useful for presenting research results clearly and concisely.

2.1 PERCENTAGES AND PROPORTIONS

Consider the following statement: "Of the 269 cases handled by the court, 167 resulted in prison sentences of five years or more." While there is nothing wrong with this statement, the same fact could have been more clearly conveyed if it had been reported as a percentage: "About 62% of all cases resulted in prison sentences of five or more years."

Percentages and proportions supply a frame of reference for reporting research results, in the sense that they standardize the raw data: percentages to the base 100 and proportions to the base 1.00. The mathematical definitions of **proportions** and **percentages** are

FORMULA 2.1

$$\text{Proportion: } p = \frac{f}{N}$$

FORMULA 2.2

$$\text{Percentage: } \% = \left(\frac{f}{N}\right) \times 100$$

where f = frequency, or the number of cases in any category
N = the number of cases in all categories

To illustrate the computation of percentages, consider the data presented in Table 2.1. How can we find the percentage of cases in the first category (sen-

TABLE 2.1 DISPOSITION OF 269 CRIMINAL CASES (fictitious data)*

Sentence	Frequency (f)	Percentage (%)	Proportion (p)
Five years or more	167	62.08	0.62
Less than five years	72	26.77	0.27
Suspended	20	7.44	0.07
Acquitted	10	3.72	0.04
	$N = 269$	100.01%	1.00

*The slight discrepancy in the totals of the percentage column is due to rounding error. See the Preface (Basic Mathematical Review) for more on rounding.

tences of five years or more)? Note that there are 167 cases in the category ($f = 167$) and a total of 269 cases in all ($N = 269$). So

$$\text{Percentage } (\%) = \left(\frac{f}{N}\right) \times 100 = \left(\frac{167}{269}\right) \times 100 = (0.6208) \times 100 = 62.08\%$$

Using the same procedures, we can also find the percentage of cases in the second category:

$$\text{Percentage } (\%) = \left(\frac{f}{N}\right) \times 100 = \left(\frac{72}{269}\right) \times 100 = (0.2677) \times 100 = 26.77\%$$

Both results could have been expressed as proportions. For example, the proportion of cases in the third category is 0.07:

$$\text{Proportion } (p) = \frac{f}{N} = \frac{20}{269} = 0.07$$

Percentages and proportions are easier to read and comprehend than frequencies. This advantage is particularly obvious when attempting to compare groups of different sizes. For example, based on the information presented in Table 2.2, which college has the higher relative number of social science majors? Because the total enrollments are so different, comparisons are difficult to make from the raw frequencies. Computing percentages eliminates the difference in size of the two campuses by standardizing both distributions to the base of 100. The same data are presented in percentages in Table 2.3.

The percentages in Table 2.3 make it easier to identify differences as well as similarities between the two colleges. College A has a much higher percentage of

TABLE 2.2 DECLARED MAJOR FIELDS ON TWO COLLEGE CAMPUSES (fictitious data)

Major	College A	College B
Business	103	3120
Natural sciences	82	2799
Social sciences	137	1884
Humanities	93	2176
	$N = 415$	$N = 9979$

Application 2.1

Not long ago, in a large social service agency, the following conversation took place between the executive director of the agency and a supervisor of one of the divisions.

> *Executive director:* Well, I don't want to seem abrupt, but I've only got a few minutes. Tell me, as briefly as you can, about this staffing problem you claim to be having.
>
> *Supervisor:* Ma'am, we just don't have enough people to handle our workload. Of the 177 full-time employees of the agency, only 50 are in my division. Yet, 6231 of the 16,722 cases handled by the agency last year were handled by my division.
>
> *Executive director (smothering a yawn):* Very interesting. I'll certainly get back to you on this matter.

How could the supervisor have presented his case more effectively? Because he wants to compare two sets of numbers (his staff versus the total staff and the workload of his division versus the total workload of the agency), proportions or percentages would be a more forceful way of presenting results. What if the supervisor had said, "Only 28.25% of the staff is assigned to my division, but we handle 37.26% of the total workload of the agency"? Is this a clearer message?

The first percentage is found by

$$\% = \left(\frac{f}{N}\right) \times 100 = \frac{50}{177} \times 100$$

$$= (.2825) \times 100 = 28.25\%$$

and the second percentage is found by

$$\% = \left(\frac{f}{N}\right) \times 100 = \left(\frac{6231}{16{,}722}\right) \times 100$$

$$= (.3726) \times 100 = 37.26\%$$

social science majors (even though the absolute number of social science majors is less than at College B) and about the same percentage of humanities majors. How would you describe the differences in the remaining two major fields? *(For practice in computing and interpreting percentages and proportions, see problems 2.1 and 2.2.)*

Here are some further guidelines on the use of percentages and proportions.

1. When working with a small number of cases (say, fewer than 20), it is usually preferable to report the actual frequencies rather than percentages or proportions. With a small number of cases, the percentages can change drastically with relatively minor changes in the size of the data set. For example, if you

TABLE 2.3 DECLARED MAJOR FIELDS ON TWO COLLEGE CAMPUSES (fictitious data)

Major	College A	College B
Business	24.82%	31.27%
Natural sciences	19.76%	28.05%
Social sciences	33.01%	18.88%
Humanities	22.41%	21.81%
	100.00%	100.01%
	(415)	(9979)

ONE STEP AT A TIME **Finding Percentages and Proportions**

Step 1: Determine the values for f (number of cases in a category) and N (number of cases in all categories). Remember that f will be the number of cases in a *specific category* (e.g., males on your campus) and N will be the number of cases in *all* categories (e.g., all students, males and females, on your campus) and that f will be smaller than N, except when the category and the entire group are the same (e.g., when all students are male). Proportions cannot exceed 1.00, and percentages cannot exceed 100.00%.

Step 2: For a proportion, divide f by N.

Step 3: For a percentage, multiply the value you calculated in step 2 by 100.

begin with a group of 10 males and 10 females (that is, 50% of each gender) and then add another female, the percentage distributions will change noticeably to 52.38% female and 47.62% male. Of course, as the number of observations increases, each additional case will have a smaller impact. If we started with 500 males and females and then added one more female, the percentage of females would change by only a tenth of a percent (from 50% to 50.10%).

2. Always report the number of observations along with proportions and percentages. This permits the reader to judge the adequacy of the sample size and, conversely, helps to prevent the researcher from lying with statistics. Statements like "Two out of three people questioned prefer courses in statistics to any other course" might impress you, but the claim would lose its gloss if you learned that only three people were tested. *You should be extremely suspicious of reports that fail to report the number of cases that were tested.*

3. Percentages and proportions can be calculated for variables at the ordinal and nominal levels of measurement, in spite of the fact that they require division. This is not a violation of the level-of-measurement guideline (see Table 1.2). Percentages and proportions do not require the division of the *scores* of the variable (as would be the case in computing the average score on a test, for example) but rather the *number of cases* in a particular category (f) of the variable by the total number of cases in the sample (N). When we make a statement like "43% of the sample is female," we are merely expressing the relative size of a category (female) of the variable (gender) in a convenient way.

2.2 RATIOS, RATES, AND PERCENTAGE CHANGE

Ratios, rates, and percentage change provide some additional ways of summarizing results simply and clearly. Although they are similar to each other, each statistic has a specific application and purpose.

Ratios. **Ratios** are especially useful for comparing categories of a variable in terms of relative frequency. Instead of standardizing the distribution of the variable to the base 100 or 1.00, as we did in computing percentages and proportions, we determine ratios by dividing the frequency of one category by the frequency in another. Mathematically, a ratio can be defined as

FORMULA 2.3

$$\text{Ratio} = \frac{f_1}{f_2}$$

Application 2.2

In Table 2.2, how many natural science majors are there compared to social science majors at College B? This question could be answered with frequencies, but a more easily understood way of expressing the answer would be with a ratio. The ratio of natural science to social science majors would be

$$\text{Ratio} = \frac{f_1}{f_2} = \frac{2799}{1884} = 1.49$$

For every social science major, there are 1.49 natural science majors at College B.

where f_1 = the number of cases in the first category
f_2 = the number of cases in the second category

To illustrate the use of ratios, suppose that you were interested in the relative sizes of the various religious denominations and found that a particular community included 1370 Protestant families and 930 Catholic families. To find the ratio of Protestants (f_1) to Catholics (f_2), divide 1370 by 930:

$$\text{Ratio} = \frac{f_1}{f_2} = \frac{1370}{930} = 1.47$$

The ratio of 1.47 means that there are 1.47 Protestant families for every Catholic family.

Ratios can be very economical ways of expressing the relative predominance of two categories. That Protestants outnumber Catholics in our example is obvious from the raw data. Percentages or proportions could have been used to summarize the overall distribution (e.g., "59.56% of the families were Protestant, 40.44% were Catholic"). In contrast to these other methods, ratios express the relative size of the categories: They tell us exactly how much one category outnumbers the other.

Ratios are often multiplied by some power of 10 to eliminate decimal points. For example, the ratio just computed might be multiplied by 100 and reported as 147 instead of 1.47. This would mean that there are 147 Protestant families for every 100 Catholic families in the community. To ensure clarity, the comparison units for the ratio are often expressed as well. Based on a unit of ones, the ratio of Protestants to Catholics would be expressed as 1.47:1. Based on hundreds, the same statistic might be expressed as 147:100. *(For practice in computing and interpreting ratios, see problems 2.1 and 2.2.)*

Rates. **Rates** provide still another way of summarizing the distribution of a single variable. Rates are defined as the number of actual occurrences of some phenomenon divided by the number of possible occurrences per some unit of time. Rates are usually multiplied by some power of 10 to eliminate decimal points. For example, the crude death rate for a population is defined as the number of deaths in that population (actual occurrences) divided by the number of people in the population (possible occurrences) per year. This quantity is then multiplied by 1000. The formula for the crude death rate can be expressed as

$$\text{Crude death rate} = \frac{\text{Number of Deaths}}{\text{Total population}} \times 1000$$

Application 2.3

In 2005, there were 2500 births in a city of 167,000. In 1965, when the population of the city was only 133,000, there were 2700 births. Is the birthrate rising or falling? Although this question can be answered from the preceding information, the trend in birthrates will be much more obvious if we compute birthrates for both years. Like crude death rates, crude birthrates are usually multiplied by 1000 to eliminate decimal points. For 1965:

$$\text{Crude birthrate} = \frac{2700}{133{,}000} \times 1000 = 20.30$$

In 1965, there were 20.30 births for every 1000 people in the city. For 2000:

$$\text{Crude birthrate} = \frac{2500}{167{,}000} \times 1000 = 14.97$$

In 2005, there were 14.97 births for every 1000 people in the city. With the help of these statistics, the decline in the birthrate is clearly expressed.

If there were 100 deaths during a given year in a town of 7000, the crude death rate for that year would be

$$\text{Crude death rate} = \frac{100}{7{,}000} \times 1{,}000 = (0.01429) \times 1{,}000 = 14.29$$

Or, for every 1000 people, there were 14.29 deaths during this particular year. In the same way, if a city of 237,000 people experienced 120 auto thefts during a particular year, the auto theft rate would be

$$\text{Auto theft rate} = \frac{120}{237{,}000} \times 100{,}000 = (0.0005063) \times 100{,}000 = 50.63$$

Or, for every 100,000 people, there were 50.63 auto thefts during the year in question. *(For practice in computing and interpreting rates, see problems 2.3 and 2.4a.)*

Percentage Change. Measuring social change, in all its variety, is an important task for all social sciences. One very useful statistic for this purpose is the **percentage change**, which tells us how much a variable has increased or decreased over a certain span of time.

To compute this statistic, we need the scores of a variable at two different points in time. The scores could be in the form of frequencies, rates, or percentages. The percentage change will tell us how much the score has changed at the later time relative to the earlier time. Using death rates as an example once again, imagine a society suffering from a devastating outbreak of disease in which the death rate rose from 16 deaths per 1000 population in 1995 to 24 deaths per 1000 in 2005. Clearly, the death rate is higher in 2005, but by how much relative to 1995?

The formula for the percent change is

FORMULA 2.4

$$\text{Percent change} = \left(\frac{f_2 - f_1}{f_1}\right) \times 100$$

where f_1 = first score, frequency, or value
f_2 = second score, frequency, or value

Application 2.4

The American family has been changing rapidly over the past several decades. One major change has been an increase in the number of married women and mothers with jobs outside the home. For example, in 1975, 36.7% of women with children under the age of 6 worked outside the home. In 2005, this percentage had risen to 68.4%.* How large has this change been?

It is obvious that the 2005 percentage is much higher, and calculating the percentage change will give us an exact idea of the magnitude of the change. The 1975 percentage is f_1 and the 2005 figure is f_2, so

$$\text{Percent change} = \left(\frac{68.4 - 36.7}{36.7}\right) \times 100$$

$$= \left(\frac{31.7}{36.7}\right) \times 100$$

$$= (.86376) \times 100$$

$$= 86.38\%$$

In the 30-year period between 1975 and 2005, the percentage of women with children younger than 6 who worked outside the home increased by 86.38%. This is an extremely large change (approaching 100%, or double the earlier percentage) in a short time frame and signals major changes in this social institution.

*U.S. Bureau of the Census. 2007. *Statistical Abstract of the United States, 2007.* Washington, DC: Government Printing Office. p. 380.

In our example, f_1 is the death rate in 1995 ($f_1 = 16$) and f_2 is the death rate in 2005 ($f_2 = 24$). The formula tells us to subtract the earlier score from the later and then divide by the earlier score. The value that results expresses the size of the change in scores ($f_2 - f_1$) relative to the score at the earlier time (f_1). The value is then multiplied by 100 to express the change in the form of a percentage:

$$\text{Percent change} = \left(\frac{24 - 16}{16}\right) \times 100 = \left(\frac{8}{16}\right) \times 100 = (.50) \times 100 = 50\%$$

The death rate in 2005 is 50% higher than in 1995. This means that the 2005 rate was equal to the 1995 rate *plus* half of the earlier score. If the rate had risen to 32 deaths per 1000, the percent change would have been 100% (the rate would have doubled), and if the death rate had fallen to 8 per 1000, the percent change would have been −50%. Note the negative sign: It means that the death rate has decreased by 50%. The 2005 rate would have been half the size of the 1995 rate.

An additional example should make the computation and interpretation of the percentage change clearer. Suppose we wanted to compare the projected population growth rates for various nations over the next 50 years. The necessary information is presented in Table 2.4, which shows the actual population for each nation in 2000 and the projected population for 2050. The "Increase/Decrease" column shows how many people will be added or lost over the 50-year time span.

Casual inspection will give us some information about population trends. For example, compare the "Increase/Decrease" column for China and the United States. These societies are projected to add roughly similar numbers of people (about 155 million for China, a little less for the United States), but, since China's 2000 population is about five times the size of the population of the United States, its percent change will be much lower (about 12% vs. almost 50%).

TABLE 2.4 PROJECTED POPULATION GROWTH FOR SIX NATIONS, 2000–2050

Nation	Population, 2000 (f_1)	Population, 2050 (f_2)	Increase/Decrease ($f_2 - f_1$)	Percent Change
China	1,268,853,000	1,424,162,000	155,309,000	12.24
United States	282,339,000	420,081,000	137,742,000	48.79
Canada	31,278,000	41,430,000	10,152,000	32.46
Mexico	99,927,000	147,908,000	47,981,000	48.02
Italy	57,719,000	50,390,000	−7,329,000	−12.70
Nigeria	114,307,000	356,524,000	242,217,000	211.90

Source: U.S. Bureau of the Census: http://www.census.gov/ipc/www/idbsum.html

Calculating percent change will make comparisons more precise. The right-hand column shows the percent change in projected population for each nation. These values were computed by subtracting the 2000 population (f_1) from the 2050 population (f_2), dividing by the 2000 population, and multiplying by 100.

Although China has the largest population of these six nations, it will grow at the slowest rate (12.24%). The United States and Mexico will increase by about 50% (in 2050, their populations will be half again larger than in 2000) and Canada will grow by about one-third. Italy will actually lose people and its population will decline by over 12%. Nigeria has by far the highest growth rate: It will add the most people and its population will increase in size by over 200%. This means that in 2050 the population of Nigeria will be more than three times its 2000 size. *(For practice in computing and interpreting percent change, see problem 2.4b.)*

ONE STEP AT A TIME Finding Ratios, Rates, and Percent Change

Ratios

Step 1: Determine the values for f_1 and f_2. The value for f_1 will be the number of cases in the first category (e.g., the number of males on your campus), and the value for f_2 will be the number of cases in the second category (e.g., the number of females on your campus).

Step 2: Divide the value of f_1 by the value of f_2.

Rates

Step 1: Determine the number of actual occurrences (e.g., births, deaths, homicides, assaults). This value will be the numerator.

Step 2: Determine the number of possible occurrences. This value will usually be the total population for the area in question.

Step 3: Divide the number of actual occurrences (step 1) by the number of possible occurrences (step 2).

Step 4: Multiply the value you calculated in step 3 by some power of 10. Conventionally, birthrates and death rates are multiplied by 1000 and crime rates are multiplied by 100,000.

Percent Change

Step 1: Determine the values for f_1 and f_2. The former will be the score at time 1 (the earlier time) and the latter will be the score at time 2 (the later time).

Step 2: Subtract f_1 from f_2.

Step 3: Divide the quantity you found in step 2 by f_1.

Step 4: Multiply the quantity you found in step 3 by 100.

2.3 FREQUENCY DISTRIBUTIONS: INTRODUCTION

Frequency distributions are tables that summarize the distribution of a variable by reporting the number of cases contained in each category of the variable. They are very helpful and commonly used ways of organizing and working with data. In fact, the construction of frequency distributions is almost always the first step in any statistical analysis.

To illustrate the usefulness of frequency distributions and to provide some data for examples, assume that the counseling center at a university is assessing the effectiveness of its services. Any realistic evaluation research would collect a variety of information from a large group of students, but, for the sake of this example, we will confine our attention to just four variables and 20 students. The data are reported in Table 2.5.

Note that, even though the data in Table 2.5 represent an unrealistically low number of cases, it is difficult to discern any patterns or trends. For example, try to ascertain the general level of satisfaction of the students from Table 2.5. You may be able to do so with just 20 cases, but it will take some time and effort. Imagine the difficulty with 50 cases or 100 cases presented in this fashion. Clearly the data need to be organized in a format that allows the researcher (and his or her audience) to understand easily the distribution of the variables.

One general rule that applies to all frequency distributions is that the categories of the frequency distribution must be exhaustive and mutually exclusive. In other words, the categories must be stated in a way that permits each case to be counted in one and only one category. This basic principle applies to the construction of frequency distributions for variables measured at all three levels of measurement.

Beyond this rule, there are only guidelines to help you construct useful frequency distributions. As you will see, the researcher has a fair amount of discretion in stating the categories of the frequency distribution (especially with vari-

TABLE 2.5 DATA FROM COUNSELING CENTER SURVEY

Student	Sex	Marital Status	Satisfaction with Services	Age
A	Male	Single	4	18
B	Male	Married	2	19
C	Female	Single	4	18
D	Female	Single	2	19
E	Male	Married	1	20
F	Male	Single	3	20
G	Female	Married	4	18
H	Female	Single	3	21
I	Male	Single	3	19
J	Female	Divorced	3	23
K	Female	Single	3	24
L	Male	Married	3	18
M	Female	Single	1	22
N	Female	Married	3	26
O	Male	Single	3	18
P	Male	Married	4	19
Q	Female	Married	2	19
R	Male	Divorced	1	19
S	Female	Divorced	3	21
T	Male	Single	2	20

ables measured at the interval-ratio level). I will identify the issues to consider as you make decisions about the nature of any particular frequency distribution. Ultimately, however, the guidelines I state are aids for decision-making, nothing more than helpful suggestions. As always, the researcher has the final responsibility for making sensible decisions and presenting his or her data in a meaningful way.

2.4 FREQUENCY DISTRIBUTIONS FOR VARIABLES MEASURED AT THE NOMINAL AND ORDINAL LEVELS

Nominal-Level Variables. For nominal-level variables, constructing frequency distribution is typically very straightforward. Count the number of times each category or score of the variable occurred and then display the frequencies in table format. Table 2.6 displays a frequency distribution for the variable "sex" from the counseling center survey. For purposes of illustration, a column for tallies has been included in this table to illustrate how the score could be counted. This column would *not* be included in the final form of the frequency distribution. Note that the table has a descriptive title, clearly labeled categories (male and female), and a report of the total number of cases at the bottom of the frequency column. These items must be included in all tables regardless of the variable or level of measurement. The meaning of the table is quite clear. There are 10 males and 10 females in the sample, a fact that is much easier to comprehend from the frequency distribution than from the unorganized data presented in Table 2.5.

For some nominal variables, the researcher might have to make some choices about the number of categories he or she wishes to report. For example, the distribution of the variable "marital status" could be reported using the categories listed in Table 2.5. Table 2.7 presents the resultant frequency distribution. Although this is a perfectly fine frequency distribution, it may be too detailed for some purposes. For example, the researcher might want to focus on unmarried as opposed to married students. That is, the researcher might not be concerned with the difference between single and divorced respondents but may want to treat both as simply "not married." In that case, these categories could be grouped together and treated as a single entity, as in Table 2.8. Notice that, when categories are collapsed like this, information and detail will be lost. This latter version of the table would not allow the researcher to discriminate between the two unmarried states.

Ordinal-Level Variables. Frequency distributions for ordinal-level variables are constructed following the same routines used for nominal-level variables. Table 2.9 reports the frequency distribution of the "satisfaction" variable from the counseling center survey. Note that a column of percentages by category has been added to this table. Such columns heighten the clarity of the table (especially with larger samples) and are common adjuncts to the basic frequency

TABLE 2.6 SEX OF RESPONDENTS, COUNSELING CENTER SURVEY

Sex	Tallies	Frequency (f)
Male	~~||||~~ ~~||||~~	10
Female	~~||||~~ ~~||||~~	10
		$N = 20$

TABLE 2.7 MARITAL STATUS OF RESPONDENTS, COUNSELING CENTER SURVEY

Status	Frequency (f)
Single	10
Married	7
Divorced	3
	$N = 20$

TABLE 2.8 MARITAL STATUS OF RESPONDENTS, COUNSELING CENTER SURVEY

Status	Frequency (f)
Married	7
Not married	13
	$N = 20$

TABLE 2.9 SATISFACTION WITH SERVICES, COUNSELING CENTER SURVEY

Satisfaction	Frequency (f)	Percentage (%)
(4) Very satisfied	4	20
(3) Satisfied	9	45
(2) Dissatisfied	4	20
(1) Very dissatisfied	3	15
	$N = 20$	100%

TABLE 2.10 SATISFACTION WITH SERVICES, COUNSELING CENTER SURVEY

Satisfaction	Frequency (f)	Percentage (%)
Satisfied	13	65
Dissatisfied	7	35
	$N = 20$	100%

distribution for variables measured at all levels. This table reports that most students were either satisfied or very satisfied with the services of the counseling center. The most common response (nearly half the sample) was "satisfied." If the researcher wanted to emphasize this major trend, the categories could be collapsed as in Table 2.10. Again, the price paid for this increased compactness is that some information (in this case, the exact breakdown of degrees of satisfaction and dissatisfaction) is lost. *(For practice in constructing and interpreting frequency distributions for nominal- and ordinal-level variables, see problem 2.5.)*

2.5 FREQUENCY DISTRIBUTIONS FOR VARIABLES MEASURED AT THE INTERVAL-RATIO LEVEL

Basic Considerations. In general, the construction of frequency distributions for variables measured at the interval-ratio level is more complex than for nominal and ordinal variables. Interval-ratio variables usually have a large number of possible scores (that is, a wide range from the lowest to the highest score). The large number of scores requires some collapsing or grouping of categories to produce reasonably compact frequency distributions. To construct frequency distributions for interval-ratio-level variables, you must decide how many categories to use and how wide these categories should be.

Application 2.5

The following list shows the ages of 50 prisoners enrolled in a work-release program. Is this group young or old? A frequency distribution will provide an accurate picture of the overall age structure.

18	60	57	27	19
20	32	62	26	20
25	35	75	25	21
30	45	67	41	30
37	47	65	42	25
18	51	22	52	30
22	18	27	53	38
27	23	32	35	42
32	37	32	40	45
55	42	45	50	47

We will use about 10 intervals to display these data. By inspection we see that the youngest prisoner is 18 and the oldest is 75. The range is thus 57. Interval size will be 57/10, or 5.7, which we can round off to either 5 or 6. Let's use a six-year interval beginning at 18. The limits of the lowest interval will be 18–23. Now we must state the limits of all other intervals, count the number of cases in each interval, and display these counts in a frequency distribution. Columns may be added for percentages, cumulative percentages, and/or cumulative frequency. The complete distribution, with a column added for percentages, is

Ages	Frequency	Percentages
18–23	10	20
24–29	7	14
30–35	9	18
36–41	5	10
42–47	8	16
48–53	4	8
54–59	2	4
60–65	3	6
66–71	1	2
72–77	1	2
	$N = 50$	100%

The prisoners seem to be fairly evenly spread across the age groups up to the 48–53 interval. There is a noticeable lack of prisoners in the oldest age groups and a concentration of prisoners in their 20s and 30s.

For example, suppose you wished to report the distribution of the variable "age" for a sample drawn from a community. Unlike the college data reported in Table 2.5, a community sample would have a very broad range of ages. If you simply reported the number of times that each year of age (or score) occurred, you could easily wind up with a frequency distribution that contained 80, 90, or even more categories. Such a large frequency distribution would not present a concise picture. The scores (years) must be grouped into larger categories to heighten clarity and ease of comprehension. How large should these categories be? How many categories should be included in the table? Although there are no hard-and-fast rules for making these decisions, they always involve a trade-off between more detail (a greater number of narrow categories) or more compactness (a smaller number of wide categories).

Constructing the Frequency Distribution. To introduce the mechanics and decision-making processes involved, we will construct a frequency distribution to display the ages of the students in the counseling center survey. Because of the narrow age range of a group of college students, we can use categories of only one year (these categories are often called *class intervals* when working with interval-ratio data). The frequency distribution is constructed by listing the ages

TABLE 2.11 AGE OF RESPONDENTS, COUNSELING CENTER SURVEY (interval width = one year of age)

Class Intervals	Frequency (*f*)
18	5
19	6
20	3
21	2
22	1
23	1
24	1
25	0
26	1
	$N = 20$

from youngest to oldest, counting the number of times each score (year of age) occurs, and then totaling the number of scores for each category. Table 2.11 presents the information and reveals a concentration or clustering of scores in the 18 and 19 class intervals.

Even though the picture presented in this table is fairly clear, assume for the sake of illustration that you desire a more compact (less detailed) summary. To do this, you will have to group scores into wider class intervals. By increasing the interval width (say, to two years), you can reduce the number of intervals and achieve a more compact expression. The grouping of scores in Table 2.12 clearly emphasizes the relative predominance of younger respondents. This trend in the data can be stressed even more by the addition of a column displaying the percentage of cases in each category.

Note that the class intervals in Table 2.12 have been stated with an apparent gap between them (that is, the class intervals are separated by a distance of one unit). At first glance, these gaps may appear to violate the principle of exhaustiveness; but, because age has been measured in whole numbers, the gaps actually pose no problem. Given the level of precision of the measurement (in whole years, as opposed to, say, 10ths of a year), no case could have a score falling between these class intervals. For these data, the set of class intervals in Table 2.12 are exhaustive and mutually exclusive. Each of the 20 respondents in the sample can be sorted into one and only one age category.

TABLE 2.12 AGE OF RESPONDENTS, COUNSELING CENTER SURVEY (interval width = two years of age)

Class Intervals	Frequency (*f*)	Percentage (%)
18–19	11	55
20–21	5	25
22–23	2	10
24–25	1	5
26–27	1	5
	$N = 20$	100%

ONE STEP AT A TIME **Finding Midpoints**

Step 1: Find the upper and lower limits of the lowest interval in the frequency distribution. For any interval, the upper limit is the highest score included in the interval and the lower limit is the lowest score included in the interval. For example, for the top set of intervals in Table 2.13, the lowest interval (0–2) includes scores of 0, 1, and 2. The upper limit of this interval is 2 and the lower limit is 0.

Step 2: Add the upper and lower limits and divide by 2. For the interval 0–2: (0 + 2)/2 = 1. The midpoint for this interval is 1.

Step 3: Midpoints for other intervals can be found by repeating steps 1 and 2 for each interval. As an alternative, you can find the midpoint for any interval by adding the value of the interval width to the midpoint of the next lower interval. For example, the lowest interval in Table 2.13 is 0–2 and the midpoint is 1. Intervals are 3 units wide (that is, they each include three scores), so the midpoint for the next higher interval (3–5) is 1 + 3, or 4. The midpoint for the interval 6–8 is 4 + 3, or 7, and so forth.

However, consider the potential difficulties if age had been measured with greater precision. If age had been measured in 10ths of a year, into which class interval in Table 2.12 would a 19.4-year-old subject be placed? You can avoid this ambiguity by always stating the limits of the class intervals at the same level of precision as the data. Thus, if age were being measured in 10ths of a year, the limits of the class intervals in Table 2.12 would be stated in 10ths of a year. For example:

17.0–18.9
19.0–20.9
21.0–22.9
23.0–24.9
25.0–26.9

To maintain mutual exclusivity between categories, do not overlap the class intervals. If you state the limits of the class intervals at the same level of precision as the data (which might be in whole numbers, tenths, hundredths, etc.) and maintain a "gap" between intervals, you will always produce a frequency distribution where each case can be assigned to one and only one category.

Midpoints. On occasion, you will need to work with the **midpoints** of the class intervals, for example, when constructing or interpreting certain graphs. Midpoints are defined as the points exactly halfway between the upper and lower limits and can be found for any interval by dividing the sum of the upper and lower limits by 2. Table 2.13 displays midpoints for two different sets of class intervals. *(For practice in finding midpoints, see problems 2.8b and 2.9b.)*

Real Limits.[1] For certain purposes, you must eliminate the "gap" between class intervals and treat a distribution as a continuous series of categories that

[1]This section is optional. It is necessary for understanding the material presented in Chapters 3 and 4 on computing measures of central tendency and dispersion for grouped data.

ONE STEP AT A TIME **Finding Real Limits***

Step 1: Find the distance (the "gap") between the stated class intervals. In Table 2.12, for example, this value is 1.

Step 2: Divide the value found in step 1 in half.

Step 3: Add the value found in step 2 to all upper stated limits and subtract it from all lower stated limits.

*This section is optional.

TABLE 2.13 MIDPOINTS

Class Interval Width = 3	
Class Intervals	Midpoints
0–2	1
3–5	4
6–8	7
9–11	10

Class Interval Width = 6	
Class Intervals	Midpoints
100–105	102.5
106–111	108.5
112–117	114.5
118–123	120.5

border each other. This is necessary for the construction of some graphs (see Section 2.7) and for computing summary statistics for variables that have been grouped into frequency distributions.

To illustrate, we'll begin with Table 2.12. Note the "gap" of one year between intervals. As we saw before, the gap is only apparent: scores are measured in whole years (i.e., 19, 21 vs. 19.5 or 21.3) and cannot fall between intervals. These types of class intervals are called **stated class limits** and they organize the scores of the variable into a series of discrete, nonoverlapping intervals.

To treat the variable as continuous, we must use the **real class limits**. To find the real limits of any class interval, divide the distance between the stated class intervals (the "gap") in half and add the result to all upper stated limits and subtract it from all lower stated limits. This process is illustrated below with the class intervals stated in Table 2.12. The distance between intervals is one, so the real limits can be found by adding 0.5 to all upper limits and subtracting 0.5 from all lower limits.

Stated Limits	Real Limits
18–19	17.5–19.5
20–21	19.5–21.5
22–23	21.5–23.5
24–25	23.5–25.5
26–27	25.5–27.5

TABLE 2.14 REAL CLASS LIMITS

Class Intervals (stated limits)	Real Class Limits
3–5	2.5–5.5
6–8	5.5–8.5
9–11	8.5–11.5
Class Intervals (stated limits)	**Real Class Limits**
100–105	99.5–105.5
106–111	105.5–111.5
112–117	111.5–117.5
118–123	117.5–123.5

Note that, with real limits, the class intervals overlap and the distribution can be seen as continuous. Table 2.14 presents additional illustrations of real limits for two different sets of class intervals. In both cases, the "gap" between the stated limits is 1. *(For practice in finding real limits, see problem 2.7c and problem 2.8d.)*

Cumulative Frequency and Cumulative Percentage. Two commonly used adjuncts to the basic frequency distribution for interval-ratio data are the **cumulative frequency** and **cumulative percentage** columns. Their primary purpose is to allow the researcher (and his or her audience) to tell at a glance how many cases fall below a given score or class interval in the distribution.

To construct a cumulative frequency column, begin with the lowest class interval (i.e., the class interval with the lowest scores) in the distribution. The entry in the cumulative frequency columns for that interval will be the same as the number of cases in the interval. For the next-higher interval, the cumulative frequency will be all cases in the interval plus all the cases in the first interval. For the third interval, the cumulative frequency will be all cases in the interval plus all cases in the first two intervals. Continue adding (or accumulating) cases until you reach the highest class interval, which will have a cumulative frequency of all the cases in the interval plus all cases in all other intervals. For the highest interval, cumulative frequency equals the total number of cases. Table 2.15 shows a cumulative frequency column added to Table 2.12.

The cumulative percentage column of Table 2.15 is quite similar to the cumulative frequency column. Begin by adding a column to the basic frequency dis-

TABLE 2.15 AGE OF RESPONDENTS, COUNSELING CENTER SURVEY

Class Intervals	Frequency (f)	Cumulative Frequency
18–19	11	11
20–21	5	16
22–23	2	18
24–25	1	19
26–27	1	20
	$N = 20$	

tribution for percentages as in Table 2.12. This column shows the percentage of all cases in each class interval. To find cumulative percentages, follow the same addition pattern explained earlier for cumulative frequency. That is, the cumulative percentage for the lowest class interval will be the same as the percentage of cases in the interval. For the next-higher interval, the cumulative percentage is the percentage of cases in the interval plus the percentage of cases in the first interval, and so on. Table 2.16 shows the age data with a cumulative percentage column added.

These cumulative columns are quite useful in situations where the researcher wants to make a point about how cases are spread across the range of scores. For example, Tables 2.15 and 2.16 show quite clearly that most students in the counseling center survey are less than 21 years of age. If the researcher wishes to impress this feature of the age distribution on his or her audience, then these cumulative columns are quite handy. Most realistic research situations will be concerned with many more than 20 cases and/or many more categories than our tables have. Since the cumulative percentage column is clearer and easier to interpret in such cases, it is normally preferred to the cumulative frequencies column.

Unequal Class Intervals. As a general rule, the class intervals of frequency distributions should be equal in size in order to maximize clarity and ease of comprehension. For example, note that all of the class intervals in Tables 2.15 and 2.16 are the same width (2 years). There are several situations, however, in which the researcher may chose to use open-ended class intervals or intervals of unequal size. Open-ended intervals have an unspecified upper or lower limit and can be used when there are a few cases with unusually high or low scores. Intervals of unequal size can be used to collapse a variable with a wide range of scores into more easily comprehended groupings. We will examine each situation separately.

Open-Ended Intervals. What would happen to the frequency distribution in Table 2.15 if we added one more student who was 47 years of age? We would now have 21 cases and there would be a large gap between the oldest respondent (now 47) and the second oldest (age 26). If we simply added the older student to the frequency distribution, we would have to include nine new class

TABLE 2.16 AGE OF RESPONDENTS, COUNSELING CENTER SURVEY

Class Intervals	Frequency (f)	Cumulative Frequency	Percentage	Cumulative Percentage
18–19	11	11	55%	55%
20–21	5	16	25%	80%
22–23	2	18	10%	90%
24–25	1	19	5%	95%
26–27	1	20	5%	100%
	$N = 20$		100%	

TABLE 2.17 AGE OF RESPONDENTS, COUNSELING CENTER SURVEY ($N = 21$)

Class Intervals	Frequency (f)	Cumulative Frequency
18–19	11	11
20–21	5	16
22–23	2	18
24–25	1	19
26–27	1	20
28 and older	1	21
	$N = 21$	

intervals (28–30, 31–32, 32–33, etc.) with zero cases in them before we got to the 46–47 interval. This would waste space and probably be unclear and confusing. An alternative way to handle the situation would be to add an "open-ended" interval to the frequency distribution, as in Table 2.17:

The open-ended interval in Table 2.17 allows us to present the information more compactly and efficiently than listing all of the empty intervals between "28–29" and "46–47." Note also that we could handle an extremely low score by adding an open-ended interval as the lowest class interval (e.g., "17 and younger"). There is a small price to pay for this efficiency (there is no information in Table 2.17 about the value of the scores included in the open-ended interval), so this technique should not be used indiscriminately.

Intervals of Unequal Size. On some variables, most scores are tightly clustered together but others are strewn across a broad range of scores. Consider, as an example, the distribution of income in the United States. In 2005, most households (a little more than 50%) reported annual incomes between \$20,000 and \$75,000 and a sizeable grouping (about 20%) earned less than that. The problem (from a statistical point of view) comes with more affluent households. Many of these cases are in the \$75,000–\$100,000 range but some have incomes in the high six- or seven- (and even eight-) -figure range. The number of very wealthy households is quite small, of course, but we must still account for these extreme cases.

If we tried to use a frequency distribution with equal intervals of, say, \$10,000 to summarize this variable, we would need 30 or 40 or more intervals to include all of the more affluent households, and many of our intervals in the higher income ranges—those over \$100,000—would have few or zero cases. In situations such as this, researchers often use intervals of unequal size to summarize the variable more efficiently. To illustrate, Table 2.18 uses unequal intervals to summarize the distribution of income in the United States.

Some of the intervals in Table 2.18 are \$10,000 wide, others are \$25,000, \$50,000 or \$150,000 wide, and two (the lowest and highest intervals) are open ended. Tables that use intervals of mixed widths might be a little confusing for the reader, but the trade-off in compactness and efficiency can be considerable. *(For practice in constructing and interpreting frequency distributions for interval-ratio level variables, see problems 2.5 to 2.9.)*

TABLE 2.18 DISTRIBUTION OF INCOME BY HOUSEHOLD, UNITED STATES, 2005

Income	Households (Frequency)	Households (Percent)
Less than $20,000	23,848,000	20.9
$20,000 to $29,999	13,642,000	11.9
$30,000 to $39,999	12,388,000	10.8
$40,000 to $49,999	11,028,000	9.6
$50,000 to $74,999	21,031,000	18.4
$75,000 to $99,999	12,734,000	11.1
$100,000 to $149,999	12,132,000	10.6
$150,000 to $199,999	4,031,000	3.5
$200,000 to $249,000	1,529,000	1.3
$250,000 and above	2,023,000	1.8
	114,386,000	99.9%

Source: U.S. Census Bureau, http://pubdb3.census.gov/macro/032006/hhinc/new06_000.htm

2.6 CONSTRUCTING FREQUENCY DISTRIBUTIONS FOR INTERVAL-RATIO-LEVEL VARIABLES: A REVIEW

We covered a lot of ground in the preceding section, so let's pause and review these principles by considering a specific research situation. The following data represent the numbers of visits received over the past year by 90 residents of a retirement community.

0	52	21	20	21	24	1	12	16	12
16	50	40	28	36	12	47	1	20	7
9	26	46	52	27	10	3	0	24	50
24	19	22	26	26	50	23	12	22	26
23	51	18	22	17	24	17	8	28	52
20	50	25	50	18	52	46	47	27	0
32	0	24	12	0	35	48	50	27	12
28	20	30	0	16	49	42	6	28	2
16	24	33	12	15	23	18	6	16	50

Listed in this format, the data are a hopeless jumble from which no one could derive much meaning. The function of the frequency distribution is to arrange and organize these data so that their meanings will be made obvious.

First, we must decide how many class intervals to use in the frequency distribution. Following the guidelines presented in the *One Step at a Time: Constructing Frequency Distributions for Interval-Ratio Variables* box, let's use about 10 intervals ($k = 10$). By inspecting the data, we can see that the lowest score is 0 and the highest is 52. The range of these scores (R) is $52 - 0$, or 52. To find the approximate interval size (i), divide the range (52) by the number of intervals (10). Since $52/10 = 5.2$, we can set the interval size at 5.

The lowest score is 0, so the lowest class interval will be 0–4. The highest class interval will be 50–54, which will include the high score of 52. All that remains is to state the intervals in table format, count the number of scores that fall into each interval, and report the totals in a frequency column. These steps have been taken in Table 2.19, which also includes columns for the percentages and cumulative percentages. Note that this table is the product of several relatively arbitrary decisions. The researcher should remain aware of this fact and inspect the frequency distribution carefully. If the table is unsatisfactory for any reason, it can be reconstructed with a different number of categories and interval sizes.

ONE STEP AT A TIME Finding Frequency Distributions for Interval-Ratio Variables

Step 1: Decide how many class intervals (k) you wish to use. One reasonable convention suggests that the number of intervals should be about 10. Many research situations may require fewer than 10 intervals ($k = 10$), and it is common to find frequency distributions with as many as 15 intervals. Only rarely will more than 15 intervals be used, since the resultant frequency distribution would be too large for easy comprehension.

Step 2: Find the range (R) of the scores by subtracting the low score from the high score.

Step 3: Find the size of the class intervals (i) by dividing R (from step 2) by k (from step 1):

$$i = R/k$$

Round the value of i to a convenient whole number. This will be the interval size or width.

Step 4: State the lowest interval so that its lower limit is equal to or below the lowest score. By the same token, your highest interval will be the one that contains the highest score. Generally, intervals should be equal in size, but unequal and open-ended intervals may be used when convenient.

Step 5: State the limits of the class intervals at the same level of precision as you have used to measure the data. Do not overlap intervals. You will thereby define the class intervals so that each case can be sorted into one and only one category.

Step 6: Count the number of cases in each class interval, and report these subtotals in a column labeled "Frequency." Report the total number of cases (N) at the bottom of this column. The table may also include a column for percentages, cumulative frequencies, and cumulative percentages.

Step 7: Inspect the frequency distribution carefully. Has too much detail been lost? If so, reconstruct the table with a greater number of class intervals (or smaller interval size). Is the table too detailed? If so, reconstruct the table with fewer class intervals (or use wider intervals). Are there too many intervals with no cases in them? If so, consider using open-ended intervals or intervals of unequal size. Remember that the frequency distribution results from a number of decisions you make in a rather arbitrary manner. If the appearance of the table seems less than optimal given the purpose of the research, redo the table until you are satisfied that you have struck the best balance between detail and conciseness.

Step 8: Give your table a clear, concise title, and number the table if your report contains more than one. All categories and columns must also be clearly labeled.

TABLE 2.19 NUMBER OF VISITS PER YEAR, 90 RETIREMENT COMMUNITY RESIDENTS

Class Intervals	Frequency (f)	Cumulative Frequency	Percentage (%)	Cumulative Percentage
0–4	10	10	11.11%	11.11
5–9	5	15	5.56%	16.67
10–14	8	23	8.89%	25.26
15–19	12	35	13.33%	38.89
20–24	18	53	20.00%	58.89
25–29	12	65	13.33%	72.22
30–34	3	68	3.33%	75.55
35–39	2	70	2.22%	77.77
40–44	2	72	2.22%	79.99
45–49	6	78	6.67%	86.66
50–54	12	90	13.33%	99.99
	$N = 90$		99.99%*	

*Percentage columns will occasionally fail to total to 100% because of rounding error. If the total is between 99.90% and 100.10%, ignore the discrepancy. Discrepancies of greater than ±0.10% may indicate mathematical errors, and the entire column should be computed again.

Now, with the aid of the frequency distribution, some patterns in the data can be discerned. There are three distinct groupings of scores in the table. Ten residents were visited rarely, if at all (the 0–4 visits per year interval). The single largest interval, with 18 cases, is 20–24. Combined with the intervals immediately above and below, this represents quite a sizeable grouping of cases (42 out of 90, or 46.66% of all cases) and suggests that the dominant visiting rate is about twice a month, or approximately 24 visits per year. The third grouping, in the 50–54 class interval (12 cases), reflects a visiting rate of about once a week. The cumulative percentage column indicates that the majority of the residents (58.89%) were visited 24 or fewer times a year.

2.7 CHARTS AND GRAPHS

Researchers frequently use charts and graphs to present their data in ways that are visually more dramatic than frequency distributions. These devices are particularly useful for conveying an impression of the overall shape of a distribution and for highlighting any clustering of cases in a particular range of scores. Many graphing techniques are available, but we will examine just four. The first two, pie and bar charts, are appropriate for discrete variables at any level of measurement. The last two, histograms and line charts (or frequency polygons), are used with both discrete and continuous interval-ratio variables but are particularly appropriate for the latter.

The sections that follow explain how to construct graphs and charts "by hand." These days, however, computer programs are almost always used to produce graphic displays. Graphing software is sophisticated and flexible but also relatively easy to use; if such programs are available to you, you should familiarize yourself with them. The effort required to learn these programs will be repaid in the quality of the final product. The SPSS for Windows section at the end of this chapter includes a demonstration of how to produce bar charts and line charts.

Pie Charts. To construct a **pie chart**, begin by computing the percentage of all cases that fall into each category of the variable. Then divide a circle (the pie) into segments (slices) proportional to the percentage distribution. Be sure that the chart and all segments are clearly labeled.

Figure 2.1 is a pie chart that displays the distribution of "marital status" from the counseling center survey. The frequency distribution (Table 2.7) is reproduced as Table 2.20, with a column added for the percentage distribution. Since a circle's circumference is 360°, we will apportion 180° (or 50%) for the first category, 126° (35%) for the second, and 54° (15%) for the last category. The pie chart visually reinforces the relative preponderance of single respondents and the relative absence of divorced students in the counseling center survey.

Bar Charts. Like pie charts, **bar charts** are relatively straightforward. Conventionally, the categories of the variable are arrayed along the horizontal axis (or *abscissa*) and frequencies, or percentages if you prefer, along the vertical axis (or *ordinate*). For each category of the variable, construct (or draw) a rectangle of constant width and with a height that corresponds to the number of cases in the category. The bar chart in Figure 2.2 reproduces the marital status data from Figure 2.1 and Table 2.20.

The chart in Figure 2.2 would be interpreted in exactly the same way as the pie chart in Figure 2.1, and researchers are free to choose between these two methods of displaying data. However, if a variable has more than four or five

FIGURE 2.1 SAMPLE PIE CHART: MARITAL STATUS OF RESPONDENTS ($N = 20$)

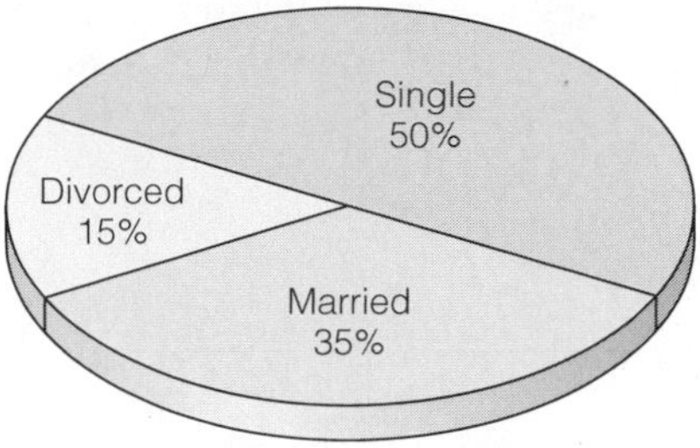

TABLE 2.20 MARITAL STATUS OF RESPONDENTS, COUNSELING CENTER SURVEY

Status	Frequency (f)	Percentage (%)
Single	10	50
Married	7	35
Divorced	3	15
	$N = 20$	100%

FIGURE 2.2 SAMPLE BAR CHART: MARITAL STATUS OF RESPONDENTS ($N = 20$)

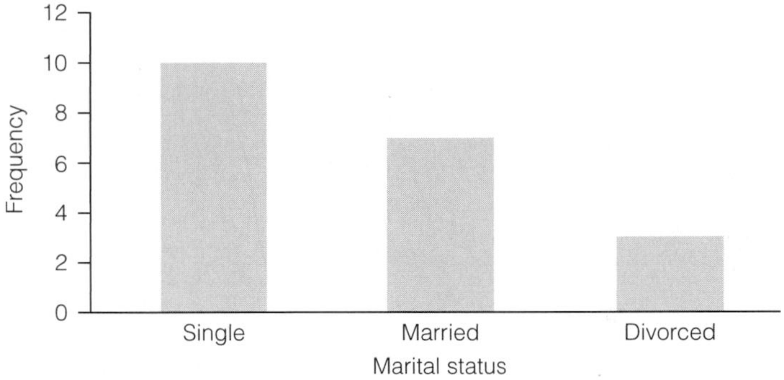

categories, the bar chart would be preferred. With too many categories, the pie chart gets very crowded and loses its visual clarity. To illustrate, Figure 2.3 uses a bar chart to display the data on visiting rates for the retirement community presented in Table 2.19. A pie chart for this same data would have had 11 different "slices," a more complex or "busier" picture than that presented by the bar chart. In Figure 2.3, the clustering of scores in the "20 to 24" range (approximately two visits a month) is readily apparent, as are the groupings in the "0 to 4" and "50 to 54" ranges.

Bar charts are particularly effective ways to display the relative frequencies for two or more categories of a variable when you want to emphasize some comparisons. Suppose, for example, that you wished to make a point about changing rates of homicide victimization for white males and females since 1955. Figure 2.4 displays the data in a dramatic and easily comprehended way. The bar chart shows that rates for males are higher than rates for females, that rates for both sexes were highest in 1975, and that rates declined after that time. *(For practice in constructing and interpreting pie and bar charts, see problems 2.5b and 2.10.)*

Histograms. **Histograms** look a lot like bar charts and, in fact, are constructed in much the same way. However, histograms use real limits rather than stated limits, and the categories or scores of the variable border each other, as if they merged into each other in a continuous series. Therefore, these graphs are most appropriate for continuous interval-ratio-level variables, but they are

FIGURE 2.3 SAMPLE BAR CHART FOR VISITS PER YEAR, RETIREMENT COMMUNITY RESIDENTS (N = 90)

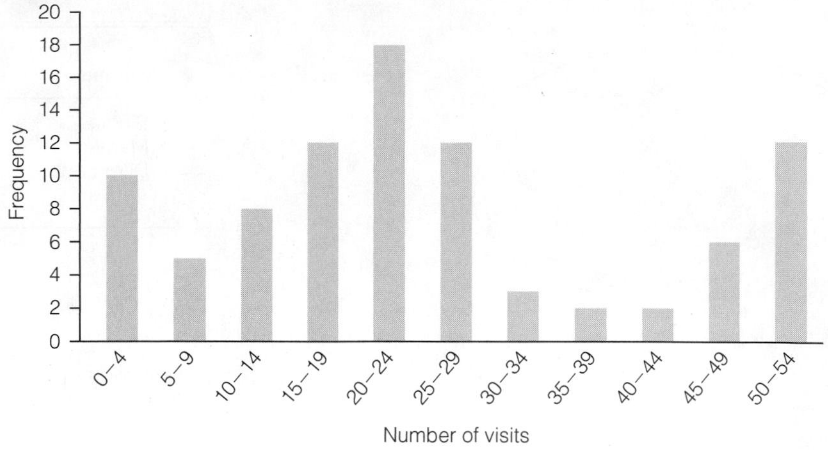

FIGURE 2.4 HOMICIDE VICTIMIZATION RATES, 1955–2003 (selected rates, per 100,000 population, whites only)

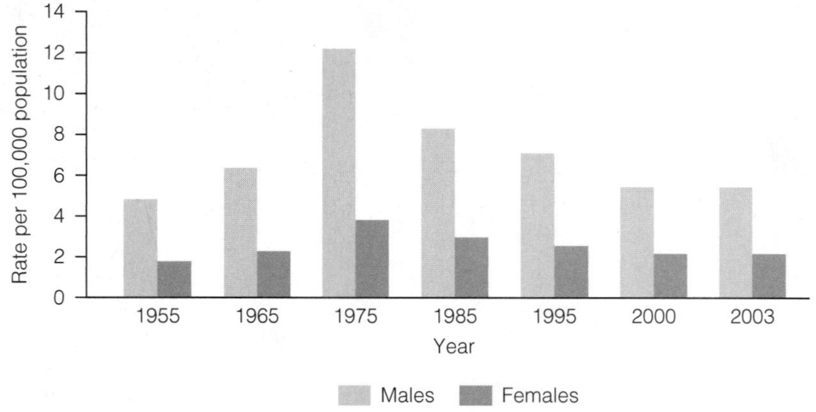

Source: U.S. Bureau of the Census. 2007. *Statistical Abstract of the United States, 2007.* Washington, D.C.: Government Printing Office. p. 195 (Available at: http://www.census.gov/prod/2006pubs/07statab/law.pdf)

commonly used for discrete interval-ratio-level variables as well. To construct a histogram from a frequency distribution, follow these steps.

1. Array the real limits of the class intervals or scores along the horizontal axis (abscissa).
2. Array frequencies along the vertical axis (ordinate).
3. For each category in the frequency distribution, construct a bar with height corresponding to the number of cases in the category and with width corresponding to the real limits of the class intervals.
4. Label each axis of the graph.
5. Title the graph.

As an example, Figure 2.5 uses a histogram to display the distribution of ages for a sample of respondents to a national public-opinion poll. The bars in the

FIGURE 2.5. AGE OF RESPONDENTS, 2006 GENERAL SOCIAL SURVEY

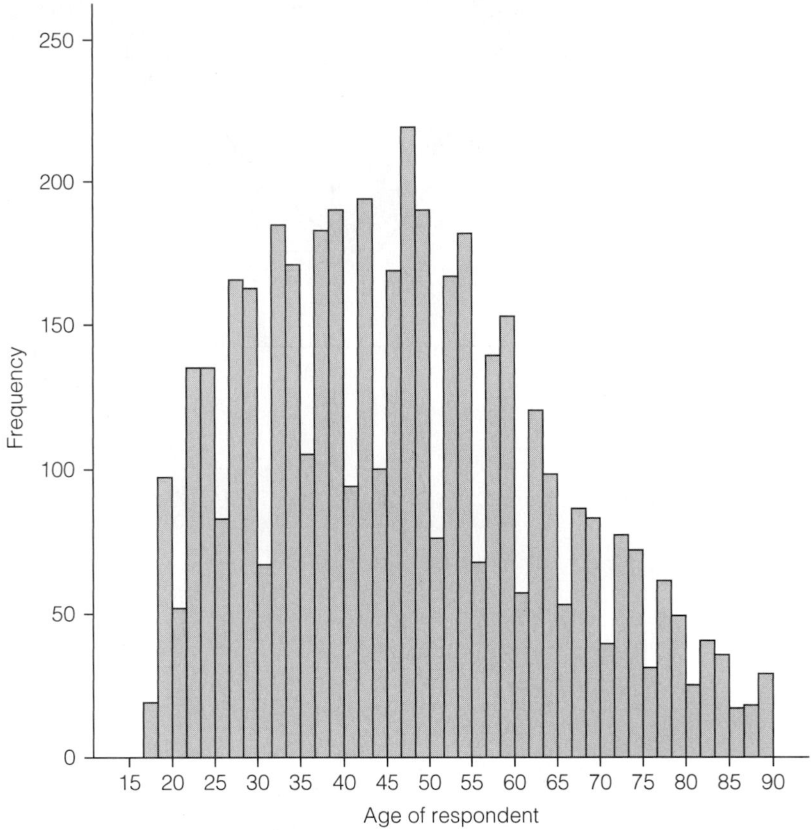

graph are 5 years wide, and their uneven heights reflect the varying number of respondents for each 5-year group. The graph peaks around age 50, and the sample has more respondents (higher bars) who are younger than 50 and fewer respondents (lower bars) older than 50. Note also that there are no people in the sample younger than age 18, the usual cutoff point for respondents to public-opinion polls.

Line Charts. Construction of a **line chart** (or **frequency polygon)** is similar to construction of a histogram. Instead of using bars to represent the frequencies, however, use a dot at the midpoint of each interval. Straight lines then connect the dots. Because the line is continuous from highest to lowest score, these graphs are especially appropriate for continuous interval-ratio-level variables but are frequently used with discrete interval-ratio-level variables. Figure 2.6 displays a line chart for the visiting data previously displayed in the bar chart in Figure 2.3.

Line charts can also be used to display trends across time. Figure 2.7 shows both marriage and divorce rates per 1000 population for the United States since 1950. Note that both rates rose until the early 1980s and have been falling since, with the marriage rate falling slightly faster.

Histograms and frequency polygons are alternative ways of displaying essentially the same message. Thus, the choice between the two techniques is left to the aesthetic pleasures of the researcher. *(For practice in constructing*

FIGURE 2.6 NUMBER OF VISITS PER YEAR, RETIREMENT COMMUNITY RESIDENTS ($N = 90$)

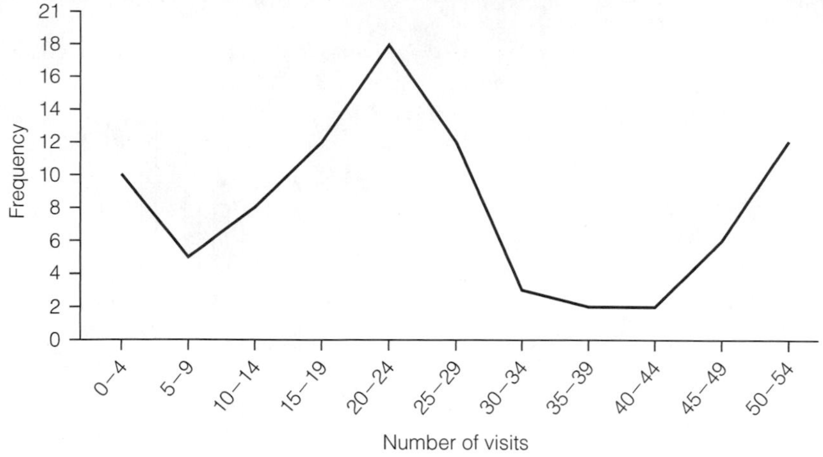

FIGURE 2.7 U.S. MARRIAGE AND DIVORCE RATES, 1950–2004 (rates per 1000 population)

Source: U.S. Bureau of the Census. 2007. *Statistical Abstract of the United States, 2007.* Washington, D.C.: Government Printing Office. p. 63. (Available at: http://www.census.gov/prod/2006pubs/07statab/law.pdf)

and interpreting histograms and line charts, see problems 2.7b, 2.8d, 2.9d, 2.11, and 2.12.)

2.8 INTERPRETING STATISTICS: USING PERCENTAGES, FREQUENCY DISTRIBUTIONS, CHARTS, AND GRAPHS TO ANALYZE CHANGING PATTERNS OF WORKPLACE SURVEILLANCE

A sizeable volume of statistical material has been introduced in this chapter, and it will be useful to conclude by focusing on meaning and interpretation. What can you say after you have calculated percentages, built a frequency distribution, or constructed a graph or chart? Remember that statistics are tools to help us analyze information and answer questions. They never speak for themselves and they always have to be understood in the context of some research question or test of hypothesis. This section provides an example of interpretation by posing and answering some questions from social science research. The interpretation (words) will be explicitly linked to the statistics (numbers) so that you will be able to see how and why conclusions are developed.

Your New Job and Workplace Surveillance. Congratulations! You have just landed a job with a major U.S. corporation and you now find yourself in the middle of the lunch hour in your cubicle. Should you log on to the Internet and spend a few minutes with your favorite role-playing game? Should you check your personal email or contact your friends about plans for the weekend? Before making a decision, consider a series of reports issued by the American Management Association (AMA). These reports suggest that the chances are growing that you may be the subject of workplace surveillance and that your email, telephone, and, more recently, your Internet use may be monitored by your employer.

Monitoring and Surveillance in 2005. Although the computer has become an important component of the workday for more and more people, your employer may feel compelled to monitor its use. Table 2.21 reports the percentage of companies that indicated they practiced a specific form of monitoring and surveillance.

Graphs are almost always a more effective method of presenting this type of information. The variable (type of monitoring and surveillance) is nominal level (the "types" are different from each other but do not form a scale), and, with nine possible scores, a bar chart would be preferred to a pie chart. Figure 2.8 shows that monitoring Internet connections was the most common form of surveillance, with about 75% of the companies practicing it. Storage and review of email messages (about 55%) and telephone monitoring (about 51%) and were also very common. This graph clearly indicates that it would be unwise to surf the net on company time or to use the phone or email for personal business.

Monitoring and Surveillance Over Time. What changes occurred in specific workplace monitoring practices between 1997 and 2005? Table 2.22 displays trends for five types of monitoring and surveillance. In 1997, the levels of monitoring were relatively low. Only about one-third of companies monitored telephone use, and no more than 16% practiced the other forms of surveillance. The levels rise quite dramatically over the period, and by 2005 three different forms of surveillance were being practiced by half or more of the companies.

Once again, these trends and patterns would be more clearly presented and appreciated in the form of a graph. Figure 2.9 presents the information in Table 2.22 in the form of a line graph. The graph clearly displays the general increase

TABLE 2.21 MONITORING AND SURVEILLANCE, 2005

Type of Monitoring and Surveillance	% Yes
Monitoring Internet connections	76%
Storage and review of e-mail messages	55%
Telephone use (time spent, numbers called)	51%
Video surveillance for security purposes	51%
Storage and review of computer files	50%
Computer use (time logged on, keystroke counts, etc.)	36%
Recording and review of telephone conversations	22%
Video recording of employee job performance	16%
Storage and review of voice mail messages	15%

FIGURE 2.8 MONITORING AND SURVEILLANCE OF EMPLOYEES, 2005

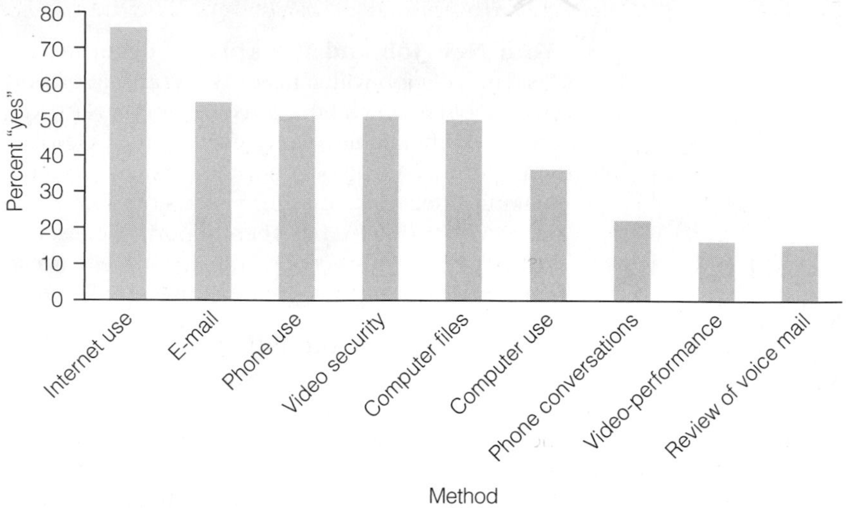

FIGURE 2.9 MONITORING AND SURVEILLANCE, 1997–2005

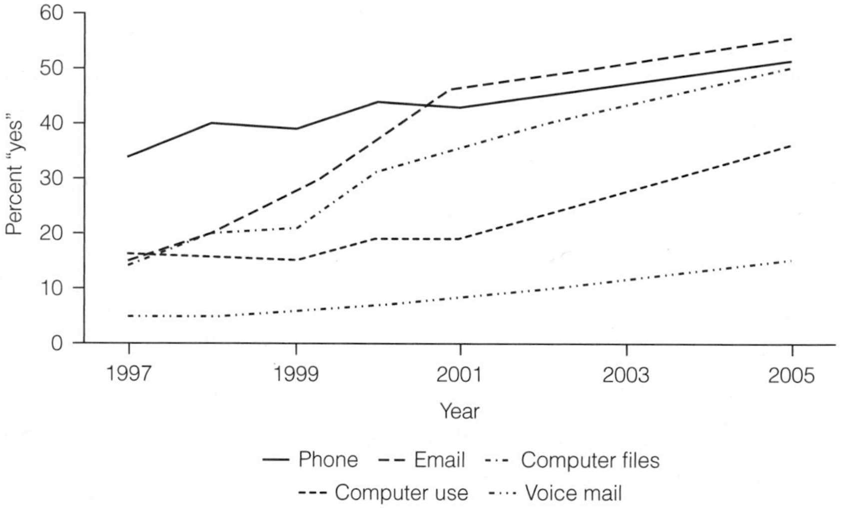

TABLE 2.22 MONITORING AND SURVEILLANCE, 1997–2005

	Percent "Yes"					
Type of Monitoring and Surveillance	1997	1998	1999	2000	2001	2005
Storage and review of voice mail messages	5	5	6	7	8	15
Storage and review of computer files	14	20	21	31	36	50
Storage and review of email messages	15	20	27	38	47	55
Telephone use (time spent, numbers called)	34	40	39	44	43	51
Computer use (time logged on, keystroke counts, etc.)	16	16	15	19	19	36

READING STATISTICS 2: Percentages, Rates, Tables, and Graphs

You will find that the statistics covered in this chapter are frequently used in the research literature of the social sciences—as well as in the popular press and the media—and one of the goals of this text is to help you develop your skills in understanding and critically analyzing these types of statistical information. Fortunately, this task is usually quite straightforward, but these statistical tools are sometimes not as simple as they appear and they can be misused. Here are some ideas to keep in mind when reading research reports that use these statistics.

First, there are many different formats for presenting results, and the tables and graphs you find in the research literature will not necessarily follow the conventions used in this text. Second, because of space limitations, tables and graphs may be presented with a minimum of detail. For example, the researcher may present a frequency distribution with only a percentage column.

Begin your analysis by examining the statistics carefully. If you are reading a table or graph, first read the title, all labels (that is, row and/or column headings), and any footnotes. These will tell you exactly what information is being presented. Inspect the body of the table or graph with the author's analysis in mind. See if you agree with the author's analysis. (You almost always will, but it never hurts to double-check and exercise your critical abilities.)

Finally, remember that most research projects analyze interrelationships among many variables. Because the tables and graphs covered in this chapter display variables one at a time, they are unlikely to be included in such research reports (or perhaps, included only as background information). Even when not reported, you can be sure that the research began with an inspection of percentages, frequency distributions, or graphs for each variable. Univariate tables and graphs display a great deal of information about the variables in a compact, easily understood format and are almost universally used as descriptive devices.

STATISTICS IN THE PROFESSIONAL LITERATURE

Social scientists rely heavily on the U.S. census for information about the characteristics and trends of change in American society, including age composition, birthrates and death rates, residential patterns, educational levels, and a host of other variables. Census data is readily available (at www.census.gov), but since they represent information about the entire population (almost 290 million people), the numbers are often large, cumbersome, and awkward to use or understand. Thus, percentages, rates, and graphs are extremely useful statistical devices when analyzing or presenting census information.

Consider, for example, a recent report on the changing U.S. family.* The purpose of the report was to present information regarding the structure of the American family and to present and discuss recent changes and trends. Consider how this report might have read if the information had been given in words and raw numbers:

> In 2003, there were about 57,320,000 million married-couple households, 13,620,000 million female-headed households, and 35,682,000 million nonfamily households. Ten years earlier, in 1993, there were 52,457,000 married-couple households, 11,692,000 female-headed households, and 28,496,000 nonfamily households.

Can you distill any meaningful understandings about American family life from these sentences? Raw information simply does not speak for itself, and these facts have to be organized or placed in some context to reveal their meaning. Thus, social scientists almost always use percentages, rates, or graphs to present this kind of info so that they can understand it themselves, assess the meaning, and convey their interpretations to others.

In contrast with the foregoing raw information, consider the following table on family trends using percentages rather than raw numbers.

U.S. Households by Type, 1970 and 2003

	Percent of All Households	
	1970	2003
Family households:		
Married couples with children	40.3%	23.3%

(*continued next page*)

READING STATISTICS 2: (*continued*)

Married couples without children	30.3%	28.2%
Other	10.6%	16.4%
Nonfamily households:		
Women living alone	5.6%	11.2%
Men living alone	11.5%	15.2%
Other	1.7%	5.6%
	100.0%	99.9%

A quick comparison of the two years reveals a dramatic decrease in the percentage of American households consisting of married couples living with children and an increase in the percentages of men and women living alone. What other trends can you see in the table?

MARTIAL STATUS OF U.S. POPULATION, 1970–2003 (15 years of age or older, male)

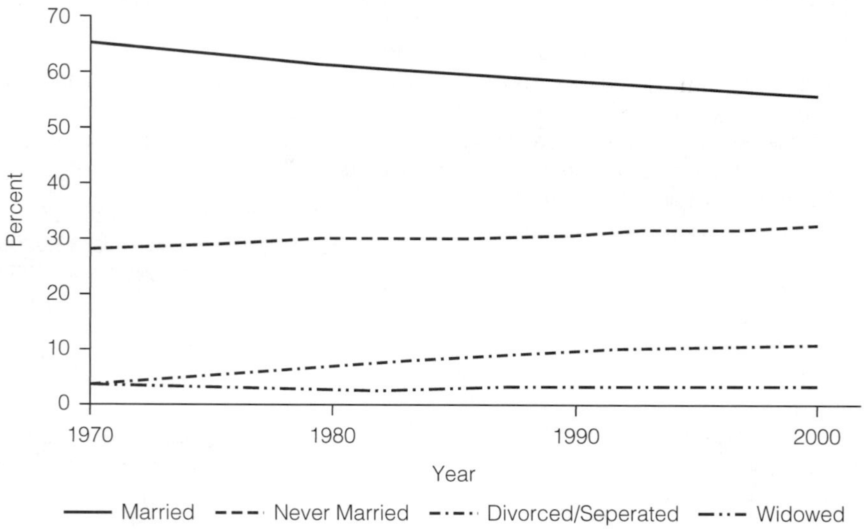

MARTIAL STATUS OF U.S. POPULATION, 1970–2003 (15 years of age or older, female)

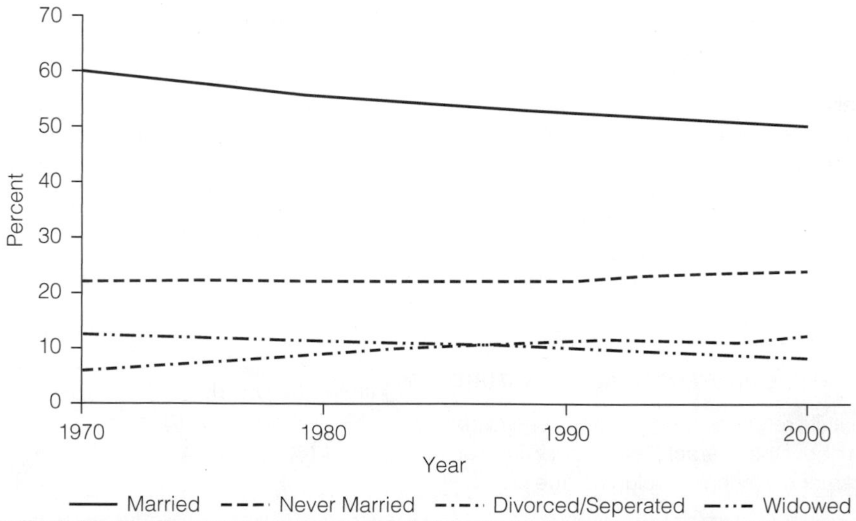

READING STATISTICS 2: *(continued)*

As we have seen, graphs are almost always more efficient and understandable ways of expressing trends. Some of the fundamental changes in American family life are presented in the two line charts on page 50, one for men and one for women. As you would expect, the graphs show essentially the same trends and, together with the frequency distribution, their message is pretty clear: The percentage of the population living in "married-couple households" is declining, and this is particularly true for married couples with children. Note also that the percentage of men and women who "never married" is increasing steadily. The author of the report attributes these changes to several factors, including people waiting longer to get married and a high divorce rate.

*Fields, Jason. 2004. "America's Families and Living Arrangements: 2003." U.S. Bureau of the Census: *Current Population Reports*. Available at http://www.census.gov/prod/2004pubs/p20-553.pdf

in monitoring and shows that monitoring of email messages, phone use, and computer files have become particularly common.

Computers may be ubiquitous features of employment in our information-age economy, but these data suggest that the potential for workplace monitoring and surveillance is also increasing. The computer on your desk is a double-edged sword. While it provides you with the necessary tools to do your job, employers are increasingly using the very same technology to watch you.

SUMMARY

1. We considered several different ways of summarizing the distribution of a single variable and, more generally, reporting the results of our research. Our emphasis throughout was on the need to communicate our results clearly and concisely. You will often find that, as you strive to communicate statistical information to others, the meanings of the information will become clearer to you as well.
2. Percentages and proportions, ratios, rates, and percentage change represent several different ways to enhance clarity by expressing our results in terms of relative frequency. Percentages and proportions report the relative occurrence of some category of a variable compared with the distribution as a whole. Ratios compare two categories with each other, and rates report the actual occurrences of some phenomenon compared with the number of possible occurrences per some unit of time. Percentage change shows the relative increase or decrease in a variable over time.
3. Frequency distributions are tables that summarize the entire distribution of some variable. Statistical analysis almost always starts with the construction and review of these tables for each variable. Columns for percentages, cumulative frequency, and/or cumulative percentages often enhance the readability of frequency distributions.
4. Pie and bar charts, histograms, and line charts (or frequency polygons) are graphic devices used to express the basic information contained in the frequency distribution in a compact and visually dramatic way.

SUMMARY OF FORMULAS

Proportions 2.1 $p = \frac{f}{N}$

Percentage 2.2 $\% = \left(\frac{f}{N}\right) \times 100$

Ratios 2.3 $\text{Ratio} = \frac{f_1}{f_2}$

Percent change 2.4 $\text{Percent change} = \left(\frac{f_2 - f_1}{f_1}\right) \times 100$

GLOSSARY

Bar chart. A graphic display device for discrete variables. Categories are represented by bars of equal width, the height of each corresponding to the number (or percentage) of cases in the category.
Class intervals. The categories used in the frequency distributions for interval-ratio variables.
Cumulative frequency. An optional column in a frequency distribution that displays the number of cases within an interval and all preceding intervals.
Cumulative percentage. An optional column in a frequency distribution that displays the percentage of cases within an interval and all preceding intervals.
Frequency distribution. A table that displays the number of cases in each category of a variable.
Frequency polygon. A graphic display device for interval-ratio variables. Class intervals are represented by dots placed over the midpoints, the height of each corresponding to the number (or percentage) of cases in the interval. All dots are connected by straight lines. Same as a **line chart.**
Histogram. A graphic display device for interval-ratio variables. Class intervals are represented by contiguous bars of equal width (equal to the class limits), the height of each corresponding to the number (or percentage) of cases in the interval.
Line chart. See **Frequency polygon.**
Midpoint. The point exactly halfway between the upper and lower limits of a class interval.
Percentage. The number of cases in a category of a variable divided by the number of cases in all categories of the variable, the entire quantity multiplied by 100.
Percent change. A statistic that expresses the magnitude of change in a variable from time 1 to time 2.
Pie chart. A graphic display device especially for discrete variables with only a few categories. A circle (the pie) is divided into segments proportional in size to the percentage of cases in each category of the variable.
Proportion. The number of cases in one category of a variable divided by the number of cases in all categories of the variable.
Rate. The number of actual occurrences of some phenomenon or trait divided by the number of possible occurrences per some unit of time.
Ratio. The number of cases in one category divided by the number of cases in some other category.
Real class limits. The class intervals of a frequency distribution when stated as continuous categories.
Stated class limits. The class intervals of a frequency distribution when stated as discrete categories.

PROBLEMS

2.1 SOC The tables that follow report the marital status of 20 respondents in two different apartment complexes. (*HINT: Make sure that you have the correct numbers in the numerator and denominator before solving the following problems. For example, problem 2.1a asks for "the percentage of respondents who are married in each complex," and the denominators will be 20 for these two fractions. Problem 2.1d, on the other hand, asks for the "percentage of the single respondents who live in Complex B," and the denominator for this fraction will be 4 + 6, or 10.*)

Status	Complex A	Complex B
Married	5	10
Unmarried ("living together")	8	2
Single	4	6
Separated	2	1
Widowed	0	1
Divorced	1	0
	20	20

a. What percentage of the respondents in each complex are married?
b. What is the ratio of single to married respondents at each complex?
c. What proportion of each sample are widowed?
d. What percentage of the single respondents live in Complex B?
e. What is the ratio of the "unmarried/living together" to the "married" at each complex?

2.2 At St. Algebra College, the numbers of males and females in the various major fields of study are as follows:

Major	Males	Females	Totals
Humanities	117	83	200
Social sciences	97	132	229
Natural sciences	72	20	92

(*continued next page*)

(*continued*)

Major	Males	Females	Totals
Business	156	139	295
Nursing	3	35	38
Education	30	15	45
Totals	475	424	899

Read each of the following problems carefully before constructing the fraction and solving for the answer. (*HINT: Be sure you place the proper number in the denominator of the fractions. For example, some problems use the total number of males or females as the denominator, whereas others use the total number of majors*)

a. What percentage of social science majors are male?

b. What proportion of business majors are female?

c. For the humanities, what is the ratio of males to females?

d. What percentage of the total student body are males?

e. What is the ratio of males to females for the entire sample?

f. What proportion of the nursing majors are male?

g. What percentage of the sample are social science majors?

h. What is the ratio of humanities majors to business majors?

i. What is the ratio of female business majors to female nursing majors?

j. What proportion of the males are education majors?

2.3 CJ The town of Shinbone, Kansas, has a population of 211,732 and experienced 47 bank robberies, 13 murders, and 23 auto thefts during the past year. Compute a rate for each type of crime per 100,000 population. (*HINT: Make sure that you set up the fraction with size of population in the denominator*)

2.4 CJ The numbers of homicides in five states and five Canadian provinces for the years 1997 and 2005 were as follows:

	1997		2005	
State	Homicides	Population	Homicides	Population
New Jersey	338	8,053,000	417	8,717,925
Iowa	52	2,852,000	38	2,966,334
Alabama	426	4,139,000	374	4,557,808
Texas	1327	19,439,000	1407	22,859,968
California	2579	32,268,000	2503	36,132,147

	1997		2005	
Province	Homicides	Population	Homicides	Population
Nova Scotia	24	936,100	20	936,100
Quebec	132	7,323,600	100	7,597,800
Ontario	178	11,387,400	218	12,558,700
Manitoba	31	1,137,900	49	1,174,100
British Columbia	116	3,997,100	98	4,257,800

Source: Statistics Canada, http://www.statcan.ca.

a. Calculate the homicide rate per 100,000 population for each state and each province for each year. Relatively speaking, which state and which province had the highest homicide rates in each year? Which society seems to have the higher homicide rate? Write a paragraph describing these results.

b. Using the rates you calculated in part a, calculate the percent change between 1997 and 2005 for each state and each province. Which states and provinces had the largest increase and decrease? Which society seems to have the largest change in homicide rates? Summarize your results in a paragraph

2.5 SOC The scores of 15 respondents on four variables are as reported next. These scores were taken from a public opinion survey called the General Social Survey, or the GSS. This data set is used for the computer exercises in this text. Small subsamples from the GSS will be used throughout the text to provide "real" data for problems. For the actual questions and other details, see Appendix G. The numerical codes for the variables are as follows:

Sex	Support for Gun Control	Level of Education	Age
Male	1 = In favor	0 = Less than HS	Actual years
Female	2 = Opposed	1 = HS	
		2 = Jr. college	
		3 = Bachelor's	
		4 = Graduate	

Case Number	Sex	Support for Gun Control	Level of Education	Age
1	2	1	1	45
2	1	2	1	48
3	2	1	3	55
4	1	1	2	32
5	2	1	3	33
6	1	1	1	28
7	2	2	0	77
8	1	1	1	50
9	1	2	0	43
10	2	1	1	48
11	1	1	4	33
12	1	1	4	35
13	1	1	0	39
14	2	1	1	25
15	1	1	1	23

a. Construct a frequency distribution for each variable. Include a column for percentages.

b. Construct pie and bar charts to display the distributions of sex, support for gun control, and level of education.

2.6 SW A local youth service agency has begun a sex education program for teenage girls who have been referred by the juvenile courts. The girls were given a 20-item test for general knowledge about sex, contraception, and anatomy and physiology upon admission to the program and again after completing the program. The scores of the first 15 girls to complete the program are as follows.

Case	Pretest	Posttest	Case	Pretest	Posttest
A	8	12	I	5	7
B	7	13	J	15	12
C	10	12	K	13	20
D	15	19	L	4	5
E	10	8	M	10	15
F	10	17	N	8	11
G	3	12	O	12	20
H	10	11			

Construct frequency distributions for the pretest and posttest scores. Include a column for percentages. (*HINT: There were 20 items on the test, so the maximum range for these scores is 20. If you use 10 class intervals to display these scores, the interval size will be 2. Since there are no scores of 0 or 1 for either test, you may state the first interval as 2–3. To make comparisons easier, both frequency distributions should have the same intervals*)

2.7 SOC Sixteen high school students completed a class to prepare them for the College Boards. Their scores were as follows.

420	345	560	650
459	499	500	657
467	480	505	555
480	520	530	589

These same 16 students were given a test of math and verbal ability to measure their readiness for college-level work. Scores are reported here in terms of the percentage of correct answers for each test.

Math Test

67	45	68	70
72	85	90	99
50	73	77	78
52	66	89	75

Verbal Test

89	90	78	77
75	70	56	60
77	78	80	92
98	72	77	82

a. Display each of these variables in a frequency distribution with columns for percentages and cumulative percentages.

b. Construct a histogram and frequency polygon for these data.

c.[2] Find the upper and lower real limits for the intervals you established.

2.8 GER Following are reported the number of times 25 residents of a community for senior citizens left their homes for any reason during the past week.

0	2	1	7	3
7	0	2	3	17
14	15	5	0	7
5	21	4	7	6
2	0	10	5	7

a. Construct a frequency distribution to display these data.

b. What are the midpoints of the class intervals?

c. Add columns to the table to display the percentage distribution, cumulative frequency, and cumulative percentages.

[2] This problem is optional.

d.[3] Find the real limits for the intervals you selected.

e. Construct a histogram and a frequency polygon to display this distribution.

f. Write a paragraph summarizing this distribution of scores.

2.9 SOC Twenty-five students completed a questionnaire that measured their attitudes toward interpersonal violence. Respondents who scored high believed that in many situations a person could legitimately use physical force against another person. Respondents who scored low believed that in no situation (or very few situations) could the use of violence be justified.

52	47	17	8	92
53	23	28	9	90
17	63	17	17	23
19	66	10	20	47
20	66	5	25	17

a. Construct a frequency distribution to display these data.

b. What are the midpoints of the class intervals?

c. Add columns to the table to display the percentage distribution, cumulative frequency, and cumulative percentage.

d. Construct a histogram and a frequency polygon to display these data.

e. Write a paragraph summarizing this distribution of scores.

2.10 PA/CJ As part of an evaluation of the efficiency of your local police force, you have gathered the following data on police response time to calls for assistance during two different years. (Response times were rounded off to whole minutes.) Convert both frequency distributions into percentages, and construct pie charts and bar charts to display the data. Write a paragraph comparing the changes in response time between the two years.

Response Time, 1995	Frequency (*f*)
21 minutes or more	35
16–20 minutes	75
11–15 minutes	180
6–10 minutes	375
Less than 6 minutes	210
	875

Response Time, 2005	Frequency (*f*)
21 minutes or more	45
16–20 minutes	95
11–15 minutes	155
6–10 minutes	350
Less than 6 minutes	250
	895

2.11 SOC Figures 2.10 through 2.12 display trends in crime in the United States over the last two decades. Write a paragraph describing each of these graphs. What similarities and differences can you observe among the three graphs? (For example, do crime rates always change in the same direction?) Note the differences in the vertical axes from chart to chart—for homicide the axis ranges from 0 to 12, while for burglary and auto theft the range is from 0 to 1600. The latter crimes

[3] This problem is optonal.

FIGURE 2.10 U.S. HOMICIDE RATES, 1984–2005 (per 100,000 population)

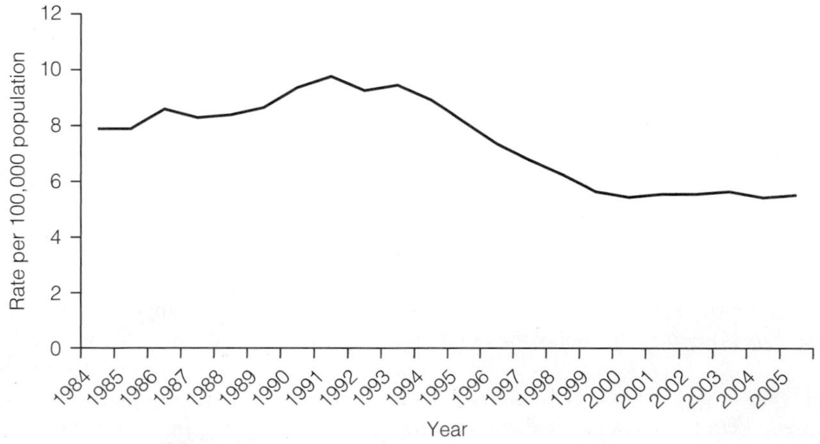

FIGURE 2.11 U.S. ROBBERY AND AGGRAVATED ASSAULT RATES, 1984–2005 (per 100,000 population)

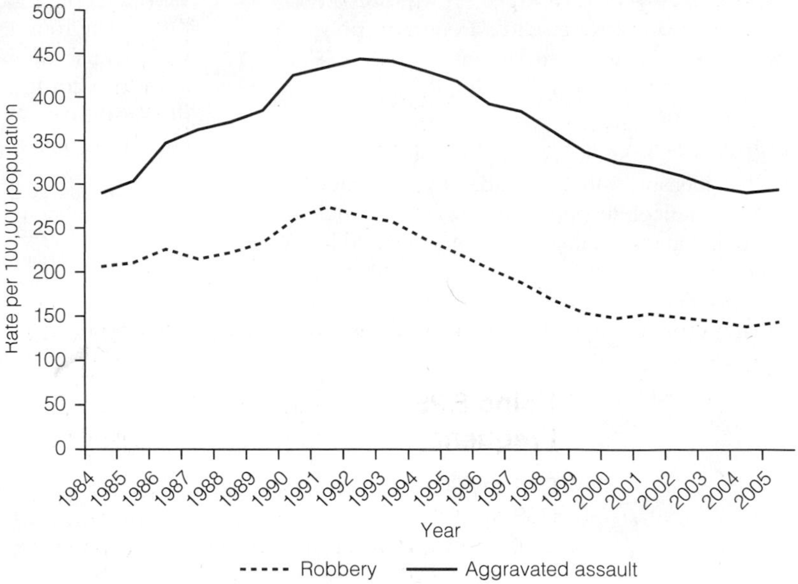

FIGURE 2.12 U.S. BURGLARY AND CAR THEFT RATES, 1984–2005 (per 100,000 population)

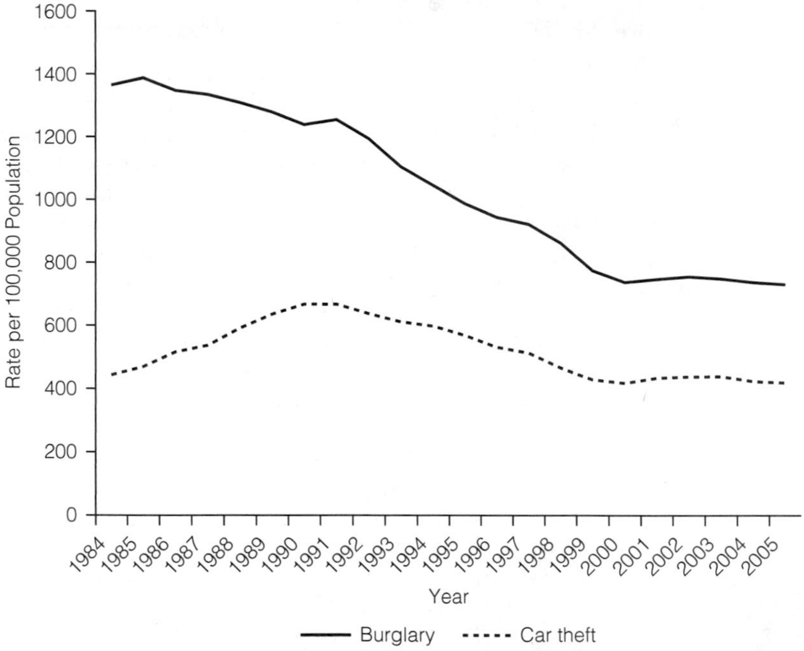

are far more common, and a scale with smaller intervals is needed to display the rates.

2.12 PA The city's Department of Transportation has been keeping track of accidents on a particularly dangerous stretch of highway. Early in the year, the city lowered the speed limit on this highway and increased police patrols. Data on the number of accidents before and after the changes are presented here. Did the changes work? Is the highway safer? Construct a line chart to display these two sets of data (use graphics software if available), and write a paragraph describing the changes.

Month	12 Months Before	12 Months After
January	23	25
February	25	21
March	20	18
April	19	12
May	15	9
June	17	10
July	24	11
August	28	15
September	23	17
October	20	14
November	21	18
December	22	20

SPSS FOR WINDOWS

Using SPSS for Windows to Produce Frequency Distributions and Graphs

Click the SPSS icon on your monitor screen to start SPSS for Windows. Load the 2006 GSS by clicking the file name on the first screen or by clicking **file, Open,** and **Data** on the **SPSS Data Editor screen.** You may have to change the drive specification to locate the 2006 GSS data supplied with this text (probably named **GSS2006.sav**). Double-click the file name to open the data set. When you see the message "SPSS Processor is Ready" on the bottom of the screen, you are ready to proceed.

SPSS DEMONSTRATION 2.1 Frequency Distributions

We produced and examined a frequency distribution for the variable *sex* in Appendix F. Use the same procedures to produce frequency distributions for the variables *age* and *marital* (marital status). From the menu bar, click **Analyze.** From the menu that drops down, click **Descriptive Statistics** and **Frequencies.** The **Frequencies** window appears, with the variables listed in alphabetical order in the left-hand box. The window may display variables by name (e.g. abany, abhlth) or by label (e.g., ABORTION IF WOMAN WANTS FOR ANY REASON). If labels are displayed, you may switch to variable names by clicking **Edit, Options,** and then making the appropriate selections on the "General" tab. Depending on the version of SPSS you are using, these changes may not take effect until you load a new data set or restart SPSS. See Appendix F and Table F.2 for further information. The variable *age* (AGE OF RESPONDENT) will be visible. Click on it to highlight it, and then click the arrow button in the middle of the screen to move *age* to the right-hand window.

Find *marital* in the left-hand box by using the slider button or the arrow keys on the right-hand border to scroll through the variable list. As an alternative, type "m"; the cursor will move to the first variable name in the list that begins with that letter. Highlight *marital* and click the arrow button in the center of the screen to move the variable name to the **Variables** box. There should now be two variable names in the box, *age* and *marital*. SPSS will process together all variables listed in the right-hand box. Click **OK** in the upper-right-hand corner, and SPSS will rush off to create the frequency distributions you requested.

The table will be in the **Output** window that will now be "closest" to you on the screen. The tables, along with other information, will be in the right-hand box of the **Output** window. To change the size of the output window, click the middle symbol (shaped like either a square or two intersecting squares) in the upper-right-hand corner of the **Output** window.

The frequency distribution for *marital* will look like this:

MARITAL STATUS

		Frequency	Percent	Valid Percent	Cumulative Percent
Valid	MARRIED	686	48.1	48.1	48.1
	WIDOWED	119	8.3	8.4	56.5
	DIVORCED	222	15.6	15.6	72.1
	SEPARATED	39	2.7	2.7	74.8
	NEVER MARRIED	359	25.2	25.2	100.0
	Total	1425	99.9	100.0	
Missing	NA	1	.1		
Total		1426	100.0		

Let's briefly examine the elements of this table. The variable description or label is printed at the top of the output ("MARITAL STATUS"). The various categories are printed on the left. Moving one column to the right, we find the actual frequencies, or the number of times each score of the variable occurred. We see that 686 of the respondents were married, 119 were widowed, and so forth. Next are two columns that report percentages. The entries in the "Percent" column are based on all respondents who were asked this question, including those coded as "NA" (No Answer), "DK" (Don't Know), or "NAP" (Not Applicable). The "Valid Percent" column eliminates all cases with missing values. Since we almost always ignore missing values, we will pay attention only to the "Valid Percent" column (even though, in this case, all respondents except one supplied this information and there is virtually no difference between the two columns). The final column is a cumulative percent column (see Table 2.16). For nominal-level variables such as *marital*, this information is not meaningful, since the order in which the categories are stated is arbitrary.

Turning to the frequency distribution for *age*, note that the table follows the same format as the table for *marital* but is much longer. This, of course, reflects the fact that *age* has many more possible scores than *marital*—so many scores, in fact, that the table is not easy to read or understand. The table needs to be made more compact by collapsing scores into fewer categories. For example, if ages were grouped into categories 10 years wide (10–19, 20–29, etc.), the number of categories would be reduced to about eight. We could collapse the categories by hand (see Section 2.5), or we could have SPSS do the work. The procedures for recoding or collapsing variables are explained next, in Demonstration 2.2.

SPSS DEMONSTRATION 2.2 Collapsing Categories with the Recode Command

The scores of interval-ratio-level variables will often have to be collapsed in order to produce readable frequency distributions. SPSS for Windows provides a number of ways to change the scores of a variable, and one of the most useful of these is the **Recode** command. We will use this command to create a new version of *age* that has fewer categories and is more suitable for display in a frequency distribution. When we are finished, we will have two different versions of the same variable in the data set: the original, interval-ratio version, with age measured in years, and a new, ordinal-level version, with collapsed categories. If we wish, the new version of age can be added to the permanent data file and used in the future.

We will collapse the values of *age* into categories of 10 years each. Since the youngest respondents are 18, let's (arbitrarily) begin with an interval of 10–19. The next interval will be 20–29, followed by 30–39, and so forth until the interval 80–89. As you can see from the frequency distribution for age, 12 individuals in the sample are coded as "89 or older." Since we do not know exactly how old these people are, we will simply include them in the 80-89 category.

As you will see, recoding requires many small steps, so please be patient and execute the commands as they are discussed.

1. In the **SPSS Data Editor** window, click **Transform** from the menu bar and then click **Recode.** A window will open that gives us two choices: **into same variable** and **into different variable,** If we choose the former (**into same variable**), the new version of the variable will replace the old version—the original version of *age* (with actual years) would disappear. We definitely do *not* want this to happen, so we will choose (click on) **into different variable,** This option will allow us to keep both the old and new versions of the variable.
2. The **Recode into Different Variable** window will open. A box containing an alphabetical list of variables will appear on the left. Use the cursor to highlight *age*, and then click on the arrow button to move the variable to the **Input Variable → Output Variable** box. The input variable is the old version of age, and the output variable is the new, recoded version we will soon create.
3. In the **Output Variable** box on the right, click in the **Name** box and type a name for the new (output) variable. I suggest *ager* (age recoded) for the new variable, but you can assign any name as long as it does not duplicate the name of some other variable in the data set and is no longer than eight characters. Click the **Change** button and the expression *age* → *ager* will appear in the **Input Variable → Output Variable** box.
4. Click on the **Old and New Values** button in the middle of the screen, and a new dialog box will open. Read down the left-hand column until you find the **Range** button. Click on the button, and the cursor will move to the small box immediately below. In these boxes we will specify the low and high points of each interval of the new variable *ager*.
5. Starting with the interval 10–19, type 10 into the left-hand **Range** dialog box, and then click on the right-hand box and type 19. In the **New Value** box in the upper-right-hand corner of the screen, click the **Value** button. Type 1 in the **Value** dialog box and then click the **Add** button directly below. The expression 10–19 → 1 will appear in the **Old → New** dialog box. This completes the first recode instruction to SPSS.
6. Continue recoding by returning to the **Range** dialog boxes on the left. Type 20 in the left-hand box and 29 in the right-hand box, and then click the **Value** button in the **New Values** box. Type 2 in the **Value** dialog box and then click the **Add** button. The expression 20–29 → 2 appears in the **Old → New** dialog box.
7. Continue this sequence of operations until all old values of *age* have been recoded into new values. The last expression in the **Old → New** dialog box should be 80–89 → 8. Now click the **Continue** button at the bottom of the screen, and you will return to the **Recode into Different Variable** dialog box. Click **OK,** and SPSS will execute the transformation.

You now have a data set with one more variable, named *ager* (or whatever name you gave the recoded variable). SPSS adds the new variable to the data set, and you can find it in the last column to the right in the data window. You can make the new variable a permanent part of the data set by saving the data file at the end of the session. If you do not wish to save the new, expanded data file, click No when you are asked if you want to save the data file. If you are using the student version of SPSS for Windows, remember that you are limited to a maximum of 50 variables, so you may not be able to save the new variable.

To produce a frequency distribution for the new variable, click **Analyze, Descriptive Statistics,** and **Frequencies.** find *ager* in the left-hand box and click the arrow key to move it to the right-hand box. Click **OK,** and you will soon be looking at the following table:

AGER

		Frequency	Percent	Valid Percent	Cumulative Percent
Valid	1.00	18	1.3	1.3	1.3
	2.00	240	16.8	16.9	18.2
	3.00	283	19.8	20.0	38.2
	4.00	285	20.0	20.1	58.3
	5.00	256	18.0	18.1	76.4
	6.00	166	11.6	11.7	88.1
	7.00	110	7.7	7.8	95.8
	8.00	59	4.1	4.2	100.0
	Total	1417	99.4	100.0	
Missing	System	9	.6		
Total		1426	100.0		

There are few respondents in their teens (category 1, or ages 10–19) and their 80s (category 8). The sample is clustered in the 30s, 40s, and 50s. Almost 60% of the sample falls into categories 3, 4, and 5, that is, are in their 30s, 40s, and 50s. We are missing data on only nine respondents for this variable. If you want to retain the data set with *ager* added, remember to save the data file at the end of your session.

SPSS DEMONSTRATION 2.3 Graphs and Charts

SPSS for Windows can produce a variety of graphs and charts, and we will use the program to produce a bar chart and a line chart (frequency polygon) in this demonstration. To conserve space, I will keep the choices as simple as possible, but you should

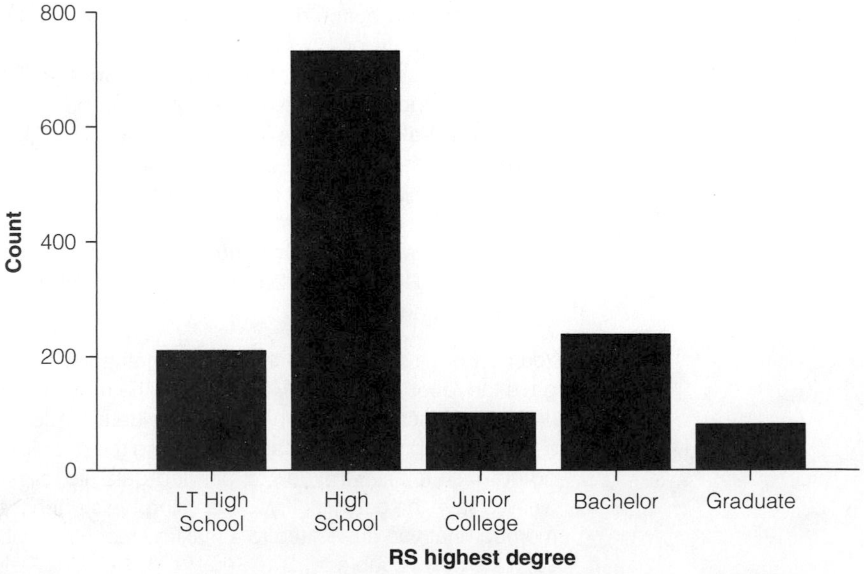

explore the options for yourself. For any questions you might have that are not answered in this demonstration, click on Help on the main menu bar.

To produce a bar chart, first click on **Graphs** on the main menu bar and then click **Bar.** The Bar Chart dialog box appears, with three choices for the type of graph we want. The **Simple** option is already highlighted, and this is the one we want, so just click **Define.** The **Define Simple Bar** dialog box appears, with variable names listed on the left. Choose *degree* from the variable list by moving the cursor to highlight this variable name. Click the arrow button in the middle of the screen to move degree to the **Category Axis** box.

Note the **Bars Represent** box above the **Category Axis** box. This dialog box gives you control over the vertical axis of the graph, which can be calibrated in frequencies, percentages, or cumulative frequencies or percentages. Let's choose **N of cases or frequencies,** the option that is already selected.

Click the **Options** button in the lower-right-hand corner. In the **Options** dialog box, make sure that the button next to the **Display groups defined by missing values** option is *not* selected (or checked). Click **Continue** and then click **OK** in the **Define Simple Bar** dialog box, and the following bar chart will be produced:

Click on any part of the graph, and the **SPSS Chart Editor** window will appear. The menu bar across the top of the window gives you a wide array of options for the final appearance of the chart, including—under the **Format** command—the color of the bars and their borders. Explore these options at your leisure using the **Help** function as necessary.

Turning to the chart itself, by far the most common level of education for this sample was "high school," with college ("bachelor") and less than high school ("LT high school") second and third most common. When you are ready, the bar chart can be printed or saved by selecting the appropriate command from the **File** menu.

The procedures for producing a line chart are very similar to those for a bar chart. Click **Graphs** and then **Line.** The **Line Charts** dialog box will appear, with the **Simple**

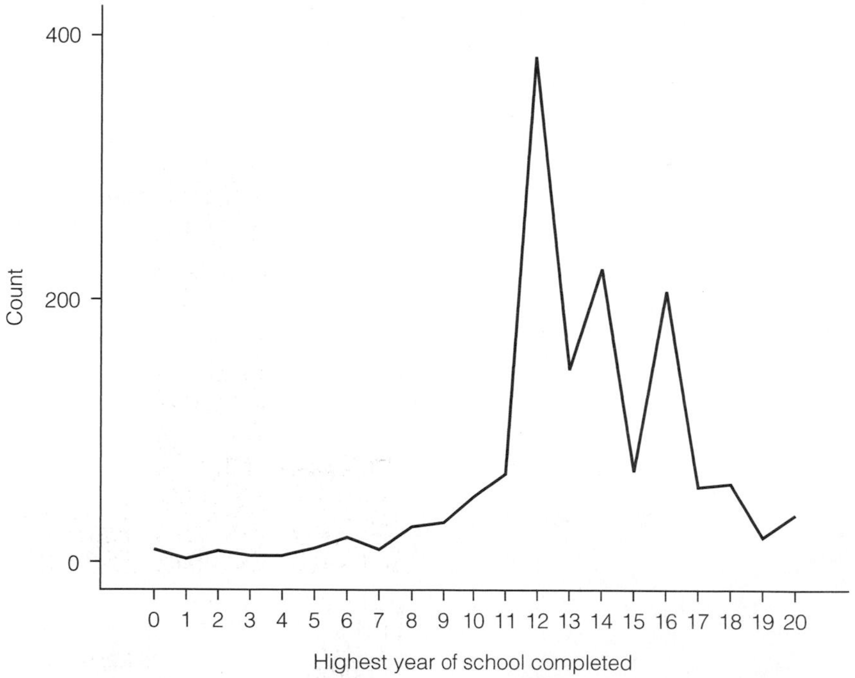

option already chosen. This is the one we want, so click **Define,** and the **Define Simple Line** dialog box will appear. Your choices here are the same as in the **Define Simple Bar** dialog box. Line charts are appropriate for interval-ratio data, so let's use *educ* (years of education) for this demonstration. You might think of this variable as a noncollapsed form of degree, and we should be able to compare the two variables quickly and easily.

Select *educ* from the variable list on the left, and click the arrow button in the middle of the screen to move *educ* to the **Category Axis** box. Note that the **Lines Represent** box is the same as in the **Simple Bar** dialog box. We want the bars to represent frequencies, and this option is already selected. Click the **Options** button, and make sure that the **Display Groups Defined by Missing Values** option is not selected (or checked). Click **Continue** and then click **OK** in the **Define Simple Line** dialog box, and the following line chart will be produced:

The line chart can be edited by clicking on any part of the chart and activating the Chart Editor window. For example, to change the color of the line, click anywhere on the line in the Chart Editor window, then click **Format** and **Color.** Click on the new color and then click **Apply.** Don't forget to **Save** or **Print** the chart if you wish.

The chart has a very high peak at 12 years of education (high school) and other peaks at 14 (AA or other two-year degrees) and 16 (bachelor or four-year degrees). This essentially duplicates the information in the bar chart for degree.

Exercises

2.1 Get frequency distributions for 5 or 10 nominal or ordinal variables, including race (*raccen1*) and religion (*relig*). Get bar charts for each variable and write a sentence or two summarizing each frequency distribution. Your description should clearly identify the most and least common scores and any other noteworthy patterns you observe.

2.2 Get frequency distributions for the two variables (*degree* and *educ*) that measure education. Use the **Recode** command to collapse *educ* into categories comparable to degree. That is, use the categories 0–11 for less than high school, 12 for high school graduates, 13–15 to approximate the number of junior college or two-year program graduates, 16 for college graduates, and 17–20 for graduate degrees. Get a frequency distribution for recoded *educ* and compare it to the frequency distribution for *degree*. Describe the three frequency distributions in a few sentences.

2.3 Get a frequency distribution for *prestg80* (respondent's occupational prestige). Find the range of this variable and use this information to determine reasonable class intervals by following the procedures described in Section 2.5. Use the **Recode** command to produce a final frequency distribution for the recoded variable, and write a sentence or two of description and interpretation.

2.4 Choose five or six more variables of interest to you and produce bar charts or line charts for each. Remember that bar charts are more appropriate for nominal and ordinal variables and line charts for interval-ratio variables. Write a sentence or two of interpretation for each chart.

3 Measures of Central Tendency

LEARNING OBJECTIVES

By the time you finish this chapter, you will be able to

1. Explain the purposes of measures of central tendency and interpret the information they convey.
2. Calculate, explain, and compare and contrast the mode, median, and mean.
3. Explain the mathematical characteristics of the mean.
4. Select an appropriate measure of central tendency according to level of measurement and skew.

3.1 INTRODUCTION

One clear benefit of frequency distributions, graphs, and charts is that they summarize the overall shape of a distribution of scores in a way that can be quickly comprehended. Almost always, however, you will need to report more detailed information about the distribution. Specifically, two additional kinds of statistics are extremely useful: some idea of the typical or average case in the distribution (for example, "The average starting salary for social workers is $39,000 per year") and some idea of how much variety or heterogeneity there is in the distribution ("In this state, starting salaries for social workers range from $31,000 per year to $42,000 per year"). The first kind of statistic, called **measures of central tendency,** is the subject of this chapter. The second kind of statistic, measures of dispersion, is presented in Chapter 4.

The three commonly used measures of central tendency—the mode, median, and mean—are all probably familiar to you. All three summarize an entire distribution of scores by describing the most common score (the mode), the score of the middle case (the median), or the average score (the mean) of that distribution. These statistics are powerful because they can reduce huge arrays of data to a single, easily understood number. Remember that the central purpose of descriptive statistics is to summarize or "reduce" data.

Even though they share a common purpose, the three measures of central tendency are quite different from each other. In fact, they will have the same value only under specific and limited conditions. As we shall see, they vary in terms of level-of-measurement considerations, and, perhaps more importantly, they also vary in terms of how they define central tendency—they will not necessarily identify the same score or case as "typical." Thus, your choice of an appropriate measure of central tendency will depend in part on the way you measure the variable and in part on the purpose of the research

3.2 THE MODE

The **mode** of any distribution of scores is the value that occurs most frequently. For example, in the set of scores 58, 82, 82, 90, 98, the mode is 82 because it occurs twice and the other scores occur only once.

TABLE 3.1 RELIGIOUS PREFERENCE (fictitious data)

Denomination	Frequency
Protestant	128
Catholic	57
Jew	10
None	32
Other	15
	N = 242

The mode is a simple statistic, most useful when you want a "quick and easy" indicator of central tendency and when you are working with nominal-level variables. In fact, the mode is the only measure of central tendency that you can use with nominal-level variables. Because these variables do not have numerical "scores," the mode of a variable measured at the nominal level would be its largest category. For example, Table 3.1 reports the religious affiliations of a fictitious sample of 242 respondents. The mode of this distribution, the single largest category, is Protestant.

If a researcher desires to report only the most popular or common value of a distribution or if the variable under consideration is nominal, then the mode is the appropriate measure of central tendency. However, keep in mind that the mode does have several limitations. First, some distributions have no mode at all (see Table 2.6). Other distributions have so many modes that the statistic loses its usefulness. For example, consider the distribution of test scores for a class of 100 students presented in Table 3.2. The distribution has modes at 55, 66, 78, 82, 90, and 97. Would reporting all six of these modes actually convey any useful information about central tendency in the distribution?

Second, the modal score of variables measured at the ordinal or interval-ratio level may not be central to the distribution as a whole. That is, *most common* does not necessarily mean "typical" in the sense of identifying the center of the distribution. For example, consider the rather unusual (but not impossible) distribution of scores on a statistics test presented in Table 3.3. The mode of the distribution is 93. Is this score very close to the majority of the scores? If the instructor summarized this distribution by reporting only the modal score, would he or she be conveying an accurate picture of the distribution as a whole? *(For practice in finding and interpreting the mode, see problems 3.1 to 3.7.)*

3.3 THE MEDIAN

Unlike the mode, the **median (Md)** always represents the exact center of a distribution of scores. The median is the score of the case that is in the exact middle of a distribution: Half the cases have scores higher than the median and half the cases have scores lower than the median. Thus, if the median family income for

TABLE 3.2 A DISTRIBUTION OF TEST SCORES

Scores (% correct)	Frequency
97	10
91	7
90	10
86	8
82	10
80	3
78	10
77	6
75	9
70	7
66	10
55	10
	N = 100

TABLE 3.3 A DISTRIBUTION OF TEST SCORES

Scores (% correct)	Frequency
58	2
60	2
62	3
64	2
66	3
67	4
68	1
69	1
70	1
93	5
	N = 24

ONE STEP AT A TIME **Finding the Median**

Step1: Array the scores in order from high score to low score

Step 2: Count the number of scores to see if *N* is odd or even.

If *N* is odd:

Step 3: The median will be the score of the middle case.

Step 4: To find the middle case, add 1 to *N* and divide by 2.

Step 5: The value you calculated in step 4 is the number of the middle case. The median is the score of this case. For example, if $N = 13$, the median will be the score of the $(13 + 1)/2$, or seventh, case.

If *N* is even:

Step 3: The median is halfway between the scores of the two middle cases.

Step 4: To find the first middle case, divide *N* by 2.

Step 5: To find the second middle case, increase the value you computed in step 4 by 1.

Step 6: Find the scores of the two middle cases. Add the scores together and divide by 2. The result is the median. For example, if $N = 14$, the median is the score halfway between the scores of the seventh and eighth cases.

a community is $45,000, half the families earn more than $45,000 and half earn less.

Before finding the median, the cases must be placed in order from the highest to the lowest score—or from the lowest to the highest. Then you can determine the median by locating the case that divides the distribution into two equal halves. The median is the score of the middle case. If five students received grades of 93, 87, 80, 75, and 61 on a test, the median would be 80, the score that splits the distribution into two equal halves.

When the number of cases (*N*) is odd, the value of the median is unambiguous because there will always be a middle case. With an even number of cases, however, there will be two middle cases and, in this situation, the median is defined as the score exactly halfway between the scores of the two middle cases.

To illustrate, assume that seven students were asked to indicate their level of support for the intercollegiate athletic program at their universities on a scale ranging from 10 (indicating great support) to 0 (no support). After arranging their

TABLE 3.4 FINDING THE MEDIAN WITH SEVEN CASES (N = odd)

Case	Score	
1	10	
2	10	
3	8	
4	7	← Md
5	5	
6	4	
7	2	

TABLE 3.5 FINDING THE MEDIAN

Case	Score	
1	10	
2	10	
3	8	
4	7	
		Md = (7 + 5)/2 = 6
5	5	
6	4	
7	2	
8	1	

responses from high to low, you can find the median by locating the case that divides the distribution into two equal halves. With a total of seven cases, the middle case would be the fourth case, since there will be three cases above and three cases below this case. Table 3.4 lists the cases in order and identifies the median. With seven cases, the median is the score of the fourth case.

Now suppose we added one more student, whose support for athletics was measured as a 1 to the sample. This would make N an even number (8) and we would no longer have a single middle case. Table 3.5 presents the new distribution of scores, and, as you can see, any value between 7 and 5 would technically satisfy the definition of a median (that is, it would split the distribution into two equal halves of four cases each). We resolve the ambiguity created by having an even number of cases by defining the median as the average of the scores of the two middle cases. In this example, the median would be defined as (7 + 5)/2, or 6.

We can now state these procedures in general terms:

- *When N is odd,* find the middle case by adding 1 to N and then dividing that sum by 2. This gives you the number of the middle case. The median is the score of that case. With an N of 7, the median is the score associated with the (7 + 1)/2, or fourth, case. If N had been 25, the median would be the score associated with the (25 + 1)/2, or 13th, case.
- *When N is even,* divide N by 2 to find the first middle case and then increase that number by 1 to find the second middle case. The median is the average of the scores of the two middle cases.[1] With an N of 8 cases, the first middle case would be the fourth case ($N/2 = 4$) and the second middle case would be the ($N/2$) + 1, or fifth, case. If N had been 142, the first middle case would have been the 71st case and the second the 72nd case. Remember that the median is defined as the average of the scores associated with the two middle cases.

The median cannot be calculated for variables measured at the nominal level because it requires that scores be ranked from high to low. Remember that the

[1] If the middle cases have the same score, that score is defined as the median. In the distribution 10, 10, 8, 6, 6, 4, 2, 1 the middle cases both have scores of 6, so the median would be defined as 6.

scores of nominal-level variables cannot be ordered or ranked: The scores are different from each other but do not form a mathematical scale of any sort. The median can be found for either ordinal or interval-ratio data but is generally more appropriate for the former. *(The median may be found for any problem at the end of this chapter.)*

3.4 OTHER MEASURES OF POSITION: PERCENTILES, DECILES, AND QUARTILES[2]

In addition to serving as a measure of central tendency, the median is a member of a class of statistics that measure position or location. The median identifies the exact middle of a distribution, but it is sometimes useful to locate other points as well. We may want to know, for example, the scores that split the distribution into thirds or fourths or the point below which a given percentage of the cases fall. A familiar application of these measures would be scores on standardized tests, which are often reported in terms of location (for example, "A score of 476 is higher than 46% of the scores").

One commonly used statistic for reporting position is the **percentie.** A percentile identifies the point below which a specific percentage of cases fall. If a score of 476 is reported as the 46th percentile, this means that 46% of the cases had scores lower than 476. To find a percentile, first arrange the scores in order. Next, multiply the number of cases (N) by the proportional value of the percentile. For example, the proportional value for the 46th percentile would be 0.46. The resultant value identifies the number of the case that marks the percentile. To find the 37th percentile for a sample of 78 cases, multiply 78 by 0.37. The result is 28.86, and the 37th percentile would be 86/100 of the distance between the scores of the 28th and 29th cases. In most cases, we would probably round off 28.86 to 29 and call the score of the 29th case the 37th percentile. The slight inaccuracy would be worth the savings in time and computational effort.

Note that, if we think in terms of percentiles, then the median is the 50th percentile, and in our example we would find the median by multiplying 78 by 0.50, finding the 39th case, and declaring the score of that case to be the 50th percentile. Notice again that we are cutting some corners here. Technically, the median would be the score halfway between the two middle cases (the 39th and 40th cases), but it is unlikely that this inaccuracy would be very significant.

Some other commonly used measures of position are **deciles** and **quartiles.** Deciles divide the distribution of scores into tenths. So, the first decile is the point below which 10% of the cases fall and is equivalent to the 10th percentile. The fifth decile is also the same as the 50th percentile, which is the same as (you guessed it) the median. Quartiles divide the distribution into quarters, and the first quartile is the same as the 25th percentile, the second quartile is the 50th percentile, and the third quartile is the 75th percentile. Any of these measures can be found by the method described earlier for percentiles. Remember that multiplying N by the proportional value of the percentile, decile, or quartile gives the number of the appropriate *case* and it's the *score* of the case that actually marks the location. Also remember that this technique cuts some (probably minor) computational corners, so use it with caution. *(Quartiles, deciles, and percentiles may be found for any ordinal or interval-ratio variable in the problems at the end of this chapter.)*

[2]This section is optional.

ONE STEP AT A TIME **Finding the Mean**

Step 1: Add up the scores (X_i).

Step 2: Divide the quantity you found in step 1 (ΣX_i) by N

3.5 THE MEAN

The mean ($\overline{X}$, read this as "ex-bar"),[3] or arithmetic average, is by far the most commonly used measure of central tendency. It reports the average score of a distribution, and its calculation is straightforward: To compute the mean, add the scores and then divide by the number of scores (N). To illustrate: A birth control clinic administered a 20-item test of general knowledge about contraception to 10 clients. The number of correct responses was 2, 10, 15, 11, 9, 16, 18, 10, 11, and 7. To find the mean of this distribution, add the scores (total = 109) and divide by the number of scores (10). The result (10.9) is the average score on the test.

The mathematical formula for the mean is

FORMULA 3.1

$$\overline{X} = \frac{\Sigma(X_i)}{N}$$

where $\overline{X}$ = the mean
$\Sigma(X_i)$ = the summation of the scores
N = the number of cases

Let's take a moment to consider the new symbols introduced in this formula. First, the symbol Σ (uppercase Greek letter sigma) is a mathematical operator just like the plus sign (+) or divide sign (÷). It stands for "the summation of" and directs us to add whatever quantities are stated immediately following it.[4] The second new symbol is X_i ("X-sub-i"), which refers to any single score—the "ith" score. If we wished to refer to a particular score in the distribution, the specific number of the score could replace the subscript. Thus, X_1 would refer to the first score, X_2 to the second, X_{26} to the 26th, and so forth. The operation of adding all the scores is symbolized as $\Sigma(X_i)$. This combination of symbols directs us to sum the scores, beginning with the first score and ending with the last score in the distribution. Thus, Formula 3.1 states in symbols what has already been stated in words (to calculate the mean, add the scores and divide by the number of scores), but in a succinct and precise way. *(For practice in computing the mean, see any of the problems at the end of this chapter.)*

Because computation of the mean requires addition and division, it should be used with variables measured at the interval-ratio level. However, researchers do calculate the mean for variables measured at the ordinal level, because the mean is much more flexible than the median and is a central feature of many interesting and powerful advanced statistical techniques. Thus, if the researcher

[3]This is the symbol for the mean of a sample. The mean of a population is symbolized with the Greek letter mu (μ—read this symbol as "mew").
[4]See the Prologue (Basic Review of Math) for further information on the summation sign and on summation operations.

plans to do any more than merely describe his or her data, the mean will probably be the preferable measure of central tendency even for ordinal-level variables.

3.6 THREE CHARACTERISTICS OF THE MEAN

The mean is the most commonly used measure of central tendency, and we will consider its mathematical and statistical characteristics in some detail. First, the mean is an excellent measure of central tendency because it acts like a fulcrum that "balances" all the scores, in the sense that the mean is the point around which all of the scores cancel out. We can express this property symbolically:

$$\sum (X_i - \overline{X}) = 0$$

This expression says that if we subtract the mean ($\overline{X}$) from each score (X_i) in a distribution and then sum the differences, the result will always be zero.

To illustrate, consider the test scores presented in Table 3.6. The mean of these five scores is 390/5, or 78. The difference between each score and the mean is listed in the right-hand column ($X_i - \overline{X}$), and the sum of these differences is zero. The total of the negative differences (−19) is exactly equal to the total of the positive differences (+19), as will always be the case. Thus, the mean "balances" the scores and is at the center of the distribution.

A second characteristic of the mean is called the "least squares" principle, a characteristic that is expressed in the statement:

$$\sum (X_i - \overline{X})^2 = \text{minimum}$$

or: the mean is the point in a distribution around which the variation of the scores (as indicated by the squared differences) is minimized. If the differences between the scores and the mean are squared and then added, the resultant sum will be less than the sum of the squared differences between the scores and any other point in the distribution.

To illustrate this principle, consider the distribution of five scores mentioned earlier: 65, 73, 77, 85, and 90. The differences between the scores and the mean have already been found. As illustrated in Table 3.7, if we square and sum these differences, we would get a total of 388. If we performed those same mathematical operations with any number other than the mean—say, the value 77—the resultant sum would be greater than 388. Table 3.7 illustrates this point by showing that the sum of the squared differences around 77 is 393, a value greater than 388.

This least-squares principle underlines the fact that the mean is closer to all of the scores than the other measures of central tendency. Also, this

TABLE 3.6 A DEMONSTRATION SHOWING THAT ALL SCORES CANCEL OUT AROUND THE MEAN

X_i	$(X_i - \overline{X})$
65	65 − 78 = −13
73	73 − 78 = −5
77	77 − 78 = −1
85	85 − 78 = 7
90	90 − 78 = 12
$\Sigma(X_i) = 390$	$\Sigma(X_i - \overline{X}) = 0$
$\overline{X} = 390/5 = 78$	

TABLE 3.7 A DEMONSTRATION SHOWING THAT THE MEAN IS THE POINT OF MINIMIZED VARIATION

X_i	$(X_i - \bar{X})$	$(X_i - \bar{X})^2$	$(X_i - 77)^2$
65	$65 - 78 = -13$	$(-13)^2 = 169$	$(65 - 77)^2 = (-12)^2 = 144$
73	$73 - 78 = -5$	$(-5)^2 = 25$	$(73 - 77)^2 = (-4)^2 = 16$
77	$77 - 78 = -1$	$(-1)^2 = 1$	$(77 - 77)^2 = (0)^2 = 0$
85	$85 - 78 = 7$	$(7)^2 = 49$	$(85 - 77)^2 = (8)^2 = 64$
90	$90 - 78 = 12$	$(12)^2 = 144$	$(90 - 77)^2 = (13)^2 = 169$
$\Sigma(X_i) = 390$	$\Sigma(X_i - \bar{X}) = 0$	$\Sigma(X_i - \bar{X})^2 = 388$	$\Sigma(X_i - 77)^2 = 393$

characteristic of the mean is important for the statistical techniques of correlation and regression, topics we take up toward the end of this text.

The final important characteristic of the mean is that every score in the distribution affects it. The mode (which is only the most common score) and the median (which deals only with the score of the middle case or cases) are not so affected. This quality is both an advantage and a disadvantage. On one hand, the mean utilizes all the available information—every score in the distribution affects the mean. On the other hand, when a distribution has a few very high or very low scores, the mean may become a very misleading measure of centrality.

To illustrate, consider Table 3.8. The five scores listed in column 1 have a mean and median of 25 (see column 2). In column 3, the scores are listed again with one score changed: 35 is changed to 3500. Look in column 4 and you will see that this change has no effect on the median, it remains at 25. This is because the median is based *only* on the score of the middle case and is not affected by changes in the scores of other cases in the distribution.

The mean, in contrast, is very much affected by the change because it takes *all* scores into account. The mean changes from 25 to 718 solely because of the one extreme score of 3500. Note also that the mean in column 4 is very different from four of the five scores listed in column 3. In this case, is the mean or the median a better representation of the scores? For distributions that have a few very high or very low scores, the mean may present a very misleading picture of the typical or central score. In these cases, the median may be the preferred measure of central tendency for interval-ratio variables. *(For practice in dealing with the effects of extreme scores on means and medians, see problems 3.11, 3.13, 3.14, and 3.15)*

TABLE 3.8 A DEMONSTRATION SHOWING THAT THE MEAN IS AFFECTED BY EVERY SCORE

1 Scores	2 Measures of Central Tendency	3 Scores	4 Measures of Central Tendency
15	$\bar{X} = 125/5 = 25$	15	$\bar{X} = 3590/5 = 718$
20	Md = 25	20	Md = 25
25		25	
30		30	
35		3500	
$\Sigma = 125$		$\Sigma = 3590$	

The general principle to remember is that, relative to the median, the mean is always pulled in the direction of extreme scores (i.e., scores that are much higher or lower than other scores). The mean and median will have the same value when and only when a distribution is symmetrical. When a distribution has some extremely high scores (this is called a *positive* ***skew***), the mean will always have a greater numerical value than the median. If the distribution has some very low scores (a *negative skew*), the mean will be lower in value than the median. Figures 3.1 to 3.3 depict three different frequency polygons that demonstrate these relationships.

These relationships between medians and means also have a practical value. For one thing, a quick comparison of the median and mean will always tell you if a distribution is skewed and the direction of the skew. If the mean is less than the median, the distribution has a negative skew. If the mean is greater than the median, the distribution has a positive skew.

FIGURE 3.1 A POSITIVELY SKEWED DISTRIBUTION (The mean is greater in value than the median)

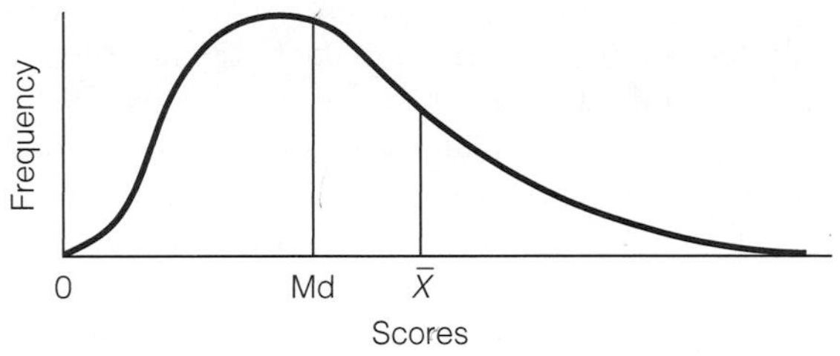

FIGURE 3.2 A NEGATIVELY SKEWED DISTRIBUTION (The mean is less than the median)

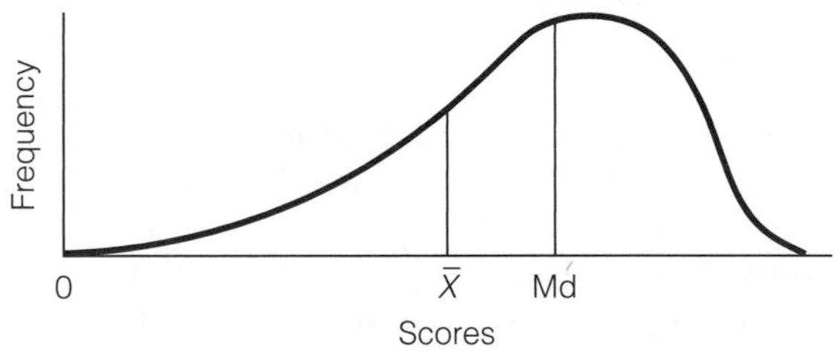

FIGURE 3.3 AN UNSKEWED, SYMMETRICAL DISTRIBUTION (The mean and median are equal)

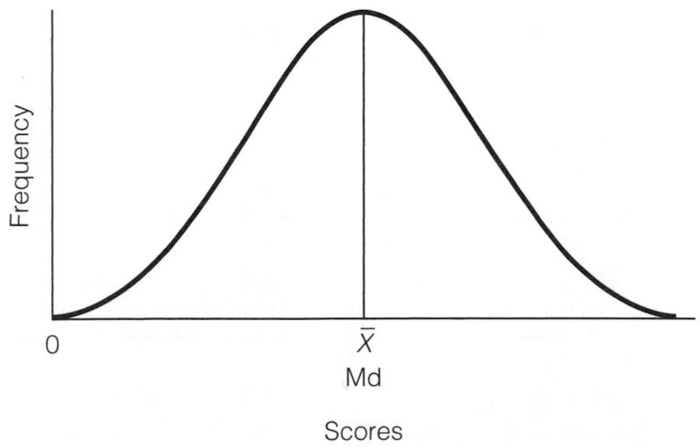

Second, these characteristics of the mean and median also provide a simple and effective way to "lie" with statistics. For example, if you want to maximize the average score of a positively skewed distribution, report the mean. Income data usually have a positive skew (there are only a few very wealthy people). If you want to impress someone with the general affluence of a mixed-income community, report the mean. If you want a lower figure, report the median.

Which measure is most appropriate for skewed distributions? This will depend on what point the researcher wishes to make but, as a rule, either both statistics or the median alone should be reported when the distribution has a few extreme scores.

3.7 COMPUTING MEASURES OF CENTRAL TENDENCY FOR GROUPED DATA[5]

In this section, we consider the techniques for computing means and medians for data presented in frequency distribution form. These techniques are useful when scores have been organized into a frequency distribution and you do not have access to the ungrouped data. Under this condition, there is no other way to calculate means or medians, but a small price is attached to using these techniques. Specifically, certain assumptions must be made about the way the scores are distributed within the intervals of the frequency distribution. These assumptions will not reflect the way in which the scores are actually distributed, and, technically, the statistics we calculate based on these techniques are only approximations of the true mean or median.

To illustrate, consider the following situation: As a new employee of a Youth Service Bureau, you are faced with the task of preparing a report on truancy for high school students identified as chronic absentees. The report is due in two weeks, leaving you no time to gather data. Fortunately, you stumble across some truancy data left behind by the person you replaced, but the data are in the form of a frequency distribution (Table 3.9), and the ungrouped data are not available. What will you do?

Computing the Mean for Grouped Data. To compute a mean for any distribution, we must find two values: the summation of the scores and the number of scores. For the distribution in Table 3.7, we know only the latter ($N = 124$). Because the data have been grouped, we do not know the exact distribution of the original scores. We know from the table that 25 students were truant between one

TABLE 3.9 TRUANCY DATA FOR 124 HIGH SCHOOL STUDENTS

Days Absent	Frequency
1–5	25
6–10	30
11–15	30
16–20	20
21–25	10
26–30	5
31–35	4
	$N = 124$

[5]This section is optional.

ONE STEP AT A TIME **Finding the Mean for Grouped Data**

Step 1: Find the midpoint (m) of each interval.

Step 2: For each interval, multiply the midpoint (m) by the number of cases in the interval (f).

Step 3: Find the sum of all the values calculated in step 2. This value is $\Sigma(fm)$.

Step 4: Divide the quantity you found in step 3 by N. The result is the approximate mean.

TABLE 3.10 COMPUTING A MEAN FOR GROUPED DATA

Days Absent	Frequency (f)	Midpoints (m)	Frequency × Midpoint ($f \times m$)
1–5	25	3	75
6–10	30	8	240
11–15	30	13	390
16–20	20	18	360
21–25	10	23	230
26–30	5	28	140
31–35	4	33	132
	$N = 124$		$\Sigma(fm) = 1567$

and five days, but we do not know exactly how many days. For all we know, these 25 students could have been truant for only one day each, or three or five days, or any combination of days between one and five, inclusive.

We can resolve this dilemma by assuming that all the scores are located at the midpoint of the interval. Then, by multiplying each midpoint by the number of cases in the interval, we can obtain an approximation of the sum of the original scores in the interval. To do this, add two columns to the original frequency distribution, one for the midpoints and one for the midpoints times the frequencies. In Table 3.10 this is done for the truancy data.

The summation of the last column can be labeled $\Sigma(fm)$ and will be approximately the same value as the summation of the original scores, or $\Sigma(X_i)$. So we have

$$\bar{X} = \frac{\Sigma(X_i)}{N} \cong \frac{\Sigma(fm)}{N}$$

where ≅ means "approximately equal to"

The final step in computing the mean for grouped data is to divide the total of the last column, $\Sigma(fm)$, by the number of cases:

FORMULA 3.2

$$\bar{X} = \frac{\Sigma(fm)}{N} = \frac{1567}{124} = 12.64$$

Thus, these 124 students were truant an average of 12.64 days over the year.

Computing the Median for Grouped Data. To locate the median, we must first find the middle case of the distribution. With 124 cases in the truancy sample,

ONE STEP AT A TIME **Finding the Median for Grouped Data**

Step 1: Add a cumulative frequency column to the frequency distribution.

Step 2: Find the middle case by multiplying *N* by 0.5.

Step 3: Find the interval in the frequency distribution that contains the middle case.

Step 4: Find the number of cases you will have to go into the interval to locate the middle case. This is given by *N*(0.5) − cfb. (cfb = cumulative frequency below the interval that contains the median).

Step 5: Divide the value you found in step 4 by *f* (the number of cases in the interval).

Step 6: Multiply the number you found in step 5 by the interval size (*i*).

Step 7: Add the number you found in step 6 to the lower real limit of the interval that contains the median. The result is an approximation of the median.

we will identify the middle case as the (*N*/2), or 62nd, case. (Technically, with an even *N*, we should find the scores of the two middle cases to locate the median. Designating only the (*N*/2) case as the median will, however, result in a significant simplification in computation without much loss in accuracy.) Our problem is first to locate this case and then to find the score associated with it. To do so, we make the assumption that the cases are evenly spaced throughout each interval in the frequency distribution. We then locate the interval that contains the middle case and, based on the assumption of equal spacing, find the score associated with that case. To use this procedure, we must use upper and lower real limits.

To locate the interval that contains the median, it is helpful to add a cumulative frequency column to the original frequency distribution, as in Table 3.11. Looking at the cumulative frequency column, we see that 55 cases are accumulated below the upper limit of the 6–10 interval and that 85 cases have been accumulated below the upper limit of the 11–15 interval. We know now that the middle, or 62nd, case is in the 11–15 interval and that the median is between 10.5 (the lower real limit of this interval) and 15.5 (the upper real limit), but we do not know the exact value.

TABLE 3.11 COMPUTING A MEDIAN FOR GROUPED DATA

Days Absent	Frequency (*f*)	Cumulative Frequency (cf)
1–5	25	25
6–10	30	55
11–15	30	85
16–20	20	105
21–25	10	115
26–30	5	120
31–35	4	124
	N = 124	

To resolve this dilemma, assume that all 30 of the cases in this interval are evenly spaced throughout the interval, with case 56 located at the lower real limit (10.5) and case 85 at the upper real limit (15.5). Case 62 (our middle case) is the seventh case of the 30 in the interval. If the scores are evenly spaced, then the 62nd case will be 7/30 of the distance from 10.5 to 15.5. Now, 7/30 of 5 (the interval size) is 1.17, so we can approximate the median at 10.5 + 1.17, or 11.67. Formula 3.3 summarizes these steps.

FORMULA 3.3

$$\text{Md} = rll + \left(\frac{N(.50) - \text{cfb}}{f}\right)i$$

where rll = real lower limit of the interval containing the median

cfb = cumulative frequency below the interval containing the median
f = number of cases in the interval containing the median
i = interval width

Applying this formula to our example, we would have

$$\text{Md} = 10.5 + \left(\frac{124(.50) - 55}{30}\right)5 = 10.5 + \left(\frac{62 - 55}{30}\right)5$$

$$= 10.5 + \left(\frac{7}{30}\right)5 = 10.5 + 1.17 = 11.67$$

Thus, half of this sample was truant fewer than 11.67 days and half more than 11.67 days.

Because the value of the median (11.67) is almost a full day lower than the value of the mean (12.64), we can also conclude that this distribution has a positive skew, or a few very high scores. This skew is reflected in the frequency distribution itself, with most cases in the lower intervals. For purposes of description, the median may be the preferred measure of central tendency for these data, because the mean is affected by the relatively few very high scores.

This procedure may seem rather formidable the first time you attempt it. With a little practice, however, it will become routine. Of course, if you do not have access to the ungrouped data, there simply is no other way to find the median. (*For practice in finding means and medians with grouped data, see problems 3.5b and 3.6b*)

3.8 CHOOSING A MEASURE OF CENTRAL TENDENCY

You should consider two main criteria when choosing a measure of central tendency. First, make sure that you know the level of measurement of the variable in question. This will generally tell you whether you should report the mode, median, or mean. Table 3.12 shows the relationship between the level of measurement and measures of central tendency. The capitalized, boldface "**YES**" identifies the most appropriate measure of central tendency for each level of measurement and the nonboldface "Yes" indicates the levels of measurement for which the measure is also permitted. An entry of "No" in the table means that the statistic cannot be computed for that level of measurement. Finally, the "Yes (?)" entry in the bottom row indicates that the mean is often used with ordinal-level variables even though, strictly speaking, this practice violates level of measurement guidelines.

TABLE 3.12 THE RELATIONSHIP BETWEEN LEVEL OF MEASURE AND MEASURES OF CENTRAL TENDENCY

Measure of Central Tendency:	Level of Measurement: Nominal	Ordinal	Interval-Ratio
Mode	**YES**	Yes	Yes
Median	No	**YES**	Yes
Mean	No	Yes (?)	**YES**

Second, consider the definitions of the three measures of central tendency and remember that they provide different types of information. They will be the same value only under certain specific conditions (that is, for symmetrical distributions with one mode), and each has its own message to report. In many circumstances, you might want to report all three. The guidelines in Table 3.13 stress both selection criteria and may be helpful when choosing a specific measure of central tendency:

Application 3.1

Ten students have been asked how many hours they spent in the college library during the past week. What is the average "library time" for these students? The hours are reported in the following list, and we will find the mode, the median, and the mean for these data.

Student	Number of Visits to the Library Last Week (X_i)
1	0
2	2
3	5
4	5
5	7
6	10
7	14
8	14
9	20
10	30
$\Sigma(X_i) =$	107

By scanning the scores, we can see that two scores, 5 and 14, occurred twice, and no other score occurred more than once. This distribution has two modes, 5 and 14.

Because the number of cases is even, the median will be the average of the two middle cases after all cases have been ranked in order. With 10 cases, the first middle case will be the (N/2), or (10/2), or fifth case. The second middle case is the (N/2) + 1, or (10/2) + 1, or sixth case. The median will be the score halfway between the scores of the fifth and sixth cases. Counting down from the top, we find that the score of the fifth case is 7 and the score of the sixth case is 10. The median for these data is (7 + 10)/2, or (17/2), or 8.5. The median, the score that divides this distribution in half, is 8.5.

The mean is found by first adding all the scores and then dividing by the number of scores. The sum of the scores is 107, so the mean is

$$\bar{X} = \frac{\Sigma(X_i)}{N} = \frac{107}{10} = 10.7$$

These 10 students spent an average of 10.7 hours in the library during the week in question.

Note that the mean is a higher value than the median. This indicates a positive skew in the distribution (a few extremely high scores). By inspection we can see that the positive skew is caused by the two students who spent many more hours (20 hours and 30 hours) in the library than the other eight students.

TABLE 3.13 CHOOSING A MEASURE OF CENTRAL TENDENCY

Use the mode when:	1. The variable is measured at the nominal level. 2. You want a quick and easy measure for ordinal and interval-ratio variables. 3. You want to report the most common score.
Use the median when:	1. The variable is measured at the ordinal level. 2. A variables measured at the interval-ratio level has a highly skewed distribution. 3. You want to report the central score. The median always lies at the exact center of a distribution.
Use the Mean when:	1. The variable is measured at the interval-ratio level (except when the variable is highly skewed). 2. You want to report the typical score. The mean is "the fulcrum that exactly balances all of the scores." 3. You anticipate additional statistical analysis.

SUMMARY

1. The three measures of central tendency presented in this chapter share a common purpose. Each reports some information about the most typical or representative value in a distribution. Appropriate use of these statistics permits the researcher to report important information about an entire distribution of scores in a single, easily understood number.
2. The mode reports the most common score and is used most appropriately with nominally measured variables.
3. The median (Md) reports the score that is the exact center of the distribution. It is most appropriately used with variables measured at the ordinal level and with variables measured at the interval-ratio level when the distribution is skewed.
4. The mean ($\overline{X}$), the most frequently used of the three measures, reports the most typical score. It is used most appropriately with variables measured at the interval-ratio level (except when the distribution is highly skewed).
5. The mean has a number of mathematical characteristics that are significant for statisticians. First, it is the point in a distribution of scores around which all other scores cancel out. Second, the mean is the point of minimized variation. Last, as distinct from the mode or median, the mean is affected by every score in the distribution and is therefore pulled in the direction of extreme scores.

SUMMARY OF FORMULAS

Mean 3.1 $$\overline{X} = \frac{\Sigma(X_i)}{N}$$

Mean for grouped data 3.2 $$\overline{X} = \frac{\Sigma(fm)}{N}$$

Median for grouped data 3.3 $$\text{Md} = rll + \left(\frac{N(.50) - \text{cfb}}{f}\right)i$$

GLOSSARY

Deciles. The points that divide a distribution of scores into 10ths.

Mean. The arithmetic average of the scores. $\bar{X}$ represents the mean of a sample, and μ, the mean of a population.

Measures of central tendency. Statistics that summarize a distribution of scores by reporting the most typical or representative value of the distribution.

Median (Md). The point in a distribution of scores above and below which exactly half of the cases fall.

Mode. The most common value in a distribution or the largest category of a variable.

Percentile. A point below which a specific percentage of the cases fall.

Quartiles. The points that divide a distribution into quarters.

Σ (uppercase Greek letter sigma). "The summation of."

Skew. The extent to which a distribution of scores has a few scores that are extremely high (positive skew) or extremely low (negative skew).

X_i ("X sub i"). Any score in a distribution.

PROBLEMS

3.1 SOC A variety of information has been gathered from a sample of college freshmen and seniors, including their region of birth, the extent to which they support legalization of marijuana (measured on a scale on which 7 = strong support, 4 = neutral, and 1 = strong opposition), the amount of money they spend each week out-of-pocket for food, drinks, and entertainment, how many movies they watched in their dorm rooms last week, their opinion of cafeteria food (10 = excellent, 0 = very bad), and their religious affiliation. Some results are presented here. Find the *most appropriate* measure of central tendency for each variable for freshmen and then for seniors. Report both the measure you selected as well as its value for each variable (e.g., "Mode = 3" or "Median = 3.5"). *(HINT: Determine the level of measurement for each variable first. In general, this will tell you which measure of central tendency is appropriate. See Section 3.7 to review the relationship between measure of central tendency and level of measurement. Also, remember that the mode is the* most common *score, and especially remember to array scores from high to low before finding the median.)*

FRESHMEN

Student	Region of Birth	Legalization	Out-of-pocket Expenses	Movies	Cafeteria Food	Religion
A	North	7	33	0	10	Protestant
B	North	4	39	14	7	Protestant
C	South	3	45	10	2	Catholic
D	Midwest	2	47	7	1	None
E	North	3	62	5	8	Protestant
F	North	5	48	1	6	Jew
G	South	1	52	0	10	Protestant
H	South	4	65	14	0	Other
I	Midwest	1	51	3	5	Other
J	West	2	43	4	6	Catholic

SENIORS

Student	Region of Birth	Legalization	Out-of-pocket Expenses	Movies	Cafeteria Food	Religion
K	North	7	65	0	1	None
L	Midwest	6	62	5	2	Protestant
M	North	7	60	11	8	Protestant
N	North	5	90	3	4	Catholic
O	South	1	62	4	3	Protestant
P	South	5	57	14	6	Protestant
Q	West	6	40	0	2	Catholic
R	West	7	49	7	9	None
S	North	3	45	5	4	None
T	West	5	85	3	7	Other
U	North	4	78	5	4	None

3.2 A variety of information has been collected for each of the nine high schools in a district. Find the most appropriate measure of central tendency for each variable and summarize this information in a paragraph. *(HINT: The level of measurement of the variable will generally tell you which measure of central tendency is appropriate. Remember to organize the scores from high to low before finding the median.)*

High 1–10, School	Enrollment	Largest Racial/ Ethnic Group	Percent College Bound	Most Popular Sport	Condition of Physical Plant (scale o 10 = high)
1	1400	White	25	Football	10
2	1223	White	77	Baseball	7
3	876	Black	52	Football	5
4	1567	Hispanic	29	Football	8
5	778	White	43	Basketball	4
6	1690	Black	35	Basketball	5
7	1250	White	66	Soccer	6
8	970	White	54	Football	9

3.3 PS You have been observing the local Democratic Party in a large city and have compiled some information about a small sample of party regulars. Find the appropriate measure of central tendency for each variable.

Respondent	Sex	Social Class	No. of Years in Party	Education	Marital Status	Number of Children
A	M	High	32	High school	Married	5
B	M	Medium	17	High school	Married	0
C	M	Low	32	High school	Single	0
D	M	Low	50	Eighth grade	Widowed	7
E	M	Low	25	Fourth grade	Married	4
F	M	Medium	25	High school	Divorced	3
G	F	High	12	College	Divorced	3
H	F	High	10	College	Separated	2

(*continued next page*)

(*continued*)

Respondent	Sex	Social Class	No. of Years in Party	Education	Marital Status	Number of Children
I	F	Medium	21	College	Married	1
J	F	Medium	33	College	Married	5
K	M	Low	37	High school	Single	0
L	F	Low	15	High school	Divorced	0
M	F	Low	31	Eighth grade	Widowed	1

3.4 SOC You have compiled the following information on each of the graduates voted "most likely to succeed" by a local high school for a 10-year period. For each variable, find the appropriate measure of central tendency.

Case	Present Income	Marital Status	Owns a BMW?	Years of Education Post–High School
A	24,000	Single	No	8
B	48,000	Divorced	No	4
C	54,000	Married	Yes	4
D	45,000	Married	No	4
E	30,000	Single	No	4
F	35,000	Separated	Yes	8
G	30,000	Married	No	3
H	17,000	Married	No	1
I	33,000	Married	Yes	6
J	48,000	Single	Yes	4

3.5 SOC For 15 respondents, data have been gathered on four variables (see the following table).

a. Find and report the appropriate measure of central tendency for each variable.

b.[6] For the variable *age*, construct a frequency distribution with interval size = 5 and the first interval set at 15–19. Find the mean and median for the grouped variable and compare with the value for the ungrouped variable. How accurate is the estimate of the mean and median based on the grouped data?

Respondent	Marital Status	Racial/Ethnic Group	Age	Attitude on Abortion Scale (High score = Strong opposition
A	Single	White	18	10
B	Single	Hispanic	20	9
C	Widowed	White	21	8
D	Married	White	30	10
E	Married	Hispanic	25	7
F	Married	White	26	7
G	Divorced	Black	19	9
H	Widowed	White	29	6
I	Divorced	White	31	10
J	Married	Black	55	5
K	Widowed	Asian American	32	4
L	Married	American Indian	28	3
M	Divorced	White	23	2
N	Married	White	24	1
O	Divorced	Black	32	9

*This scale is constructed so that a high score indicates strong opposition to abortion under any circumstances.

3.6 SOC Following are four variables for 30 cases from the General Social Survey. Age is reported in years. The variable *happiness* consists of answers to the question "Taken all together, would you say that you are (1) very happy (2) pretty happy, or (3) not too happy?" Respondents were asked how many sex partners they had over the past five

[6] This problem is optional.

years. Responses were measured on the following scale: 0–4 = actual numbers; 5 = 5–10 partners; 6 = 11–20 partners; 7 = 21–100 partners; 8 = more than 100.

a. For each variable, find the appropriate measure of central tendency and write a sentence reporting this statistical information as you would in a research report.

b.[7] For the variable *age*, construct a frequency distribution with interval size = 5 and the first interval set at 20–24. Compute the mean and median for the grouped data and compare with the values computed for the ungrouped data. How accurate are the estimates based on the grouped data?

Respondent	Age	Happiness	Number of Partners	Religion
1	20	1	2	Protestant
2	32	1	1	Protestant
3	31	1	1	Catholic
4	34	2	5	Protestant
5	34	2	3	Protestant
6	31	3	0	Jew
7	35	1	4	None
8	42	1	3	Protestant
9	48	1	1	Catholic
10	27	2	1	None
11	41	1	1	Protestant
12	42	2	0	Other
13	29	1	8	None
14	28	1	1	Jew
15	47	2	1	Protestant
16	69	2	2	Catholic
17	44	1	4	Other
18	21	3	1	Protestant
19	33	2	1	None
20	56	1	2	Protestant
21	73	2	0	Catholic
22	31	1	1	Catholic
23	53	2	3	None
24	78	1	0	Protestant
25	47	2	3	Protestant
26	88	3	0	Catholic
27	43	1	2	Protestant
28	24	1	1	None
29	24	2	3	None
30	60	1	1	Protestant

3.7 Find the appropriate measure of central tendency for each variable displayed in Table 2.5. Report each statistic as you would in a formal research report.

3.8 SOC The following table lists the median family incomes for 13 Canadian provinces and territories in 2000 and 2004. Compute the mean and median for each year and compare the two measures of central tendency. Which measure of central tendency is greater for each year? Are the distributions skewed? In which direction?

Province or Territory	2000	2004
Newfoundland and Labrador	38,800	46,100
Prince Edward Island	44,200	51,300
Nova Scotia	44,500	51,500
New Brunswick	43,200	49,700
Quebec	47,700	54,400
Ontario	55,700	62,500
Manitoba	47,300	54,100
Saskatchewan	45,800	53,500
Alberta	55,200	66,400
British Columbia	49,100	55,900
Yukon	56,000	67,800
Northwest Territories	61,000	79,800
Nunavut	37,600	49,900

3.9 SOC The administration is considering a total ban on student automobiles. You have conducted a poll on this issue of 20 fellow students and 20 of the neighbors who live around the campus and have calculated scores for your respondents. On the scale you used, a high score indicates strong opposition to the proposed ban. The scores are presented here for both groups. Calculate an appropriate measure of central tendency and compare the two groups in a sentence or two.

Students		Neighbors	
10	11	0	7
10	9	1	6
10	8	0	0
10	11	1	3
9	8	7	4
10	11	11	0
9	7	0	0
5	1	1	10
5	2	10	9
0	10	10	0

3.10 SW As the head of a social services agency, you believe that your staff of 20 social workers is very much overworked compared to 10 years ago. The case loads for each worker are reported below for each of the two years in question. Has the average caseload increased? What measure of central tendency is most appropriate to answer this question? Why?

1997		2007	
52	55	42	82
50	49	75	50

(continued next page)

[7]This problem is optional

(*continued*)

1997		2007	
57	50	69	52
49	52	65	50
45	59	58	55
65	60	64	65
60	65	69	60
55	68	60	60
42	60	50	60
50	42	60	60

3.11 SOC The following table lists the approximate number of cars per 100 population for eight nations in 1999. Compute the mean and median for these data. Which is greater in value? Is there a positive skew in the data? How do you know?

Nation	Number of Cars per 100 Population
United States	50
Canada	45
France	46
Germany	51
Japan	39
Mexico	10
Sweden	44
United Kingdom	37

Source: http://www.nationmaster.com

3.12 SW For the test scores first presented in problem 2.6 and reproduced here, compute a median and mean for both the pretest and posttest. Interpret these statistics.

Case	Pretest	Posttest
A	8	12
B	7	13
C	10	12
D	15	19
E	10	8
F	10	17
G	3	12
H	10	11
I	5	7
J	15	12
K	13	20
L	4	5
M	10	15
N	8	11
O	12	20

3.13 SOC A sample of 25 freshmen at a major university completed a survey that measured their degree of racial prejudice (the higher the score, the greater the prejudice).

a. Compute the median and mean scores for these data.

10	43	30	30	45
40	12	40	42	35
45	25	10	33	50
42	32	38	11	47
22	26	37	38	10

b. These same 25 students completed the same survey during their senior year. Compute the median and mean for this second set of scores, and compare them to the earlier set. What happened?

10	45	35	27	50
35	10	50	40	30
40	10	10	37	40
40	15	30	20	43
23	25	30	40	10

3.14 PA The following table presents the annual person-hours of time lost due to traffic congestion for a group of cities for 2003. This statistic is a measure of traffic congestion.

City	Annual person-hours of time lost to traffic congestion per year
Baltimore	27
Boston	25
Buffalo	6
Chicago	31
Cleveland	6
Dallas	35
Detroit	30
Houston	36
Kansas City	9
Los Angeles	50
Miami	29
Minneapolis	23
New Orleans	10
New York	23
Philadelphia	21
Pittsburgh	8
Phoenix	26
San Antonio	18
San Diego	28
San Francisco	37
Seattle	25
Washington, DC	34

Source: U.S. Bureau of the Census. 2007. *Statistical Abstract of the United States: 2007*. p. 688. (Available at http:// www .census.gov/ compendia/statab/)

a. Calculate the mean and median of this distribution.
b. Compare the mean and median. Which is the higher value? Why?
c. If you removed Los Angeles from this distribution and recalculated, what would happen to the mean? To the median? Why?
d. Report the mean and median as you would in a formal research report.

3.15 Professional athletes are threatening to strike because they claim that they are underpaid. The team owners have released a statement that says, in part, "The average salary for players was $1.2 million last year." The players counter by issuing their own statement that says, in part, "The average player earned only $753,000 last year." Is either side necessarily lying? If you were a sports reporter and had just read Chapter 3 of this text, what questions would you ask about these statistics?

SPSS for Windows

Using *SPSS for Windows* for Measures of Central Tendency and Percentiles

Start *SPSS for Windows* by clicking the SPSS icon on your monitor screen. Load the 2006 GSS, and when you see the message "SPSS Processor is Ready" on the bottom of the "closest" screen, you are ready to proceed.

SPSS DEMONSTRATION 3.1 Producing Measures of Central Tendency

The only procedure in SPSS that will produce all three commonly used measures of central tendency (mode, median, and mean) is **Frequencies.** We used this procedure to produce frequency distributions in Demonstrations 2.1 and 2.2 and in Appendix F. Here we will use **Frequencies** to calculate measures of central tendency for three variables: *relig* (religious denomination), *attend* (frequency of attendance at religious services), and *age*

The three variables vary in level of measurement, and we could request only the appropriate measure of central tendency for each variable. That is, we could request the mode for *relig* (nominal), the median for *attend* (ordinal), and the mean for *age* (interval-ratio). While this would be reasonable, it's actually more convenient to get all three measures for each variable and ignore the irrelevant output. Statistical packages typically generate more information than necessary, and it is common to disregard some of the output.

To produce modes, medians, and means, begin by clicking **Analyze** from the menu bar and then click **Descriptive Statistics** and **Frequencies.** In the **Frequencies** dialog box, find the variable names in the list on the left and click the arrow button in the middle of the screen to move the names (*age, attend,* and *relig*) to the **Variables** box on the right.

To request specific statistics, click the **Statistics** button at the bottom of the **Frequencies** dialog box, and the **Frequencies: Statistics** dialog box will open. Find the **Central Tendency** box on the right and click **Mean, Median,** and **Mode.** Click **Continue,** and you will be returned to the **Frequencies** dialog box, where you might want to click the **Display Frequency Tables** box. When this box is *not* checked, SPSS will not produce frequency distribution tables, and only the statistics we request (mode, median, and mean) will appear in the **Output** window. Click **OK,** and SPSS will produce the following output:

Statistics

		AGE OF RESPONDENT	HOW OFTEN R ATTENDS RELIGIOUS SERVICES	RS RELIGIOUS PREFERENCE
N	Valid	1417	1419	1417
	Missing	9	7	9
Mean		46.88	3.52	1.88
Median		45.00	3.00	1.00
Mode		48	0	1

Looking only at the most appropriate measures for each variable, the mode for *relig* ("RS RELIGIOUS PREFERENCE") is "1" (see the bottom line of output). What does the value mean? You can find out by consulting either Appendix G in this the text or the online code book. To use the latter, click **Utilities** and then click **Variables** and find *relig* in the variable list on the left. Either way, you will find that a score of 1 indicates Protestant. This was the most common religious affiliation in the sample and, thus, is the mode.

The median for *attend* ("HOW OFTEN R ATTENDS RELIGIOUS SERVICES") is a value of "3.00." Again, use either Appendix G or the online code book, and you will see that the category associated with this score is "several times a year." This means that the middle case in this distribution of 1417 cases has a score of 3 (or that the middle case is in the interval "several times a year").

The output for *age* indicates that the respondents were, on the average, 46.88 years old. Because age is an interval-ratio variable that has been measured in a defined unit (years), the value of the mean is numerically meaningful, and we do not need to consult the code book to interpret its meaning.

Note that SPSS did not hesitate to compute means and medians for the two variables that were not interval-ratio. The program cannot distinguish between numerical codes (such as the scores for *relig*) and actual numbers—to SPSS, all numbers are the same. Also, SPSS cannot screen your commands to see if they are statistically appropriate. If you request nonsensical statistics (average sex or median religious denomination), SPSS will carry out your instructions without hesitation. The blind willingness of SPSS simply to obey commands makes it easy for you, the user, to request statistics that are completely meaningless. Computers don't care about meaning; they just crunch numbers.

In this case, it was more convenient for us to produce statistics indiscriminately and then ignore the ones that are nonsensical. This will not always be the case, and the point of all this, of course, is to caution you to use SPSS wisely. As the manager of your local computer center will be quick to remind you, computer resources (including paper) are not unlimited.

Before closing this demonstration, note that the mean for *age* is slightly greater than the median. Recalling Section 3.6, this indicates that the variable has a positive skew, or a few extremely high (old) cases. Verify this by doing a line chart for *age:* Click **Graphs** and then click **Line** and then **Define.** The **Define Simple Line** dialog box appears, with variable names listed on the left. Choose *age* from the variable list and click the arrow button in the middle of the screen to move the variable name to the **Category Axis** box. Click the **Options** button in the lower-left-hand corner and make sure that the button next to the **Display groups defined by missing values** option is *not* selected (or checked). Click **Continue** and then click **OK,** and the line chart for *age* will be produced. Note how the chart peaks in the 30s and 40s age groups and then gradually declines into the 70s and 80s. Although this figure is not as smooth as Figure 3.1, it does indicate that *age* is skewed in a positive direction (i.e., has a few very high scores).

SPSS DEMONSTRATION 3.2 The Descriptives Command

The **Descriptives** command in *SPSS for Windows* is designed to provide summary statistics for continuous interval-ratio-level variables. By default (i.e., unless you tell it otherwise), **Descriptives** produces the mean, the minimum and maximum scores (i.e., the lowest and highest scores), and the standard deviation (which is discussed in Chapter 4). Unlike **Frequencies,** this procedure will not produce frequency distributions.

To illustrate the use of **Descriptives,** let's run the procedure for *age educ* (HIGHEST YEAR OF SCHOOL COMPLETED), and *tvhours* (HOURS PER DAY WATCHING TV). Click **Analyze, Descriptive Statistics,** and **Descriptives.** The **Descriptives** dialog box will open. This dialog box looks just like the **Frequencies** dialog box and works in the same way. Find the variable names in the list on the left and, once they are highlighted, click the arrow button in the middle of the screen to transfer them to the **Variables** box on the right. Click **OK,** and the following output will be produced:

Descriptive Statistics

	N	Minimum	Maximum	Mean	Std. Deviation
AGE OF RESPONDENT	1417	18	89	46.88	17.086
HIGHEST YEAR OF SCHOOL COMPLETED	1424	0	20	13.26	3.191
HOURS PER DAY WATCHING TV	618	0	24	3.00	2.404
Valid N (listwise)	616				

On the average, the sample is 46.88 years of age (this duplicates the **Frequencies** output in Demonstration 3.1), has over 13 years of education, and watches three hours of TV daily.

SPSS DEMONSTRATION 3.3 Finding Percentiles, Deciles, and Quartiles

The **Frequencies** procedure can be used to find percentiles, deciles, and quartiles for any variable (see Section 3.4). These statistics are especially useful for continuous variables with broad ranges of scores. One of the few variables that fits this description in the GSS data file is *age* and we will once again use this variable for our illustrations.

Click **Analyze, Descriptive Statistics,** and **Frequencies** to get the **Frequencies** dialog box. Click the **Statistics** button to get the **Frequencies: Statistics** dialog box. For purposes of illustration, let's find quartiles, deciles, and the 23rd and 47th percentiles. In the **Percentile Values** box, check the boxes next to **Quartiles** and next to **Cut points for 10 equal groups.** The latter instruction will find deciles (ten equal groups) and could be changed to split the distribution in other ways (e.g., into thirds).

To get specific percentiles, check the box next to **Percentiles,** type the desired percentile values in the box to the right (23 and 47), and click the **Add** button after each value. Click **Continue,** and you will return to the **Frequencies** dialog box, where you might want to click the **Display Frequency Tables** box. Recall that when this box is *not* checked, SPSS will not produce frequency distribution tables. Click **OK,** and SPSS will produce the following output:

AGE OF RESPONDENT

N	Valid	1417
	Missing	9
Percentiles	10	25.00
	20	30.00

(continued next page)

(continued)

AGE OF RESPONDENT

N	Valid	1417
	Missing	9
Percentiles	23	32.00
	25	33.00
	30	36.00
	40	41.00
	47	44.00
	50	45.00
	60	50.00
	70	56.00
	75	59.00
	80	62.00
	90	72.00

The output prints the deciles, quartiles, and the percentiles we requested in order. Find the 10th percentile (noted as 10 in the column labeled "Percentiles"). The score associated with this percentile is 25 (noted as a value of 25.00). This indicates that 10% of the sample were younger than 25. Similarly, 20% were younger than age 30, 23% (the percentile we requested) were younger than 32, 25% (the first quartile) were younger than 33, and so forth. Note that the fifth decile, 2nd quartile, and 50th percentile are all associated with the age of 45, which is also the value of the median.

Exercises

3.1 Use the **Frequencies** command to get all three measures of central tendency for *marital*, *income06*, *race*, *region*, *polviews*, *cappun*, and five more variables of your own choosing. Select the most appropriate measure of central tendency for each variable (*Hint: Use the level of measurement of the variable as your criteria*) and write a sentence or two reporting each measure.

3.2 Use the **Descriptives** command to get means for *partnrs5* and *sexfreq*. Note that these variables are ordinal in level of measurement (see Appendix G) and, therefore, calculation of a mean is not fully justified. Such violations are common, however, in social science research. Write a sentence or two reporting and explaining the mean of each variable. Make sure you understand the coding scheme for each variable (see Appendix G) to help interpret the values.

3.3 Revise the command you wrote in exercise 3.1 to find the quartiles and deciles for each of the ordinal and interval-ratio-level variables in the list.

4 Measures of Dispersion

LEARNING OBJECTIVES

By the end of this chapter, you will be able to

1. Explain the purpose of measures of dispersion and the information they convey.
2. Compute and explain the index of qualitative variation (IQV), the range (R), the interquartile range (Q), the standard deviation (s), and the variance (s^2).
3. Select an appropriate measure of dispersion and correctly calculate and interpret the statistic.
4. Describe and explain the mathematical characteristics of the standard deviation.

4.1 INTRODUCTION

Chapters 2 and 3 presented a variety of ways of describing a variable, including frequency distributions, graphs, and measures of central tendency. For a complete description of a distribution of scores, we must combine these with **measures of dispersion**, the subject of this chapter. While measures of central tendency describe the typical, average, or central score, measures of dispersion describe variety, diversity, or heterogeneity of a distribution of scores.

The importance of the concept of **dispersion** might be easier to grasp if we consider a brief example. Suppose that the director of public safety wants to evaluate two ambulance services that have contracted with the city to provide emergency medical aid. As a part of the investigation, she has collected data on the response time of both services to calls for assistance. Data collected for the past year show that the mean response time is 7.4 minutes for Service A and 7.6 minutes for Service B. These averages, or means (calculated by adding up the response times to all calls and dividing by the number of calls), are very similar and provide no basis for judging one service as more or less efficient than the other. Measures of dispersion, however, can reveal substantial differences between distributions even when the measures of central tendency are the same. For example, consider Figure 4.1, which displays the distribution of response times for the two services in the form of frequency polygons, or line charts (see Chapter 2).

Compare the shapes of these two figures. Note that the line chart for Service B is much flatter than that for Service A. This is because the scores for Service B are more spread out, or more diverse, than the scores for Service A. In other words, Service B was much more variable in response time and had more scores in the high and low ranges and fewer in the middle. Service A was more consistent in its response time, and its scores are more clustered, or grouped, around the mean. Both distributions have essentially the same *average* response time, but there is considerably more *variation*, or dispersion, in the response times for Service B. If you were the director of public safety, would you be more likely to select an ambulance service that was always on the scene of an emergency in about the same amount of time (Service A) or one that was sometimes

FIGURE 4.1 RESPONSE TIME FOR TWO AMBULANCE SERVICES

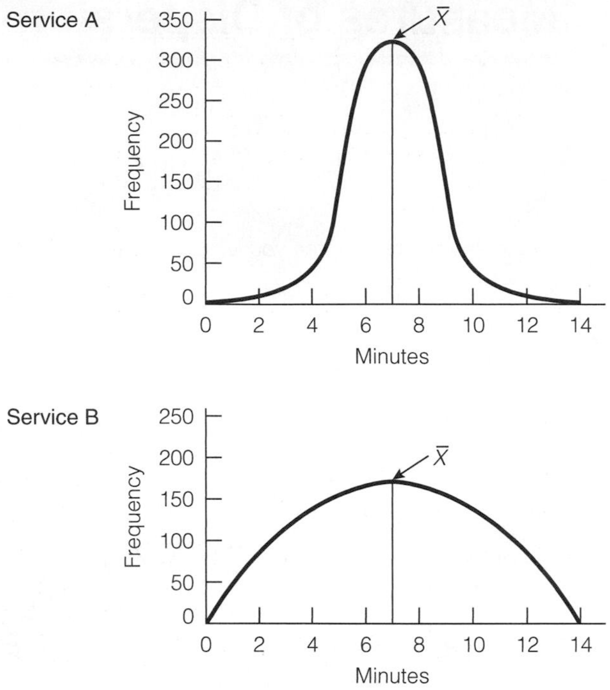

very slow and sometimes very quick to respond (Service B)? Note that if you had not considered dispersion, a possibly important difference in the performance of the two ambulance services might have gone unnoticed.

Keep the two shapes in Figure 4.1 in mind as visual representations of the concept of dispersion. The greater clustering of scores around the mean in the distribution for Service A indicates *less* dispersion, and the flatter curve of the distribution for Service B indicates *more* variety, or dispersion. Any of the measures of dispersion discussed in this chapter will decrease in value as the scores become less dispersed and the distribution becomes more peaked (that is, as the distribution looks more and more like Service A's) and increase in value as the scores become more dispersed and the distribution becomes flatter (that is, as the distribution looks more and more like Service B's).

These ideas and Figure 4.1 may give you a general notion of what is meant by dispersion, but the concept is not easily described in words alone. In this chapter we introduce some of the more common measures of dispersion, each of which provides a quantitative indication of the variety in a set of scores. We begin with the index of qualitative variation, mention two measures—the range and the interquartile range—briefly, and devote most of our attention to the standard deviation and the variance.

4.2 THE INDEX OF QUALITATIVE VARIATION (IQV)

We begin our consideration of measures of dispersion with the **index of qualitative variation (IQV)**. This statistics is rarely used in the professional research literature, but, since it is the only measure of dispersion available for nominal-

TABLE 4.1 RACIAL AND ETHNIC GROUPS IN THE UNITED STATES

	Percent of Total Population		
Group	2000	2025	2050
Non-Hispanic white	71%	62%	53%
Black	12%	13%	13%
Native American	1%	1%	1%
Asian American	4%	6%	9%
Hispanic	12%	18%	24%
Total =	100%	100%	100%

level variables (although it can be used with any variable that has been grouped into a frequency distribution), it deserves at least a brief consideration.

The IQV is essentially the ratio of the amount of variation actually observed in a distribution of scores to the maximum variation that could exist in that distribution. The index varies from 0.00 (no variation) to 1.00 (maximum variation).

To illustrate the logic of this statistic, consider the idea that the United States will grow more racially, ethnically, and linguistically diverse in the decades to come. Table 4.1 presents some information about the relative size of some groups for 2000 and some projections for 2025 and 2050 in the form of three separate frequency distributions, one for each year. Note that the values in the table are percentages instead of frequencies. This will greatly simplify computations. If there were no diversity in U.S. society (e.g., if everyone were white), the IQV would be 0.00. At the other extreme, if Americans were distributed equally across the five categories (i.e., if each group comprised exactly 20% of the population), the IQV would achieve its maximum value of 1.00.

By inspection, you can see that the United States was quite diverse in 2000, not at all close to the extreme of zero diversity (or an IQV of 0.00). Whites comprised 71% of the population, but two other groups (blacks and Hispanics) were also quite sizeable at about 12% of the population. In 2025 and again in 2050, the percentage of whites declines and the proportional share of other groups, especially Hispanic Americans, increases. By 2050, the population will be much closer to maximum variation (an IQV of 1.00) than it was in 2000. Let us see how the IQV substantiates these observations. The computational formula for the IQV is

FORMULA 4.1

$$\text{IQV} = \frac{k(N^2 - \Sigma f^2)}{N^2(k - 1)}$$

where k = the number of categories

N = the number of cases

Σf^2 = the sum of the squared frequencies

To use this formula, the sum of the squared frequencies must first be computed. It is most convenient to do this by adding a column to the frequency distribution for the squared frequencies and then summing this column. This procedure is illustrated in Table 4.2.

TABLE 4.2 FINDING THE SUM OF THE SQUARED FREQUENCIES

	2000		2025		2050	
Group	*f*	f^2	*f*	f^2	*f*	f^2
Non-Hispanic white	71	5041	62	3844	53	2809
Black	12	144	13	169	13	169
Native American	1	1	1	1	1	1
Asian American	4	16	6	36	9	81
Hispanic	12	144	18	169	24	576
Total =	100	5346	100	4219	100	3636

For each year, the sum of the frequency column is *N* and the sum of the squared frequency ($\sum f^2$) is the total of the second column. Substituting these values into Formula 4.1 for the year 2000, we would have an IQV of 0.58:

$$\text{IQV} = \frac{k(N^2 - \Sigma f^2)}{N^2(k-1)} = \frac{(5)(100^2 - 5{,}346)}{(100^2)(4)} = \frac{5(10{,}000 - 5{,}346)}{(10{,}000)(4)} = \frac{(5)(4654)}{40{,}000} = \frac{23{,}270}{40{,}000} = .58$$

Since the values of *k* and *N* are the same for all three years, the IQV for the remaining years can be found by simply changing the values for $\sum f^2$. For the year 2025,

$$\text{IQV} = \frac{k(N^2 - \Sigma f^2)}{N^2(k-1)} = \frac{(5)(100^2 - 4{,}219)}{(100^2)(4)} = \frac{5(10{,}000 - 4{,}219)}{(10{,}000)(4)} = \frac{(5)(5{,}781)}{40{,}000} = \frac{28{,}905}{40{,}000} = .72$$

and, similarly, for the year 2050,

$$\text{IQV} = \frac{k(N^2 - \Sigma f^2)}{N^2(k-1)} = \frac{(5)(100^2 - 3{,}636)}{(100^2)(4)} = \frac{5(10{,}000 - 3{,}636)}{(10{,}000)(4)} = \frac{(5)(6{,}634)}{40{,}000} = \frac{31{,}820}{40{,}000} = .80$$

Thus, the IQV, in a quantitative and precise way, substantiates our earlier impressions.

ONE STEP AT A TIME Finding the Index of Qualitative Variation (IQV)

If there is more than one column of frequencies (as in Table 4.1), complete these steps for each column in the table.

Step 1: Add a column to the frequency distribution and label the new column "f^2."

Step 2: Square each frequency and place the result in the new column you created in Step 1.

Step 3: Find the sum of the frequency (*f*) column and multiply this sum by itself. This value is N^2.

Step 4: Find the sum of the squared frequency column. This value is Σf^2.

Step 5: Subtract the sum of the squared frequencies (Σf^2) from N^2.

Step 6: Multiply the value you found in step 5 by the number of categories (*k*).

Step 7: Multiply N^2 by the number of categories minus 1 ($k - 1$).

Step 8: Divide the quantity you found in step 6 by the quantity you found in step 7. The result is the IQV.

The IQV of .58 for the year 2000 means that the distribution of frequencies shows almost 60% of the maximum variation possible. By 2050, the variation has increased to 80% of the maximum variation possible for this distribution of categories. United States society was already quite heterogeneous in 2000 and will grow increasingly diverse as the century progresses. *(For practice in calculating and interpreting the IQV, see problems 4.1 and 4.2.)*

4.3 THE RANGE (*R*) AND INTERQUARTILE RANGE (*Q*)

The **range (*R*)** is defined as the distance between the highest and lowest scores in a distribution:

FORMULA 4.2

$$R = \text{High score} - \text{Low score}$$

The range is quite easy to calculate (high score minus low score) and is perhaps most useful to gain a quick and general notion of variability while scanning many distributions.

Unfortunately, the range has some important limitations, related to the fact that it is based on only two scores (the highest and lowest scores). First, almost any sizeable distribution will contain some scores that are atypically high and/or low compared to most of the scores (for example, see Table 3.3). Thus, R might exaggerate the amount of dispersion for most of the scores in the distribution. Also, R yields no information about the variation of the scores between the highest and lowest scores.

The **interquartile range (*Q*)** is a kind of range. It avoids some of the problems associated with R by considering only the middle 50% of the cases in a distribution. To find Q, arrange the scores from highest to lowest and then divide the distribution into quarters (as distinct from halves when locating the median). The first quartile (Q_1) is the point below which 25% of the cases fall and above which 75% of the cases fall. The second quartile (Q_2) divides the distribution into halves (thus, Q_2 is equal in value to the median). The third quartile (Q_3) is the point below which 75% of the cases fall and above which 25% of the cases fall. Thus, if line LH represents a distribution of scores, the quartiles are located as shown:

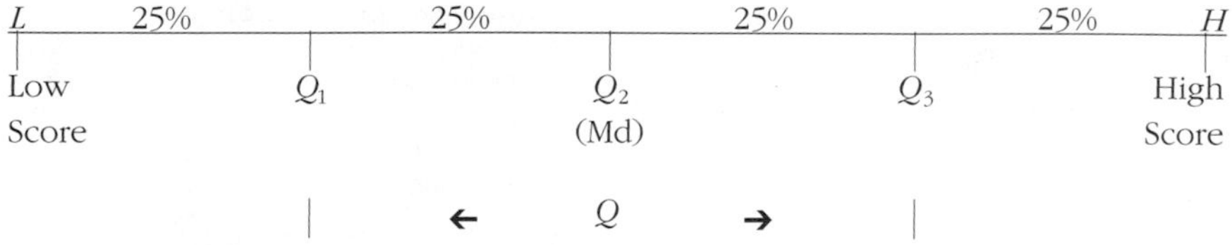

The interquartile range is defined as the distance from the third to the first quartile, as stated in Formula 4.3:

FORMULA 4.3

$$Q = Q_3 - Q_1$$

The interquartile range essentially extracts the middle 50% of the distribution and, like R, is based on only two scores. Unlike the range, Q avoids the problem of being based on the most extreme scores, but it also fails to yield any information about the variation of the scores other than the two upon which it is based.

ONE STEP AT A TIME **Finding the Range (*R*)**

Step 1: Find the high and low scores.

Step 2: Subtract the low score from the high score. The result is the range (*R*).

4.4 COMPUTING THE RANGE AND INTERQUARTILE RANGE[1]

Table 4.3 presents per capita school expenditures for 20 states. What are the range and interquartile range of these data? Note that the scores have already been ordered from high to low. This makes the range easy to calculate and is necessary for finding the interquartile range. Of these 20 states, Illinois spent the most per capita on public education ($2234) and Arizona spent the least ($1137). The range is therefore 2234 − 1137, or $1097 ($R$ = $1097).

To find Q, we must locate the first and third quartiles (Q_1 and Q_3). We will define these points in terms of the scores associated with certain cases, as we did when finding the median. Q_1 is determined by multiplying N by (.25). Since (20) × (.25) is 5, Q_1 is the score associated with the fifth case, counting up from the lowest score. The fifth case is Idaho, with a score of 1244. So $Q_1 = 1244$. The case that lies at the third quartile (Q_3) is given by multiplying N by (.75),

TABLE 4.3 PER CAPITA EXPENDITURES ON PUBLIC SCHOOLS IN SELECTED STATES, 2004

Rank	State	Expenditure
20 (highest)	Illinois	2234
19	New Jersey	1903
18	Wyoming	1896
17	Michigan	1880
16	Maine	1716
15	Texas	1696
14	Ohio	1667
13	California	1657
12	New Hampshire	1632
11	Virginia	1631
10	Pennsylvania	1617
9	Oregon	1384
8	Nebraska	1373
7	Louisiana	1367
6	Florida	1286
5	Idaho	1244
4	North Carolina	1233
3	Alabama	1221
2	Mississippi	1178
1 (lowest)	Arizona	1137

Source: U.S. Bureau of the Census, 2007. *Statistical Abstract of the United States, 2007*, p. 161. Available at http://www.census.gov/compendia/statab/

[1] This section is optional.

ONE STEP AT A TIME **Finding the Interquartile Range (*Q*)**

Step 1: Find the case that lies at the first quartile (Q_1) by multiplying *N* by 0.25. If the result is not a whole number, round off to the nearest whole number. This is the number of the case that marks the first quartile. Note the *score* of this case.

Step 2: Find the case that lies at the third quartile (Q_3) by multiplying *N* by 0.75. If the result is not a whole number, round off to the nearest whole number. This is the number of the case that marks thethird quartile. Note the *score* of this case.

Step 3: Subtract the score of the case at the first quartile (Q_1)—see step 1—from the score of the case at the third quartile (Q_3)—see step 2. The result is the interquartile range (*Q*).

and (20) × (.75) = 15th case. The 15th case, again counting up from the lowest score, is Texas, with a score of 1696 ($Q_3 = 1696$). Therefore

$$Q = Q_3 - Q_1$$
$$Q = 1696 - 1244$$
$$Q = 452$$

In most situations, the locations of Q_1 and Q_3 will not be as obvious as they are when $N = 20$. For example, if *N* had been 157, then Q_1 would be (157)(.25), or the score associated with the 39.25th case, and Q_3would be (157)(.75), or the score associated with the 117.75th case. Since fractions of cases are impossible, these numbers present some problems. The easy solution to this difficulty is to round off and take the score of the closest case to the numbers that mark the quartiles. Thus, Q_1 would be defined as the score of the 39th case and Q_3 as the score of the 118th case.

The more accurate solution would be to take the fractions of cases into account. For example, Q_1 could be defined as the score that is one-quarter of the distance between the scores of the 39th and 40th cases, and Q_3 could be defined as the score that is three-quarters of the distance between the scores of the 117th and 118th cases. (This procedure could be analogous to defining the median—which is also Q_2—as halfway between the two middle scores when *N* is even.) In most cases, the differences in the values of *Q* for these two methods would be quite small. *(For practice in finding and interpreting Q, see problem 4.5. The range may be found for any of the problems at the end of this chapter except 4.1 and 4.2.)*

4.5 THE STANDARD DEVIATION AND VARIANCE

A basic limitation of both *Q* and *R* is that they are based on only two scores. They do not use all the scores in the distribution, and, in this sense, they do not capitalize on all the available information. Also, neither statistic provides any information on how far the scores are from each other or from some central point, such as the mean. How can we design a measure of dispersion that would correct these faults? We can begin with some specifications. A good measure of dispersion should:

1. Use all the scores in the distribution. The statistic should use all the information available.

2. Describe the average or typical deviation of the scores. The statistic should give us an idea about how far the scores are from each other or from the center of the distribution.
3. Increase in value as the distribution of scores becomes more diverse. This would be a very handy feature because it would permit us to tell at a glance which distribution was more variable: The higher the numerical value of the statistic, the greater the dispersion.

One way to develop a statistic to meet these criteria would be to start with the distances between each score and the mean. The distances between the scores and the mean $(X_i - \overline{X})$ are called **deviations,** and this quantity will increase in value as the scores increase in their variety or heterogeneity. If the scores are more clustered around the mean (remember the graph for Service A in Figure 4.1), the deviations would be small. If the scores are more spread out, or more varied (like the scores for Service B in Figure 4.1), the deviations would be greater in value. How can we use the deviations of the scores around the mean to develop a useful statistic?

One course of action would be to use the sum of the deviations—$\Sigma(X_i - \overline{X})$—as the basis for a statistic, but, as we saw in Section 3.6, the sum of deviations will always be zero. To illustrate, consider a distribution of five scores: 10, 20, 30, 40, and 50. If we sum the deviations of the scores from the mean, we would always wind up with a total of zero, regardless of the amount of variety in the scores:

Scores (X_i)	Deviations $(X_i - \overline{X})$
10	$(10 - 30) = -20$
20	$(20 - 30) = -10$
30	$(30 - 30) = 0$
40	$(40 - 30) = 10$
50	$(50 - 30) = 20$
$\Sigma(X_i) = 150$	$\Sigma(X_i - \overline{X}) = 0$

$$\overline{X} = 150/5 = 30$$

Still, the sum of the deviations are a logical basis for a statistic that measures the amount of variety in a set of scores, and statisticians have developed two ways around the fact that the positive deviations always equal the negative deviations. Both solutions eliminate the negative signs. The first does so by using the absolute values, that is, by ignoring signs, when summing the deviations. This is the basis for a statistic called the **average deviation,** a measure of dispersion that is rarely used and will not be mentioned further.

The second solution squares each of the deviations. This makes all values positive because a negative number multiplied by a negative number becomes positive. For example: $(-20) \times (-20) = 400$. In the preceding example, the sum of the squared deviations would be $(400 + 100 + 0 + 100 + 400)$, or 1000. Thus, a statistic based on the sum of the squared deviations will have some of the properties we want in a good measure of dispersion.

Before we finish designing our measure of dispersion, we must deal with another problem. The sum of the squared deviations will increase with sample size: The larger the *number* of scores, the greater the value of the measure. This would make it very difficult to compare the relative variability of distributions based on samples of different size. We can solve this problem by dividing the sum of the squared deviations by N (sample size) and thus standardizing for samples of different sizes.

These procedures yield a statistic known as the **variance,** which is symbolized as s^2. The variance is used primarily in inferential statistics, although it is a central concept in the design of some measures of association. For purposes of describing the dispersion of a distribution, a closely related statistic, called the **standard deviation** (symbolized as s), is typically used, and this statistic is our focus for the remainder of the chapter.

The formulas for the variance and standard deviation are

FORMULA 4.4

$$s^2 = \frac{\Sigma(X_i - \overline{X})^2}{N}$$

FORMULA 4.5

$$s = \sqrt{\frac{\Sigma(X_i - \overline{X})^2}{N}}$$

Strictly speaking, Formulas 4.2 and 4.3 are for the variance and standard deviation of a population. Slightly different formulas, with $N - 1$ instead of N in the denominator, should be used when we are working with random samples rather than entire populations. This is an important point because many of the electronic calculators and statistical software packages you might be using (including SPSS) use "$N - 1$" in the denominator and, thus, produce results that are at least slightly different from Formulas 4.2 and 4.3. The size of the difference will decrease as sample size increases, but the problems and examples in this chapter use small samples and the differences between using N and $N - 1$ in the denominator can be considerable in such cases. Some calculators offer the choice of "$N - 1$" or "N" in the denominator. If you use the latter, the values calculated for the standard deviation should match the values in this text.

To compute the standard deviation, it is advisable to construct a table such as Table 4.4 to organize computations. The five scores used in the previous example are listed in the left-hand column, the deviations are in the middle column, and the squared deviations are in the right-hand column.

The total of the last column in Table 4.4 is the sum of the squared deviations and should be substituted into the numerator of Formula 4.3. To finish solving the formula, divide the sum of the squared deviations by N and take the square root of the result. To find the variance, square the standard deviation. For our example problem, the variance is $s^2 = (14.14)^2 = 200$:

TABLE 4.4 COMPUTING THE STANDARD DEVIATION

Scores (X_i)	Deviations $(X_i - \overline{X})$	Deviations Squared $(X_i - \overline{X})^2$
10	$(10 - 30) = -20$	$(-20)^2 = 400$
20	$(20 - 30) = -10$	$(-10)^2 = 100$
30	$(30 - 30) = 0$	$(-0)^2 = 0$
40	$(40 - 30) = 10$	$(-10)^2 = 100$
50	$(50 - 30) = 20$	$(-20)^2 = 400$
$\Sigma(X_i) = 150$	$\Sigma(X_i - \overline{X}) = 0$	$\Sigma(X_i - \overline{X})^2 = 1000$

ONE STEP AT A TIME Finding the Standard Deviation (s) and the Variance (s^2)

Step 1: Construct a computing table like Table 4.4, with columns for the scores (X_i), the deviations ($X_i - \overline{X}$), and the deviations squared $(X_i - \overline{X})^2$.

Step 2: Place the scores (X_i) in the left-hand column. Add up the scores and divide by N to find the mean.

Step 3: Find the deviations ($X_i - \overline{X}$) by subtracting the mean from each score, one at a time. Note each deviation in the second column of the table.

Step 4: Add up the deviations. The sum must equal zero (within rounding error). If the sum of the deviations does not equal zero, you have made a computational error and need to repeat steps 2 and 3.

Step 5: Square each deviation and place the result in the third column.

Step 6: Add up the squared deviations listed in the third column and transfer this sum to the numerator in Formula 4.5.

Step 7: Divide the sum of the squared deviations (see step 6) by N.

Step 8: Take the square root of the quantity you computed in step 7. This is the standard deviation.

To Find the Variance (s^2):

Step 1: Square the value of the standard deviation (s). See step 8.

$$s = \sqrt{\frac{\Sigma(X_i - \overline{X})^2}{N}}$$

$$s = \sqrt{\frac{1000}{5}}$$

$$s = \sqrt{200}$$

$$s = 14.14$$

4.6 COMPUTING THE STANDARD DEVIATION: AN ADDITIONAL EXAMPLE

An additional example will help to clarify the procedures for computing and interpreting the standard deviation. A researcher is comparing the student bodies of two campuses. One college is located in a small town, and almost all the students reside on campus. The other is located in a large city, and the students are almost all part-time commuters. The researcher wishes to compare the age structure of the two campuses and has compiled the information presented in Table 4.5. Which student body is older and more diverse on age? (Needless to say, these very small groups are much too small for serious research and are used here only to simplify computations.)

We see from the means that the students from the residential campus are quite a bit younger than the students from the urban campus (19 vs. 23 years of age). Which group is more diverse on this variable? Computing the standard deviation will answer this question. To solve Formula 4.3, substitute the sum of the right-hand column ("Deviations Squared") in the numerator and N (5 in this case) in the denominator:

$$\text{Residential Campus:} \quad s = \sqrt{\frac{\Sigma(X_i - \overline{X})^2}{N}} = \sqrt{\frac{4}{5}} = \sqrt{.8} = .89$$

$$\text{Urban Campus:} \quad s = \sqrt{\frac{\Sigma(X_i - \overline{X})^2}{N}} = \sqrt{\frac{88}{5}} = \sqrt{17.6} = 4.20$$

TABLE 4.5 COMPUTING THE STANDARD DEVIATION FOR TWO CAMPUSES

Residential Campus		
Ages (X_i)	Deviations ($X_i - \bar{X}$)	Deviations Squared $(X_i - \bar{X})^2$
18	$(18 - 19) = -1$	$(-1)^2 = 1$
19	$(19 - 19) = 0$	$(0)^2 = 0$
20	$(20 - 19) = 1$	$(1)^2 = 1$
18	$(18 - 19) = -1$	$(-1)^2 = 1$
20	$(20 - 19) = 1$	$(1)^2 = 1$
$\Sigma(X_i) = 95$	$\Sigma(X_i - \bar{X}) = 0$	$\Sigma(X_i - \bar{X})^2 = 4$
	$\bar{X} = \dfrac{\Sigma(X_i)}{N} = \dfrac{95}{5} = 19$	

Urban Campus		
Ages (X_i)	Deviations ($X_i - \bar{X}$)	Deviations Squared $(X_i - \bar{X})^2$
20	$(20 - 23) = -3$	$(-3)^2 = 9$
22	$(22 - 23) = -1$	$(-1)^2 = 1$
18	$(18 - 23) = -5$	$(-5)^2 = 25$
25	$(25 - 23) = 2$	$(2)^2 = 4$
30	$(30 - 23) = 7$	$(7)^2 = 49$
$\Sigma(X_i) = 115$	$\Sigma(X_i - \bar{X}) = 0$	$\Sigma(X_i - \bar{X})^2 = 88$
	$\bar{X} = \dfrac{\Sigma(X_i)}{N} = \dfrac{115}{5} = 23$	

The higher value of the standard deviation for the urban campus means that it is much more diverse. As you can see by scanning the scores, the students at the residential college are within a narrow age range ($R = 20 - 18 = 2$), whereas the students at the urban campus are more mixed and include students of age 25 and 30 ($R = 30 - 18 = 12$) *(For practice in computing and interpreting the standard deviation, see any of the problems at the end of this chapter except 4.1 and 4.2. Problems with smaller data sets, such as 4.3 to 4.5, are recommended for practicing computations until you are comfortable with these procedures.)*

4.7 COMPUTING THE STANDARD DEVIATION FROM GROUPED DATA[2]

When data have been grouped into a frequency distribution, we face the same problem in computing the standard deviation that we faced with the mean and median: The exact value of the scores is no longer known. We resolve this problem by using the midpoints of the intervals as approximations of the original scores. In other words, instead of subtracting the scores from the mean to find the deviations, we'll subtract the midpoints from the mean. This procedure assumes that the scores in each interval of the frequency distribution are clustered

[2]This section is optional.

TABLE 4.6 COMPUTING A MEAN FOR GROUPED DATA

Correct	Frequency (f)	Midpoints (m)	Frequency × Midpoint ($f \times m$)
0–2	1	1	1
3–5	2	4	8
6–8	3	7	21
9–11	4	10	40
12–14	3	13	39
15–17	2	16	32
18–20	2	19	38
	17		$\Sigma(fm) = 179$

$$\overline{X} = \frac{\Sigma(fm)}{N} = \frac{179}{17} = 10.53$$

at the midpoint and that the value we compute will be an approximation of the actual standard deviation.

To illustrate this procedure, we will use the data presented in Table 4.6, which reports the number of quiz questions missed in a class of 17 students. As you can see from the frequency column, some students got only a few items correct, two students got 18–20 items correct, and most got between 6 and 14 items correct. Table 4.6 also demonstrates the computation of the mean for grouped data (see Section 3.7). The mean is 10.53, and we will use this value to compute the standard deviation.

The formula for computing the standard deviation for grouped data is:

FORMULA 4.6

$$s = \sqrt{\frac{\Sigma f(m - \overline{X})^2}{N}}$$

where f = the number of cases in an interval
m = the midpoint of that interval
$\overline{X}$ = the mean
N = the total number of cases

Formula 4.4 tells us to subtract the value of the mean ($\overline{X}$) from the midpoint (m) of each interval, square the result, and then multiply that quantity by the number of cases in the interval (f). Next, divide by N, and, finally, find the square root of the resultant quantity. It will be helpful to organize computations in table format, as in Table 4.7.

TABLE 4.7 COMPUTING THE STANDARD DEVIATION FOR GROUPED DATA

Correct	f	m	$m - \overline{X}$	$(m - \overline{X})^2$	$f(m - \overline{X})^2$
0–2	1	1	1 − 10.53 = −9.53	90.82	90.82
3–5	2	4	4 − 10.53 = −6.53	42.64	85.28
6–8	3	7	7 − 10.53 = −3.53	12.46	37.38
9–11	4	10	10 − 10.53 = −0.53	0.28	1.12
12–14	3	13	13 − 10.53 = 2.47	6.10	18.30
15–17	2	16	16 − 10.53 = 5.47	29.92	59.84
18–20	2	19	19 − 10.53 = 8.47	71.74	143.48
	17				$\Sigma f(m - \overline{X})^2 = 436.22$

ONE STEP AT A TIME **Finding the Standard Deviation (s) and the Variance (s^2) from Grouped Data**

Step 1: If necessary, construct a computing table like Table 4.6 and find the mean by dividing the sum of the frequency (f) column by N.

Step 2: Construct a computing table like Table 4.7, with columns for the intervals, frequencies (f), midpoints (m), and so forth.

Step 3: Subtract the mean from the midpoint of each interval. Note the result in a column labeled "$m - \bar{X}$."

Step 4: Square the value you found in step 3 and note the result in a column labeled "$(m - \bar{X})^2$."

Step 5: Multiply the value you found in step 4 by the frequency in that interval. Note the result in a column labeled "$f(m - \bar{X})^2$."

Step 6: Add up the values in the "$f(m - \bar{X})^2$" column and transfer this sum to the numerator in Formula 4.6.

Step 7: Divide the sum of the "$f(m - \bar{X})^2$" column (see step 6) by N.

Step 8: Take the square root of the quantity you computed in step 7. This is the standard deviation.

To Find the Variance (s^2) for Grouped Data:

Step 1: Square the value of the standard deviation (s). See step 8.

To solve Formula 4.4, substitute the sum of the far right-hand column—$\Sigma f(m - \bar{X})^2$—into the numerator under the square root sign, divide by N, and take the square root of the resultant sum:

$$s = \sqrt{\frac{\Sigma f(m - \bar{X})^2}{N}} = \sqrt{\frac{436.22}{17}} = \sqrt{25.66} = 5.07$$

These 17 students averaged 10.53 items correct, and the distribution has a standard deviation of 5.07. Remember that these are approximations of the actual mean and standard deviations, based on the assumption that the scores in each interval are clustered at the midpoint. Of course, it would be preferable to have the exact values. But if the data has been grouped into a frequency distribution, the procedures presented here and in Section 3.7 are the only ways to approximate the mean and standard deviation. (*For practice in computing descriptive statistics with grouped data, see problems 4.8b and 4.14b.*)

4.8 INTERPRETING THE STANDARD DEVIATION

It is very possible that the meaning of the standard deviation (i.e., why we calculate it) is not completely obvious to you at this point. You might be asking: "Once I've gone to the trouble of calculating the standard deviation, what do I have?" The meaning of this measure of dispersion can be expressed in three ways. The first and most important involves the normal curve, and we will defer this interpretation until the next chapter.

A second way of thinking about the standard deviation is as an index of variability that increases in value as the distribution becomes more variable. In other words, the standard deviation is higher for more diverse distributions and lower for less diverse distributions. The lowest value the standard deviation can have is zero, and this would occur for distributions with no dispersion (i.e., if

Application 4.1

At a local preschool, 10 children were observed for 1 hour, and the number of aggressive acts committed by each was recorded in the following list. What is the standard deviation of this distribution? We will use Formula 4.5 to compute the standard deviation. If you use the preprogrammed function on a hand calculator to check these computations, remember to choose "divide by N" not "divide by $N - 1$."

NUMBER OF AGGRESSIVE ACTS

(X_i)	$(X_i - \overline{X})$	$(X_i - \overline{X})^2$
1	$1 - 4 = -3$	9
3	$3 - 4 = -1$	1
5	$5 - 4 = 1$	1
2	$2 - 4 = -2$	4
7	$7 - 4 = 3$	9

(continued next column)

(continued)

(X_i)	$(X_i - \overline{X})$	$(X_i - \overline{X})^2$
11	$11 - 4 = 7$	49
1	$1 - 4 = -3$	9
8	$8 - 4 = 4$	16
2	$2 - 4 = -2$	4
0	$0 - 4 = -4$	16
$\Sigma(X_i) = 40$	$\Sigma(X_i - \overline{X}) = 0$	$\Sigma(X_i - \overline{X})^2 = 118$

$$\overline{X} = \frac{\Sigma(X_i)}{N} = \frac{40}{10} = 4.0$$

Substituting into Formula 4.5, we have

$$s = \sqrt{\frac{\Sigma(X_i - \overline{X})^2}{N}} = \sqrt{\frac{118}{10}} = \sqrt{11.8} = 3.44$$

The standard deviation for these data is 3.44.

every single case in the sample had exactly the same score). Thus, 0 is the lowest value possible for the standard deviation (although there is no upper limit).

A third way to get a feel for the meaning of the standard deviation is by comparing one distribution with another. We already did this when comparing the two ambulance services in Figure 4.1 and the residential and urban campuses in Section 4.6. You might also do this when comparing one group against another (e.g., men vs. women, blacks vs. whites) or the same variable at two different times. For example, suppose we found that the ages of the students on a particular campus had changed over time as indicated by the following summary statistics:

1975	2005
$\overline{X} = 21$	$\overline{X} = 25$
$s = 1$	$s = 3$

In 1975, students were, on the average, 21 years of age. By 2005, the average age had risen to 25. Clearly, the student body has grown older, and, according to the standard deviation, it has also grown more diverse in terms of age. The lower standard deviation for 1975 indicates that the distribution of ages in that year would be more clustered around the mean (remember the distribution for Service A in Figure 4.1), whereas, in 2005 the distribution would be flatter and more spread out, like the distribution for Service B in Figure 4.1. In other words, compared to 2005, the students in 1975 were more similar to each other and more clustered in a narrower age range. The standard deviation is extremely useful for making comparisons of this sort between distributions of scores.

Application 4.2

Five western and five eastern states were compared as part of a study of traffic safety. The states differ in population, so comparisons are made in terms of the rate of fatal accidents (number of fatal accidents per 100,000 licensed drivers). Columns for the computation of the standard deviation have already been added to the tables below. Which group of states is most variable? Computations for both the mean and standard deviation are shown.

FATALITIES PER 100,000 LICENSED DRIVERS FOR 2004, EASTERN STATES

State	Fatalities (X_i)	Deviations ($X_i - \overline{X}$)	Deviations Squared $(X_i - \overline{X})^2$
Pennsylvania	17.67	17.67 − 12.92 = 4.75	22.56
New York	13.27	13.27 − 12.92 = 0.35	.12
New Jersey	12.60	12.60 − 12.92 = −.32	.10
Connecticut	10.80	10.80 − 12.92 = −2.12	4.49
Massachusetts	10.25	10.25 − 12.92 = −2.67	7.13
	$\Sigma(X_i) = 64.59$	$\Sigma(X_i - \overline{X}) = 0.01$	$\Sigma(X_i - \overline{X})^2 = 34.40$

Source: U.S. Bureau of the Census, 2007. *Statistical Abstract of the United States, 2007.* U.S. Government Printing Office: Washington, D.C., pp. 687, 691.

$$\overline{X} = \frac{\Sigma(X_i)}{N} = \frac{64.59}{5} = 12.92$$

$$s = \sqrt{\frac{\Sigma(X_i - \overline{X})^2}{N}} = \sqrt{\frac{34.40}{5}} = \sqrt{6.90} = 2.63$$

FATALITIES PER 100,000 LICENSED DRIVERS FOR 2001, WESTERN STATES

State	Fatalities (X_i)	Deviations ($X_i - \overline{X}$)	Deviations Squared $(X_i - \overline{X})^2$
Wyoming	43.16	43.16 − 23.03 = 20.13	405.22
Nevada	25.52	25.52 − 23.03 = 2.49	6.20
California	18.10	18.10 − 23.03 = −4.93	24.31
Oregon	17.36	17.36 − 23.03 = −5.67	32.15
Washington	11.01	11.01 − 23.03 = −12.02	144.48
	$\Sigma(X_i) = 115.15$	$\Sigma(X_i - \overline{X}) = 0.00$	$\Sigma(X_i - X)^2 = 612.36$

Source: U.S. Bureau of the Census, 2007. *Statistical Abstract of the United States, 2007.* U.S. Government Printing Office: Washington, D.C., pp. 687, 691.

$$\overline{X} = \frac{\Sigma(X_i)}{N} = \frac{115.15}{5} = 23.03$$

$$s = \sqrt{\frac{\Sigma(X_i - \overline{X})^2}{N}} = \sqrt{\frac{612.36}{5}} = \sqrt{122.47} = 11.07$$

With such small groups, you can tell by simply inspecting the scores that the western states have higher fatality rates. This impression is confirmed by both the median, which is 12.06 for the five eastern states (New Jersey is the middle case) and 18.10 for the western states (California is the middle case), and the mean (12.92 for the eastern states and 23.03 for the western states). For both groups, the mean is greater than the median, indicating a positive skew. Note also that the skew is much greater for the western states (that is, for the western states, the mean is much higher than the median). This is caused by Wyoming, which has a much higher score than any other state. However, even with Wyoming removed, the western states still have higher rates. With Wyoming removed, the mean of the remaining four western states is 18.00.

The five western states are also much more variable than the eastern states. The range for the western states is 32.15 ($R = 43.16 - 11.01 = 32.15$), much higher than the range for the eastern states of 7.42 ($R = 17.67 - 10.25 = 7.42$). Similarly, the standard deviation for the western states (11.07) is more than four times greater in value than the standard deviation for the eastern states (2.63). The greater dispersion in the western states is also largely a reflection of Wyoming's extremely high score.

In summary, the five western states average higher fatality rates and are also more variable than the five eastern states.

READING STATISTICS 3: Measures of Central Tendency and Dispersion

As was the case with frequency distributions, measures of central tendency and dispersion may not be presented in research reports in the professional literature. Given the large number of variables included in a typical research project and the space limitations in journals and other media, there may not be room for the researcher to describe each variable fully. Furthermore, virtually all research reports focus on relationships between variables rather than the distribution of single variables. In this sense, univariate descriptive statistics will be irrelevant to the main focus of the report.

This does not mean, of course, that univariate descriptive statistics are irrelevant to the research project. Nor do I mean to imply that researchers do not calculate and interpret these statistics. In fact, measures of central tendency and dispersion will be calculated and interpreted for virtually every variable, in virtually every research project. However, these statistics are less likely to be included in final research reports than the more analytical statistical techniques to be presented in the remainder of this text. Furthermore, some statistics (for example, the mean and standard deviation) serve a dual function. They not only are valuable descriptive statistics but also form the basis for many analytical techniques. Thus, they may be reported in the latter role if not in the former.

When included in research reports, measures of central tendency and dispersion will most often be presented in some summary form for all relevant variables—often in the form of a table. Means and standard deviations for many variables might, for example, be presented in the following table format.

Variable	$\bar{X}$	s	N
Age	33.2	1.3	1078
Number of children	2.3	.7	1078
Years married	7.8	1.5	1052
Income	55,786	1500	987
.	.	.	.
.	.	.	.
.	.	.	.
.	.	.	.

These tables describe the overall characteristics of the sample succinctly and clearly. If you inspect the table carefully, you will have a good sense of the nature of the sample on the traits relevant to the project. Note that the number of cases varies from variable to variable in the preceding table. This is normal in social science research and is caused by missing data or incomplete information on some of the cases.

Statistics in the Professional Literature

Professors Bill McCarthy, Diane Felmlee, and John Hagan were concerned with the effects of friendship networks on involvement in crime for teenagers. Specifically, they hypothesized that friendships with females would provide better social control and less motivation for criminal behavior than friendships with males. While the popular media has raised awareness of "mean girls" and female aggression in recent years, these researchers believed that teens with strong relationships with female peers would be less involved in delinquency and would have lower scores on a number of other "risk factors" associated with deviant behavior.

McCarty, Felmlee, and Hagan also believed that the relationship between friendships and delinquency would be affected by the social context in which

(continued)

4.9 INTERPRETING STATISTICS: THE CENTRAL TENDENCY AND DISPERSION OF INCOME IN THE UNITED STATES

In this installment of "Interpreting Statistics," we examine the changing distribution of income in the United States. We will use the latest information from the Bureau of the Census to answer several questions: Is average income rising or falling? Does the average American receive more or less income than in the past? Is the distribution of income becoming more unequal (i.e., are the rich getting richer)? We can answer these questions by looking at changes in measures of central tendency and dispersion. Changes in the measures of central tendency will tell us about changes in the average income (mean income) and income for

READING STATISTICS 3: *(continued)*

those friendships arose. They had access to two samples of teenagers: (1) males and females who lived at home and attended school and (2) a group of homeless youth who spent their days and nights on the streets. The researchers believed that female friendships that developed in the less conventional, homeless context would have weaker effects on involvement in delinquency. The researchers tested their ideas on a sample of 563 youths who lived in Toronto, Canada, and attended high school and a sample of street youth from the same city.

In this installment of "Reading Statistics," we review some of the descriptive statistics reported by the researchers. As is usually the case, they report this information to give the reader some information about background and general characteristics of the respondents. The actual hypotheses are tested with more advanced statistics.

Means and Standard Deviations on Selected Variables for Four Samples

	Females				Males			
	School		Street		School		Street	
	$\bar{X}$	s	$\bar{X}$	s	$\bar{X}$	s	$\bar{X}$	s
Friendship Variables:								
No. of friends	2.50	0.83	2.31	0.80	2.60	0.86	2.34	1.10
Confide in friends	4.00	1.25	4.18	1.53	4.00	1.36	4.07	1.82
Proportion of friends arrested	0.51	1.00	2.60	1.54	1.00	1.20	2.76	1.53
Background Variables:								
Parental attachment	13.35	4.14	7.01	4.01	12.10	4.06	7.36	4.06
Positive school experience	11.05	2.00	9.00	3.00	10.24	2.42	8.39	3.01
Dependent Variable								
Property crime	4.88	35.10	36.23	95.42	5.41	23.78	71.68	132.32

There is little difference among the four groups in total number of friends or on their average scores on a scale that measures the degree to which they can confide in friends. Comparing males with females, we see that females in both the school and street samples were less likely to have deviant friends and be involved in property crimes. The researchers also found that context mattered: Males and females from the street sample are more likely to have deviant friends and are much more likely to be involved in property crime than the males and females from the school sample. Also, "street" males and females had lower scores on scales that measured their attachment to parents and how positive their school experiences were.

Want to find out more? See the following source.

McCarthy, Bill, Felmlee, Diane, and Hagan John. 2004. "Girl Friends are Better: Gender, Friends, and Crime Among School and Street Youth." *Criminology*, 42: 805–835.

the average American (median income). The standard deviation would be the preferred measure of dispersion for an interval-ratio-level variable such as income, but, unfortunately, the Census Bureau does not provide this information. Instead, we will measure dispersion with a statistic that is a variation of the interquartile range.

Before considering the data, we should keep in mind that virtually any distribution of income will be positively skewed (see Figure 3.1). That is, in any group, community, state, or nation, the incomes of most people will be grouped

FIGURE 4.2 MEAN AND MEDIAN INCOME OF HOUSEHOLDS, 1967–2005 (2005 dollars)

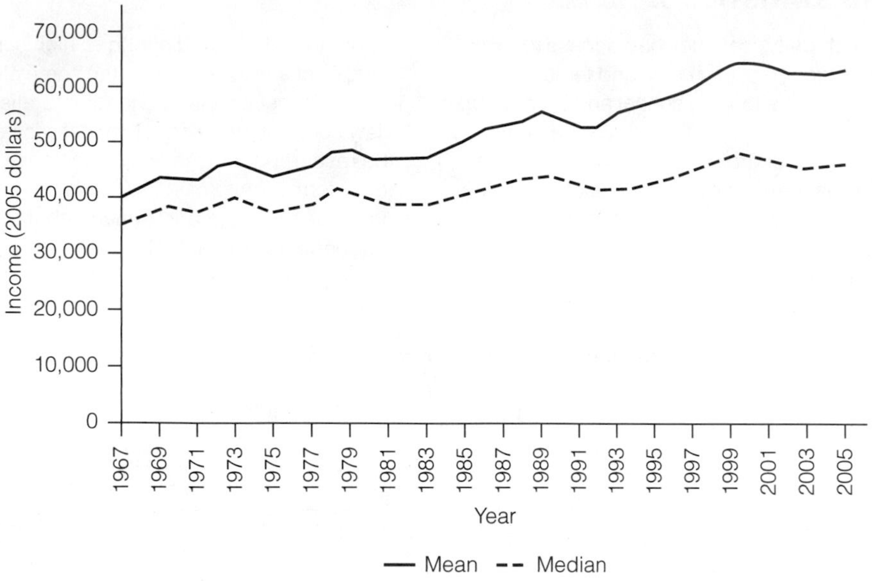

around the mean or median but some people—the wealthiest members of the group—will have incomes far higher than the average. Of course, some people will also have incomes far below the mean or median and their incomes may even approach zero. However, low incomes will have less effect on the mean than very high incomes. The lowest income cannot be less than zero, but the highest incomes can be in the millions or even billions. Since the mean uses *all* the scores, including the very highest, and is pulled in the direction of extreme scores relative to the median, mean income will always be greater than median income.

Figure 4.2 shows the changes in mean and median income for the United States over almost four decades. Please note that incomes are expressed in 2005 dollars to eliminate any changes in the mean and median caused by inflation over the years. Without this adjustment, recent incomes would appear to be much higher than older incomes, not because of increased buying power and well-being, but rather because of the changing value of the dollar. Also, Figure 4.2 is based on total income for entire households, not individual income.

The line labeled "Median" in Figure 4.2 shows that, expressed in 2005 dollars, the income of the average American household was about 35,000 in 1967. This value gradually trends upward, with some noticeable declines in the mid-1970s, early 1980s, and early 1990s, all periods of recession. There is a pronounced increase in median income during the boom economy of the 1990s, but then we see a decline and leveling off as the economy once again fell into recession. By 2005, median income for the average American household had reached $46,326, almost $11,000 higher than in 1967. This indicates an increase in standard of living over this time period.

The pattern of the mean income is almost identical to that of the median. It rises and falls in almost exactly the same ways and at the same times. Notice, however, that the mean is always higher than the median, a reflection of the characteristic positive skew of income data. Also notice that the size of the gap

FIGURE 4.3 80TH AND 20TH PERCENTILE HOUSEHOLD INCOMES, 1967–2005 (2005 dollars)

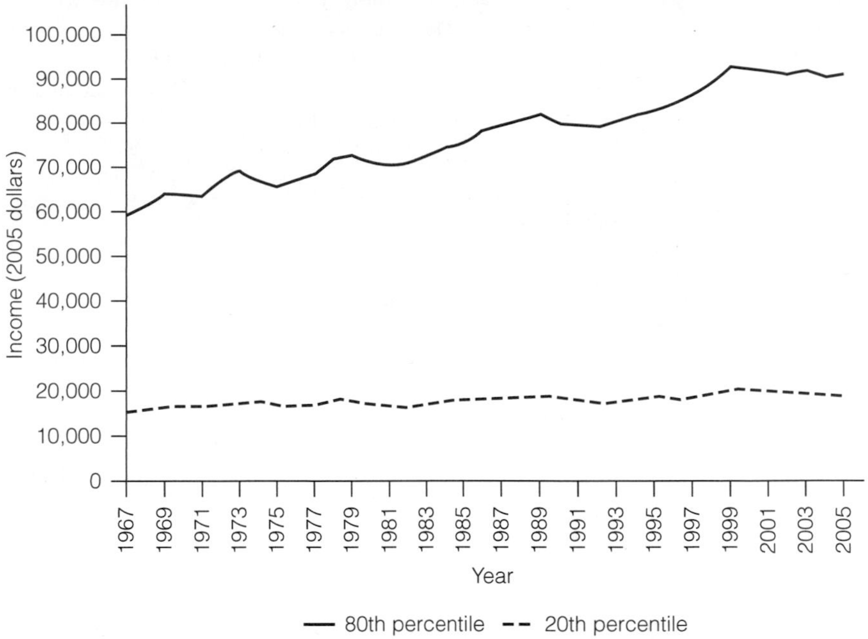

between the mean and the median increases over time, indicating that the degree of skew is increasing. In other words, the difference between the median and mean increases because the underlying distribution is becoming more skewed—the income of the rich is increasing relative to the income of the typical American household.

An increasing skew means that the scores are becoming more varied, and this means, in turn, that all measures of dispersion computed on the distribution of income between 1967 and 2005 will increase in value. The Census Bureau does not report standard deviations, but it does provide a statistic similar to the interquartile range (see Section 4.3). This measure of dispersion is based on the 80th and 20th percentiles rather than Q_3 and Q_1 (or the 75th and the 25th percentiles). The 20th percentile is the point below which 20% of all household incomes fell, and the 80th percentile is the point below which 80% of all incomes fell. Both are mapped in Figure 4.3. The distance between the 80th and 20th percentiles is a kind of range. You can see by inspecting Figure 4.3 that the distance between the 20th and 80th percentiles is increasing, and this verifies the conclusion that American household incomes are growing more dispersed, or variable.

In 1967, the 20th percentile was at about $15,000. In other words, in this year 20% of households made less than $15,000 and 80% made more. The 20th percentile moves up slightly over the years, but in 2005 it was still slightly less than $20,000. This indicates that the financial situation of lower-income Americans had improved only slightly over the time period.

In contrast, the line marking the 80th percentile increases dramatically, rising about $30,000 over the time period from the high $50,000s to the low $90,000s. This indicates that more affluent Americans received a much higher level of income at the end of the time period, which is very consistent with the increasing positive skew in Figure 4.2.

Taken together, these statistics and graphs show increases in the income of the average American (median) and the average income for all Americans (mean). The increase was much more dramatic for more affluent Americans (80th percentile) than for the less affluent (20th percentile). Thus, people with modest incomes continued to have modest incomes, and, consistent with the ancient folk wisdom, the rich got richer.

SUMMARY

1. Measures of dispersion summarize information about the heterogeneity, or variety, in a distribution of scores. When combined with an appropriate measure of central tendency, these statistics convey a large volume of information in just a few numbers. While measures of central tendency locate the central points of the distribution, measures of dispersion indicate the amount of diversity in the distribution.
2. The index of qualitative variation (IQV) can be computed for any variable that has been organized into a frequency distribution. It is the ratio of the amount of variation observed in the distribution to the maximum variation possible in the distribution. The IQV is most appropriate for variables measured at the nominal level.
3. The range (R) is the distance from the highest to the lowest score in the distribution. The interquartile range (Q) is the distance from the third to the first quartile (the "range" of the middle 50% of the scores). These two ranges can be used with variables measured at either the ordinal or interval-ratio level.
4. The standard deviation (s) is the most important measure of dispersion because of its central role in many more advanced statistical applications. The standard deviation has a minimum value of zero (indicating no variation in the distribution) and increases in value as the variability of the distribution increases. It is used most appropriately with variables measured at the interval-ratio level.
5. The variance (s^2) is used primarily in inferential statistics and in the design of some measures of association.

SUMMARY OF FORMULAS

Index of Qualitative Variation	4.1	$\text{IQV} = \dfrac{k(N^2 - \Sigma f^2)}{N^2(k-1)}$
Range	4.2	$R = \text{High score} - \text{Low score}$
Interquartile Range	4.3	$Q = Q_3 - Q_1$
Variance	4.4	$s^2 = \dfrac{\Sigma(X_i - \overline{X})^2}{N}$
Standard Deviation	4.5	$s = \sqrt{\dfrac{\Sigma(X_i - \overline{X})^2}{N}}$
Standard Deviation, grouped data	4.6	$s = \sqrt{\dfrac{\Sigma f(m - \overline{X})^2}{N}}$

GLOSSARY

Average deviation (AD). The average of the absolute deviations of the scores around the mean.

Deviations. The distances between the scores and the mean.

Dispersion. The amount of variety, or heterogeneity, in a distribution of scores.

Index of qualitative variation (IQV). A measure of dispersion for variables that have been organized into frequency distributions.

Interquartile range (Q). The distance from the third quartile to the first quartile.

Measures of dispersion. Statistics that indicate the amount of variety, or heterogeneity, in a distribution of scores.

Range (*R*). The highest score minus the lowest score.

Standard deviation. The square root of the squared deviations of the scores around the mean, divided by *N*. The most important and useful descriptive measure of dispersion; *s* represents the standard deviation of a sample; Σ represents the standard deviation of a population.

Variance. The squared deviations of the scores around the mean divided by *N*. A measure of dispersion used primarily in inferential statistics and also in correlation and regression techniques; s^2 represents the variance of a sample; Σ^2 represents the variance of a population.

PROBLEMS

4.1 SOC The marital status of residents of four apartment complexes is reported here. Compute the index of qualitative variation (IQV) for each neighborhood. Which is the most heterogeneous of the four? Which is the least? *(HINT: It may be helpful to organize your computations as in Table 4.2.)*

Complex A

Marital Status	Frequency
Single	26
Married	31
Divorced	12
Widowed	5
	N = 74

Complex B

Marital Status	Frequency
Single	10
Married	12
Divorced	8
Widowed	7
	N = 37

Complex C

Marital Status	Frequency
Single	20
Married	30
Divorced	2
Widowed	1
	N = 53

Complex D

Marital Status	Frequency
Single	52
Married	3
Divorced	20
Widowed	10
	N = 85

4.2 The following table shows the religious preference of college students living in two small dormitories. Compute the index of qualitative variation for the table. Which dorm is more diverse in terms of this variable?

Status	Dorm A	Dorm B
Protestant	20	10
Catholic	7	17
Jew	3	8
None	10	3
Other	7	10
	47	48

4.3 Compute the range and standard deviation of the following 10 scores. *(HINT: It will be helpful to organize your computations as in Tables 4.4.)*

10, 12, 15, 20, 25, 30, 32, 35, 40, 50

4.4 Compute the range and standard deviation of the following 10 test scores.

77, 83, 69, 72, 85, 90, 95, 75, 55, 45

4.5 SOC In problem 3.8, you computed mean and median income for 13 Canadian provinces and territories in two separate years. Now compute the standard deviation and range for each year, and, taking account of the two measures of central tendency and the two measures of dispersion, write a paragraph summarizing the distributions. What do the measures of dispersion add to what you already knew about central tendency? Did the provinces become more or less variable over the period? The scores are reproduced here.

Province or Territory	2000	2004
Newfoundland and Labrador	38,800	46,100
Prince Edward Island	44,200	51,300
Nova Scotia	44,500	51,500
New Brunswick	43,200	49,700
Quebec	47,700	54,400
Ontario	55,700	62,500
Manitoba	47,300	54,100
Saskatchewan	45,800	53,500
Alberta	55,200	66,400
British Columbia	49,100	55,900
Yukon	56,000	67,800
Northwest Territories	61,000	79,800
Nunavut	37,600	49,900

4.6 SOC Data on several variables measuring overall heath and well-being for five nations are reported here for 2005, with projections to 2010. Are these nations becoming more or less diverse on these variables? Calculate the mean, range, and standard deviation for each year for each variable. Summarize the results in a paragraph.

	Life Expectancy (years)		Infant Mortality Rate*		Fertility Rate#	
	2005	2010	2005	2010	2005	2010
Canada	80	81	4.8	4.5	1.6	1.6
United States	78	78	6.5	6.2	2.1	2.1
Mexico	75	76	20.9	17.8	2.5	2.3
Columbia	72	73	21.0	17.8	2.6	2.4
Japan	81	82	3.3	3.2	1.4	1.4
China	72	74	24.2	19.4	1.7	1.8
Sudan	59	61	62.5	55.2	4.9	4.2
Kenya	53	59	60.7	53.5	5.0	4.3
Italy	80	80	5.9	5.4	1.3	1.3
Germany	79	79	4.2	4.0	1.4	1.4

Source: U.S. Bureau of the Census, 2007. *Statistical Abstract of the United States, 2007*, p. 837. Available at http://www.census.gov/compendia/statab/

* Number of deaths of children under one year of age per 1000 live births.

\# Average number of children per female.

4.7 SOC Labor force participation rates (percent employed), percent high school graduates, and mean income for males and females in 10 states are reported here. Calculate a mean and a standard deviation for both groups for each variable and describe the differences. Are males and females unequal on any of these variables? How great is the gender inequality?

	Labor Force Participation		% High School Graduates		Mean Income	
State	Male	Female	Male	Female	Male	Female
A	74	54	65	67	35,623	27,345
B	81	63	57	60	32,345	28,134
C	81	59	72	76	35,789	30,546
D	77	60	77	75	38,907	31,788
E	80	61	75	74	42,023	35,560
F	74	52	70	72	34,000	35,980
G	74	51	68	66	25,800	19,001
H	78	55	70	71	29,000	26,603
I	77	54	66	66	31,145	30,550
J	80	75	72	75	34,334	29,117

4.8 a. Compute the standard deviation for the pretest and posttest scores that were used in problems 2.6 and 3.12. The scores are reproduced here. Taking into account all of the information you have on these variables, write a paragraph describing how the sample changed from test to test. What does the standard deviation add to the information you already had?

Case	Pretest	Posttest
A	8	12
B	7	13
C	10	12
D	15	19
E	10	8
F	10	17
G	3	12
H	10	11
I	5	7
J	15	12
K	13	20
L	4	5
M	10	15
N	8	11
O	12	20

b. Frequency distributions for both pretest and posttest scores are presented here. Use this information to compute a mean and standard deviation for each group, and compare these values with the actual mean and standard deviation you computed earlier. How accurate are the approximate mean and standard deviation as compared to the actual statistics?

Pretest

Scores	Frequency (*f*)
0–4	2
5–9	4
10–14	7
15–19	2
20–24	0
	N = 15

Posttest

Scores	Frequency (*f*)
0–4	0
5–9	3
10–14	7
15–19	3
20–24	2
	N = 15

4.9 In problem 3.11, you computed measures of central tendency for the number of cars per 100 population for eight nations. The scores are reproduced here. Compute the standard deviation for this variable, and write a paragraph summarizing the mean, median, and standard deviation.

Nation	Number of Cars per 100 Population
United States	50
Canada	45
France	46
Germany	51
Japan	39
Mexico	10
Sweden	44
United Kingdom	37

4.10 CJ Per capita expenditures for police protection for 20 cities are reported here for 1995 and 2000. Compute a mean and standard deviation for each year, and describe the differences in expenditures for the five-year period.

City	1995	2000
A	180	210
B	95	110
C	87	124
D	101	131
E	52	197
F	117	200
G	115	119
H	88	87
I	85	125
J	100	150
K	167	225
L	101	209
M	120	201
N	78	141
O	107	94
P	55	248
Q	78	140
R	92	131
S	99	152
T	103	178

4.11 Compute the range and standard deviation for the data presented in problem 3.14. The data are reproduced here. What would happen to the value of the standard deviation if you removed Los Angeles from this distribution and recalculated? Why?

City	Annual Person-Hours of Time Lost to Traffic Congestion per Year
Baltimore	27
Boston	25
Buffalo	6
Chicago	31
Cleveland	6
Dallas	35

(continued next column)

(continued)

City	Annual Person-Hours of Time Lost to Traffic Congestion per Year
Detroit	30
Houston	36
Kansas City	9
Los Angeles	50
Miami	29
Minneapolis	23
New Orleans	10
New York	23
Philadelphia	21
Pittsburgh	8
Phoenix	26
San Antonio	18
San Diego	28
San Francisco	37
Seattle	25
Washington, DC	34

4.12 SOC Listed here are the rates of abortion per 100,000 women for 20 states in 1973 and 1975. Describe what happened to these distributions over the two-year period. Did the average rate increase or decrease? What happened to the dispersion of this distribution? What happened between 1973 and 1975 that might explain these changes in central tendency and dispersion? *(HINT: It was a Supreme Court decision.)*

State	1973	1975
Maine	3.5	9.5
Massachusetts	10.0	25.7
New York	53.5	40.7
Pennsylvania	12.1	18.5
Ohio	7.3	17.9
Michigan	18.7	20.3
Iowa	8.8	14.7
Nebraska	7.3	14.3
Virginia	7.8	18.0
South Carolina	3.8	10.3
Florida	15.8	30.5
Tennessee	4.2	19.2
Mississippi	0.2	0.6
Arkansas	2.9	6.3
Texas	6.8	19.1
Montana	3.1	9.9
Colorado	14.4	24.6
Arizona	6.9	15.8
California	30.8	33.6
Hawaii	26.3	31.6

Source: United States Bureau of the Census, *Statistical Abstracts of the United States: 1977* (98th edition). Washington, D.C., 1977.

4.13 SW One of your goals as the new chief administrator of a large social service bureau is to equalize work loads within the various divisions of the agency. You have gathered data on caseloads per worker within each division. Which division comes closest to the ideal of an equalized workload? Which is farthest away?

A	B	C	D
50	60	60	75
51	59	61	80
55	58	58	74
60	55	59	70
68	56	59	69
59	61	60	82
60	62	61	85
57	63	60	83
50	60	59	65
55	59	58	60

4.14 a. Compute the standard deviation for both sets of data presented in problem 3.13 and reproduced here. Compare the standard deviation computed for freshmen with the standard deviation computed for seniors. What happened? Why? Does this change relate at all to what happened to the mean over the four-year period? How? What happened to the shapes of the underlying distributions?

Freshmen

10	43	30	30	45
40	12	40	42	35
45	25	10	33	50
42	32	38	11	47
22	26	37	38	10

Seniors

10	45	35	27	50
35	10	50	40	30
40	10	10	37	10
40	15	30	20	43
23	25	30	40	10

b. Frequency distributions for both freshman and seniors are presented here. Use this information to compute a mean and standard deviation for each group, and compare these values with the actual mean and standard deviation you computed earlier. How accurate are the approximate mean and standard deviation as compared to the actual statistics?

Freshman

Scores	Frequency (f)
10–19	5
20–29	3
30–39	8
40–49	8
50–59	1
	$N = 25$

Seniors

Scores	Frequency (f)
10–19	7
20–29	4
30–39	6
40–49	6
50–59	2
	$N = 25$

4.15 At St. Algebra College, the math department ran some special sections of the freshman math course using a variety of innovative teaching techniques. Students were randomly assigned to either the traditional sections or the experimental sections, and all students were given the same final exam. The results of the final are summarized here. What was the effect of the experimental course?

Traditional	Experimental
$\bar{X} = 77.8$	$\bar{X} = 76.8$
$s = 12.3$	$s = 6.2$
$N = 478$	$N = 465$

4.16 You're the governor of the state and must decide which of four metropolitan police departments will win the annual award for efficiency. The performance of each department is summarized in monthly arrest statistics as reported here. Which department will win the award? Why?

	Departments			
	A	B	C	D
$\bar{X} =$	601.30	633.17	592.70	599.99
$s =$	2.30	27.32	40.17	60.23

SPSS for Windows

Using *SPSS for Windows* to Produce Measures of Dispersion

Start *SPSS for Windows* by clicking the SPSS icon on your monitor screen, and load the 2006 GSS data set.

SPSS DEMONSTRATION 4.1 Producing the Range and the Standard Deviation

Most of the statistics discussed in this chapter are available from either the **Frequencies** or **Descriptives** procedures that you are already familiar with. In this demonstration, we will use **Descriptives** to find the range and standard deviation for *age* (YEARS OF AGE), *educ* (HIGHEST YEAR OF SCHOOL COMPLETED), and *tvhours* (HOURS PER DAY WATCHING TV). These three variables were also used in SPSS Demonstration 3.2.

From the main menu, click **Analyze, Descriptive Statistics,** and **Descriptives.** The **Descriptives** dialog box will open. Use the cursor to find the names of the three variables in the list on the left, and click the right arrow button to transfer them to the **Variables** box. Click **OK**, and SPSS will produce the same output we analyzed in Demonstration 3.2. Now, however, we will consider dispersion rather than central tendency. The output looks like this:

Descriptive Statistics

	N	Minimum	Maximum	Mean	Std. Deviation
AGE OF RESPONDENT	1417	18	89	46.88	17.086
HIGHEST YEAR OF SCHOOL COMPLETED	1424	0	20	13.26	3.191
HOURS PER DAY WATCHING TV	618	0	24	3.00	2.404
Valid N (listwise)	616				

The standard deviation for each variable is reported in the column labeled "Std. Deviation," and the range can be computed from the values given in the Minimum and Maximum columns. The standard deviation for *age* was about 17.1 years, and the youngest and oldest respondents were 18 and 89, respectively. For *educ*, the standard deviation was 3.2 and scores ranged from zero to 20. Respondents with scores of 20 have completed four years of formal education beyond the bachelor's level. The standard deviation is 2.4 for *tvhours*, and scores ranged from 0 hours of television watching to 24 (the maximum possible in a single day).

At this point, the range is probably easier to understand and interpret than the standard deviation. As we saw in Section 4.8, the latter is more meaningful when we have a point of comparison. For example, suppose we were interested in the variable *tvhours* and how television-viewing habits have changed over the years. The **Descriptives** output for 2006 shows that people watched an average of 3.0 hours a day, with a standard deviation of 2.4. Suppose that a sample from 1986 showed an average of 3.7 hours of television viewing a day, with a standard deviation of 1.1. You could conclude that television watching had, on the average, decreased over the 20-year period but that Americans had also become much more diverse in their viewing habits.

SPSS DEMONSTRATION 4.2 Using the COMPUTE Command to Create an "Attitude Toward Abortion" Scale

SPSS provides a variety of ways to transform and manipulate variables. In the SPSS exercises at the end of Chapter 2, the **Recode** command was introduced as a way of changing the values associated with a variable. In this demonstration, we will use the **Compute** command to create new variables and summary scales.

Let's begin by considering the two questions from the 2006 GSS that measure attitudes toward abortion, *abany* and *abrape*. The items present two different situations under which an abortion might be desired and ask the respondent to react to each situation independently. The first item, a*bany*, asks if an abortion should be possible for "any reason" at all, and *abrape* asks if a legal abortion should be available for the more specific reason that the pregnancy is the result of rape. Since these two situations are distinct, each item should be analyzed in its own right. Suppose, however, that you wanted to create a summary scale that indicated a person's *overall* feelings about abortion.

One way to do this would be to add the scores of the two variables together. This would create a new variable, which we will call *abscale*, with three possible scores. If a respondent was consistently "pro-abortion" and answered "yes" (coded as "1") to both items, the respondent's score on the summary variable would be 2. A score of 3 would occur when a respondent answered "yes" to one item and "no" to the other. This might be labeled an "intermediate" or "moderate" position. The final possibility would be a score of 4, if the respondent answered "no" to both items. This would be a consistent "antiabortion" position. The following table summarizes the scoring possibilities.

If Response on *abrape* Is	and	Response on *abany* Is	Score on *abscale* Will Be
1 (Yes)		1 (Yes)	2 (pro-abortion)
1 (Yes)		2 (No)	3 (moderate)
2 (No)		1 (Yes)	3 (moderate)
2 (No)		2 (No)	4 (antiabortion)

The new variable, *abscale*, summarizes each respondent's overall position on the issue. Once created, *abscale* could be analyzed, transformed, and manipulated exactly like a variable actually recorded in the data file.

To use the **Compute** command, click **Transform** and then **Compute** from the main menu. The **Compute Variable** window will appear. Find the **Target Variable** box in the upper-left-hand corner of this window. The first thing we need to do is to assign a name to the new variable we are about to compute (*abscale*) and to type that name in this box. Next, we need to tell SPSS how to compute the new variable. In this case, *abscale* will be computed by adding the scores of *abany* and *abrape*. Find *abany* in the variable list on the left and click the arrow button in the middle of the screen to transfer the variable name to the **Numeric Expression** box. Next, click the plus sign (+) on the calculator pad under the **Numeric Expression** box, and the sign will appear next to *abany*. Finally, highlight *abrape* in the variable list and click the arrow button to transfer the variable name to the **Numeric Expression** box.

The expression in the **Numeric Expression** box should now read

abany + *abrape*

Click **OK**, and *abscale* will be created and added to the data set. If you want to keep this new variable permanently, click **Save** from the **File** menu, and the updated data set with *abscale* added will be saved to disk. If you are using the student version of SPSS, remember that your data set is limited to 50 variables.

We now have three variables that measure attitudes toward abortion—two items referring to specific situations, and a more general, summary item. It is always a good idea to check the frequency distribution for computed and recoded variables to make sure that the computations were carried out correctly. Use the **Frequencies** procedure (click **Analyze, Descriptive Statistics**, and **Frequencies**) to get tables for *abany*, *abrape*, and *abscale*. Your output will look like this:

ABORTION IF WOMAN WANTS FOR ANY REASON

		Frequency	Percent	Valid Percent	Cumulative Percent
Valid	YES	276	19.4	43.8	43.8
	NO	354	24.8	56.2	100.0
	Total	630	44.2	100.0	
Missing	NAP	776	54.4		
	DK	18	1.3		
	NA	2	.1		
	Total	796	55.8		
Total		1426	100.0		

PREGNANT AS RESULT OF RAPE

		Frequency	Percent	Valid Percent	Cumulative Percent
Valid	YES	490	34.4	79.4	79.4
	NO	127	8.9	20.6	100.0
	Total	617	43.3	100.0	
Missing	NAP	776	54.4		
	DK	31	2.2		
	NA	2	.1		
	Total	809	56.7		
Total		1426	100.0		

abscale

		Frequency	Percent	Valid Percent	Cumulative Percent
Valid	2.00	268	18.8	44.2	44.2
	3.00	219	15.4	36.1	80.2
	4.00	120	8.4	19.8	100.0
	Total	607	42.6	100.0	
Missing	System	819	57.4		
Total		1426	100.0		

Note that the level of approval varies quite a bit for the two specific situations. For *abany*, about 44% of the respondents approve of abortion, but the level of approval rises to almost 80% when the pregnancy is the result of rape (*abrape*). Looking at the combined scores on *abscale*, we see that about 44% of the sample approved of the legal right to an abortion in both cases (scored 2), and about 20% disapproved in both cases (scored 4). Over a third of the sample scored a 3, meaning that they approved in one situation but not in the other.

Note that, out of the total sample of over 1400 people, only 630 and 617 respondents answered the original two abortion items. Remember that no respondent is given the entire GSS, and the vast majority of the "missing cases" received a form of the GSS that did not include these two items. Now look at *abscale*, the summary scale, and note that even fewer cases ($N = 607$) are included in the summary scale than in either of the two original items. When SPSS executes a **Compute** statement, it automatically eliminates any cases that are missing scores on *any* of the constituent items. If these cases were not eliminated, a variety of errors and misclassifications could result. For example, if cases with missing scores were included, a person who scored a 2 ("antiabortion") on *abany* and then failed to respond to *abrape* would have a total score of

2 on *abscale*. Thus, this case would be treated as "pro-abortion" when the only information we have indicates that this respondent is "antiabortion." To eliminate this kind of error, cases with missing scores on any of the constituent variables are deleted from calculations.

Exercises

4.1 Use **Descriptives** to produce univariate descriptive statistics for *educ* (respondents' education), *paeduc* (respondent's father's years of education), *prestg80* (respondent's occupational prestige), and *papres80* (respondent's father's occupational prestige).

a. Compare the statistics for *prestg80* and *papres80*. Describe the difference in occupation prestige between the two generations. Are the respondents higher or lower in prestige than their fathers? Is the sample more or less homogeneous on this variable than their fathers?

b. Compare the statistics for *educ* and *paeduc*. Describe the differences between the two generations. Are the respondents more or less educated than their fathers? Is the sample more or less homogeneous on this variable than their fathers?

4.2 Use the **Compute** command to create a summary scale for *fepresch* and *fefam*. Get univariate descriptive statistics for the summary scale and for *fepresch* and *fefam*. Write a few sentences summarizing these tables using our description of the distributions for *abany*, *abrape*, and *abindex* as a model.

5 The Normal Curve

LEARNING OBJECTIVES

By the end of this chapter, you will be able to

1. Define and explain the concept of the normal curve.
2. Convert empirical scores to Z scores and use Z scores and the normal curve table (Appendix A) to find areas above, below, and between points on the curve.
3. Express areas under the curve in terms of probabilities.

5.1 INTRODUCTION

The **normal curve** is a concept of great importance in statistics. In combination with the mean and standard deviation, the normal curve can be used to construct precise descriptive statements about empirical distributions. In addition, as we shall see in Part II, it is also central to the theory that underlies inferential statistics. Thus, this chapter concludes our treatment of descriptive statistics in Part I and lays important groundwork for Part II.

The normal curve is a theoretical model, a kind of frequency polygon, or line chart, that is unimodal (i.e., has a single mode, or peak), perfectly smooth, and symmetrical (unskewed), so its mean, median, and mode are all exactly the same value. It is bell shaped, and its tails extend infinitely in both directions. Of course, no empirical distribution has a shape that perfectly matches this ideal model, but many variables (e.g., test results from large classes, standardized test scores such as the GRE) are close enough to permit the assumption of normality. In turn, this assumption makes possible one of the most important uses of the normal curve—the description of empirical distributions based on our knowledge of the theoretical normal curve.

The crucial point about the normal curve is that distances along the abscissa (horizontal axis) of the distribution, when measured in standard deviations from the mean, always encompass the same proportion of the total area under the curve. In other words, on any normal curve, the distance from any given point to the mean (when measured in standard deviations) will cut off exactly the same proportion of the total area.

To illustrate, Figures 5.1 and 5.2 present two hypothetical distributions of IQ scores for fictional groups of males and females, both normally distributed (or nearly so), such that:

Males	Females
$\bar{X} = 100$	$\bar{X} = 100$
$s = 20$	$s = 10$
$N = 1000$	$N = 1000$

Figures 5.1 and 5.2 are drawn with two scales on the horizontal axis, or abscissa, of the graph. The upper scale is stated in "IQ units" and the lower scale in

FIGURE 5.1 IQ SCORES FOR A GROUP OF MALES

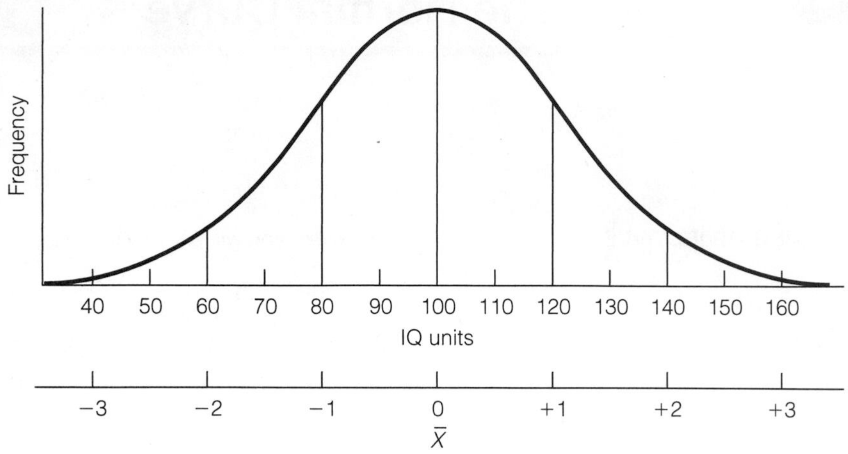

FIGURE 5.2 IQ SCORES FOR A GROUP OF FEMALES

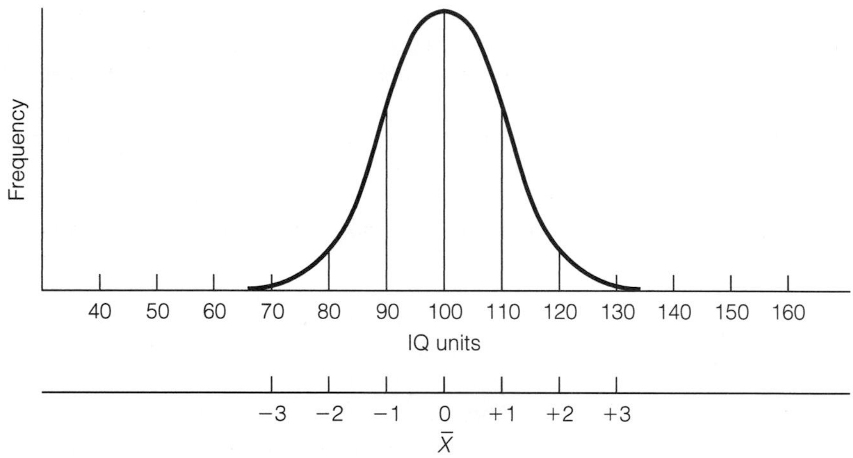

standard deviations from the mean. These scales are interchangeable and we can easily shift from one to the other. For example, for the males, an IQ score of 120 is one standard deviation (remember that, for the male group, $s = 20$) above the mean and an IQ of 140 is two standard deviations above (to the right of) the mean. Scores to the left of the mean are marked as negative values because they are less than the mean. An IQ of 80 is one standard deviation below the mean, an IQ score of 60 is two standard deviations less than the mean, and so forth. Figure 5.2 is marked in a similar way, except that, since its standard deviation is a different value ($s = 10$), the markings occur at different points. For the female sample, one standard deviation above the mean is an IQ of 110, one standard deviation below the mean is an IQ of 90, and so forth.

Recall that, on any normal curve, distances along the abscissa, when measured in standard deviations, always encompass exactly the same proportion of

FIGURE 5.3 AREAS UNDER THE THEORETICAL NORMAL CURVE

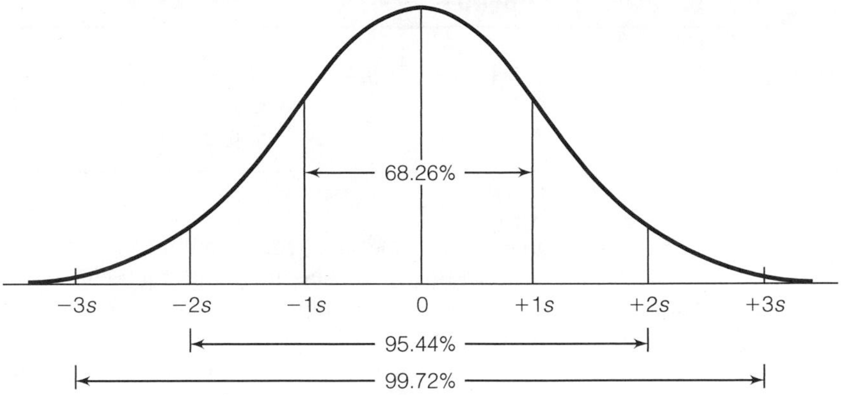

the total area under the curve. Specifically, the distance between one standard deviation above the mean and one standard deviation below the mean (or ±1 standard deviation) encompasses exactly 68.26% of the total area under the curve. This means that in Figure 5.1, 68.26% of the total area lies between the score of 80 (−1 standard deviation) and 120 (+1 standard deviation). The standard deviation for females is 10, so the same percentage of the area (68.26%) lies between the scores of 90 and 110. As long as an empirical distribution is normal, 68.26% of the total area will always be encompassed between ±1 standard deviation—regardless of the trait being measured and the number values of the mean and standard deviation.

It will be useful to familiarize yourself with the following relationships between distances from the mean and areas under the curve:

Between	Lies
±1 standard deviation	68.26% of the area
±2 standard deviations	95.44% of the area
±3 standard deviations	99.72% of the area

These relationships are displayed graphically in Figure 5.3.

The relationship between distance from the mean and area allows us to describe empirical distributions that are at least approximately normal. The position of individual scores can be described with respect to the mean, the distribution as a whole, or any other score in the distribution.

The areas between scores can also be expressed, if desired, in numbers of cases rather than percentage of total area. For example, a normal distribution of 1000 cases will contain about 683 cases (68.26% of 1000 cases) between ±1 standard deviation of the mean, about 954 between ±2 standard deviations, and about 997 between ±3 standard deviations. Thus, for any normal distribution, only a few cases will be farther away from the mean than ±3 standard deviations.

5.2 COMPUTING *Z* SCORES

To find the percentage of the total area (or number of cases) above, below, or between scores in an empirical distribution, the original scores must first be expressed in units of the standard deviation, or converted into ***Z* scores**. The

ONE STEP AT A TIME **Finding *Z* Scores**

Step 1: Subtract the value of the mean ($\bar{X}$) from the value of the score (X_i).

Step 2: Divide the quantity found in Step 1 by the value of the standard deviation (s). The result is the Z-score equivalent for this raw score.

original scores could be in any unit of measurement (feet, IQ, dollars), but Z scores always have the same values for their mean (0) and standard deviation (1).

Think of converting the original scores into Z scores as a process of changing value scales—similar to changing from meters to yards, kilometers to miles, or gallons to liters. These units are different but equally valid ways of expressing distance, length, or volume. For example, a mile is equal to 1.61 kilometers, so two towns that are 10 miles apart are also 16.1 kilometers apart and a "5k" race covers about 3.10 miles. Although you may be more familiar with miles than kilometers, either unit works perfectly well as a way of expressing distance.

In the same way, the original (or "raw") scores and Z scores are two equally valid but different ways of measuring distances under the normal curve. In Figure 5.1, for example, we could describe a particular score in terms of IQ units ("John's score was 120") or standard deviations ("John scored one standard deviation above the mean").

When we compute Z scores, we convert the original units of measurement (IQ scores, inches, dollars, etc.) to Z scores and, thus, "standardize" the normal curve to a distribution that has a mean of 0 and a standard deviation of 1. The mean of the empirical normal distribution will be converted to 0, its standard deviation to 1, and all values will be expressed in Z-score form. The formula for computing Z scores is

FORMULA 5.1

$$Z = \frac{X_i - \bar{X}}{s}$$

This formula will convert any score (X_i) from an empirical normal distribution into the equivalent Z score. To illustrate with the men's IQ data (Figure 5.1), the Z-score equivalent of a raw score of 120 would be

$$Z = \frac{120 - 100}{20} = +1.00$$

The Z score of positive 1.00 indicates that the original score lies one standard deviation unit above (to the right of) the mean. A negative score would fall below (to the left of) the mean. *(For practice in computing Z scores, see any of the problems at the end of this chapter.)*

5.3 THE NORMAL CURVE TABLE

The theoretical normal curve has been very thoroughly analyzed and described by statisticians. The areas related to any Z score have been precisely determined and organized into a table format. This **normal curve table,** or Z-score table, is presented as Appendix A in this text; for purposes of illustration, a small portion of it is reproduced here as Table 5.1.

TABLE 5.1 AN ILLUSTRATION OF HOW TO FIND AREAS UNDER THE NORMAL CURVE USING APPENDIX A

(a) Z	(b) Area Between Mean and Z	(c) Area Beyond Z
0.00	0.0000	0.5000
0.01	0.0040	0.4960
0.02	0.0080	0.4920
0.03	0.0120	0.4880
⋮	⋮	⋮
1.00	0.3413	0.1587
1.01	0.3438	0.1562
1.02	0.3461	0.1539
1.03	0.3485	0.1515
⋮	⋮	⋮
1.50	0.4332	0.0668
1.51	0.4345	0.0655
1.52	0.4357	0.0643
1.53	0.4370	0.0630
⋮	⋮	⋮

The normal curve table consists of three columns, with Z scores in the left-hand column a, areas between the Z score and the mean of the curve in the middle column b, and areas beyond the Z score in the right-hand column c. To find the area between any Z score and the mean, go down the column labeled "Z" until you find the Z score. For example, go down column a either in Appendix A or in Table 5.1 until you find a Z score of +1.00. The entry in column b shows that the "area between mean and Z" is 0.3413. The table presents all areas in the form of proportions, but we can easily translate these into percentages by multiplying them by 100 (see Chapter 2). We could say either "a proportion of 0.3413 of the total area under the curve lies between a Z score of 1.00 and the mean" or "34.13% of the total area lies between a score of 1.00 and the mean."

To illustrate further, find the Z score of 1.50 either in column a of Appendix A or the abbreviated table presented in Table 5.1. This score is 1½ standard deviations to the right of the mean and corresponds to an IQ of 130 for the men's IQ data. The area in column b for this score is 0.4332. This means that a proportion of 0.4332—or a percentage of 43.32%—of all the area under the curve lies between this score and the mean.

The third column in the table presents "Areas Beyond Z." These are areas above positive scores or below negative scores. This column will be used when we want to find an area above or below certain Z scores, an application that will be explained in Section 5.4.

To conserve space, the normal curve table in Appendix A includes only positive Z scores. Since the normal curve is perfectly symmetrical, however, the area between the score and the mean—column b—for a negative score will be exactly the same as those for a positive score of the same numerical value. For example, the area between a Z score of −1.00 and the mean will also be 34.13%, exactly the same as the area we found previously for a score of +1.00.

As will be repeatedly demonstrated later, however, the sign of the Z score is extremely important and should be carefully noted.

For practice in using Appendix A to describe areas under an empirical normal curve, verify that the following Z scores and areas are correct for the men's IQ distribution. For each IQ score, the equivalent Z score is computed using Formula 5.1, and then Appendix A is consulted to find areas between the score and the mean. ($\bar{X} = 100$, $s = 20$ throughout.)

IQ Score	Z Score	Area Between Z and the Mean
110	+0.50	19.15%
125	+1.25	39.44%
133	+1.65	45.05%
138	+1.90	47.13%

The same procedures apply when the Z-score equivalent of an actual score happens to be a minus value (that is, when the raw score lies below the mean).

IQ Score	Z Score	Area Between Z and the Mean
93	−0.35	13.68%
85	−0.75	27.34%
67	−1.65	45.05%
62	−1.90	47.13%

Remember that the areas in Appendix A will be the same for Z scores of the same numerical value regardless of sign. The area between the score of 138 (+1.90) and the mean is the same as the area between 62 (−1.90) and the mean. *(For practice in using the normal curve table, see any of the problems at the end of this chapter.)*

5.4 FINDING TOTAL AREA ABOVE AND BELOW A SCORE

To this point, we have seen how the normal curve table can be used to find areas between a Z score and the mean. The information presented in the table can also be used to find other kinds of areas in empirical distributions that are at least approximately normal in shape. For example, suppose you need to determine the total area below the scores of two male subjects in the distribution described in Figure 5.1. The first subject has a score of 117 ($X_1 = 117$), which is equivalent to a Z score of +0.85:

$$Z_1 = \frac{X_i - \bar{X}}{s} = \frac{117 - 100}{20} = \frac{17}{20} = +0.85$$

The plus sign of the Z score indicates that the score should be placed above (to the right of) the mean. To find the area below a positive Z score, the area between the score and the mean—given in column b—must be added to the area below the mean. As we noted earlier, the normal curve is symmetrical (unskewed) and its mean will be equal to its median. Therefore, the area below the mean (just like the median) will be 50%. Study Figure 5.4 carefully. We are interested in the shaded area.

By consulting the normal curve table, we find that the area between the score and the mean in column b is 30.23% of the total area. The area below a

FIGURE 5.4 FINDING THE AREA BELOW A POSITIVE *Z* SCORE

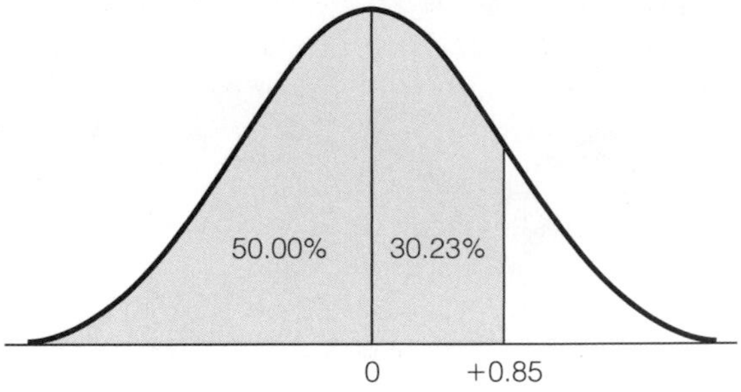

Z score of +0.85 is therefore 80.23% (50.00% + 30.23%). This subject scored higher than 80.23% of the persons tested.

The second subject has an IQ score of 73 ($X_2 = 73$), which is equivalent to a *Z* score of −1.35:

$$Z_2 = \frac{X_i - \bar{X}}{s} = \frac{73 - 100}{20} = -\frac{27}{20} = -1.35$$

To find the area below a negative score, we use column c, labeled "Area Beyond *Z*." The area of interest is depicted in Figure 5.5, and we must determine the size of the shaded area. The area beyond a score of −1.35 is given as 0.0885, which we can express as 8.85%. The second subject ($X_2 = 73$) scored higher than 8.85% of the tested group.

ONE STEP AT A TIME Finding Areas Above and Below Positive and Negative *Z* Scores

Step 1: Compute the *Z* score. Note whether the score is positive or negative.

Step 2: Find the *Z* score in column a of the normal curve table (Appendix A) and do one of the following:

To Find the Total Area Below a Positive *Z* Score:

Step 3: Add the column b area for this *Z* score to .5000, or multiply the column b area by 100 and add it to 50.00%.

To Find the Total Area Above a Positive *Z* Score:

Step 4: Look in column c for this *Z* score. This value is the area above the score expressed as a proportion. To express the area as a percentage, multiply the column c area by 100.

To Find the Total Area Below a Negative *Z* Score:

Step 5: Look in column c for this *Z* score. This value is the area below the score expressed as a proportion. To express the area as a percentage, multiply the column c area by 100.

To Find the Total Area above a Negative *Z* Score:

Step 6: Add the column b area for this *Z* score to .5000, or multiply the column b area by 100 and add it to 50.00%. The result is the total area above this score.

FIGURE 5.5 FINDING THE AREA BELOW A NEGATIVE *Z* SCORE

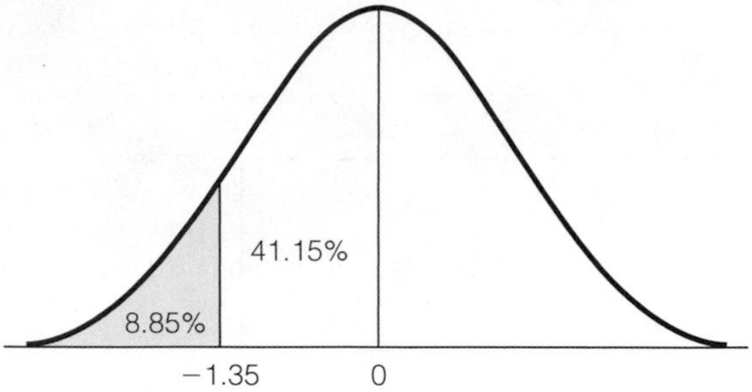

In the foregoing examples, we applied the techniques for finding the area below a score. Essentially the same techniques are used to find the area above a score. If we need to determine the area above an IQ score of 108, for example, we would first convert to a *Z* score,

$$Z = \frac{X_i - \overline{X}}{s} = \frac{108 - 100}{20} = \frac{8}{20} = +0.40$$

and then proceed to Appendix A. The shaded area in Figure 5.6 represents the area in which we are interested. The area above a positive score is found in the "Area Beyond *Z*" column, and, in this case, the area is 0.3446, or 34.46%.

These procedures are summarized in Table 5.2 and in the ONE STEP AT A TIME: Finding Areas Above and Below Positive and Negative *Z* Scores box. To find the total area above a positive *Z* score or below a negative *Z* score, go down the "*Z*" column of Appendix A until you find the score. The area you are seeking will be in the "Area Beyond *Z*" column, or column c. To find the total area below a positive *Z* score or above a negative score, locate the score and then add the area in the "Area Between Mean and *Z*" in column b to either .5000 (for proportions) or 50.00 (for percentages). These techniques might be confusing at first, and you will find it helpful to draw the curve and shade in the areas in which you are interested. *(For practice in finding areas above or below Z scores, see problems 5.1 to 5.7.)*

FIGURE 5.6 FINDING THE AREA ABOVE A POSITIVE *Z* SCORE

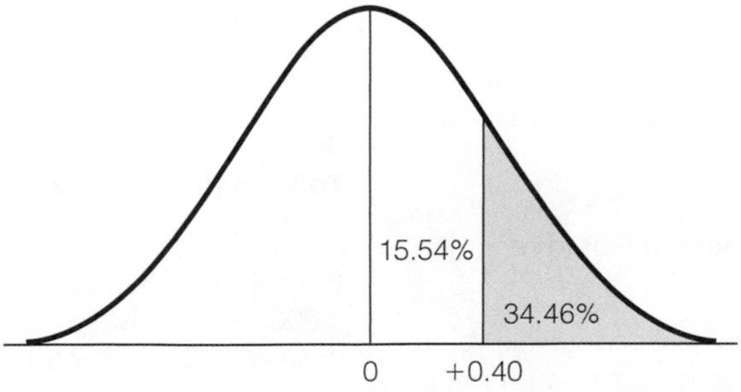

TABLE 5.2 FINDING AREAS ABOVE AND BELOW POSITIVE AND NEGATIVE SCORES

	When the Z Score Is	
To Find Area:	Positive	Negative
Above Z	Look in column c	Add column b area to .5000 or to 50.00%
Below Z	Add column b area to .5000 or to 50.00%	Look in column c

5.5 FINDING AREAS BETWEEN TWO SCORES

On occasion, you will need to determine the area between two scores rather than the total area above or below one score. When the scores are on opposite sides of the mean, the area between the scores can be found by adding the areas between each score and the mean. Using the men's IQ data as an example, if we wished to know the area between the IQ scores of 93 and 112, we would convert both scores to Z scores, find the area between each score and the mean from Appendix A, and add these two areas together. The first IQ score of 93 converts to a Z score of -0.35:

$$Z_1 = \frac{X_i - \overline{X}}{s} = \frac{93 - 100}{20} = -\frac{7}{20} = -0.35$$

The second IQ score (112) converts to $+0.60$:

$$Z_2 = \frac{X_i - \overline{X}}{s} = \frac{112 - 100}{20} = \frac{12}{20} = 0.60$$

Both scores are placed on Figure 5.7. We are interested in the total shaded area. The total area between these two scores is 13.68% + 22.57%, or 36.25%. Therefore, 36.25% of the total area (or about 363 of the 1000 cases) lies between the IQ scores of 93 and 112.

When the scores of interest are on the same side of the mean, a different procedure must be followed to determine the area between them. For example, if we were interested in the area between the scores of 113 and 121, we would begin by converting these scores into Z scores:

FIGURE 5.7 FINDING THE AREA BETWEEN TWO SCORES

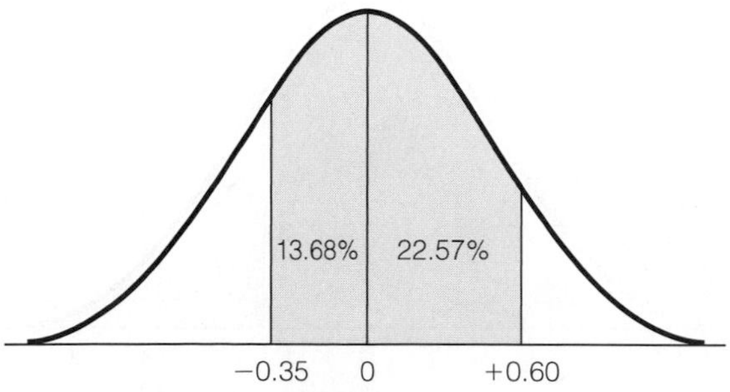

ONE STEP AT A TIME **Finding Areas Between *Z* Scores**

Step 1: Compute the *Z* scores for both raw scores. Note whether the scores are positive or negative and do one of the following:

If the Scores Are on the *Same* Side of the Mean

Step 2: Find the areas between each score and the mean in column b. Subtract the smaller area from the larger area. Multiply this value by 100 to express it as a percentage.

If the Scores Are on *Opposite* Sides of the Mean

Step 3: Find areas between each score and the mean in column b. Add the two areas together to get the total area between the scores. Multiply this value by 100 to express it as a percentage.

$$Z_1 = \frac{X_i - \overline{X}}{s} = \frac{113 - 100}{20} = \frac{13}{20} = +0.65$$

$$Z_2 = \frac{X_i - \overline{X}}{s} = \frac{121 - 100}{20} = \frac{21}{20} = +1.05$$

The scores are noted in Figure 5.8; we are interested in the shadowed area. To find the area between two scores on the same side of the mean, find the area between each score and the mean (given in column b of Appendix A) and then subtract the smaller area from the larger. Between the *Z* score of +0.65 and the mean lies 24.22% of the total area. Between +1.05 and the mean lies 35.31% of the total area. Therefore, the area between these two scores is 35.31% − 24.22%, or 11.09% of the total area. The same technique would be followed if both scores had been below the mean. The procedures for finding areas between two scores are summarized in Table 5.3 and in the ONE STEP AT A TIME: Finding Areas Between *Z* Scores box. *(For practice in finding areas between two scores, see problems 5.3, 5.4, 5.6 to 5.9.)*

FIGURE 5.8 FINDING THE AREA BETWEEN TWO SCORES

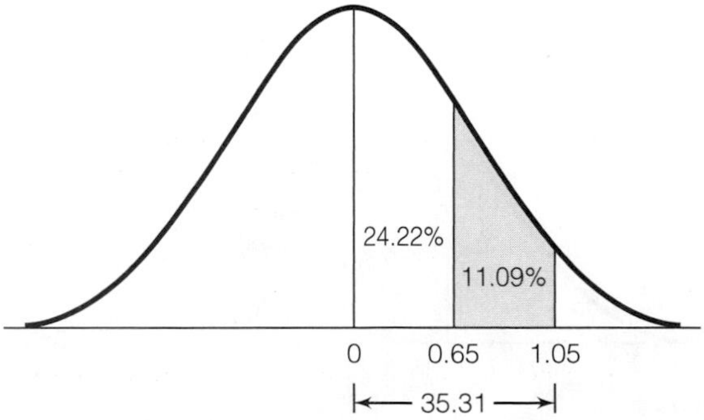

Application 5.1

You have just received your score on a test of intelligence. If your score was 78 and you know that the mean score on the test was 67 with a standard deviation of 5, how does your score compare with the distribution of all test scores?

If you can assume that the test scores are normally distributed, you can compute a Z score and find the area below or above your score. The Z-score equivalent of your raw score would be

$$Z = \frac{X_i - \overline{X}}{s} = \frac{78 - 67}{5} = \frac{11}{5} = +2.20$$

Turning to Appendix A, we find that the "Area Between Mean and Z" for a Z score of 2.20 is 0.4861, which could also be expressed as 48.61%. Since this is a positive Z score, we need to add this area to 50.00% to find the total area below. Your score is higher than 48.61 + 50.00, or 98.61%, of all the test scores. You did pretty well!

TABLE 5.3 FINDING AREAS BETWEEN SCORES

Situation	Procedure
Scores are on the SAME side of the mean	Find areas between each score and the mean in column b. Subtract the smaller area from the larger area.
Scores are on OPPOSITE sides of the mean	Find areas between each score and the mean in column b. Add the two areas together.

5.6 USING THE NORMAL CURVE TO ESTIMATE PROBABILITIES[1]

To this point, we have thought of the theoretical normal curve as a way of describing the percentage of total area above, below, and between scores in an empirical distribution. We have also seen that these areas can be converted into the number of cases above, below, and between scores. In this section, we introduce the idea that the theoretical normal curve may also be thought of as a distribution of probabilities. Specifically, we may use the properties of the theoretical normal curve (Appendix A) to estimate the probability that a case randomly selected from an empirical normal distribution will have a score that falls in a certain range. In terms of techniques, these probabilities will be found in exactly the same way as areas were found. Before we consider these mechanics, however, let us examine what is meant by the concept of probability.

Although we are rarely systematic or rigorous about it, we all attempt to deal with probabilities every day, and, indeed, we base our behavior on our estimates of the likelihood that certain events will occur. We constantly ask (and answer) questions such as: What is the probability of rain? Of drawing to an inside straight in poker? Of the worn-out tires on my car going flat? Of passing a test if I don't study?

[1] A more detailed and mathematical treatment of probability is available at the Web site for this text.

Application 5.2

All sections of Biology 101 at a large university were given the same final exam. Test scores were distributed normally, with a mean of 72 and a standard deviation of 8. What percentage of student scored between 60 and 69 (a grade of D) and what percentage scored between 70 and 79 (a grade of C)? The first two scores are both below the mean. Using Table 5.2 as a guide, we must first compute Z scores, find areas between each score and the mean, and then subtract the smaller area from the larger.

$$Z_1 = \frac{X_i - \overline{X}}{s} = \frac{60 - 72}{8} = -\frac{12}{8} = -1.50$$

$$Z_2 = \frac{X_i - \overline{X}}{s} = \frac{69 - 72}{8} = -\frac{3}{8} = -0.38$$

Using column b, we see that the area between $Z = -1.50$ and the mean is .4332 and the area between $Z = -0.38$ and the mean is .1480. Subtracting the smaller from the larger (.4332 − .1480) gives .2852. Changing to percentage format, we can say that 28.52% of the students earned a D on the test.

To find the percentage of students who earned a C, we must add column b areas together, since the scores (70 and 79) are on opposite sides of the mean (see Table 5.2):

$$Z_1 = \frac{X_i - \overline{X}}{s} = \frac{70 - 72}{8} = -\frac{2}{8} = -0.25$$

$$Z_2 = \frac{X_i - \overline{X}}{s} = \frac{79 - 72}{8} = \frac{7}{8} = 0.88$$

Using column b, we see that the area between $Z = -0.25$ and the mean is .0987 and the area between $Z = 0.88$ and the mean is .3106. Therefore, the total area between these two scores is .0987 + .3106, or 0.4093. Translating to percentages again, we can say that 40.93% of the students earned a C on this test.

To estimate the probability of an event, we must first be able to define what would constitute a "success." The preceding examples contain several different definitions of a success (that is, rain, drawing a certain card, flat tires, and passing grades). To determine a probability, a fraction must be established, with the numerator equaling the number of events that would constitute a success and the denominator equaling the total number of possible events where a success could theoretically occur:

$$\text{Probability} = \frac{\#\ successes}{\#\ events}$$

To illustrate, assume that we wish to know the probability of selecting a specific card—say, the king of hearts—in one draw from a well-shuffled deck of cards. Our definition of a success is quite specific (drawing the king of hearts); and with the information given, we can establish a fraction. Only one card satisfies our definition of success, so the number of events that would constitute a success is 1; this value will be the numerator of the fraction. There are 52 possible events (that is, 52 cards in the deck), so the denominator will be 52. The fraction is thus 1/52, which represents the probability of selecting the king of hearts on one draw from a well-shuffled deck of cards. Our probability of success is 1 out of 52.

We can leave this fraction as it is, or we can express it in several other ways. For example, we can express it as an odds ratio by inverting the fraction, showing that the odds of selecting the king of hearts on a single draw are 52:1 (or

ONE STEP AT A TIME Finding Probabilities

Step 1: Compute the Z score (or scores). Note whether the score is positive or negative.

Step 2: Find the Z score (or scores) in column a of the normal curve table (Appendix A).

Step 3: Find the area above or below the score (or between the scores) as you would normally (see the three previous "One Step at a Time" boxes in this chapter) and express the result as a proportion. Typically, probabilities are expressed as a value between 0.00 and 1.00 rounded to two digits beyond the decimal point.

fifty-two to one). We can express the fraction as a proportion by dividing the numerator by the denominator. For our example, the corresponding proportion is .0192, which is the proportion of all possible events that would satisfy our definition of a success. In the social sciences, probabilities are usually expressed as proportions, and we will follow this convention throughout the remainder of this section. Using p to represent "probability," the probability of drawing the king of hearts (or any specific card) can be expressed as

$$p\,(\text{king of hearts}) = \frac{\#\ successes}{\#\ events} = \frac{1}{52} = .0192$$

As conceptualized here, probabilities have an exact meaning: Over the long run, the events that we define as successes will bear a certain proportional relationship to the total number of events. The probability of .0192 for selecting the king of hearts in a single draw really means that, over thousands of selections of one card at a time from a full deck of 52 cards, the proportion of successful draws would be .0192. Or, for every 10,000 draws, 192 would be the king of hearts, and the remaining 9,808 selections would be other cards. Thus, when we say that the probability of drawing the king of hearts in one draw is .0192, we are essentially applying to a single draw our knowledge of what would happen over thousands of draws.

Like proportions, probabilities range from 0.00 (meaning that the event has absolutely no chance of occurrence) to 1.00 (a certainty). As the value of the probability increases, the likelihood that the defined event will occur also increases. A probability of .0192 is close to zero, and this means that the event (drawing the king of hearts) is unlikely, or improbable.

These techniques can be used to establish simple probabilities in any situation in which we can specify the number of successes and the total number of events. For example, a single die has six sides, or faces, each with a different value, ranging from 1 to 6. The probability of getting any specific number (say, a 4) in a single roll of a die is therefore

$$p\,(\text{rolling a 4}) = \frac{1}{6} = .1667$$

Combining this way of thinking about probability with our knowledge of the theoretical normal curve allows us to estimate the likelihood of selecting a case that has a score within a certain range. For example, suppose we wished to estimate the probability that a randomly chosen subject from the distribution of men's IQ scores would have an IQ score between 95 and the mean score of 100. Our definition of a success here would be the selection of any subject with a score in the specified range. Normally, we would next establish a fraction with the numerator equal to the number of subjects with scores in the defined range and the denominator equal to the total number of subjects. However, if the empirical distribution is normal in form, we can skip this step, since the probabilities, in proportion form, are already stated in Appendix A. That is, the areas in Appendix A can be interpreted as probabilities.

To determine the probability that a randomly selected case will have a score between 95 and the mean, we would convert the original score to a Z score:

$$Z = \frac{X_i - \overline{X}}{s} = \frac{95 - 100}{20} = -\frac{5}{20} = -0.25$$

Using Appendix A, we see that the area between this score and the mean is 0.0987. This is the probability we are seeking. The probability that a randomly selected case will have a score between 95 and 100 is 0.0987 (or, rounded off, 0.1, or 1 out of 10). In the same fashion, the probability of selecting a subject from any range of scores can be estimated. Note that the techniques for estimating probabilities are exactly the same as those for finding areas. The only new information introduced in this section is the idea that the areas in the normal curve table can also be thought of as probabilities.

To consider an additional example, what is the probability that a randomly selected male will have an IQ less than 123? We will find probabilities in exactly the same way we found areas. The score (X_i) is above the mean and, following the directions in Table 5.2, we will find the probability we are seeking by adding the area in column b to 0.5000. First, we find the Z score:

$$Z = \frac{X_i - \overline{X}}{s} = \frac{123 - 100}{20} = \frac{23}{20} = +1.15$$

Next, look in column b of Appendix A to find the area between this score and the mean. Then add the area (0.3749) to 0.5000. The probability of selecting a male with an IQ of less than 123 is 0.3749 + 0.5000, or 0.8749. Rounding this value to .88, we can say that the odds are .88 (very high) that we will select a male with an IQ score in this range. Technically, remember that this probability expresses what would happen over the long run: For every 100 males selected from this group over an infinite number of trials, 88 would have IQ scores less than 123 and 12 would not.

Let me close by stressing a very important point about probabilities and the normal curve. The probability is very high that any case randomly selected from a normal distribution will have a score close in value to that of the mean. The shape of the normal curve is such that most cases are clustered around the mean and decline in frequency as we move farther away—either to the right or to the left—from the mean value. In fact, given what we know about the normal curve, the probability that a randomly selected case will have a score within ± 1 standard deviation of the mean is 0.6826. Rounding off, we can say that 68 out of 100 cases—or about two-thirds of all cases—selected over the long run will have a score between ± 1 standard deviation, or Z score, of the mean. The probabilities are that any randomly selected case will have a score close in value to the mean.

Application 5.3

The distribution of scores on a biology final exam used in Application 5.2 had a mean of 72 and a standard deviation of 8. What is the probability that a student selected at random will have a score less than 61? More than 80? Less than 98? To answer these questions, we must first calculate Z scores and then consult Appendix A. We are looking for probabilities, so we will leave the areas in proportion form. The Z score for a score of 61 is

$$Z_1 = \frac{X_i - \overline{X}}{s} = \frac{61 - 72}{8} = -\frac{11}{8} = -1.38$$

This score is a negative value (below, or to the left of, the mean) and we are looking for the area below. Using Table 5.1 as a guide, we see that we must use column c to find area below a negative score. This area is .0838. Rounding off, we can say that the odds of selecting a student with a score less than 61 are only 8 out of 100. This low value tells us this would be an unlikely event.

The Z score for the score of 80 is

$$Z_2 = \frac{X_i - \overline{X}}{s} = \frac{80 - 72}{8} = \frac{8}{8} = 1.00$$

The Z score is positive, and to find the area above (greater than) 80, we look in column c (see Table 5.1). This value is .1587. The odds of selecting a student with a score greater than 80 is roughly 16 out of 100, about twice as likely as selecting a student with a score of less than 61.

The Z score for the score of 98 is

$$Z_3 = \frac{X_i - \overline{X}}{s} = \frac{98 - 72}{8} = \frac{26}{8} = 3.25$$

To find the area below a positive Z score, we add the area between the score and the mean (column b) to .5000 (see Table 5.1). This value is .4994 + .5000, or .9994. It is extremely likely that a randomly selected student will have a score less than 98. Remember that scores more than ±3 standard deviations from the mean are very rare.

In contrast, the probability of the case having a score beyond 3 standard deviations from the mean is very small. Look in column c ("Area Beyond Z") for a Z score of 3.00 and you will find the value .0014. Adding the areas in the upper tail (beyond +3.00) to the area in the lower tail (beyond −3.00), gives us .0014 + .0014, for a total of .0028. The probability of selecting a case with a very high score or a very low score is .0028. If we randomly select cases from a normally distributed variable, we would select cases with Z scores beyond ±3.00 only 28 times out of every 10,000 trials.

The general point to remember is that cases with scores close to the mean are common and cases with scores that have scores far above or below the mean are rare. This relationship is central for an understanding of inferential statistics in Part II. *(For practice in using the normal curve table to find probabilities, see problems 5.8 to 5.10 and 5.13.)*

SUMMARY

1. The normal curve, in combination with the mean and standard deviation, can be used to construct precise descriptive statements about empirical distributions that are normally distributed. This chapter also lays some important groundwork for Part II.
2. To work with the theoretical normal curve, raw scores must be transformed into their equivalent Z scores. Z scores allow us to find areas under the theoretical normal curve (Appendix A).

3. We considered three uses of the theoretical normal curve: finding total areas above and below a score, finding areas between two scores, and expressing these areas as probabilities. This last use of the normal curve is especially germane because inferential statistics are centrally concerned with estimating the probabilities of defined events in a fashion very similar to the process introduced in Section 5.6.

SUMMARY OF FORMULAS

Z scores 5.1 $$Z = \frac{X_i - \overline{X}}{s}$$

GLOSSARY

Normal curve. A theoretical distribution of scores that is symmetrical, unimodal, and bell shaped. The standard normal curve always has a mean of 0 and a standard deviation of 1.

Normal curve table. Appendix A; a detailed description of the area between a *Z* score and the mean of any standardized normal distribution.

***Z* scores.** Standard scores; the way scores are expressed after they have been standardized to the theoretical normal curve.

PROBLEMS

5.1 Scores on a quiz were normally distributed and had a mean of 10 and a standard deviation of 3. For each of the following scores, find the *Z* score and the percentage of area above and below the score.

X_i	*Z* Score	% Area Above	% Area Below
5			
6			
7			
8			
9			
11			
12			
14			
15			
16			
18			

5.2 Assume that the distribution of a college entrance exam is normal, with a mean of 500 and a standard deviation of 100. For each of the following scores, find the equivalent *Z* score, the percentage of the area above the score, and the percentage of the area below the score.

X_i	*Z* Score	% Area Above	% Area Below
650			
400			
375			
586			
437			
526			
621			
498			
517			
398			

5.3 The senior class has been given a comprehensive examination to assess educational experience. The mean on the test was 74 and the standard deviation was 10. What percentage of the students had scores

a. between 75 and 85? ______
b. between 80 and 85? ______
c. above 80? ______
d. above 83? ______
e. between 80 and 70? ______
f. between 75 and 70? ______
g. below 75? ______
h. below 77? ______
i. below 80? ______
j. below 85? ______

5.4 For a normal distribution where the mean is 50 and the standard deviation is 10, what percentage of the area is

a. between the scores of 40 and 47? ________
b. above a score of 47? ________
c. below a score of 53? ________
d. between the scores of 35 and 65? ________
e. above a score of 72? ________
f. below a score of 31 and above a score of 69? ________
g. between the scores of 55 and 62? ________
h. between the scores of 32 and 47? ________

5.5 At St. Algebra College, the 200 freshmen enrolled in introductory biology took a final exam on which their mean score was 72 and their standard deviation was 6. The following table presents the grades of 10 students. Convert each into a *Z* score and determine the *number of people* who scored higher or lower than each of the 10 students. *(HINT: Multiply the appropriate proportion by N and round the result.)*

X_i	*Z* Score	Number of Students Above	Number of Students Below
60			
57			
55			
67			
70			
72			
78			
82			
90			
95			

5.6 If a distribution of test scores is normal, with a mean of 78 and a standard deviation of 11, what percentage of the area lies
a. below 60? ________
b. below 70? ________
c. below 80? ________
d. below 90? ________
e. between 60 and 65? ________
f. between 65 and 79? ________
g. between 70 and 95? ________
h. between 80 and 90? ________
i. above 99? ________
j. above 89? ________
k. above 75? ________
l. above 65? ________

5.7 A scale measuring prejudice has been administered to a large sample of respondents. The distribution of scores is approximately normal, with a mean of 31 and a standard deviation of 5. What percentage of the sample had scores
a. below 20? ________
b. below 40? ________
c. between 30 and 40? ________
d. between 35 and 45? ________
e. above 25? ________
f. above 35? ________

5.8 The average burglary rate for a jurisdiction has been 311 per year with a standard deviation of 50. What is the probability that next year the number of burglaries will be
a. less than 250? ________
b. less than 300? ________
c. more than 350? ________
d. more than 400? ________
e. between 250 and 350? ________
f. between 300 and 350? ________
g. between 350 and 375? ________

5.9 For a math test on which the mean was 59 and the standard deviation was 4, what is the probability that a student randomly selected from this class will have a score
a. between 55 and 65? ________
b. between 60 and 65? ________
c. above 65? ________
d. between 60 and 50? ________
e. between 55 and 50? ________
f. below 55? ________

5.10 SOC On the scale mentioned in problem 5.7, if a score of 40 or more is considered "highly prejudiced," what is the probability that a person selected at random will have a score in that range?

5.11 The local police force gives all applicants an entrance exam and accepts only those applicants who score in the top 15% on this test. If the mean score this year is 87 and the standard deviation is 8, would an individual with a score of 110 be accepted?

5.12 After taking the state merit examinations for the positions of social worker and employment counselor, you receive the following information on the tests and on your performance. On which of the tests did you do better?

Social Worker	Employment Counselor
$\overline{X} = 118$	$\overline{X} = 27$
$s = 17$	$s = 3$
Your score = 127	Your score = 29

5.13 In a distribution of scores with a mean of 35 and a standard deviation of 4, which event is more likely: that a randomly selected score will be between 29 and 31 or that a randomly selected score will be between 40 and 42?

5.14 To be accepted into an honor society, students must have GPAs in the top 10% of the school. If the mean GPA is 2.78 and the standard deviation is .33, which of the following GPAs would qualify?
3.20, 3.21, 3.25, 3.30, 3.35

SPSS for Windows

Using SPSS for Windows to Transform Raw Scores into *Z* Scores

SPSS DEMONSTRATION 5.1 Computing *Z Scores*

The **Descriptives** program introduced at the end of Chapter 4 can also be used to compute *Z* scores for any variable. These *Z* scores are then available for further operations and may be used in other tasks. SPSS will create a new variable consisting of the transformed scores of the original variable. The program uses the letter *Z* and the first seven letters of the variable name to designate the normalized scores of a variable.

In this demonstration, we will have SPSS compute *Z* scores for *age*. First, load the 2006 GSS data set and then click **Analyze, Descriptive Statistics,** and **Descriptives.** Find *age* in the variable list and click the arrow to move the variable to the **Variable(s):** window. Find the "Save standardized values as variables" option below the variable list and click the box next to it. With this option selected for **Descriptives,** SPSS will compute *Z* scores for all variables listed in the **Variable(s):** window. Click **OK,** and SPSS will produce the usual set of descriptive statistics for *age*. It will also add to the data set the new variable (called *zage*), which contains the standardized scores for *age*. To verify this, run the **Descriptives** command again, and you will find *zage* in the variable list. Transfer *zage* to the **Variable(s):** window with *age*, click **OK,** and the following output will be produced:

Descriptive Statistics

	N	Minimum	Maximum	Mean	Std. Deviation
AGE OF RESPONDENT	1417	18	89	46.88	17.086
Zscore: AGE OF RESPONDENT	1417	-1.69058	2.46491	.0000000	1.00000000
Valid N (listwise)	1417				

Like any set of *Z* scores, *zage* has a mean of zero and a standard deviation of 1. The new variable, *zage*, can be treated just like any other variable and used in any SPSS procedure.

If you would like to inspect the scores of *zage*, use the **Case Summaries** procedure. Click **Analyze, Reports,** and then **Case Summaries.** Move both *age* and *zage* to the **Variable(s):** window. Find the **Limit cases to first** option at the bottom of the window. This option can be used to set the number of cases included in the output. For this exercise, let's set a limit of 10 cases. Make sure the box to the left of the option is checked, and type 10 in the box to the right. Click **OK**, and the following output will be produced:

Case Summaries(a)

		AGE OF RESPONDENT	Zscore: AGE OF RESPONDENT
1		50	.18232
2		50	.18232
3		20	-1.57352
4		23	-1.39794
5		32	-.87119
6		81	1.99668
7		47	.00673
8		60	.76760
9		18	-1.69058
10		24	-1.33941
Total	N	10	10

[a] Limited to first 10 cases.

Scan the list of scores, and note that the scores that are close in value to the mean of *age* (46.88) are very close to the mean of *zage* (0.00), and the further away the score is from 46.88, the greater the numerical value of the *Z* score. Also note that, of course, scores below the mean (less than 46.88) have negative signs and scores above the mean (greater than 46.88) have positive signs.

We should note that SPSS calculates *Z* scores using Formula 5.1 but typically performs the calculations at a very high level of precision. Still, we can verify the *Z* scores produced by SPSS are the same as those we would compute by hand. To illustrate, consider a raw score of 50, the age of the first two cases in the list. With a mean of 46.88 and a standard deviation of 17.09, this score would translate to a Z score of 0.18:

$$Z = \frac{X_i - \overline{X}}{s} = \frac{50 - 46.88}{17.09} = \frac{3.12}{17.09} = 0.18$$

This score is essentially the same as that computed by SPSS (0.18232).

Exercises

5.1 Compute *Z* scores for *prestg80, papres80,* and *tvhours.* Use the **Case Summaries** procedure to display the normalized and "raw" scores for each variable for 10 cases.

5.2 Use the **Graphs** command to get simple line charts of *age* and then *zage.* How close are these charts to smooth, bell-shaped, normal curves?

PART I CUMULATIVE EXERCISES

1. To what extent do people apply religion to the problems of everyday living? Fifteen people have responded to a series of questions including the following:

1. What is your religion?
   1. Protestant
   2. Catholic
   3. Jewish
   4. None
   5. Other (Muslim, Hindu, etc.)
2. On a scale of 1 to 10 (with 10 being the highest), how strong would you say your faith is?
3. How many times a day do you pray?
4. When things aren't going well, my religion is my major source of comfort.
   1. Strongly agree
   2. Slightly agree
   3. Neither agree nor disagree
   4. Slightly disagree
   5. Strongly disagree
5. How old are you?

Case	Religion	Strength	Pray	Comfort	Age
1	1	2	0	2	30
2	2	8	1	1	67
3	1	8	3	1	45
4	1	6	0	1	43
5	3	9	3	1	32
6	3	3	0	3	18
7	4	0	0	5	52
8	5	9	6	1	37
9	1	5	0	2	54
10	2	8	2	1	55
11	2	3	0	5	33
12	1	6	1	3	45
13	1	8	2	2	37
14	1	7	2	1	50
15	2	9	1	1	25

For each variable, construct a frequency distribution and calculate appropriate measures of central tendency and dispersion. Write a sentence summarizing each variable.

2. A survey measuring attitudes toward interracial dating was administered to 1000 people. The survey asked the following questions:

1. What is your age?
2. What is your sex?
   1. Male
   2. Female

3. Marriages between people of different racial groups just don't work out and should be banned by law.
   1. Strongly agree
   2. Agree
   3. Undecided
   4. Disagree
   5. Strongly disagree
4. How many years of schooling have you completed?
5. Which of the following categories best describes the place where you grew up?
   1. Large city
   2. Medium-size city
   3. Suburbs of a city
   4. Small town
   5. Rural area
6. What is your marital status?
   1. Married
   2. Separated or divorced
   3. Widowed
   4. Never married

The scores of 20 respondents are reproduced here.

Case	Age	Sex	Attitude on Interracial Dating	Years of School	Area	Marital Status
1	17	1	5	12	1	4
2	25	2	3	12	2	1
3	55	2	3	14	2	1
4	45	1	1	12	3	1
5	38	2	1	10	3	1
6	21	1	1	16	5	1
7	29	2	2	16	2	2
8	30	2	1	12	4	1
9	37	1	1	12	2	1
10	42	2	3	18	5	4
11	57	2	4	12	2	3
12	24	2	2	12	4	1
13	27	1	2	18	3	2
14	44	1	1	15	1	1
15	37	1	1	10	5	4
16	35	1	1	12	4	1
17	41	2	2	15	3	1
18	42	2	1	10	2	4
19	20	2	1	16	1	4
20	21	2	1	16	1	4

a. For each variable, construct a frequency distribution and select and calculate an appropriate measure of central tendency and a measure of dispersion. Summarize each variable in a sentence.

b. For all 1000 respondents, the mean age was 34.70 with a standard deviation of 3.4 years. Assuming the distribution of age is approximately normal, compute Z scores for each of the first 10 respondents and determine the percentage of the area below (younger than) each respondent.

3. The following data set is taken from the General Social Survey. Abbreviated versions of the questions along with the meanings of the codes are also presented. See Appendix G for the codes and the complete question wordings. For each variable, construct a frequency distribution and select and calculate an appropriate measure of central tendency and a measure of dispersion. Summarize each variable in a sentence.

1. How many children have you ever had? (Values are actual numbers.)
2. Respondent's educational level:
   0. Less than HS
   1. HS
   2. Jr. college
   3. Bachelor's degree
   4. Graduate school
3. Race:
   1. White
   2. Black
   3. Other
4. "It is sometimes necessary to discipline a child with a good, hard spanking."
   1. Strongly agree
   2. Agree
   3. Disagree
   4. Strongly disagree
5. Number of hours of TV watched per day. (Values are actual numbers of hours.)
6. What is your religious preference?
   1. Protestant
   2. Catholic
   3. Jewish
   4. None
   5. Other

Case	Number of Children	Years of School	Race	Attitude on Spanking	TV Hours	Religion
1	3	1	1	3	3	1
2	2	0	1	4	1	1
3	4	2	1	2	3	1
4	0	3	1	1	2	1
5	5	1	1	3	2	1
6	1	1	1	3	3	1
7	9	0	1	1	6	1
8	6	1	2	3	4	1
9	4	3	1	1	2	4
10	2	1	3	1	1	1
11	2	0	1	2	4	1
12	4	1	2	1	5	2
13	0	1	1	3	2	2
14	2	1	1	4	2	1
15	3	1	2	3	4	1
16	2	0	1	2	2	1
17	2	1	1	2	2	1
18	0	3	1	3	2	1
19	3	0	1	3	5	2

20	2	1	2	1	10	1
21	2	1	1	3	4	1
22	1	0	1	3	5	1
23	0	2	1	1	2	2
24	0	1	1	2	0	4
25	2	4	1	1	1	2

Part II Inferential Statistics

The six chapters in this part cover the techniques and concepts of inferential statistics. Generally speaking, these applications allow us to learn about large groups (populations) from small, carefully selected subgroups (samples). These statistical techniques are powerful and extremely useful. They are used to poll public opinion, to research the potential market for new products, to project the winners of elections, to test the effects of new drugs, and in hundreds of other ways both inside and outside of the social sciences.

Chapter 6 includes a brief description of the technology of sampling, or how subgroups are selected so as to justify making inferences to populations. That section is intended to give you a general overview of the process and some insight into the actual selection of cases for the sample, not as a comprehensive or detailed treatment of the subject. The most important part of this chapter, however, concerns the sampling distribution, the single most important concept in inferential statistics. The sampling distribution is normal in shape, and it is the key link between populations and samples.

There are two main applications in inferential statistics, and Chapter 7 covers the first: using statistical information from a sample (e.g., a mean or a proportion) to estimate the characteristics of a population. The technique, called *estimation,* is most commonly used in public opinion polling and election projection.

Chapters 8 through 11 cover the second application of inferential statistics: hypothesis testing. Most of the relevant concepts for this material are introduced in Chapter 8, and each chapter covers a different situation in which hypothesis testing is done. For example, Chapter 9 presents the techniques used when we are comparing information from two different samples or groups (e.g., men vs. women), while Chapter 10 covers applications involving more than two groups or samples (e.g., Republicans vs. Democrats vs. Independents).

Hypothesis testing is one of the more challenging aspects of statistics for beginning students, and I have included an abundance of learning aids to ease the chore of assimilating this material. Hypothesis testing is also one of the most common and important statistical applications to be found in social science research. Mastery of this material is essential for developing the capability to read the professional literature.

6

Introduction to Inferential Statistics
Sampling and the Sampling Distribution

LEARNING OBJECTIVES

By the end of this chapter, you will be able to

1. Explain the purpose of inferential statistics in terms of generalizing from a sample to a population.
2. Define and explain the basic techniques of random sampling.
3. Explain and define these key terms: population, sample, parameter, statistic, representative, EPSEM.
4. Differentiate between the sampling distribution, the sample, and the population.
5. Explain the two theorems presented.

6.1 INTRODUCTION

One of the goals of social science research is to test our theories and hypotheses using many different people, groups, societies, and historical eras. Obviously, we can have the greatest confidence in theories that have stood up to testing against the greatest variety of cases and social settings. A major problem we often face in social science research is that the most appropriate populations for our test of theory are very large. For example, a theory concerning political party preference among U.S. citizens would be most suitably tested using the entire electorate, but it is impossible to interview every member of this group (about 100 million people). Indeed, even for theories that could be reasonably tested with smaller populations–such as a local community or the student body at a university–the logistics of gathering data from every single case in the population are staggering to contemplate.

If it is too difficult or expensive to research entire populations, how can we reasonably test our theories? To deal with this problem, social scientists select samples, or subsets of cases, from the populations of interest. Our goal in inferential statistics is to learn about the characteristics of a population (often called **parameters**) based on what we can learn from the sample. Two applications of inferential statistics are covered in this text. In estimation procedures, covered in Chapter 7, a "guess" of the population parameter is made, based on what is known about the sample. In hypothesis testing, covered in Chapters 8 through 11, the validity of a hypothesis about the population is tested against sample outcomes. In this chapter, we look briefly at sampling (the techniques for selecting cases for a sample) and then introduce a key concept in inferential statistics: the sampling distribution.

6.2 PROBABILITY SAMPLING: BASIC CONCEPTS

Social scientists have developed a variety of techniques for selecting samples from populations. In this chapter, we review the basic procedures for selecting probability samples, the only type of sample that fully supports the use of inferential statistical techniques to generalize to populations. These types of samples are often described as *random,* and you may be more familiar with this terminology. Because of its greater familiarity, I will often use the term *random sample* in the following chapters. The term *probability sample* is preferred, however, because, in everyday language, *random* is often used to mean "by coincidence" or to give a connotation of unpredictability. As you will see, probability samples are selected by techniques that are careful and methodical and leave no room for haphazardness. Interviewing the people you happen to meet in a mall one afternoon may be "random" in some sense of the word, but this technique will not result in a sample that could support inferential statistics.

Before considering probability sampling, let me point out that social scientists often use nonprobability samples. For example, social scientists studying small-group dynamics or the structure of attitudes or personal values might use the students enrolled in their classes as subjects. These "convenience" samples are very useful for a number of purposes (e.g., exploring ideas or pretesting survey forms before embarking on a more ambitious project) and are typically less costly and easier to assemble. The major limitation of these samples is that results cannot be generalized beyond the group being tested. If a theory of prejudice, for example, has been tested only on the students who happen to have been enrolled in a particular section of introductory sociology at a particular university, we cannot conclude that the theory would be true for other types of people. Therefore, we cannot place a lot of confidence in theories tested on nonprobability samples, even when the evidence is very strong.

When constructing a probability sample, our goal is to select cases so that the final sample is **representative** of the population from which it was drawn. A sample is representative if it reproduces the important characteristics of the population. For example, if the population consists of 60% females and 40% males, the sample should have about the same proportional makeup. In other words, a representative sample is very much like the population—only smaller. It is crucial for inferential statistics that samples be representative. if they are not, generalizing to the population becomes, at best, extremely hazardous.

How can we assure ourselves that our samples are representative? Unfortunately, it is not possible to guarantee that samples will be representative. However, we can maximize the chances of a representative sample by following the principle of EPSEM (the "**e**qual **p**robability of **se**lection **m**ethod"), the fundamental principle of probability sampling. To follow the EPSEM principle, we select the sample so that every element or case in the population has an equal probability of being selected. Following the rule of EPSEM will make it extremely likely that we will achieve our goal of selecting a representative sample.

Remember that the EPSEM selection technique and the representativeness of the final sample are two different things. In other words, the fact that a sample is selected according to EPSEM does not guarantee that it will be an exact representation of the population. The probability is very high that an EPSEM sample will be representative; but, just as a perfectly honest coin will sometimes show 10 heads in a row when flipped, an EPSEM sample will occasionally present an inaccurate picture of the population. One of the great strengths of inferential statistics is that they allow the researcher to estimate the probability of this type of error and to interpret results accordingly.

6.3 EPSEM SAMPLING TECHNIQUES

This section presents four techniques for selecting random samples. The most basic technique is presented first; the remaining techniques are refinements of or variations on the basic technique.

Simple Random Sampling. The most basic EPSEM sampling technique produces a **simple random sample.** To use this technique, we need a list of all cases in the population and a system of selection that will guarantee that every case has an equal chance of being chosen for the sample. The selection process could be based on a variety of operations (for example, drawing cards from a well-shuffled deck, flipping coins, throwing dice, drawing numbers from a hat). Cases are often selected by using tables of random numbers. These tables are lists of numbers that have no pattern to them (that is, they are random); an example of such a table is available at the Web site for this text.

To use the table of random numbers, first assign a unique identification number to each case in the population. Then pick numbers from the table one at a time and select the cases whose identification numbers match the numbers selected from the table. This procedure will produce an EPSEM sample because the numbers in the table are in random order and any number is just as likely as any other number. Stop selecting cases when you have reached your desired sample size; if an identification number is selected more than once, ignore the repeats.[1]

Systematic Random Sampling. Following the foregoing procedures will produce random samples as long as you select from a complete list of the population. However, this technique can be very cumbersome when the list of cases is long. Consider the situation when your population numbers 10,000 cases. It is perfectly possible that the first case you select will come from the front of the list, the second from the back, the third from the front again, and so on—leading to a great deal of paper shuffling and a fair amount of confusion. To save time and money in such a situation, researchers often use a technique called **systematic sampling,** where only the first case is randomly selected. Thereafter, every *k*th case is selected. For example, if you are drawing from a list of 10,000 and desire a sample of 200, select the first case randomly and every 10,000/200th, or 50th, case thereafter. If you randomly start with case 13, then your second case will be case 63, your third case 113, and so on, until you reach the end of the list.

Note that systematic sampling does not strictly conform to the criterion of EPSEM. That is, once the first case has been selected, the other cases no longer have an equal probability of being chosen. In our example, cases other than the 13th, the 63rd, the 113th, and so on will not be selected for the sample. In general, this increased probability of error is very slight as long as the list from which cases are chosen is random, or at least noncyclical with respect to the traits you wish to measure. For example, there might be a problem if you are concerned with ethnicity and are drawing your sample from an alphabetical list, because there is a tendency for certain ethnic names to begin with the same letter (for example, the Irish prefix *O*). Therefore, when using systematic sampling, pay careful attention to the nature of the population list as well as your sampling technique.

---

[1]Ignoring identification numbers when they are repeated is called *sampling without replacement.* Technically, this practice compromises the randomness of the selection process. However, if the sample is a small fraction of the total population, we will be unlikely to select the same case twice, and ignoring repeats will not bias our conclusions.

Stratified Random Sampling. A third type of EPSEM sampling produces a **stratified sample.** This technique is very desirable because it guarantees that the sample will be representative on the selected traits. To apply this technique, you first *stratify* (or divide) the population list into sublists according to some relevant trait and then sample from the sublists. If you select a number of cases from each sublist proportional to the numbers for that characteristic in the population, the sample will be representative of the population.

For example, suppose you are drawing a sample of 300 of your classmates and you wish to have proportional representation from every major field on campus. If only 10% of the student body is majoring in zoology, random and systematic sampling could result in a sample with very few (or even no) zoologists. If, however, you first divide the population into sublists by major, you can use the EPSEM technique to select exactly 30 zoologists from the appropriate sublist. Following the same procedure with other majors will create a sample that is, by definition, representative of the population on this characteristic. Thus, stratified samples are guaranteed to meet the all-important criterion of representativeness (at least for the traits that are used to stratify the samples). The major limitation of stratified random sampling is that the exact composition of the population is often unknown. If we have no information about the population, we will be unable to establish a scheme for stratification and determine how many cases should be taken from each sublist.

Cluster Sampling. To this point, sampling techniques have been presented as straightforward processes of randomly selecting cases from a list or sublists of the population. However, sampling is rarely so uncomplicated, and the major difficulty almost always centers on what might appear, at first glance, to be the easiest part: finding a list of the population. For many of the populations of interest to the social sciences, there are no complete, up-to-date lists. There is no list of United States citizens, no list of the residents of any given state, and no complete, up-to-date list of residents of your local community. Devices such as telephone books might appear to be complete lists of local residents. However, they omit unlisted numbers and are very likely to be outdated.

Social scientists have devised several ways of dealing with the limitations imposed by the scarcity of lists. Probably the most significant of these is **cluster sampling,** which involves selecting groups of cases (clusters) rather than single cases. The clusters are often based on geography, and the selection of clusters often proceeds in stages. For example, you might draw a cluster sample of your city or town by first numbering all of the voting precincts within the political boundaries. Next, you would use an EPSEM technique to select a sample of precincts. The second stage of selection would involve numbering the blocks within each of the selected precincts and, following EPSEM, selecting a sample of blocks. A third stage might involve the selection of households within each selected block. When these stages are completed, you would have a sample that had a very high probability of being representative of the entire city without ever using a list of city residents.

Unfortunately, a cluster sample is somewhat less likely to be representative of the population than a simple random sample of comparable size. In part, this lower accuracy is a result of the multiple selection stages. With a simple random sample, the sample is drawn in one selection from the list of the population. In a multistage cluster sample, each stage in the selection process (e.g., first the precincts, then the blocks, and then the households) has a probability of error. That is, each time we sample, we run the risk of selecting an unrepresentative sample. In simple random sampling, we run this risk once; with cluster sampling we will run the risk anew at each stage.

Although we have to treat inferences to populations based on cluster samples with some additional caution, we usually have no alternative method of sampling. While it may be extremely difficult (or even impossible) to construct an accurate list of an entire city population, all you need to compile a cluster sample is a map (or a list of voting precincts, census tracts, and so forth).

Summary. In closing our consideration of sampling techniques, let us return to a major point. The purpose of inferential statistics is to acquire knowledge about populations based on the information derived from samples of that population. Each of the statistics to be presented in the following chapters in Part II requires that samples be selected according to EPSEM. While even the most painstaking and sophisticated sampling techniques are no guarantee, the probability is high that EPSEM samples will be representative of the populations from which they are selected.

6.4 THE SAMPLING DISTRIBUTION

Once we have selected a probability sample according to some EPSEM procedure, what do we know? On one hand, we can gather a great deal of information from the cases in the sample. On the other hand, we know nothing about the population. Indeed, if we had information about the population, we probably wouldn't need the sample. Remember that we use inferential statistics to learn more about populations, and information from the sample is important primarily insofar as it allows us to generalize to the population.

When we use inferential statistics, we generally measure some variable (e.g., age, political party preference, or opinions about abortion) in the sample and then use the information from the sample to learn more about that variable in the population. In Part I of this text, you learned that three types of information are generally necessary to adequately characterize a variable: (1) the shape of its distribution, (2) some measure of central tendency, and (3) some measure of dispersion. Clearly, all three kinds of information can be gathered (or computed) on a variable from the cases in the sample. Just as clearly, none of the information is available for the population. Except in rare situations (for example, IQ tests are designed so that the scores of a population will be approximately normal in distribution, and the distribution of income is almost always positively skewed), nothing can be known about the exact shape of the distribution of a variable in the population. The means and standard deviations of variables in the population are also unknown. Let me remind you that if we had this information for the population, inferential statistics would be unnecessary.

In statistics, we link information from the sample to the population with the **sampling distribution:** *the theoretical, probabilistic distribution of a statistic for all possible samples of a certain sample size (N).* That is, the sampling distribution is the distribution of a statistic (e.g., a mean or a proportion) based on every conceivable combination of cases from the population. A crucial point about the sampling distribution is that its characteristics are based on the laws of probability, not on empirical information, and are very well known. In fact, the sampling

FIGURE 6.1 RELATIONSHIPS BETWEEN THE SAMPLE, THE SAMPLING DISTRIBUTION, AND THE POPULATION

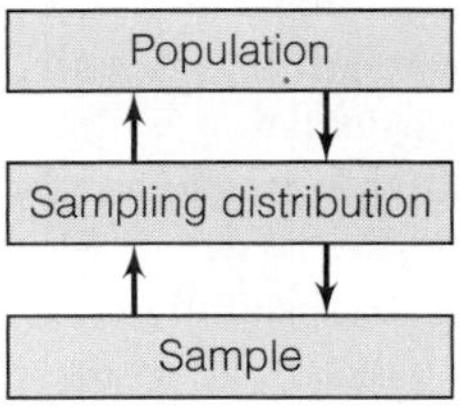

distribution is the central concept in inferential statistics, and a prolonged examination of its characteristics is certainly in order.

The Three Distributions Used in Inferential Statistics. As illustrated by Figure 6.1, the general strategy of all applications of inferential statistics is to move between the sample and the population via the sampling distribution. Thus, three separate and distinct distributions of a variable are involved in every application of inferential statistics:

1. The sample distribution of the variable, which is empirical (i.e., it exists in reality) and is known. The shape, central tendency, and dispersion of the variable can be ascertained for the sample. Remember, however, that information from the sample is important primarily insofar as it allows the researcher to learn about the population.

2. The population distribution of the variable, which, while empirical, is unknown. Amassing information about or making inferences to the population is the sole purpose of inferential statistics.

3. The sampling distribution of the variable, which is nonempirical, or theoretical. Because of the laws of probability, a great deal is known about this distribution. Specifically, the shape, central tendency, and dispersion of the distribution can be deduced and, therefore, the distribution can be adequately characterized.

The utility of the sampling distribution is implied by its definition. Because it includes the statistics from all possible sample outcomes, the sampling distribution enables us to estimate the probability of any particular sample outcome, a process that will occupy our attention for the next five chapters.

Constructing a Sampling Distribution. The sampling distribution is theoretical, which means that it is never obtained in reality by the researcher. However, to understand better the structure and function of the distribution, let's consider an example of how one might be constructed. Suppose we wanted to gather some information about the age of a particular community of 10,000 individuals. We draw an EPSEM sample of 100 residents, ask all 100 respondents their age, and use those individual scores to compute a mean age of 27. This score is noted on the graph in Figure 6.2. Note that this sample is one of countless possible combinations of 100 people taken from this population of 10,000, and the statistics (the mean of 27) is one of millions of possible sample outcomes.

Now replace the 100 respondents in the first sample and draw another sample of the same size ($N = 100$) and again compute the average age. Assume that the mean for the second sample is 30 and note this sample outcome on Figure 6.2. This second sample is another of the countless possible combinations

FIGURE 6.2 CONSTRUCTING A SAMPLE DISTRIBUTION

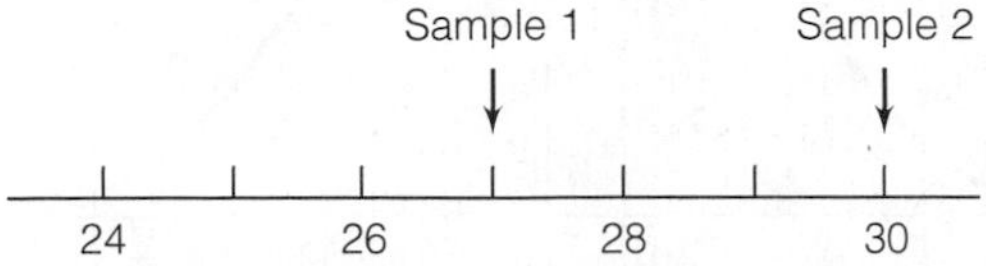

of 100 people taken from this population of 10,000, and the sample mean of 30 is another of the millions of possible sample statistics. Replace the respondents from the second sample and draw still another sample, calculate and note the mean, replace this third sample, and draw a fourth sample, continuing these operations an infinite number of times, calculating and noting the mean of each sample. Now try to imagine what Figure 6.2 would look like after tens of thousands of individual samples had been collected and the mean had been computed for each sample. What shape, mean, and standard deviation would this distribution of sample means have after we had collected all possible combinations of 100 respondents from the population of 10,000?

For one thing, we know that each sample will be at least slightly different from every other sample, since it is very unlikely that we will sample exactly the same 100 people twice. Since each sample will almost certainly be a unique combination of individuals, each sample mean will be at least slightly different in value. We also know that even though the samples are chosen according to EPSEM, they will not all be representative of the population. For example, if we continue taking samples of 100 people long enough, we will eventually choose a sample that includes only the very youngest residents. Such a sample would have a mean much lower than the true population mean. Likewise, some of our samples will include only senior citizens and will have means that are much higher than the population mean. Common sense suggests, however, that such nonrepresentative samples will be rare and that most sample means will cluster around the true population value.

To illustrate further, assume that we somehow come to know that the true mean age of the population of 10,000 individuals is 30. Since, as we have just seen, most of the sample means will also be approximately 30, the sampling distribution of these sample means should peak at 30. Some of the samples will be unrepresentative and their means will "miss the mark," but the frequency of such misses should decline as we get farther away from the population mean of 30. That is, the distribution should slope to the base as we move away from the population value–sample means of 29 or 31 should be common; means of 20 or 40 should be rare. Since the samples are random, their means should miss an equal number of times on either side of the population value, and the distribution itself should therefore be roughly symmetrical. In other words, the sampling distribution of all possible sample means should be approximately normal and will resemble the distribution presented in Figure 6.3. Recall from Chapter 5 that, on any normal curve, cases close to the mean (say, within ±1 standard

FIGURE 6.3 A SAMPLING DISTRIBUTION OF SAMPLE MEANS

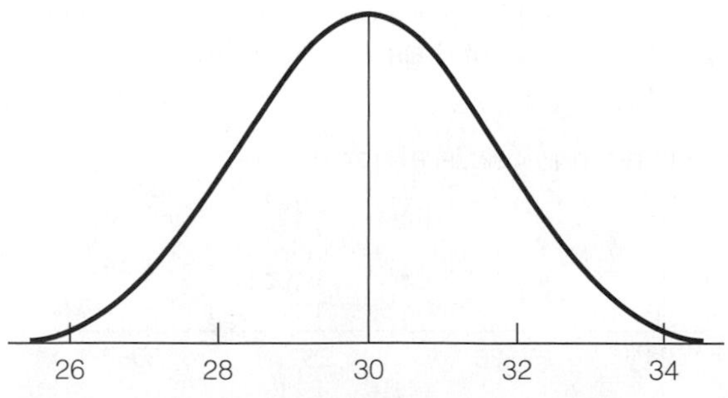

deviation) are common and cases far away from the mean (say, beyond ±3 standard deviations) are rare.

Two Theorems. These commonsense notions about the shape of the sampling distribution and other very important information about central tendency and dispersion are stated in two theorems. The first of these theorems states that

> **If repeated random samples of size N are drawn from a normal population with mean μ and standard deviation σ, then the sampling distribution of sample means will be normal, with a mean μ and a standard deviation of $\sigma/\sqrt{N}$.**

To translate: if we begin with a trait that is normally distributed across a population (IQ scores, for example) and take an infinite number of equal-sized random samples from that population, then the sampling distribution of sample means will be normal. If we know that the variable is distributed normally in the population, we can assume that the sampling distribution will be normal.

The theorem tells us more than the shape of the sampling distribution, however. It also defines its mean and standard deviation. In fact, it says that the mean of the sampling distribution will be exactly the same value as the mean of the population. That is, if we know that the mean IQ of the entire population is 100, then we know that the mean of any sampling distribution of sample mean IQ scores will also be 100. Exactly why this should be so is not a matter that can be fully explained at this level. Recall, however, that most sample means will cluster around the population value over the long run. Thus, the fact that these two values are equal should have intuitive appeal. As for dispersion, the theorem says that the standard deviation of the sampling distribution, also called the **standard error of the mean,** will be equal to the standard deviation of the population divided by the square root of N (symbolically: $\sigma/\sqrt{N}$).

If the mean and standard deviation of a normally distributed variable in a population are known, the theorem allows us to compute the mean and standard deviation of the sampling distribution.[2] Thus, we will know exactly as much about the sampling distribution (shape, central tendency, and dispersion) as we ever knew about any empirical distribution.

The first theorem requires that the variable be normally distributed in the population. What happens when the distribution of the variable in question is unknown or is known not to be normal in shape (like income, which always has a positive skew)? These eventualities (very common, in fact) are covered by a second theorem, called the **Central Limit Theorem:**

> **If repeated random samples of size N are drawn from *any* population, with mean μ and standard deviation σ, then, as N becomes large, the sampling distribution of sample means will approach normality, with mean μ and standard deviation $\sigma/\sqrt{N}$.**

To translate: The sampling distribution of sample means will become normal in shape as sample size increases for *any* variable, even when the variable is not normally distributed across the population. When N is large, the mean of the

---

[2]In the typical research situation, the values of the population mean and standard deviation are, of course, unknown. However, these values can be estimated from sample statistics, as we shall see in the chapters that follow

sampling distribution will equal the population mean, and its standard deviation (or the standard error of the mean) will be equal to $\sigma/\sqrt{N}$.

The Central Limit Theorem is important because it removes the condition that the variable be normally distributed in the population. Whenever sample size is large, we can assume that the sampling distribution is normal, with a mean equal to the population mean and a standard deviation equal to $\sigma/\sqrt{N}$. Thus, even if we are working with a variable that is known to have a skewed distribution (like income), we can still assume a normal sampling distribution.

The issue remaining, of course, is to define what is meant by a large sample. A good rule of thumb is that if sample size (N) is 100 or more, the Central Limit Theorem applies, and you can assume that the sampling distribution of sample statistics is normal in shape. When N is less than 100, you must have good evidence of a normal population distribution before you can assume that the sampling distribution is normal. Thus, a normal sampling distribution can be ensured by the expedient of using fairly large samples.

6.5 THE SAMPLING DISTRIBUTION: AN ADDITIONAL EXAMPLE

Developing an understanding of the sampling distribution—what it is and why it's important—is often one of the more challenging tasks for beginning students of statistics. It may be helpful to list briefly the most important points about the sampling distribution:

1. Its definition: *The sampling distribution is the distribution of a statistic (such as a mean or a proportion) for all possible sample outcomes of a certain size.*
2. Its shape: *normal* (see Chapter 5 and Appendix A).
3. Its central tendency and dispersion: *The mean of the sampling distribution is the same value as the mean of the population. The standard deviation of the sampling distribution—or the standard error—is equal to the population standard deviation divided by the square root of N (see the theorems).*
4. The role of the sampling distribution in inferential statistics: *It links the sample with the population* (see Figure 6.1).

To reinforce these points, let's consider an additional example of how the sampling distribution works together with the sample and the population. Consider the General Social Survey (GSS), the database used for SPSS exercises in this text. The GSS has been administered to randomly selected samples of adult Americans since 1972 and explores a broad range of characteristics and issues, including confidence in the Supreme Court, attitudes about assisted suicide, number of siblings, and level of education. The GSS has its limits, of course, but it has proven to be a very valuable resource for testing theory and for learning more about American society. Focusing on this survey, let's review the roles played by the population, the sample, and the sampling distribution when we use this database.

We'll start with the population, or the group we are actually interested in and want to learn more about. In the case of the GSS, the population consists of all adult (older than 18) Americans, which includes almost 225 million people. Clearly, we can never interview all of these people and learn what they are like or what they are thinking about abortion, capital punishment, gun control, sex education in the public schools, or any other issue. We should also note that this information is worth having. It could help inform public debates, provide some basis in fact for the discussion of many controversial issues (e.g., the polls show consistently that the majority of Americans favor some form of gun control), and assist people in clarifying their personal beliefs. If the information is valuable, what can be done to learn more about this huge population?

This brings us to the sample, a carefully chosen subset of the population. The General Social Survey is administered to about 3000 people, each of whom is chosen by a sophisticated technology based on the principle of EPSEM. A key point to remember is that samples chosen by this method are very likely to be representative of the populations from which they were selected. Whatever is true of the sample will also be true of the population (with some limits and qualifications, of course).

The respondents are contacted at home and asked for background information (religion, gender, years of education, and so on) as well as their opinions and attitudes. When all of this information is collated, the GSS database includes information (shape, central tendency, dispersion) on hundreds of variables (age, level of prejudice, marital status) for the people in the sample. So we have a lot of information about the variables for the *sample* (the 3000 or so people who actually responded to the survey), but no information about these variables for the *population* (the 225 million adult Americans). How do we get from the known sample to the unknown population? This is the central question of inferential statistics; the answer, as you hopefully realize by now, is "by using the sampling distribution."

Remember that, unlike the sample and the population, the sampling distribution is theoretical, and, because of the theorems presented earlier in this chapter, we know its shape, central tendency, and dispersion. For any variable from the GSS, the theorems tell us that:

- The sampling distribution will be normal in shape because the sample is "large" (N is much greater than 100). This will be true regardless of the shape of the variable in the population.
- The mean of the sampling distribution will be the same value as the mean of the population. If *all* adult Americans have completed an average of 13.5 years of schooling ($\mu = 13.5$), the mean of the sampling distribution will also be 13.5.
- The standard deviation (or standard error) of the sampling distribution is equal to the population standard deviation (σ) divided by the square root of N.

Thus, the theorems tell us the statistical characteristics of this distribution (shape, central tendency, and dispersion), and this information allows us to link the sample to the population.

How does the sampling distribution link the sample to the population? The fact that the sampling distribution will be normal when N is large is crucial. This means that more than two-thirds (68%) of all samples will be within ± 1 Z score of the mean of the sampling distribution (which is the same value as the mean of the population), about 95% are within ± 2 Z scores, and so forth. We do not (and cannot) know the actual value of the mean of the sampling distribution, but we do know that the probabilities are very high that our sample statistic is approximately equal to this parameter. Similarly, the theorems give us crucial information about the mean and standard error of the sampling distribution that we can use, as you will see in the chapters that follow, to link information from the sample to the population.

To summarize, our goal is to infer information about the population (in the case of the GSS: all adult Americans). When populations are too large to test (and contacting 225 million adult Americans is far beyond the capacity of even the most energetic pollster), we use information from randomly selected samples, carefully drawn from the population of interest, to estimate the characteristics of the population. In the case of the GSS, the full sample consists of about 3000 adult Americans who have responded to the questions on the survey. The sampling distribution, the theoretical distribution whose characteristics are defined by the theorems, links the known sample to the unknown population.

6.6 SYMBOLS AND TERMINOLOGY

In the following chapters, we will be working with three entirely different distributions. Furthermore, we will be concerned with several different kinds of sampling distributions–including the sampling distribution of sample means and the sampling distribution of sample proportions.

To distinguish clearly among these various distributions, we will often use symbols. The symbols for the means and standard deviations of samples and populations have already been introduced in Chapters 3 and 4. For quick reference, Table 6.1 introduces in summary form some of the symbols that will be used for the sampling distribution. Basically, the sampling distribution is denoted with Greek letters that are subscripted according to the sample statistic of interest.

Note that the mean and standard deviation of a sample are denoted with English letters ($\overline{X}$ and s), while the mean and standard deviation of a population are denoted with the Greek-letter equivalents (μ and σ). Proportions calculated on samples are symbolized as *P*-sub-*s* (*s* for sample), while population proportions are denoted as *P*-sub-*u* (*u* for "universe" or population). The symbols for the sampling distribution are Greek letters with English-letter subscripts. The mean and standard deviation of a sampling distribution of sample means are $\mu_{\overline{X}}$ ("mu-sub-*x*-bar") and $\sigma_{\overline{X}}$ ("sigma-sub-*x*-bar"). The mean and standard deviation of a sampling distribution of sample proportions are μ_p ("mu-sub-*p*") and σ_p ("sigma-sub-*p*").

TABLE 6.1 SYMBOLS FOR MEANS AND STANDARD DEVIATIONS OF THREE DISTRIBUTIONS

	Mean	Standard Deviation	Proportion
1. Samples	$\overline{X}$	s	P_s
2. Populations	μ	σ	P_u
3. Sampling distributions			
of means	$\mu_{\overline{X}}$	$\sigma_{\overline{X}}$	
of proportions	μ_p	σ_p	

SUMMARY

1. Since populations are almost always too large to test, a fundamental strategy of social science research is to select a sample from the defined population and then to use information from the sample to generalize to the population. This is done either by estimation or by hypothesis testing.
2. Several techniques are commonly used for selecting random samples. Each of these techniques involves selecting cases for the sample according to EPSEM. Even the most rigorous technique, however, cannot guarantee that the sample will be representative. One of the great strengths of inferential statistics is that the probability that the sample is nonrepresentative can be estimated.
3. The sampling distribution, the central concept in inferential statistics, is a theoretical distribution of all possible sample outcomes. Since its overall shape, mean, and standard deviation are known (under the conditions specified in the two theorems), the sampling distribution can be adequately characterized and utilized by researchers.
4. The two theorems that were introduced in this chapter state that when the variable of interest is normally distributed in the population or when sample size is large, the sampling distribution will be normal in shape, its mean will be equal to the population mean, and its standard deviation (or standard error) will be equal to the population standard deviation divided by the square root of *N*.
5. All applications of inferential statistics involve generalizing from the sample to the population by means of the sampling distribution. Both estimation procedures and hypothesis testing incorporate the three distributions, and it is crucial that you develop a clear understanding of each distribution and its role in inferential statistics.

GLOSSARY

Central Limit Theorem. A theorem that specifies the mean, standard deviation, and shape of the sampling distribution, given that the sample is large.

Cluster sample. A method of sampling by which geographical units are randomly selected and all cases within each selected unit are tested.

EPSEM. The **E**qual **P**robability of **SE**lection **M**ethod for selecting samples. Every element or case in the population must have an equal probability of selection for the sample.

μ. The mean of a population.

$\mu_{\bar{X}}$. The mean of a sampling distribution of sample means.

μ_p. The mean of a sampling distribution of sample proportions.

Parameter. A characteristic of a population.

P_s. (*P*-sub-*s*) Any sample proportion.

P_u. (*P*-sub-*u*) Any population proportion.

Representative. The quality a sample is said to have if it reproduces the major characteristics of the population from which it was drawn.

Sampling distribution. The distribution of a statistic for all possible sample outcomes of a certain size. Under conditions specified in two theorems, the sampling distribution will be normal in shape, with a mean equal to the population value and a standard deviation equal to the population standard deviation divided by the square root of *N*.

Simple random sample. A method for choosing cases from a population by which every case and every combination of cases has an equal chance of being included.

Standard error of the mean. The standard deviation of a sampling distribution of sample means.

Stratified sample. A method of sampling by which cases are selected from sublists of the population.

Systematic sampling. A method of sampling by which the first case from a list of the population is randomly selected. Thereafter, every *k*th case is selected.

PROBLEMS

6.1 Imagine that you had to gather a random sample ($N = 300$) of the student body at your school. How would you acquire a list of the population? Would the list be complete and accurate? What procedure would you follow in selecting cases (that is, simple or systematic random sampling)? Would cluster sampling be an appropriate technique (assuming that no list was available)? Describe in detail how you would construct a cluster sample.

6.2 This exercise is extremely tedious and hardly ever works out the way it ought to, mostly because not many people have the patience to draw an "infinite" number of even very small samples. However, if you want a more concrete and tangible understanding of sampling distributions and the two theorems presented in this chapter, then this exercise may have a significant payoff. At the end of this problem are listed the ages of a population of college students ($N = 50$). By a random method (such as a table of random numbers), draw at least 50 samples of size 2 (that is, 50 pairs of cases), compute a mean for each sample, and plot the means on a frequency polygon. (Incidentally, this exercise will work better if you draw 100 or 200 samples and/or use larger samples than $N = 2$.)

a. The curve you've just produced is a sampling distribution. Observe its shape; after 50 samples, it should be approaching normality. What is your estimate of the population mean (μ) based on the shape of the curve?

b. Calculate the mean of the sampling distribution ($\mu_{\bar{X}}$). Be careful to do this by summing the sample means (not the scores) and dividing by the number of samples you've drawn. Now compute the population mean (μ). These two means should be very close in value because $\mu_{\bar{X}} = \mu$ by the Central Limit Theorem.

c. Calculate the standard deviation of the sampling distribution (use the means as scores) and the standard deviation of the population. Compare these two values. You should find that $\sigma_{\bar{X}} = \sigma/\sqrt{N}$.

d. If none of the preceding exercises turned out as they should have, it is for one or more of the following reasons:

1. You didn't take enough samples. You may need as many as 100 or 200 (or more) samples to see the curve begin to look "normal."
2. Sample size (2) is too small. An N of 5 or 10 would work much better.
3. Your sampling method is not truly random and/or the population is not arranged in random fashion.

17	20	20	19	20
18	21	19	20	19
19	22	19	23	19
20	23	18	20	20
22	19	19	20	20
23	17	18	21	20
20	18	20	19	20
22	17	21	21	21
21	20	20	20	22
18	21	20	22	21

SPSS for Windows

Using SPSS for Windows to Draw Random Samples

DEMONSTRATION 6.1 Estimating Average Age

SPSS for Windows includes a procedure for drawing random samples from a database. We can use this procedure to illustrate some points about sampling and to convince the skeptics in the crowd that properly selected samples will produce statistics that are close approximations of the corresponding population values or parameters. For purposes of this demonstration, the 2006 GSS sample will be treated as a population and its characteristics will be treated as parameters.

The following instructions will calculate a mean for *age* for three random samples of different sizes drawn from the 2006 GSS sample. The actual average age of the sample (which will be the parameter or μ) is 46.88 (see Demonstration 5.1). The samples are roughly 10%, 25%, and 50% of the population size, and the program selects them by an EPSEM procedure. Therefore, these samples may be considered "simple random samples."

As a part of this procedure we also request the "standard error of the mean," or S.E. MEAN. This is the standard deviation of the sampling distribution ($\sigma_{\bar{X}}$) for a sample of this size. This statistic will be of interest because we can expect our sample means to be within this distance of the population value or parameter.

With the 2006 GSS loaded, click **Data** from the menu bar of the Data Editor and then click **Select Cases.** The **Select Cases** window appears and presents a number of different options. To select random samples, click the button next to "Random sample of cases" and then click on the **Sample** button. The **Select Cases: Random Sample** window will open. We can specify the size of the sample in two different ways. If we use the first option, we can specify that the sample will include a certain percentage of cases in the database. The second option allows us to specify the exact number of cases in the sample. Let's use the first option and request a 10% sample by typing 10 into the box on the first line. Click **Continue,** and then click **OK** on the **Select Cases** window. The sample will be selected and can now be processed.

To find the mean age for the 10% sample, click **Analyze, Descriptive Statistics**, and then **Descriptives.** Find *age* in the variable list and transfer it to the **Variable(s):** window. On the **Descriptives** menu, click the **Options** button and select S.E. MEAN in addition to the usual statistics. Click **Continue** and then **OK,** and the requested statistics will appear in the output window.

Now, to produce a 25% sample, return to the **Select Cases** window by clicking **Data** and **Select Cases.** Click the **Reset** button at the bottom of the window and then click **OK** and the full data set ($N = 1417$) will be restored. Repeat the procedure we followed for selecting the 10% sample. Click the button next to "Random sample of cases" and then click on the **Sample** button. The **Select Cases: Random Sample** window will open. Request a 25% sample by typing 25 in the box, click **Continue** and **OK,** and the new sample will be selected.

Run the **Descriptives** procedure for the 25% sample (don't forget S.E. MEAN) and note the results. Finally, repeat these steps for a 50% sample. The results are summarized here:

1 Sample %	2 Sample Size	3 Sample Mean	4 Standard Error	5 Sample Mean ± Standard Error	6 Sample Mean − Population Mean
10%	133	47.61	1.46	46.15–49.07	0.73
25%	361	46.89	0.91	45.98–47.80	0.01
50%	653	46.99	0.67	46.32–47.66	0.11

Let's look at this table column by column. The first column lists the sample percent and column 2 lists the actual number of cases in each sample. Sample means and standard errors (or standard deviations of the sampling distribution) are listed in columns 3 and 4. Look at column 4 and note that standard error decreases as sample size increases. This should reinforce the commonsense notion that larger samples will provide more accurate estimates of population values.

To produce column 5, I subtracted and then added the value of the standard error to the mean of each sample. Note that all three intervals listed in column 5 include the value of the population mean (46.88). The important point here is that all three sample means were close to (within 1 standard error of) the population mean. Finally, column 6 shows the distance between the sample means and the population mean. The mean of the smallest (10%) sample is furthest from the population mean. The other two sample means are quite close to the population mean, and the mean of the 25% sample is almost exactly equal to the population value. We would expect the mean of the largest (50%) sample to be closest to the population mean, but remember that we are dealing with chance and probabilities here. At any rate, the most important point is that all three sample means are within 1 standard error of the population mean.

This demonstration should reinforce one of the main points of this chapter: Statistics calculated on samples that have been selected according to the principle of EPSEM will (almost always) be reasonable approximations of their population counterparts.

Exercise

6.1 Following the procedures in Demonstration 6.1, select three samples from the 2006 GSS database: 15%, 30%, and 60%. Get descriptive statistics for *age* (don't forget to get the standard error), and use the results to complete the following table:

1 Sample %	2 Sample Size	3 Sample Mean	4 Standard Error	5 Sample Mean ± Standard Errorfdas	6 Sample Mean Mean-Population
15%	_____	_____	_____	_____	_____
30%	_____	_____	_____	_____	_____
60%	_____	_____	_____	_____	_____

Summarize these results. What happens to standard error as sample size increases? Why? How accurate are the estimates (sample means)? Are all sample means within a standard error of 46.88? How does the accuracy of the estimates change as sample size changes?

7 Estimation Procedures

LEARNING OBJECTIVES

By the end of this chapter, you will be able to

1. Explain the logic of estimation and the role of the sample, sampling distribution, and the population.
2. Define and explain the concepts of bias and efficiency.
3. Construct and interpret confidence intervals for sample means and sample proportions.
4. Explain the relationships between confidence level, sample size, and the width of the confidence interval.

7.1 INTRODUCTION

The object of this branch of inferential statistics is to estimate population values or parameters from statistics computed from samples. Although the techniques presented in this chapter may be new to you, you are certainly familiar with their most common applications: public-opinion polls and election projections. Polls and surveys on every conceivable issue—from the sublime to the trivial—have become a staple of the mass media and popular culture. The techniques you will learn in this chapter are essentially the same as those used by the most reputable, sophisticated, and scientific pollsters.

There are two kinds of estimation procedures. First, a **point estimate** is a sample statistic that is used to estimate a population value. For example, a newspaper story that reports that 74% of a sample of randomly selected Americans support capital punishment, is reporting a point estimate. The second kind of estimation procedure involves **confidence intervals,** which consist of a range of values (an interval) instead of a single point. Rather than estimating a specific figure as in a point estimate, an interval estimate might be phrased as "between 71% and 77% of Americans approve of capital punishment." In this latter estimate, we are estimating that the population value falls between 71% and 77%, but we do not specify its exact value.

7.2 BIAS AND EFFICIENCY

Both point and interval estimation procedures are based on sample statistics. Which of the many available sample statistics should be used? Estimators can be selected according to two criteria: **bias** and **efficiency.** Estimates should be based on sample statistics that are unbiased and relatively efficient. We cover each of these criteria separately.

Bias. An estimator is unbiased if the mean of its sampling distribution is equal to the population value of interest. We know from the theorems presented in Chapter 6 that sample means conform to this criterion. The mean of the sampling distribution of sample means (which we will note symbolically as $\mu_{\bar{X}}$) is the $\mu_{\bar{X}}$ same as the population mean (μ).

Sample proportions (p_s) are also unbiased. That is, if we calculate sample proportions from repeated random samples of size N and then array them in a line chart, the sampling distribution of sample proportions will have a mean (μ_p) equal to the population proportion (p_s). Thus, if we are concerned with coin flips and sample honest coins 10 at a time ($N = 10$), the sampling distribution will have a mean equal to 0.5, which is the probability that an honest coin will be heads (or tails) when flipped. All statistics other than sample means and sample proportions are biased (that is, have sampling distributions with means not equal to the population value).[1]

Knowing that sample means and proportions are unbiased allows us to determine the probability that they lie within a given distance of the population values we are trying to estimate. To illustrate, consider a specific problem. Assume that we wish to estimate the average income of a community. A random sample of 500 households is taken ($N = 500$), and a sample mean of $45,000 is computed. In this example, the population mean is the average income of *all* households in the community and the sample mean is the average income for the 500 households that happened to be selected for our sample. Note that we do not know the value of the population mean (μ)—if we did, we wouldn't need the sample—but it is μ that we are interested in. The sample mean of $45,000 is important and interesting primarily insofar as it can give us information about the population mean.

The two theorems presented in Chapter 6 give us a great deal of information about the sampling distribution of all possible sample means in this situation. Because N is large, we know that the sampling distribution is normal and that its mean is equal to the population mean. We also know that all normal curves contain about 68% of the cases (the cases here are sample means) within ± 1 Z, 95% of the cases within ± 2 Z's, and more than 99% of the cases within ± 3 Z's of the mean. Remember that we are discussing the sampling distribution here—the distribution of all possible sample outcomes, or, in this instance, sample means. Thus, the probabilities are very good (approximately 68 out of 100 chances) that our sample mean of $45,000 is within ± 1 Z, excellent (95 out of 100) that it is within ± 2 Z's, and overwhelming (99 out of 100) that it is within ± 3 Z's of the mean of the sampling distribution (which is the same value as the population mean). These relationships are graphically depicted in Figure 7.1.

If an estimator is unbiased, it is probably an accurate estimate of the population parameter (μ in this case). However, in less than 1% of the cases, a sample mean will be more than ± 3 Z's away from the mean of the sampling distribution (very inaccurate) by random chance alone. We literally have no idea if our particular sample mean of $45,000 is in this small minority. We do know, however, that the odds are high that our sample mean is considerably closer

---

[1] In particular, the sample standard deviation (s) is a biased estimator of the population standard deviation (σ). As you might expect, there is less dispersion in a sample than in a population and, as a consequence, s will underestimate σ. As we shall see, however, sample standard deviation can be corrected for this bias and still serve as an estimate of the population standard deviation for large samples.

FIGURE 7.1 AREAS UNDER THE SAMPLING DISTRIBUTION OF SAMPLE MEANS

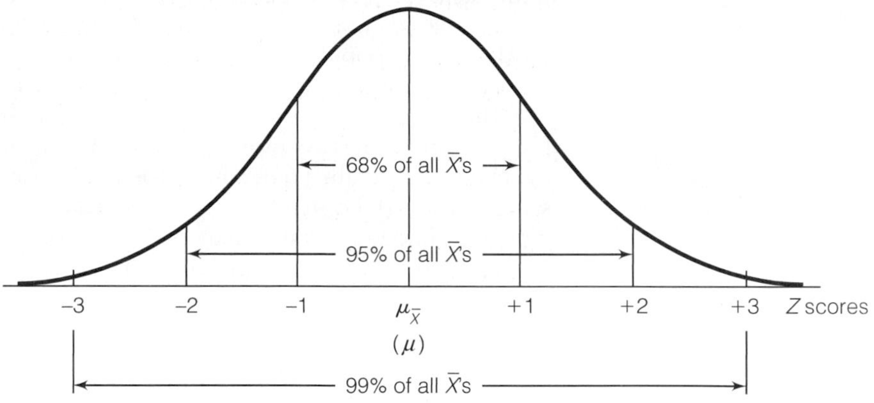

than ±3 *Z*'s to the mean of the sampling distribution and, thus, to the population mean.

Efficiency. The second desirable characteristic of an estimator is efficiency, which is the extent to which the sampling distribution is clustered about its mean. Efficiency, or clustering, is essentially a matter of dispersion, as we saw in Chapter 4 (see Figure 4.1). The smaller the standard deviation of a sampling distribution, the greater the clustering and the higher the efficiency. Remember that the standard deviation of the sampling distribution of sample means, or the standard error of the mean, is equal to the population standard deviation divided by the square root of *N*. Therefore, the standard deviation of the sampling distribution is an inverse function of $N (\sigma_{\overline{X}} = \sigma/\sqrt{N})$. As sample size increases, $\sigma_{\overline{X}}$ will decrease. We can improve the efficiency (or decrease the standard deviation of the sampling distribution) for any estimator by increasing sample size.

An example should make this clearer. Consider two samples of different sizes:

Sample 1	Sample 2
$\overline{X} = \$45{,}000$	$\overline{X} = \$45{,}000$
$N_1 = 100$	$N_2 = 1000$

Both sample means are unbiased, but which is the more efficient estimator? Consider sample 1 and assume, for the sake of illustration, that the population standard deviation (σ) is \$500.[2] In this case, the standard deviation of the sampling distribution of all possible sample means with an *N* of 100 would be $\sigma/\sqrt{N}$, or $500/\sqrt{100}$, or \$50.00. For sample 2, the standard deviation of all possible sample means with an *N* of 1000 would be much smaller. Specifically, it would be equal to $500/\sqrt{1000}$, or \$15.81.

The sampling distribution based on sample 2 is much more clustered than the sampling distribution based on sample 1. In fact, sampling distribution 2

[2] In reality, of course, the value of σ would be unknown.

contains 68% of all possible sample means within ±15.81 of μ, while sampling distribution 1 requires a much broader interval of ±50.00 to do the same. The estimate based on a sample with 1000 cases is much more likely to be close in value to the population parameter than is an estimate based on a sample of 100 cases. Figures 7.2 and 7.3 illustrate these relationships graphically.

The key point to remember here is that the standard deviation of all sampling distributions is an inverse function of N: the larger the sample, the greater the clustering and the higher the efficiency. In part, these relationships between sample size and the standard deviation of the sampling distribution do nothing more than underscore our commonsense notion that much more confidence can be placed in large samples than in small (as long as both have been randomly selected).

7.3 ESTIMATION PROCEDURES: INTRODUCTION

The procedure for constructing a point estimate is straightforward. Draw an EPSEM sample, calculate either a proportion or a mean, and estimate that the population parameter is the same as the sample statistic. Remember that the larger the sample, the greater the efficiency and the more likely that the estimator is approximately the same as the population value. Also remember that,

FIGURE 7.2 A SAMPLING DISTRIBUTION WITH N = 100 AND $\sigma_{\bar{X}}$ = \$50.00

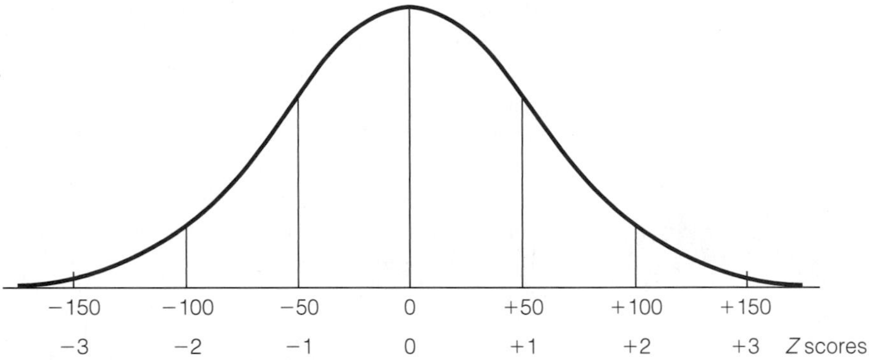

FIGURE 7.3 A SAMPLING DISTRIBUTION WITH N = 1000 AND $\sigma_{\bar{X}}$ = \$15.81

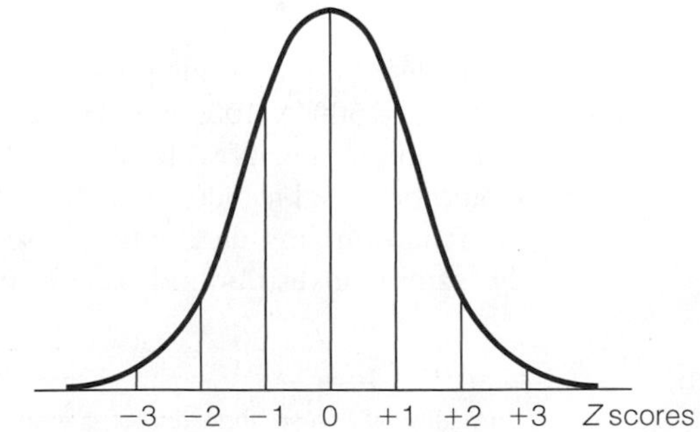

no matter how rigid the sampling procedure or how large the sample, there is always some chance that the estimator is very inaccurate.

Compared to point estimates, interval estimates are more complicated but safer because we are more likely to include the population parameter when we guess a range of values. The first step in constructing an interval estimate is to decide on the risk you are willing to take of being wrong. An interval estimate is wrong if it does *not* include the population parameter. This probability of error is called **alpha** (symbolized by Greek letter α). The exact value of alpha will depend on the nature of the research situation, but a 0.05 probability is commonly used. Setting alpha equal to 0.05, also called using the 95% **confidence level,** means that over the long run the researcher is willing to be wrong only 5% of the time. Or, to put it another way, if an infinite number of intervals were constructed at this alpha level (and with all other things being equal), 95% of them would contain the population value and 5% would not. In reality, of course, only one interval is constructed, and, by setting the probability of error very low, we are setting the odds in our favor that the interval will include the population value.

The second step is to picture the sampling distribution, divide the probability of error equally into the upper and lower tails of the distribution, and then find the corresponding *Z* score. For example, if we decided to set alpha equal to 0.05, we would place half (0.025) of this probability in the lower tail and half in the upper tail of the distribution. The sampling distribution would thus be divided as illustrated in Figure 7.4.

We need to find the *Z* score that marks the beginnings of the shaded areas. In Chapter 5, we learned how to calculate *Z* scores and find areas under the normal curve. Here, we will reverse that process. We need to find the *Z* score beyond which lies a proportion of .0250 of the total area. To do this, go down column c of Appendix A until you find this proportion (.0250). The associated *Z* score is 1.96. Since the curve is symmetrical and we are interested in both the upper and lower tails, we designate the *Z* score that corresponds to an alpha of .05 as ± 1.96 (see Figure 7.5).

We now know that 95% of all possible sample outcomes fall within ± 1.96 *Z*-score units of the population value. In reality, of course, there is only one sample outcome, but, if we construct an interval estimate based on ± 1.96 *Z*s,

FIGURE 7.4 THE SAMPLING DISTRIBUTION WITH ALPHA (α) EQUAL TO 0.05

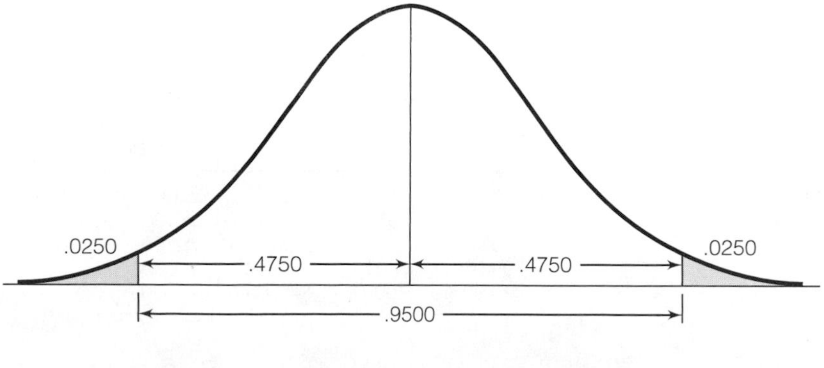

FIGURE 7.5 FINDING THE *Z* SCORE THAT CORRESPONDS TO AN ALPHA (α) OF 0.05

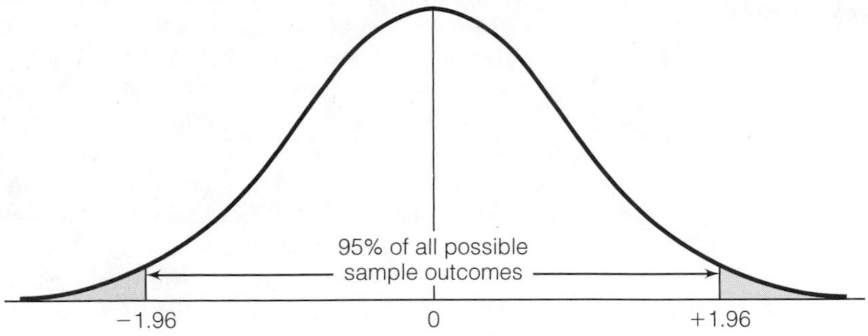

the probabilities are that 95% of all such intervals will trap the population value. Thus, we can be 95% confident that our interval contains the population value.

Besides the 95% level, there are four other commonly used confidence levels: the 90% level ($\alpha = .10$), the 99% level ($\alpha = .01$), the 99.9% level ($\alpha = .001$), and the 99.99% level ($\alpha = .0001$). To find the corresponding *Z* scores for these levels, follow the procedures outlined earlier for an alpha of 0.05. Table 7.1 summarizes all the information you will need.

You should turn to Appendix A and confirm for yourself that the *Z* scores in Table 7.1 do indeed correspond to these alpha levels. As you do, note that, in the cases where alpha is set at 0.10 and 0.01, the precise areas we seek do not appear in the table. For example, with an alpha of 0.10, we would look in column c ("Area beyond") for the area .0500. Instead we find an area of .0505 ($Z = \pm 1.64$) and an area of .0495 ($Z = \pm 1.65$). The *Z* score we are seeking is somewhere between these two other scores. When this condition occurs, take the larger of the two scores as *Z*. This will make the interval as wide as possible under the circumstances and is thus the most conservative course of action. In the case of an alpha of 0.01, we encounter the same problem (the exact area .0050 is not in the table); resolve it the same way, and take the larger score as *Z*. Finally, Table 7.1 includes the *Z* scores we will use for the two lowest alpha levels ($\alpha = .001$ and $.0001$), but Appendix A is not detailed enough to show these values. *(For practice in finding Z scores for various levels of confidence, see problem 7.3.)*

The third step is actually to construct the confidence interval. In the sections that follow, we illustrate how to construct an interval estimate first with sample means and then with sample proportions.

TABLE 7.1 *Z* SCORES FOR VARIOUS LEVELS OF ALPHA (α)

Confidence Level	Alpha (α)	$\alpha/2$	*Z* Score
90%	.10	.0500	±1.65
95%	.05	.0250	±1.96
99%	.01	.0050	±2.58
99.9%	.001	.0005	±3.29
99.99%	.0001	.00005	±3.90

7.4 INTERVAL ESTIMATION PROCEDURES FOR SAMPLE MEANS (LARGE SAMPLES)

The formula for constructing a confidence interval based on sample means is given in Formula 7.1:

FORMULA 7.1

$$\text{c.i.} = \bar{X} \pm Z\left(\frac{\sigma}{\sqrt{N}}\right)$$

where c.i. = confidence interval

$\bar{X}$ = the sample mean

Z = the Z score as determined by the alpha level

$\frac{\sigma}{\sqrt{N}}$ = the standard deviation of the sampling distribution or the standard error of the mean

As an example, suppose you wanted to estimate the average IQ of a community and had randomly selected a sample of 200 residents, with a sample mean IQ of 105. Assume that the population standard deviation for IQ scores is about 15, so we can set σ equal to 15. If we are willing to run a 5% chance of being wrong and set alpha at 0.05, the corresponding Z score will be 1.96. These values can be substituted directly into Formula 7.1, and an interval can be constructed:

$$\text{c.i.} = \bar{X} \pm Z\left(\frac{\sigma}{\sqrt{N}}\right)$$

$$\text{c.i.} = 105 \pm 1.96\left(\frac{15}{\sqrt{200}}\right)$$

$$\text{c.i.} = 105 \pm 1.96\left(\frac{15}{14.14}\right)$$

$$\text{c.i.} = 105 \pm (1.96)(1.06)$$

$$\text{c.i.} = 105 \pm 2.08$$

That is, our estimate is that the average IQ for the population in question is somewhere between 102.92 (105 − 2.08) and 107.08 (105 + 2.08). Since 95% of all possible sample means are within ±1.96 Z's (or 2.08 IQ units in this case) of the mean of the sampling distribution, the odds are very high that our interval will contain the population mean. In fact, even if the sample mean is as far off as ±1.96 Z's (which is unlikely), our interval will still contain $\mu_{\bar{X}}$ and, thus, μ. Only if our sample mean is one of the few that is more than ±1.96 Z's from the mean of the sampling distribution will we have failed to include the population mean.

Note that in our earlier example, the value of the population standard deviation was supplied. Needless to say, it is unusual to have such information about a population. In the great majority of cases, we will have no knowledge of σ. In such cases, however, we can estimate σ with s, the sample standard deviation. Unfortunately, s is a biased estimator of σ, and the formula must be changed slightly to correct for the bias. For larger samples, the bias of s will not affect the interval very much. The revised formula for cases in which σ is unknown is

FORMULA 7.2

$$\text{c.i.} = \bar{X} \pm Z\left(\frac{s}{\sqrt{N-1}}\right)$$

In comparing this formula with Formula 7.1, note that there are two changes. First, σ is replaced by s and, second, the denominator of the last term is the square root of $N - 1$ rather than the square root of N. The latter change is the correction for the fact that s is biased.

Let me stress here that the substitution of s for σ is permitted only for large samples (that is, samples with 100 or more cases). For smaller samples, when

Application 7.1

A study of the leisure activities of Americans was conducted on a sample of 1000 households. The respondents identified television viewing as a major form of recreation. If the sample reported an average of 6.2 hours of television viewing a day, what is the estimate of the population mean? The information from the sample is

$$\bar{X} = 6.2$$
$$s = 0.7$$
$$N = 1000$$

If we set alpha at 0.05, the corresponding Z score will be ± 1.96, and the 95% confidence interval will be

$$\text{c.i.} = \bar{X} \pm Z\left(\frac{s}{\sqrt{N-1}}\right)$$
$$\text{c.i.} = 6.2 \pm 1.96\left(\frac{0.7}{\sqrt{1000-1}}\right)$$
$$\text{c.i.} = 6.2 \pm 1.96\left(\frac{0.7}{31.61}\right)$$
$$\text{c.i.} = 6.2 \pm (1.96)(0.02)$$
$$\text{c.i.} = 6.2 \pm 0.04$$

Based on this result, we would estimate that the population spends an average of $6.2 \pm .04$ hours per day viewing television. The lower limit of our interval estimate $(6.2 - 0.04)$ is 6.16, and the upper limit $(6.2 + 0.04)$ is 6.24. Thus, another way to state the interval would be

$$6.16 \leq \mu \leq 6.24$$

The population mean is greater than or equal to 6.16 and less than or equal to 6.24. Because alpha was set at the .05 level, this estimate has a 5% chance of being wrong (that is, of not containing the population mean).

the value of the population standard deviation is unknown, the standardized normal distribution summarized in Appendix A cannot be used in the estimation process. To construct confidence intervals from sample means with samples smaller than 100, we must use a different theoretical distribution, called the Student's t distribution, to find areas under the sampling distribution. We defer the presentation of the t distribution until Chapter 8 and confine our attention here to estimation procedures for large samples only.

Let us close this section by working through a sample problem with Formula 7.2. Average income for a random sample of a particular community is \$45,000, with a standard deviation of \$200. What is the 95% interval estimate of the population mean, μ?

Given that

$$\bar{X} = \$45{,}000$$
$$s = \$200$$
$$N = 500$$

and using an alpha of 0.05, the interval can be constructed:

$$\text{c.i.} = \bar{X} \pm Z\left(\frac{s}{\sqrt{N-1}}\right)$$
$$\text{c.i.} = 45{,}000 \pm 1.96\left(\frac{200}{\sqrt{499}}\right)$$
$$\text{c.i.} = 45{,}000 \pm 17.55$$

The average income for the community as a whole is between \$44,982.45 (45,000 − 17.55) and \$45,017.55 (45,000 + 17.55). Remember that this interval has only a 5% chance of being wrong (that is, of not containing the population

ONE STEP AT A TIME **Constructing Confidence Intervals for Sample Means**

Step 1: Select an alpha level. Commonly used alpha levels are 0.10, 0.05, 0.01, 0.001, and 0.0001; the 0.05 level is particularly common.

Step 2: Divide the value of alpha in half. For example, if alpha is 0.05, half of alpha would be 0.025. Find this value in column c of Appendix A to find the Z score that corresponds to your selected alpha level. If alpha = 0.05, $Z = \pm 1.96$.

NOTE: If you are using the conventional alpha level of 0.05, the Z score will always be ± 1.96 and you can omit the first two steps. For other commonly used alpha levels, see Table 7.1.

Step 3: Substitute the sample values into the proper formula. If the value of the population standard deviation (σ) is known, use Formula 7.1. If the value of the population standard deviation (σ) is not known, use Formula 7.2.

Solving Formula 7.1

Step 4: Find the square root of N.

Step 5: Divide the value you found in step 4 into sigma (σ).

Step 6: Multiply the value you found in step 5 by Z.

Step 7: The value you found in step 6 is the width of the confidence interval. Insert this value in Formula 7.1 following the ± sign.

Step 9: State the confidence interval in a sentence or two that identifies the:

a. sample mean
b. upper and lower limits of the interval
c. alpha level or confidence level
d. sample size (N)

Solving Formula 7.2

Step 10: Find the square root of $N - 1$.

Step 11: Divide the value you found in step 10 into s.

Step 12: Multiply the value you found in step 11 by Z.

Step 13: The value you found in step 12 is the width of the confidence interval. Insert this value in Formula 7.2 following the ± sign.

Step 14: State the confidence interval in a sentence or two that identifies the:

a. sample mean
b. upper and lower limits of the interval
c. alpha level or confidence level
d. sample size (N)

mean). *(For practice with confidence intervals for sample means, see problems 7.1, 7.4–7.7, 7.18, and 7.19a–7.19c.)*

7.5 INTERVAL ESTIMATION PROCEDURES FOR SAMPLE PROPORTIONS (LARGE SAMPLES)

Estimation procedures for sample proportions are essentially the same as those for sample means. The major difference is that, since proportions are different statistics, we must use a different sampling distribution. In fact, again based on the Central Limit Theorem, we know that sample proportions have sampling distributions that are normal in shape with means (μ_p) equal to the population value (P_u) and standard deviations (σ_p) equal to $\sqrt{\frac{P_u(1 - P_u)}{N}}$. The formula for constructing confidence intervals based on sample proportions is

FORMULA 7.3

$$\text{c.i.} = P_s \pm Z\sqrt{\frac{P_u(1 - P_u)}{N}}$$

The values for P_s and N come directly from the sample, and the value of Z is determined by the confidence level, as was the case with sample means. This leaves one unknown in the formula, P_u—the same value we are trying to estimate. This dilemma can be resolved by setting the value of P_u at 0.5. Since the second term in the numerator under the radical $(1 - P_u)$ is the reciprocal of P_u, the entire

Application 7.2

If 45% of a random sample of 1000 Americans reports that walking is their major physical activity, what is the estimate of the population value? The sample information is

$$P_s = 0.45$$
$$N = 1000$$

Note that the percentage of "walkers" has been stated as a proportion. If we set alpha at 0.05, the corresponding Z score will be ± 1.96, and the interval estimate of the population proportion will be

$$\text{c.i.} = P_s \pm Z\sqrt{\frac{P_u(1 - P_u)}{N}}$$

$$\text{c.i.} = 0.45 \pm 1.96\sqrt{\frac{(0.5)(0.5)}{1000}}$$

$$\text{c.i.} = 0.45 \pm 1.96\sqrt{0.00025}$$
$$\text{c.i.} = 0.45 \pm (1.96)(0.016)$$
$$\text{c.i.} = 0.45 \pm 0.03$$

We can now estimate that the proportion of the population for which walking is the major form of physical exercise is between 0.42 and 0.48. That is, the lower limit of the interval estimate is (0.45 − 0.03) or 0.42, and the upper limit is (0.45 + 0.03) or 0.48. We may also express this result in percentages and say that between 42% and 48% of the population walk as their major form of physical exercise. This interval has a 5% chance of not containing the population value.

ONE STEP AT A TIME Constructing Confidence Intervals for Sample Proportions

Step 1: Select an alpha level. Commonly used alpha levels are 0.10, 0.05, 0.01, 0.001, and 0.0001; the 0.05 level is particularly common.

Step 2: Divide the value of alpha in half. For example, if alpha is 0.05, half of alpha would be 0.025. Find this value in column b of Appendix A to find the Z score that corresponds to your selected alpha level. If alpha = 0.05, Z = ±1.96.

NOTE: If you are using the conventional alpha level of 0.05, the Z score will always be ±1.96 and you can omit the first two steps. For other commonly used alpha levels, see Table 7.1.

Step 3: Substitute the sample values into Formula 7.3.

Step 4: Substitute a value of 0.5 for P_u. This will make the numerator of the fraction under the square root sign 0.25.

Step 5: Divide *N* into 0.25.

Step 6: Find the square root of the value you found in step 5.

Step 7: Multiply this value by the value of *Z*.

Step 8: The value you found in step 7 is the width of the confidence interval. Insert this value in Formula 7.3 following the ± sign.

Step 9: State the confidence interval in a sentence or two that identifies the:

a. sample proportion
b. upper and lower limits of the interval
c. alpha level or confidence level
d. sample size (*N*)

expression will always have a value of 0.5 × 0.5, or 0.25, which is the maximum value this expression can attain. That is, if we set P_u at any value other than 0.5, the expression $P_u(1 - P_u)$ will decrease in value. If we set P_u at 0.4, for example, the second term $(1 - P_u)$ would be 0.6, and the value of the entire expression would decrease to 0.24. Setting P_u at 0.5 ensures that the expression $P_u(1 - P_u)$

Application 7.3

A total of 1609 adult Canadians were randomly selected to participate in a study of attitudes towards homosexuality and same-sex marriages. Some results are reported here. What is the level of support in the population? The sample information, expressed in terms of the proportion agreeing, is

"Gays and lesbians should have the same rights as heterosexuals"

$$P_s = 0.72$$
$$N = 1609$$

"Marriage should be expanded to include same-sex unions."

$$P_s = 0.60$$
$$N = 1609$$

For the first item, the confidence interval estimate to the population at the 95% confidence level is

$$\text{c.i.} = P_s \pm Z\sqrt{\frac{P_u(1 - P_u)}{N}}$$
$$\text{c.i.} = .72 \pm 1.96\sqrt{\frac{(0.5)(0.5)}{1609}}$$
$$\text{c.i.} = .72 \pm 1.96\sqrt{.00016}$$
$$\text{c.i.} = .72 \pm (1.96)(.013)$$
$$\text{c.i.} = .72 \pm .03$$

Expressing these results in terms of percentages, we can conclude that between 69% and 75% of adult Canadians support equal rights for gays and lesbians. This estimate is based on a sample of 1609 and is constructed at the 95% confidence level.

For the second survey item, the confidence interval estimate to the population at the 95% confidence level is

$$\text{c.i.} = P_s \pm Z\sqrt{\frac{P_u(1 - P_u)}{N}}$$
$$\text{c.i.} = .60 \pm 1.96\sqrt{\frac{(0.5)(0.5)}{1609}}$$
$$\text{c.i.} = .60 \pm 1.96\sqrt{.00016}$$
$$\text{c.i.} = .60 \pm (1.96)(.013)$$
$$\text{c.i.} = .60 \pm .03$$

Again, expressing results in terms of percentages, we can conclude that between 57% and 63% of adult Canadians support same-sex marriages. This estimate is based on a sample of 1609 and is constructed at the 95% confidence level. (Note that the width of the second confidence interval is exactly the same as the first width of the confidence level. This is because we are using the same values for *Z* score and the sample size in both estimates.)

will be at its maximum possible value and, consequently, the interval will be at maximum width. This is the most conservative solution possible to the dilemma posed by having to assign a value to P_u in the estimation equation.

To illustrate these procedures, assume that you wish to estimate the proportion of students at your university who missed at least one day of classes because of illness last semester. Out of a random sample of 200 students, 60 reported that they had been sick enough to miss classes at least once during the previous semester. The sample proportion on which we will base our estimate is thus 60/200, or 0.30. At the 95% level, the interval estimate will be

$$\text{c.i.} = P_s \pm Z\sqrt{\frac{P_u(1 - P_u)}{N}}$$
$$\text{c.i.} = 0.30 \pm 1.96\sqrt{\frac{(0.5)(0.5)}{200}}$$
$$\text{c.i.} = 0.30 \pm 1.96\sqrt{\frac{0.25}{200}}$$
$$\text{c.i.} = 0.30 \pm (1.96)(0.04)$$
$$\text{c.i.} = 0.30 \pm 0.08$$

Based on this sample proportion of 0.30, you would estimate that the proportion of students who missed at least one day of classes because of illness was between 0.22 and 0.38. The estimate could, of course, also be phrased in percentages by reporting that between 22% and 38% of the student body was affected by illness at least once during the past semester. *(For practice with confidence intervals for sample proportions, see problems 7.2, 7.8–7.12, and 7.19d–7.19g.)*

7.6 A SUMMARY OF THE COMPUTATION OF CONFIDENCE INTERVALS

To this point, we have covered the construction of confidence intervals for sample means and sample proportions. In both cases, the procedures assume large samples (N greater than 100). The procedures for constructing confidence intervals for small samples are not covered in this text. Table 7.2 presents the three formulas for confidence intervals organized by the situations in which they are used. For sample means when the population standard deviation is known, use formula 7.1. When the population standard deviation is unknown (which is the usual case), use formula 7.2. For sample proportions, always use formula 7.3.

7.7 CONTROLLING THE WIDTH OF INTERVAL ESTIMATES

The width of a confidence interval for either sample means or sample proportions can be partly controlled by manipulating two terms in the equation. First, the confidence level can be raised or lowered and, second, the interval can be widened or narrowed by gathering samples of different size. The researcher alone determines the risk he or she is willing to take of being wrong (that is, of not including the population value in the interval estimate). The exact confidence level (or alpha level) will depend, in part, on the purpose of the research. For example, if potentially harmful drugs were being tested, the researcher would naturally demand very high levels of confidence (99.99% or even 99.999%). On the other hand, if intervals are being constructed only for loose "guesstimates," then much lower confidence levels can be tolerated (such as 90%).

The relationship between interval size and confidence level is that intervals widen as confidence levels increase. This should make intuitive sense. Wider intervals are more likely to trap the population value; hence, more confidence can be placed in them.

TABLE 7.2 CHOOSING FORMULAS FOR CONFIDENCE INTERVALS

If the sample statistic is a	and	Use formula
mean	the population standard deviation is known	7.1 $\text{c.i.} = \overline{X} \pm Z\left(\frac{\sigma}{\sqrt{N}}\right)$
mean	the population standard deviation is unknown	7.2 $\text{c.i.} = \overline{X} \pm Z\left(\frac{s}{\sqrt{N-1}}\right)$
proportion		7.3 $\text{c.i.} = P_s \pm Z\sqrt{\frac{P_u(1-P_u)}{N}}$

To illustrate this relationship, let us return to the example where we estimated the average income for a community. In this problem, we were working with a sample of 500 residents, and the average income for this sample was \$45,000, with a standard deviation of \$200. We constructed the 95% confidence interval and found that it extended 17.55 around the sample mean (that is, the interval was \$45,000 ± 17.55).

If we had constructed the 90% confidence interval (a lower confidence level) for these sample data, the Z score in the formula would have decreased to ±1.65, and the interval would have been narrower:

$$\text{c.i.} = \overline{X} \pm Z\left(\frac{s}{\sqrt{N-1}}\right)$$

$$\text{c.i.} = 45{,}000 \pm 1.65\left(\frac{200}{\sqrt{499}}\right)$$

$$\text{c.i.} = 45{,}000 \pm 1.65(8.95)$$

$$\text{c.i.} = 45{,}000 \pm 14.77$$

On the other hand, if we had constructed the 99% confidence interval, the Z score would have increased to ±2.58, and the interval would have been wider:

$$\text{c.i.} = \overline{X} \pm Z\left(\frac{s}{\sqrt{N-1}}\right)$$

$$\text{c.i.} = 45{,}000 \pm 2.58\left(\frac{200}{\sqrt{499}}\right)$$

$$\text{c.i.} = 45{,}000 \pm 2.58(8.95)$$

$$\text{c.i.} = 45{,}000 \pm 23.09$$

At the 99.9% confidence level, the Z score would be ±3.29, and the interval would be wider still:

$$\text{c.i.} = \overline{X} \pm Z\left(\frac{s}{\sqrt{N-1}}\right)$$

$$\text{c.i.} = 45{,}000 \pm 3.29\left(\frac{200}{\sqrt{499}}\right)$$

$$\text{c.i.} = 45{,}000 \pm 3.29(8.95)$$

$$\text{c.i.} = 45{,}000 \pm 29.45$$

We can readily see the increase in interval size in Table 7.3. Although I have used sample means to illustrate the relationship between interval width and confidence level, exactly the same relationships apply to sample proportions. (*To explore further the relationship between alpha and interval width, see problem 7.13.*)

Sample size bears the opposite relationship to interval width. As sample size increases, interval width decreases. Larger samples give more precise (narrower) estimates. Again, an example should make this clearer. In Table 7.4, confidence intervals for four samples of various sizes are constructed and then grouped together for purposes of comparison. The sample data are the same as in Table 7.3, and the confidence level is 95% throughout. The relationships illustrated in Table 7.4 also hold true, of course, for sample proportions. *(To explore further the relationship between sample size and interval width, see problem 7.14.)*

TABLE 7.3 INTERVAL ESTIMATES FOR FOUR CONFIDENCE LEVELS ($\bar{X}$ = \$45,000, s = \$200, N = 500 throughout)

Alpha	Confidence Level	Interval	Interval Width
.10	90%	\$45,000 ± 14.77	\$29.54
.05	95%	\$45,000 ± 17.55	\$35.10
.01	99%	\$45,000 ± 23.09	\$46.18
.001	99.9%	\$45,000 ± 29.45	\$58.90

TABLE 7.4 INTERVAL ESTIMATES FOR FOUR DIFFERENT SAMPLES ($\bar{X}$ = \$45,000, s = \$200, alpha = 0.05 throughout)

Sample 1 (N = 100)	Sample 2 (N = 500)
c.i. = \$45,000 ± 1.96(200/$\sqrt{99}$) c.i. = \$45,000 ± 39.40	c.i. = \$45,000 ± 1.96(200/$\sqrt{499}$) c.i. = \$45,000 ± 17.55
Sample 3 (N = 1000)	**Sample 4 (N = 10,000)**
c.i. = \$45,000 ± 1.96(200/$\sqrt{999}$) c.i. = \$45,000 ± 12.40	c.i. = \$45,000 ± 1.96(200/$\sqrt{9,999}$) c.i. = \$45,000 ± 3.92

Sample	N	Interval Width
1	100	\$78.80
2	500	\$35.10
3	1,000	\$24.80
4	10,000	\$ 7.84

Notice that the decrease in interval width (or increase in precision) does not bear a constant, or linear, relationship with sample size. For example, sample 2 is five times larger than sample 1, but the interval is not five times as narrow. This is an important relationship because it means that N might have to be increased many times over to improve the accuracy of an estimate appreciably. Since the cost of a research project is a direct function of sample size, this relationship implies a point of diminishing returns in estimation procedures. A sample of 10,000 will cost about twice as much as a sample of 5000, but estimates based on the larger sample will not be twice as precise.

7.8 INTERPRETING STATISTICS: PREDICTING THE ELECTION OF THE PRESIDENT AND JUDGING HIS OR HER PERFORMANCE

The statistical techniques covered in this chapter have become a part of everyday life in the United States and in many other societies. In politics, for example, estimation techniques are used to track public sentiment, measure how citizens perceive the performance of our leaders, and project the likely winners of upcoming elections. We should acknowledge, before proceeding, that these applications of estimation techniques can be controversial. Many people wonder if the easy availability of polls makes politicians overly sensitive to the whims of public sentiment. Others are concerned that election projections might work against people's readiness to participate fully in the democratic process and, in particular, cast their votes on election day. These are serious concerns, of course, but we can do little more than acknowledge them in this text and hope that you have the opportunity to pursue them in other, more appropriate settings.

In this installment of Interpreting Statistics, we examine the accuracy of election projections for the 2004 and 2000 presidential elections, and we'll also examine the polls that have measured the approval ratings of incumbent U.S. presidents since the middle of the 20th century. Both kinds of polls use the same formulas introduced in this chapter to construct confidence intervals (although the random samples were assembled according to a complex and sophisticated technology that is beyond the scope of this text).

"Too Close to Call." Pollsters have become very accurate in predicting the outcomes of presidential elections, but remember that 95% confidence intervals are accurate only to within ±3 percentage points. This means that the polls cannot identify the likely victor in very close races. Both the 2004 and 2000 elections were very close, and the polls indicated a statistical dead heat right up to the end of the campaigns. Table 7.5 shows CNN's "poll of polls"—or averages of polls from various sources—for the final weeks of the 2004 campaign and the final breakdown of votes for the two major candidates. The polls included by CNN were based on sample sizes of about 1000.

The final polls in October and November show that the race was very close and that the difference between the candidates was so small that a winner could not be projected. For example, in the November 1 poll, Bush's support could have been as low as 45% (48% − 3%) and Kerry's could have been as high as 49% (46% + 3%). When the confidence intervals overlap, the race is said to be "too close to call" and "a statistical dead heat."

The 2000 presidential election between President Bush and Senator Al Gore of Tennessee was even closer than the 2004 election. When the ballots were finally counted (and recounted), the candidates were in a virtual tie for the popular vote, separated by only a few thousand votes out of more than 100 million votes cast. Democrat Al Gore received 48.3% of the popular vote, while Republican George Bush garnered 48.1%. After a lengthy and intense battle over the electoral votes of Florida, the Supreme Court decided the election in favor of Bush. Table 7.6 shows the actual vote and the final election projections made by the Gallup Polls. These estimates are based on sample sizes ranging between 700 and 2300.

TABLE 7.5 POLLING RESULTS AND ACTUAL VOTE, 2004 PRESIDENTIAL ELECTION

	Actual Vote	
	Bush	Kerry
	51%	48%
	Election Projections	
Date of Poll	% of Population Estimated to Vote for Bush	% of Population Estimated to Vote for Kerry
November 1	48	46
October 25	49	46
October 18	50	45

READING STATISTICS 4: Public-Opinion Polls

You are most likely to encounter the estimation techniques covered in this chapter in the mass media in the form of public-opinion polls, election projections, and the like. Professional polling firms use interval estimates, and responsible reporting by the media will usually emphasize the estimate itself (for example, "In a survey of the American public, 57% of the respondents said that they approve of the president's performance") but also will report the width of the interval ("This estimate is accurate to within ±3%" or "Figures from this poll are subject to a sampling error of ±3%"), the alpha level (usually as the confidence level of 95%), and the size of the sample ("1458 households were surveyed").

Election projections and voter analyses, at least at the presidential level, have been common since the middle of the 20th century and are discussed further in Section 7.8. More recently, public opinion polling has been used to gauge reactions to everything from the newest movies to the hottest gossip to the president's conduct during the latest crisis. Newsmagazines routinely report poll results as an adjunct to news stories, and similar stories are regular features of TV news and newspapers. I would include an example or two of these applications here, but polls have become so pervasive that you can do your own example. Just pick up a newsmagazine or newspaper and leaf through it casually, and I bet you'll find at least one poll. Read the story and see if you can identify the population, the confidence interval width, the sample size, and the confidence level. Bring the news item to class and dazzle your instructor.

As a citizen as well as a social scientist, you should be extremely suspicious of polls that do not include such vital information as sample size or interval width. You should also check to see how the sample was assembled. Samples selected in a nonrandom fashion cannot be regarded as representative of the American public or, for that matter, of any population other than the sample itself. You might encounter such nonscientific polls when local TV news or sports programs ask viewers to call in and register their opinions about some current controversy. The "samples" assembled in this way are not representative, and the results should be regarded as for entertainment only or, perhaps, as a basis for discussion. They should not be regarded as reflections of public opinion in general. You should, of course, read all polls and surveys critically and analytically, but you should place confidence only in polls that are based

(continued next page)

TABLE 7.6 POLLING RESULTS AND ACTUAL VOTE, 2000 PRESIDENTIAL ELECTION

	Actual Vote	
	Bush	Gore
	48%	48%
	Election Projections	
Date of Poll	% of Population Estimated to Vote for Bush	% of Population Estimated to Vote for Gore
November 6	48	46
November 1	47	43
October 22	46	44

READING STATISTICS 4: *(continued)*

on samples selected according to the rule of EPSEM (see Chapter 6) from some defined population.

Advertisements, commercials, and reports published by partisan groups sometimes report statistics that seem to be estimates of the population values. Often, however, such estimates are based on woefully inadequate sampling sizes and biased sampling procedures, and the data are collected under circumstances that evoke a desired response. "Person in the street" (or shopper in the grocery store) interviews have a certain folksy appeal but must not be accorded the same credibility as surveys conducted by reputable polling firms.

Public Opinion Surveys in the Professional Literature

For the social sciences, probably the single most important consequence of the growth in opinion polling is that many nationally representative databases are now available for research purposes. These high-quality databases are often available at no cost or for a nominal fee, and they make it possible to conduct "state-of-the-art" research without the expense and difficulty of collecting data yourself. This is an important development because we can now test our theories against very high-quality data, and our conclusions will therefore have a stronger empirical basis. Our research efforts will have greater credibility with our colleagues, with policy makers, and with the public at large.

One of the more important and widely used databases of this sort is called the General Social Survey, or the GSS. Since 1972, the National Opinion Research Council has questioned a nationally representative sample of Americans about a wide variety of issues and concerns. Since many of the questions are asked every year, the GSS has been offering a longitudinal record of American sentiment and opinion about a large variety of topics for more than three decades. Each year, new topics of current concern are added and explored, and the variety of information available continues to expand. Like other nationally representative samples, the GSS sample is chosen by a complex probability design that resembles cluster sampling (see Chapter 6). Sample size varies from 1400 to over 4000, and estimates based on samples this large will be accurate to within about ±3% (see Table 7.5 and Section 7.7). The computer exercises in this text are based on the 2006 GSS; this database is described more fully in Appendix F.

Although these races were too close for the pollsters to identify a likely winner, note that in both years the polls were within the ±3% margin of error that is associated with interval estimates based on the 95% confidence level and using 1000–2000 respondents.

The Ups and Downs of Presidential Popularity. Once you get to be president, you are not free of polls and confidence intervals. Since the middle of the 20th century, pollsters have tracked the president's popularity by asking randomly selected samples of adult Americans if they approve or disapprove of the way the president is handling his job. President Bush's approval ratings (the percent of the sample that said "approve" when asked, "Do you approve or disapprove of the way George W. Bush is handling his job as president?") are presented in Figure 7.6. For purposes of clarity, point estimates are used for this graph, but remember that the confidence interval estimate would range about ±3% (at the 95% confidence level) around these points.

The single most dramatic feature of this graph is the huge increase in approval that followed the 9/11 terrorist attacks on the World Trade Center and the

Pentagon in 2001. This burst of support reflects the increased solidarity, strong emotions, and high levels of patriotism with which Americans responded to the attacks. The president's approval rating reached an astounding 90% shortly after the attacks but then, inevitably, began to trend down as the society (and politics) gradually returned to the daily routine of everyday business.

By the spring of 2003, the president's approval rating had fallen back into the high 50s, much lower than the peaks of fall, 2001, but still higher than the historical average. President Bush's approval received another boost with the invasion of Iraq in March 2003, an echo of the "rally" effect that followed 9/11, but then began to sink to prewar levels and below, reflecting the substantial reluctance of many Americans to support the war effort. The level of approval continued to decline through the fall of 2006 and into the spring of 2007; the dissatisfaction of the American public was also reflected by the victory of the Democratic Party in both the U.S. House and Senate elections in November 2006.

Returning to the controversies surrounding public opinion polls, many critics would argue that it is not a good thing to judge the president so continuously. To some extent, these approval ratings expose the president (and other leaders and politicians whose performance is similarly measured) to the whims of public sentiment and, at some level, could pressure him or her to cater to popular opinion and shape ("spin") his or her image to maintain support. On the other hand, information like that presented in Figure 7.6 supplies some interesting and useful insights into the affairs, not only of the political institution, but of the society as a whole. These approval ratings provide convincing evidence that, contrary to the expectations of some, Americans responded to the attacks of 9/11 with strong solidarity and cohesion. The numbers also show the debilitating effects of a long and controversial war on public morale and confidence in the leadership of the nation.

FIGURE 7.6 APPROVAL RATINGS FOR PRESIDENT BUSH, FEBRUARY 2001 TO MAY 2007

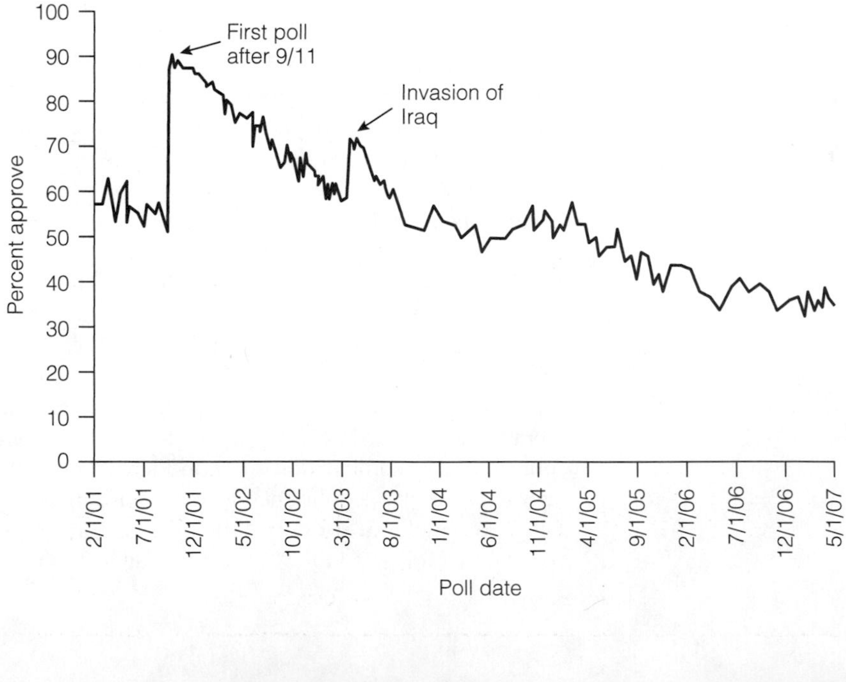

READING STATISTICS 5: Using Representative Samples to Track National Trends

In Reading Statistics 4, I mentioned nationally representative databases such as the General Social Survey (GSS) and the Gallup poll. Among many other uses, this information can be employed to track national trends in opinion and behavior and shifts in U.S. culture over time. In a recent article, Professor Amy Butler used the GSS to track changes in same-sex partnering between 1988 and 2002. She hypothesized that the percentage of same-sex sexual relationships would increase over the time period, in part because of cultural, moral, and legal changes in U.S. society. The item used in the GSS to measure the dependent variable was "Have your sex partners in the last 12 months been: exclusively male, both male and female, or exclusively female?"

As shown in Figure 1, the rates of same sex partnering fluctuated but generally increased over the time period, with females increasing more than males. Note also that, according to these data, same-sex partnering is relatively rare and peaked at about 4% of the population (for men in 1998). Because of these low percentages, the values on the vertical axis of the graph range from zero to 4.5%.

After conducting a variety of statistical tests, Butler was able to link the increases shown in Figure 1 to changing cultural patterns. She notes that while genetic factors may account for sexual preferences, fluctuations in actual same-sex partnering over time and between cultures (or at least the willingness to reveal these relationships to public opinion pollsters) are more likely to be due to cultural, political, and legal factors.

Figure 1 uses point estimates of the percentage of Americans who have had same-sex relationships in the past year. Butler was concerned about changes over time, so she did not construct interval estimates for each year. Had she done so, she would have converted the percentages to proportions and used Formula 7.3 to find the upper and lower limits of the interval. Although not germane to her analysis, you should recognize that, for each year, the sample proportions are very likely to be within ±3 percentage points of the population parameter (the proportion of *all* adult Americans who had same-sex sexual relationships within the past year).

Finally, we should note that these rates are so low that the usual ±3 percentage points of a 95% confidence interval would take us into negative numbers for most of the yearly values in Figure 1. Statistically, there are ways to deal with such situations, but these are beyond the scope of this text. Commonsensically, values of less than zero are impossible for this variable, and we should simply ignore the lower end of the confidence interval.

Want to learn more about the changing sexual behaviors of Americans? See the following citation.

Butler, Amy. 2005. "Gender Differences in the Prevalence of Same-Sex Partnering: 1988–2002." *Social Forces*, 84(1):421–449.

FIGURE 1 SEX OF SEXUAL PARTNER IN PREVIOUS YEAR

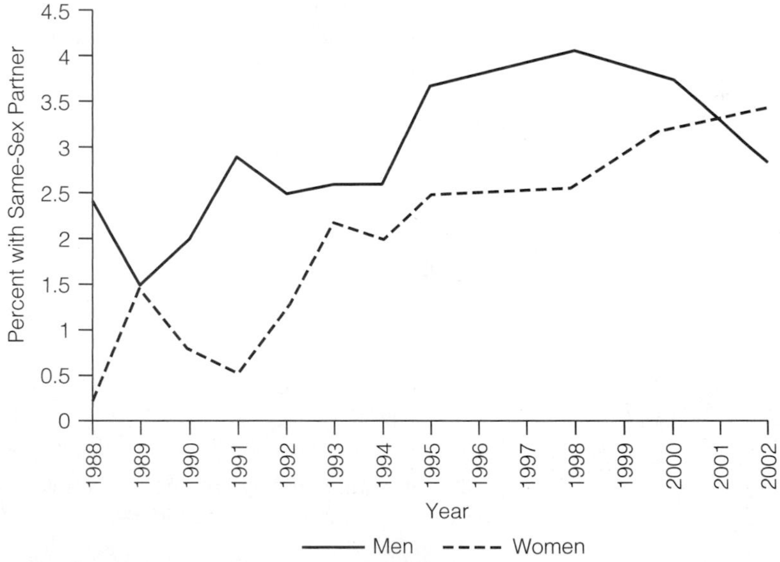

SUMMARY

1. Population values can be estimated with sample values. With point estimates, a single sample statistic is used to estimate the corresponding population value. With confidence intervals, we estimate that the population value falls within a certain range of values.
2. Estimates based on sample statistics must be unbiased and relatively efficient. Of all the sample statistics, only means and proportions are unbiased. The means of the sampling distributions of these statistics are equal to the respective population values. Efficiency is largely a matter of sample size. The greater the sample size, the lower the value of the standard deviation of the sampling distribution, the more tightly clustered the sample outcomes will be around the mean of the sampling distribution, and the more efficient the estimate.
3. With point estimates, we estimate that the population value is the same as the sample statistic (either a mean or a proportion). With interval estimates, we construct a confidence interval, a range of values into which we estimate that the population value falls. The width of the interval is a function of the risk we are willing to take of being wrong (the alpha level) and the sample size. The interval widens as our probability of being wrong decreases and as sample size decreases.

SUMMARY OF FORMULAS

Confidence interval for a sample mean, large samples, population standard deviation known:

$$7.1 \quad \text{c.i.} = \bar{X} \pm Z\left(\frac{\sigma}{\sqrt{N}}\right)$$

Confidence interval for a sample mean, large samples, population standard deviation unknown:

$$7.2 \quad \text{c.i.} = \bar{X} \pm Z\left(\frac{s}{\sqrt{N-1}}\right)$$

Confidence interval for a sample proportion, large samples:

$$7.3 \quad \text{c.i.} = P_s \pm Z\sqrt{\frac{P_u(1-P_u)}{N}}$$

GLOSSARY

Alpha (α). The probability of error, or the probability that a confidence interval does not contain the population value. Alpha levels are usually set at 0.10, 0.05, 0.01, or 0.001.

Bias. A criterion used to select sample statistics as estimators. A statistic is unbiased if the mean of its sampling distribution is equal to the population value of interest.

Confidence interval. An estimate of a population value in which a range of values is specified.

Confidence level. A frequently used alternate way of expressing alpha, the probability that an interval estimate will not contain the population value. Confidence levels of 90%, 95%, 99%, and 99.9% correspond to alphas of 0.10, 0.05, 0.01, and 0.001, respectively.

Efficiency. The extent to which the sample outcomes are clustered around the mean of the sampling distribution.

Point estimate. An estimate of a population value where a single value is specified.

PROBLEMS

7.1 For each set of sample outcomes below, construct the 95% confidence interval for estimating μ, the population mean.

a. $\bar{X} = 5.2$, $s = 0.7$, $N = 157$
b. $\bar{X} = 100$, $s = 9$, $N = 620$
c. $\bar{X} = 20$, $s = 3$, $N = 220$
d. $\bar{X} = 1020$, $s = 50$, $N = 329$
e. $\bar{X} = 7.3$, $s = 1.2$, $N = 105$
f. $\bar{X} = 33$, $s = 6$, $N = 220$

7.2 For each of the following sets of sample outcomes, construct the 99% confidence interval for estimating P_u.

a. $P_s = .14$, $N = 100$ **b.** $P_s = .37$, $N = 522$ **c.** $P_s = .79$, $N = 121$

d. $P_s = .43$, $N = 1049$ **e.** $P_s = .40$, $N = 548$ **f.** $P_s = .63$, $N = 300$

7.3 For each of the following confidence levels, determine the corresponding Z score.

Confidence Level	Alpha	Area Beyond Z	Z score
95%	.05	.0250	±1.96
94%			
92%			
97%			
98%			
99.9%			

7.4 SW You have developed a series of questions to measure "burnout" in social workers. A random sample of 100 social workers working in greater metropolitan Shinbone, Kansas, has an average score of 10.3, with a standard deviation of 2.7. What is your estimate of the average burnout score for the population as a whole? Use the 95% confidence level.

7.5 SOC A researcher has gathered information from a random sample of 178 households. For each of the following variables, construct confidence intervals to estimate the population mean. Use the 90% level.

a. An average of 2.3 people resides in each household. Standard deviation is 0.35.

b. There was an average of 2.1 television sets ($s = 0.10$) and 0.78 telephones ($s = 0.55$) per household.

c. The households averaged 6.0 hours of television viewing per day ($s = 3.0$).

7.6 SOC A random sample of 100 television programs contained an average of 2.37 acts of physical violence per program. At the 99% level, what is your estimate of the population value?

$$\bar{X} = 2.37$$
$$s = 0.30$$
$$N = 100$$

7.7 SOC A random sample of 429 college students was interviewed about a number of matters.

a. They reported that they had spent an average of \$178.23 on textbooks during the previous semester. If the sample standard deviation for these data is \$15.78, construct an estimate of the population mean at the 99% level.

b. They also reported that they had visited the health clinic an average of 1.5 times a semester. If the sample standard deviation is 0.3, construct an estimate of the population mean at the 99% level.

c. On the average, the sample had missed 2.8 days of classes per semester because of illness. If the sample standard deviation is 1.0, construct an estimate of the population mean at the 99% level.

d. On the average, the sample had missed 3.5 days of classes per semester for reasons other than illness. If the sample standard deviation is 1.5, construct an estimate of the population mean at the 99% level.

7.8 CJ A random sample of 500 residents of Shinbone, Kansas, shows that exactly 50 of the respondents had been the victims of violent crime over the past year. Estimate the proportion of victims for the population as a whole, using the 90% confidence level. *(HINT: Calculate the sample proportion P_s before using Formula 7.3. Remember that proportions are equal to frequency divided by N.)*

7.9 SOC The survey mentioned in problem 7.5 found that 25 of the 178 households consisted of unmarried couples who were living together. What is your estimate of the population proportion? Use the 95% level.

7.10 PA A random sample of 324 residents of a community revealed that 30% were very satisfied with the quality of trash collection. At the 99% level, what is your estimate of the population value?

7.11 SOC A random sample of 1496 respondents of a major metropolitan area was questioned about a number of issues. Construct estimates to the population at the 90% level for each of the results reported next. Express the final confidence interval in percentages (e.g., "between 40 and 45% agreed that premarital sex was always wrong").

a. When asked to agree or disagree with the statement "Explicit sexual books and magazines lead to rape and other sex crimes," 823 agreed.

b. When asked to agree or disagree with the statement "Hand guns shoulbe outlawed," 650 agreed.

c. 375 of the sample agreed that marijuana should be legalized.
d. 1023 of the sample said that they had attended church or synagogue at least once within the past month.
e. 800 agreed that public elementary schools should have sex education programs starting in the fifth grade.

7.12 SW A random sample of 100 patients treated in a program for alcoholism and drug dependency over the past 10 years was selected. It was determined that 53 of the patients had been readmitted to the program at least once. At the 95% level, construct an estimate of the population proportion.

7.13 For the following sample data, construct four different interval estimates of the population mean, one each for the 90%, 95%, 99%, 99.9%, and 99.99% levels. What happens to the interval width as confidence level increases? Why?

$$\bar{X} = 100$$
$$s = 10$$
$$N = 500$$

7.14 For each of the following three sample sizes, construct the 95% confidence interval. Use a sample proportion of 0.40 throughout. What happens to interval width as sample size increases? Why?

$$P_s = 0.40$$
Sample A: $N = 100$
Sample B: $N = 1000$
Sample C: $N = 10{,}000$

7.15 PS Two individuals are running for mayor of Shinbone. You conduct an election survey a week before the election and find that 51% of the respondents prefer candidate A. Can you predict a winner? Use the 99% level. *(HINT: In a two-candidate race, what percentage of the vote would the winner need? Does the confidence interval indicate that candidate A has a sure margin of victory? Remember that while the population parameter is probably (α = .01) in the confidence interval, it may be anywhere in the interval.)*

$$P_s = 0.51$$
$$N = 578$$

7.16 SOC The World Values Survey is administered periodically to random samples from societies around the globe. Listed here are the number of respondents in each nation who said that they are "very happy." Compute sample proportions, and construct confidence interval estimates for each nation at the 95% level.

Nation	Year	Number "very happy"	Sample Size	Confidence Interval
Great Britain	1998	496	1495	
Japan	1995	505	1476	
Brazil	1996	329	1492	
Nigeria	1995	695	1471	
China	1995	338	1493	

7.17 SOC The fraternities and sororities at St. Algebra College have been plagued by declining membership over the past several years and want to know if the incoming freshman class will be a fertile recruiting ground. Not having enough money to survey all 1600 freshmen, they commission you to survey the interests of a random sample. You find that 35 of your 150 respondents are "extremely" interested in social clubs. At the 95% level, what is your estimate of the number of freshmen who would be extremely interested? *(HINT: The high and low values of your final confidence interval are proportions. How can proportions also be expressed as numbers?)*

7.18 SOC You are the consumer-affairs reporter for a daily newspaper. Part of your job is to investigate the claims of manufacturers, and you are particularly suspicious of a new economy car that the manufacturer claims will get 78 miles per gallon. After checking the mileage figures for a random sample of 120 owners of this car, you find an average miles per gallon of 75.5, with a standard deviation of 3.7. At the 99% level, do your results tend to confirm or refute the manufacturer's claims?

7.19 SOC The results listed next are from a survey given to a random sample of the American public. For each sample statistic, construct a confidence interval estimate of the population parameter at the 95% confidence level. Sample size (N) is 2987 throughout.
a. The average occupational prestige score was 43.87, with a standard deviation of 13.52.
b. The respondents reported watching an average of 2.86 hours of TV per day, with a standard deviation of 2.20.

c. The average number of children was 1.81, with a standard deviation of 1.67.

d. Of the 2987 respondents, 876 identified themselves as Catholic.

e. Five hundred thirty-five of the respondents said that they had never married.

f. The proportion of respondents who said they voted for Bush in the 2004 presidential election was 0.52.

g. When asked about capital punishment, 2425 of the respondents said that they favored the death penalty for murder.

SPSS for Windows

Using *SPSS for Windows* Procedures to Produce Confidence Intervals

SPSS DEMONSTRATION 7.1 Generating Statistical Information for Use in Constructing Confidence Intervals

SPSS does not provide any programs specifically for constructing confidence intervals, although some of the procedures we cover in future chapters do include confidence intervals as part of the output. Rather than make use of these programs, I want this section to confirm something you may already suspect: Fancy computer programs are not always helpful or even particularly useful. The arithmetic required by estimation procedures (let's face it) is not particularly difficult, and you could probably complete the formulas faster by hand than by SPSS.

On the other hand, who wants to do all the calculations it would take to get the summary statistics on which the estimates will be based? Calculating the mean age for the GSS sample would require the addition of almost 1500 numbers. The dimensions of that task should suggest the proper role of the computer. Once you know the mean age, the rest of the calculations for confidence intervals are not very formidable. So how do you get the sample statistics? For interval-ratio variables in this problem, use **Descriptives** to get sample means, standard deviations, and sample sizes. For nominal variables, use **Frequencies** to produce frequency distributions with a column for percentages.

To illustrate, we'll get sample statistics for *educ* (years of education), an interval-ratio level variable, and *marital* (marital status), a nominal-level variable. The output for *educ* will look like this:

Descriptive Statistics

	N	Minimum	Maximum	Mean	Std. Deviation
HIGHEST YEAR OF SCHOOL COMPLETED	1424	0	20	13.26	3.191
Valid N (listwise)	1424				

To construct the confidence interval at the 95% level for *educ*, substitute the values in the SPSS output into Formula 7.2:

$$\text{c.i.} = \overline{X} \pm Z\left(\frac{s}{\sqrt{N-1}}\right)$$

$$\text{c.i.} = 13.26 \pm 1.96\left(\frac{3.19}{\sqrt{1423}}\right)$$

$$\text{c.i.} = 13.26 \pm (1.96)(0.09)$$

$$\text{c.i.} = 13.26 \pm 0.18$$

We estimate that Americans, on the average, have comleted between 13.08 (13.26 − .18) and 13.44 (13.26 + .18) years of school.

For *marital*, the output will look like this:

Marital Status

		Frequency	Percent	Valid Percent	Cumulative Percent
Valid	MARRIED	686	48.1	48.1	48.1
	WIDOWED	119	8.3	8.4	56.5
	DIVORCED	222	15.6	15.6	72.1
	SEPARATED	39	2.7	2.7	74.8
	NEVER MARRIED	359	25.2	25.2	100.0
	Total	1425	99.9	100.0	
Missing	NA	1	.1		
Total		1426	100.0		

We can estimate the population parameter for any or all of the categories of the variable. Let's estimate the proportion of the population that is married. Don't forget to change percentages to proportions. Substituting values into Formula 7.3 and using the 95% confidence level, we would have:

We can estimate the population parameter for any or all of the categories of the variable. Let's estimate the proportion of the population that is married. Don't forget to change percentages to proportions. Substituting values into Formula 7.3 and using the 95% confidence level, we would have:

$$\text{c.i.} = P_s \pm Z\sqrt{\frac{P_u(1 - P_u)}{N}}$$

$$\text{c.i.} = 0.48 \pm 1.96\sqrt{\frac{(0.5)(0.5)}{1425}}$$

$$\text{c.i.} = 0.48 \pm (1.96)(0.013)$$

$$\text{c.i.} = 0.48 \pm 0.03$$

Changing back to percentages, we can estimate that between 45% (48% − 3%) and 51% (48 + 3%) of Americans are married.

Exercises

7.1 Use the **Descriptives** command to calculate means and standard deviations for *papres80*, *age*, and *income06*. Use the output to formulate confidence interval estimates for the population values. Express the confidence intervals in words, as if you were reporting results in a newspaper story. *(NOTE: Your estimate based on age will not reflect the total American population. The GSS sample is restricted to people age 18 and older, and the mean of the sample is much higher than the mean of the population as a whole.)*

7.2 Use the **Frequencies** command to get sample proportions (convert the percentages in the frequency distributions) to estimate the population parameter for each of the following: proportion with college degree (*degree*), proportion black (*racecen1*), and proportion favoring gun control (*gunlaw*). Use the 95% confidence level. Express the confidence intervals in words, as if you were reporting results in a newspaper story.

8

Hypothesis Testing I
The One-Sample Case

LEARNING OBJECTIVES

By the end of this chapter, you will be able to

1. Explain the logic of hypothesis testing.
2. Define and explain the conceptual elements involved in hypothesis testing, especially the null hypothesis, the sampling distribution, the alpha level, and the test statistic.
3. Explain what it means to "reject the null hypothesis" or "fail to reject the null hypothesis."
4. Identify and cite examples of situations in which one-sample tests of hypotheses are appropriate.
5. Test the significance of single-sample means and proportions using the five-step model and correctly interpret the results.
6. Explain the difference between one- and two-tailed tests and specify when each is appropriate.
7. Define and explain Type I and Type II errors and relate each to the selection of an alpha level.

8.1 INTRODUCTION

Chapter 7 introduced the techniques for estimating population parameters from sample statistics. In Chapters 8 through 11, we investigate a second application of inferential statistics, called **hypothesis testing** or **significance testing.** In this chapter, the techniques for hypothesis testing in the one-sample case are introduced. These procedures could be used in situations such as the following:

1. A researcher has selected a sample of 789 older residents of a particular state and has asked them if they have been victimized by crime over the past year. He also has information on the percentage of the entire population of the state that was victimized by crime during the same time period. He wonders if senior citizens, as represented by this sample, are more likely to be victimized than the population in general.
2. Are the GPAs of college athletes different from the GPAs of the student body as a whole? To investigate, the academic performance of a random sample of 235 student athletes from a large state university are compared with the GPA of all students at the university.
3. A sociologist has been hired to assess the effectiveness of a rehabilitation program for alcoholics in her city. The program serves a large area, and she does not have the resources to test every single client. Instead, she draws a random sample of 127 people from the list of all clients and questions them on a variety of issues. She notices that, on the average, the people in her sample miss fewer days of work each year than the city as a whole. Her research question is "Are alcoholics treated by the program more reliable than workers in general?"

In each of these situations, we have randomly selected samples (of senior citizens, athletes, or treated alcoholics) that we want to compare to a population (the entire state, student body, or city). Note that we are not interested in the sample per se but in the larger group from which it was selected (*all* senior citizens in the state, *all* athletes on this campus, or *all* people who have completed the treatment program). Specifically, we want to know if the groups represented by the samples are different from the populations on a specific trait or variable (victimization rates, GPAs, or absenteeism).

Of course, it would be better if we could include all senior citizens, athletes, or treated alcoholics rather than these smaller samples. However, as we have seen, researchers usually do not have the resources necessary to test everyone in a large group and must use random samples instead. In these situations, conclusions will be based on a comparison of a single sample (representing the larger group) and the population. For example, if we found that the rate of victimization for the sample of senior citizens was higher than the rate for the state as a whole, we might conclude that "senior citizens are significantly more likely to be crime victims." The word *significantly* is a key word: It means that the difference between the sample's victimization rate and the population's rate is very unlikely to be caused by random chance alone. In other words, *all* senior citizens (not just the 789 people in the sample) have a higher victimization rate than the state as a whole. On the other hand, if we found little difference between the GPAs of the sample of athletes and the student body as a whole, we might conclude that athletes (*all* athletes on this campus, not just the athletes in the sample) are essentially the same as other students in terms of academic achievement.

Thus, we can use samples to represent larger groups (senior citizens, athletes, or treated alcoholics) and compare and contrast the characteristics of the sample with those of the population and be extremely confident in our conclusions. Remember, however, that the EPSEM procedure for drawing random samples does not guarantee representativeness; thus, there will always be a small amount of uncertainty in our conclusions. One of the great advantages of inferential statistics is that we will be able to estimate the probability of error and evaluate our decisions accordingly.

8.2 AN OVERVIEW OF HYPOTHESIS TESTING

We'll begin with a general overview of hypothesis testing, using the third research situation mentioned earlier as an example, and introduce the more technical considerations and proper terminology throughout the remainder of the chapter. Let's examine this situation in some detail. First of all, the main question here is "Are people treated in this program more reliable workers than people in the community in general?" In other words, what the researcher would really like to do is compare information gathered from *all* clients (the *population* of alcoholics treated in the program) with information about the entire metropolitan area. If she had information for both of these groups (all clients and the entire metro population), she could answer the question easily, completely, and finally.

The problem is that the researcher does not have the time and or money to gather information on the thousands of people who have been treated by the program. Instead, she has drawn a random sample, following the rule of EPSEM, of 127 clients from agency records. The absentee rates for the sample and the community are:

Community	Sample of Treated Alcoholics
$\mu = 7.2$ days per year	$\bar{X} = 6.8$ days per year
$\sigma = 1.43$	$N = 127$

We can see that there is a difference in rates of absenteeism and that the average rate of absenteeism for the sample is lower than the rate for the community. Although it's tempting, we can't make any conclusions yet because we are working with a random sample of the population we are interested in, not the population itself (all people treated in the program).

Figure 8.1 should clarify these relationships. The community is symbolized by the largest circle because it is the largest group. The population of all treated alcoholics is also symbolized by a large circle because it is a sizeable group, although only a small fraction of the community as a whole. The random sample of 127, the smallest of the three groups, is symbolized by the smallest circle.

The labels on the arrows connecting the circles summarize the major questions and connections in this research situation. As we noted earlier, the main question the researcher wants to answer is "Does the population of *all* treated alcoholics have different absentee rates than the community as a whole?" The population of treated alcoholics, too large to test, is represented by the randomly selected sample of 127. The main question related to the sample concerns the cause of the observed difference between its mean of 6.8 and the community mean of 7.2. There are two possible explanations for this difference, and we consider them one at a time.

The first explanation is that the difference between the community mean of 7.2 days and the sample mean of 6.8 days reflects a real difference in absentee rates between the population of all treated alcoholics and the community. The

FIGURE 8.1 A TEST OF HYPOTHESIS FOR SINGLE-SAMPLE MEANS

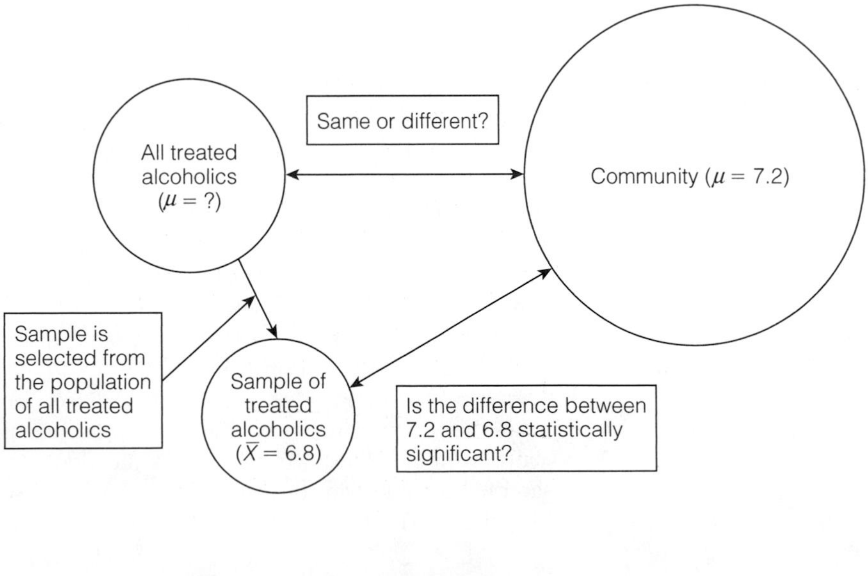

difference is "statistically significant," in the sense that it is very unlikely to have occurred by random chance alone. If this explanation is true, the population of all treated alcoholics is different from the community and the sample did *not* come from a population with a mean absentee rate of 7.2 days.

The second explanation is called the **null hypothesis** (symbolized as $\boldsymbol{H}_0$, or H-sub-zero). It states that the observed difference between sample and community means was caused by mere random chance. In other words, there is no important difference between treated alcoholics and the community as a whole, and the difference between the sample mean of 6.8 days and the community mean of 7.2 days is trivial and due to random chance. If the null hypothesis is true, the population of treated alcoholics is just like everyone else and has a mean absentee rate of 7.2 days.

Which explanation is correct? We cannot answer this question with absolute (100%) certainty as long as we are working with a sample rather than the entire group. We can, however, set up a decision-making procedure so conservative that one of these two explanations can be chosen, with the knowledge that the probability of choosing the incorrect explanation is very low.

This decision-making process, in broad outline, begins with the assumption that the second explanation—the null hypothesis—is correct. Symbolically, the assumption that the mean absentee rate for all treated alcoholics is the same as the rate for the community as a whole can be stated as

$$H_0\text{: } \mu = 7.2 \text{ days per year}$$

Remember that this μ refers to the mean for *all* treated alcoholics, not just the 127 in the sample. This assumption, $\mu = 7.2$, can be tested statistically.

If the null hypothesis ("*The population of treated alcoholics is not different from the community as a whole and has a* μ *of 7.2*") is true, then the probability of getting the observed sample outcome ($\bar{X} = 6.8$) can be found. Let us add an objective decision rule in advance. If the odds of getting the observed difference are less than .05 (5 out of 100, or 1 in 20), we will reject the null hypothesis. If this explanation were true, a difference of this size (7.2 days vs. 6.8 days) would be a very rare event, and in hypothesis testing we always bet *against* rare events.

How can we estimate the probability of the observed sample outcome ($\bar{X} = 6.8$) if the null hypothesis is correct? This value can be determined by using our knowledge of the sampling distribution of all possible sample outcomes. Looking back at the information we have and applying the Central Limit Theorem (see Chapter 6), we can assume that the sampling distribution is normal in shape, has a mean of 7.2 (because $\mu_{\bar{X}} = \mu$) and a standard deviation of $1.43/\sqrt{127}$ because $\sigma_{\bar{X}} = \sigma/\sqrt{N}$. We also know that the standard normal distribution can be interpreted as a distribution of probabilities (see Chapter 5) and that the particular sample outcome noted earlier ($\bar{X} = 6.8$) is one of thousands of possible sample outcomes. The sampling distribution, with the sample outcome noted, is depicted in Figure 8.2.

Using our knowledge of the standardized normal distribution, we can add further useful information to this sampling distribution of sample means. Specifically, with Z scores, we can depict the decision rule stated previously: Any sample outcome with probability less than 0.05 will cause us to reject the null hypothesis. The probability of 0.05 can be translated into an area and divided

FIGURE 8.2 THE SAMPLING DISTRIBUTION OF ALL POSSIBLE SAMPLE MEANS

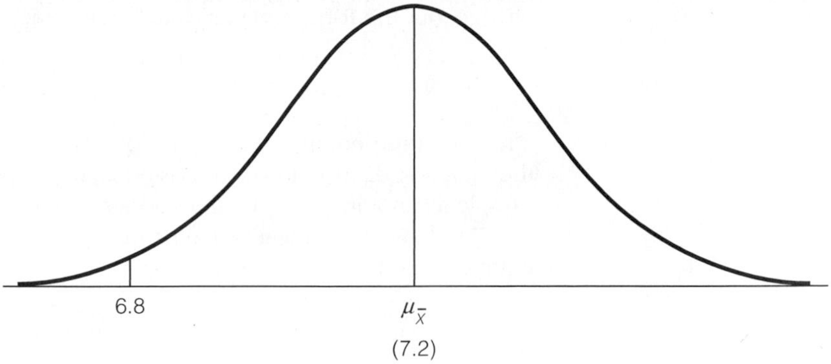

equally into the upper and lower tails of the sampling distribution. Using Appendix A, we find that the Z-score equivalent of this area is ± 1.96. (To review finding Z scores from areas or probabilities, see Section 7.3). The areas and Z scores are depicted in Figure 8.3.

The decision rule can now be rephrased. Any sample outcome falling in the shaded areas depicted in Figure 8.3 has a probability of occurrence of less than 0.05. Such an outcome would be a rare event and would cause us to reject the null hypothesis.

All that remains is to translate our sample outcome into a Z score to see where it falls on the curve. To do this, we use the standard formula for locating any particular raw score under a normal distribution. When we use known or empirical distributions, this formula is expressed as

$$Z = \frac{X_i - \bar{X}}{s}$$

Or, to find the equivalent Z score for any raw score, subtract the mean of the distribution from the raw score and divide by the standard deviation of the distribution.

FIGURE 8.3 THE SAMPLING DISTRIBUTION OF ALL POSSIBLE SAMPLE MEANS

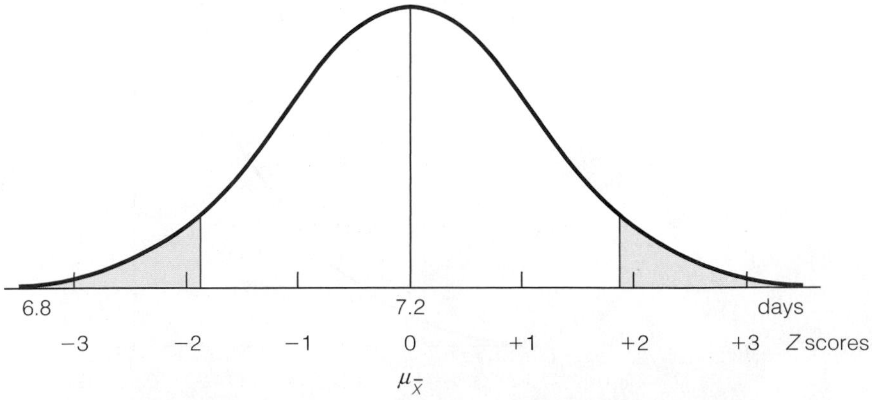

Since we are now concerned with the sampling distribution of all sample means rather than an empirical distribution, the symbols in the formula will change, but the form remains exactly the same:

FORMULA 8.1

$$Z = \frac{\bar{X} - \mu}{\sigma/\sqrt{N}}$$

Or, to find the equivalent Z score for any sample mean, subtract the mean of the sampling distribution, which is equal to the population mean, or μ, from the sample mean and divide by the standard deviation of the sampling distribution.

Recalling the data given on this problem, we can now find the Z-score equivalent of the sample mean:

$$Z = \frac{6.8 - 7.2}{1.43/\sqrt{127}}$$

$$Z = \frac{-0.40}{0.127}$$

$$Z = -3.15$$

In Figure 8.4, this Z score of -3.15 is noted on the distribution of all possible sample means, and we see that the sample outcome does fall in the shaded area. If the null hypothesis is true, this particular sample outcome has a probability of occurrence of less than 0.05. The sample outcome ($\bar{X} = 6.8$, or $Z = -3.15$) would be rare if the null hypothesis was true, and the researcher may therefore reject the null hypothesis. The sample of 127 treated alcoholics comes from a population that is significantly different from the community on the trait of absenteeism. Or, to put it another way, the sample does not come from a population that has a mean of 7.2 days of absences.

Keep in mind that our decisions in significance testing are based on information gathered from random samples. On rare occasions, an EPSEM sample may not be representative of the population from which it was selected. The decision-making process just outlined has a very high probability of resulting in correct decisions, but we always face an element of risk as long as we must work with samples rather than populations. That is, the decision to reject the null hypothesis might be incorrect if this sample happens to be one of the few that is unrepresentative of the population of alcoholics treated in this program. One important strength of hypothesis testing is that we can estimate the proba-

FIGURE 8.4 THE SAMPLING DISTRIBUTION OF SAMPLE MEANS, WITH THE SAMPLE OUTCOME ($\bar{X} = 6.8$) NOTED IN Z SCORES

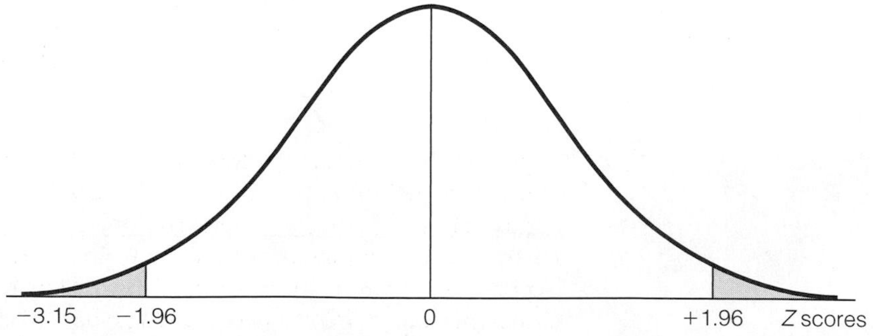

bility of making an incorrect decision. In the example at hand, the null hypothesis was rejected and the probability that this decision was incorrect is 0.05—the decision rule established at the beginning of the process. To say that the probability of rejecting the null hypothesis incorrectly is 0.05 means that, if we repeated this same test an infinite number of times, we would incorrectly reject the null hypothesis only 5 times out of every 100.

8.3 THE FIVE-STEP MODEL FOR HYPOTHESIS TESTING

All of the formal elements and concepts used in hypothesis testing were introduced in the preceding discussion. This section presents their proper names and introduces a **five-step model** for organizing all hypothesis testing.

Step 1. Making assumptions and meeting test requirements

Step 2. Stating the null hypothesis

Step 3. Selecting the sampling distribution and establishing the critical region

Step 4. Computing the test statistic

Step 5. Making a decision and interpreting the results of the test

We now look at each step individually, using the problem from Section 8.2 as an example throughout.

Step 1. Making Assumptions and Meeting Test Requirements. Any application of statistics requires that certain assumptions be made. Specifically, three assumptions about the testing situation and the variables involved have to be satisfied when conducting a test of hypothesis with a single-sample mean. First, we must be sure that we are working with a random sample, one that has been selected according to the rules of EPSEM (see Chapter 6). Second, to justify computation of a mean, we must assume that the variable being tested is interval-ratio in level of measurement. Finally, we must assume that the sampling distribution of all possible sample means is normal in shape so that we may use the standardized normal distribution to find areas under the sampling distribution. We can be sure that this assumption is satisfied by using large samples (see the Central Limit Theorem in Chapter 6).

Usually, we will state these assumptions in abbreviated form as a mathematical model for the test. For example:

Model: Random sampling
Level of measurement is interval-ratio
Sampling distribution is normal

Step 2. Stating the Null Hypothesis (H_0). The null hypothesis is always a statement of "no difference," but its exact form will vary depending on the test being conducted. In the single-sample case, the null hypothesis states that the sample comes from a population with a certain characteristic. In our example, the null hypothesis is that the population of treated alcoholics is "no different" from the community as a whole, that their average days of absenteeism is also 7.2, and that

ONE STEP AT A TIME

Testing the Significance of the Difference Between a Sample Mean and a Population Mean: Computing Z(obtained) and Interpreting Results

Use these procedures if the population standard deviation (σ) is known or sample size (N) is greater than 100. See Section 8.6 for procedures when σ is unknown and N is less than 100.

Step 4: Computing Z(obtained)

Use Formula 8.1 to compute the test statistic.

Step 1: Find the square root of N.

Step 2: Divide the square root of N into the population standard deviation (σ).

Step 3: Subtract the population mean (μ) from the sample mean ($\bar{X}$).

Step 4: Divide the quantity you found in step 3 by the quantity you found in step 2. This value is *Z*(obtained).

Step 5: Making a Decision and Interpreting the Test Result

Step 5: Compare the *Z*(obtained) you computed in step 4 to your *Z*(critical). If *Z*(obtained) is *in* the critical region, *reject* the null hypothesis. If *Z*(obtained) is *not in* the critical region, *fail to reject* the null hypothesis.

Step 6: Interpret the decision to reject or fail to reject the null hypothesis in the terms of the original question. For example, our conclusion for the example problem used in Section 8.3 was "Treated alcoholics miss significantly fewer days of work than the community as a whole."

the difference between 7.2 and the sample mean of 6.8 is caused by random chance. As we saw previously, the null hypothesis would be stated as

$$H_0: \mu = 7.2$$

where μ refers to the mean of the population of treated alcoholics. The null hypothesis is the central element in any test of hypothesis because the entire process is aimed at rejecting or failing to reject the H_0.

Usually, the researcher believes there is a significant difference and desires to reject the null hypothesis. At this point in the five-step model, the researcher's belief is stated in a **research hypothesis (H_1),** a statement that directly contradicts the null hypothesis. Thus, the researcher's goal in hypothesis testing is often to gather evidence for the research hypothesis by rejecting the null hypothesis.

The research hypothesis can be stated in several ways. One form would simply assert that the population from which the sample was selected did *not* have a certain characteristic or, in terms of our example, had a mean that was *not* equal to a specific value:

$$(H_1: \mu \neq 7.2)$$

where $\neq$ means "not equal to"

Symbolically, this statement asserts that the sample does not come from a population with a mean of 7.2, or that the population of all treated alcoholics is different from the community as a whole.

The research hypothesis is enclosed in parentheses to emphasize that it has no formal standing or role in the hypothesis-testing process (except, as we shall see in the next section, in choosing between one-tailed and two-tailed tests). It serves as a reminder of what the researcher believes to be the truth.

Step 3. Selecting the Sampling Distribution and Establishing the Critical Region. The sampling distribution is, as always, the probabilistic yardstick against which a particular sample outcome is measured. By assuming that the null hypothesis is true (and *only* by this assumption), we can attach values to the mean and standard deviation of the sampling distribution and thus measure the probability of any specific sample outcome. There are several different sampling distributions, but for now we will confine our attention to the sampling distribution described by the standard normal curve, as summarized in Appendix A.

The **critical region** consists of the areas under the sampling distribution that include unlikely sample outcomes. Prior to the test of hypothesis, we must define what we mean by *unlikely*. That is, we must specify in advance those sample outcomes so unlikely that they will lead us to reject the H_0. This decision rule will establish the critical region, or **region of rejection.** The word *region* is used because we are describing those areas under the sampling distribution that contain unlikely sample outcomes. In our earlier example, this area corresponded to a Z score of ± 1.96, called **Z(critical),** that was graphically displayed in Figure 8.3. The shaded area is the critical region. Any sample outcome for which the Z-score equivalent fell in this area (that is, below -1.96 or above $+1.96$) would have caused us to reject the null hypothesis.

By convention, the size of the critical region is reported as alpha (α), the proportion of all of the area included in the critical region. In our example, our **alpha level** was 0.05. Other commonly used alphas are 0.10, 0.01, 0.001, and 0.0001.

In abbreviated form, all the decisions made in this step are as follows. The critical region is noted by the Z scores that mark its beginnings.

$$\text{Sampling distribution} = Z \text{ distribution}$$
$$\alpha = 0.05$$
$$Z(\text{critical}) = \pm 1.96$$

(For practice in finding Z(critical) scores, see problem 8.1a.)

Step 4. Computing the Test Statistic. To evaluate the probability of the sample outcome, the sample value must be converted into a Z score. Solving the equation for Z-score equivalents is called computing the **test statistic,** and the resultant value will be referred to as **Z(obtained),** in order to differentiate the test statistic from the score that marks the beginning of the critical region. In our example, we found a Z(obtained) of -3.15. *(For practice in computing obtained Z scores for means, see problems 8.1c, 8.2 to 8.7, and 8.15e and f.)*

Step 5. Making a Decision and Interpreting the Results of the Test. Finally, the test statistic is compared with the critical region. If the test statistic falls into the critical region, our decision will be to reject the null hypothesis. If the test statistic does not fall into the critical region, we fail to reject the null hypothesis. In our example, the two values were

$$Z(\text{critical}) = \pm 1.96$$
$$Z(\text{obtained}) = -3.15$$

and we saw that the Z(obtained) fell in the critical region (see Figure 8.4). Our decision was to reject the null hypothesis ("Treated alcoholics have a mean absentee rate of 7.2 days"). When we reject the null hypothesis, we are saying that treated alcoholics do *not* have a mean absentee rate of 7.2 days and that there

TABLE 8.1 MAKING A DECISION IN STEP 5 AND INTERPRETING THE RESULTS OF THE TEST

Situation	Decision	Interpretation
The test statistic is *in* the critical region	Reject the null hypothesis (H_0)	The difference is statistically significant
The test statistic is *not* in the critical region	Fail to reject the null hypothesis (H_0)	The difference is not statistically significant

is a difference between them and the community. We can also say that the difference between the sample mean of 6.8 and the community mean of 7.2 is statistically significant, or unlikely to be caused by random chance alone.

Note that, in order to complete step 5, you have to do two things. First you make a decision about the null hypothesis: If the test statistic falls in the critical region, reject H_0. If the test statistic does not fall in the critical region, we fail to reject the H_0. The decision in step 5 is summarized in Table 8.1.

Second, you need to say what our decision means. In this case, the null hypothesis was rejected, which means that there is a significant difference between the mean of the sample and the mean for the entire community and, therefore, we can conclude that treated alcoholics are different from the community as a whole.

This five-step model will serve us as a framework for decision making throughout the hypothesis-testing chapters. The exact nature and method of expression for our decisions will be different for different situations. However, familiarity with the five-step model will assist you in mastering this material by providing a common frame of reference for all significance testing.

8.4 ONE-TAILED AND TWO-TAILED TESTS OF HYPOTHESIS

The five-step model for hypothesis testing is fairly rigid, and the researcher has little room for making choices. Follow the steps one by one as specified in Section 8.3, and you cannot make a mistake. Nonetheless, the researcher must still make two crucial choices. First, he or she must decide between a one-tailed and a two-tailed test. Second, an alpha level must be selected. In this section we discuss the former decision; in Section 8.5 we discuss the latter.

Choosing a One- or Two-Tailed Test. The choice between a one- and two-tailed test is based on the researcher's expectations about the population from which the sample was selected. These expectations are reflected in the research hypothesis (H_1), which is contradictory to the null hypothesis and usually states what the researcher believes to be "the truth." In most situations, the researcher will wish to support the research hypothesis by rejecting the null hypothesis.

The format for the research hypothesis may take either of two forms, depending on the relationship between what the null hypothesis states and what the researcher believes to be the truth. The null hypothesis states that the population has a specific characteristic. In the example that has served us throughout this chapter, the null stated, in symbols, "All treated alcoholics have the *same* absentee rate (7.2 days) as the community." The researcher might believe that the population of treated alcoholics actually has *less* absenteeism (their population mean is *lower than* the value stated in the null hypothesis) or *more* absenteeism (their population mean is *greater than* the value stated in the null hypothesis), or he or she might be unsure about the direction of the difference.

If the researcher is unsure about the direction, the research hypothesis would state only that the population mean is "not equal" to the value stated in the null hypothesis. The research hypothesis stated in Section 8.3 ($\mu \neq 7.2$) was in this format. This is called a **two-tailed test** of significance because it means that the researcher will be equally concerned with the possibility that the true population value is greater than the value specified in the null hypothesis and the possibility that the true population value is less than the value specified in the null hypothesis.

In other situations, the researcher might be concerned only with differences in a specific direction. If the direction of the difference can be predicted or if the researcher is concerned only with differences in one direction, a **one-tailed test** can be used. A one-tailed test may take one of two forms, depending on the researcher's expectations about the direction of the difference. If the researcher believes that the true population value is greater than the value specified in the null hypothesis, the research hypothesis would reflect that belief. In our example, if we had predicted that treated alcoholics had *higher* absentee rates than the community (or averaged *more* days of absenteeism than 7.2), our research hypothesis would have been

$$(H_1: \mu > 7.2)$$

where > signifies "greater than"

If we predicted that treated alcoholics had lower absentee rates than the community (or averaged *fewer* days of absenteeism than 7.2), our research hypothesis would have been

$$(H_1: \mu < 7.2)$$

where < signifies "less than"

One-tailed tests are often appropriate when programs designed to solve a problem or improve a situation are being evaluated. If the program for treating alcoholics made them *less* reliable workers with higher absentee rates, for example, the program would be considered a failure, at least on that criterion. In this situation, the researcher may well focus only on outcomes that would indicate that the program is a success (i.e., when treated alcoholics have lower rates) and conduct a one-tailed test with a research hypothesis in the form: H_1: $\mu < 7.2$. Or consider the evaluation of a program designed to reduce unemployment. The evaluators would be concerned only with outcomes that show a decrease in the unemployment rate. If the rate shows no change or if unemployment increases, the program is a failure, and both of these outcomes might be considered equally negative by the researchers. Thus, the researchers could legitimately use a one-tailed test that stated that unemployment rates for graduates of the program would be less than (<) rates in the community.

One- vs. Two-Tailed Tests and Step 3 of the Five-Step Model. In terms of the five-step model, the choice of a one-tailed or two-tailed test determines what we do with the critical region under the sampling distribution in step 3. As you recall, in a two-tailed test, we split the critical region equally into the upper and lower tails of the sampling distribution. In a one-tailed test, we place the entire critical area in one tail of the sampling distribution. If we believe that the population characteristic is greater than the value stated in the null hypothesis (if the H_1 includes the > symbol), we place the entire critical region in the upper tail. If we believe that the characteristic is less than the value stated in the null hypothesis (if the H_1 includes the < symbol), the entire critical region goes in the lower tail.

For example, in a two-tailed test with alpha equal to 0.05, the critical region begins at $Z(\text{critical}) = \pm 1.96$. In a one-tailed test at the same alpha level, the Z(critical) is $+1.65$ if the upper tail is specified and -1.65 if the lower tail is specified. Table 8.2 summarizes the procedures to follow in terms of the nature of the research hypothesis. The difference in placing the critical region is graphically summarized in Figure 8.5, and the critical Z scores for the most common alpha levels are given in Table 8.3 for both one- and two-tailed tests.

Note that the critical Z values for one-tailed tests at all values of alpha are closer to the mean of the sampling distribution. Thus, a one-tailed test is more likely to reject the H_0 without changing the alpha level (assuming that we have specified the correct tail). One-tailed tests are a way of statistically both having and eating your cake and should be used whenever (1) the direction of the difference can be confidently predicted or (2) the researcher is concerned only with differences in one tail of the sampling distribution. An example should clarify these procedures.

A Test of Hypothesis Using a One-tailed Test. A sociologist has noted that sociology majors seem more sophisticated, charming, and cosmopolitan than the rest of the student body. A "Sophistication Scale" test has been administered to the entire student body and to a random sample of 100 sociology majors, and these results have been obtained:

Student Body	Sociology Majors
$\mu = 17.3$	$\bar{X} = 19.2$
$\sigma = 7.4$	$N = 100$

TABLE 8.2 ONE- VS. TWO-TAILED TESTS, $\alpha = .05$

If the Research Hypothesis Uses	The Test Is	And Concern Is with	Z(critical) =
$\neq$	Two-tailed	Both tails	± 1.96
$>$	One-tailed	Upper tail	$+1.65$
$<$	One-tailed	Lower tail	-1.65

TABLE 8.3 FINDING CRITICAL Z SCORES FOR ONE-TAILED TESTS (single-sample means)

		One-Tailed Value	
Alpha	Two-Tailed Value	Upper Tail	Lower Tail
.10	± 1.65	$+1.29$	-1.29
.05	± 1.96	$+1.65$	-1.65
.01	± 2.58	$+2.33$	-2.33
.001	± 3.29	$+3.10$	-3.10
.0001	± 3.90	$+3.70$	-3.70

FIGURE 8.5 ESTABLISHING THE CRITICAL REGION, ONE-TAILED TESTS VERSUS TWO-TAILED TESTS ($\alpha = 0.05$)

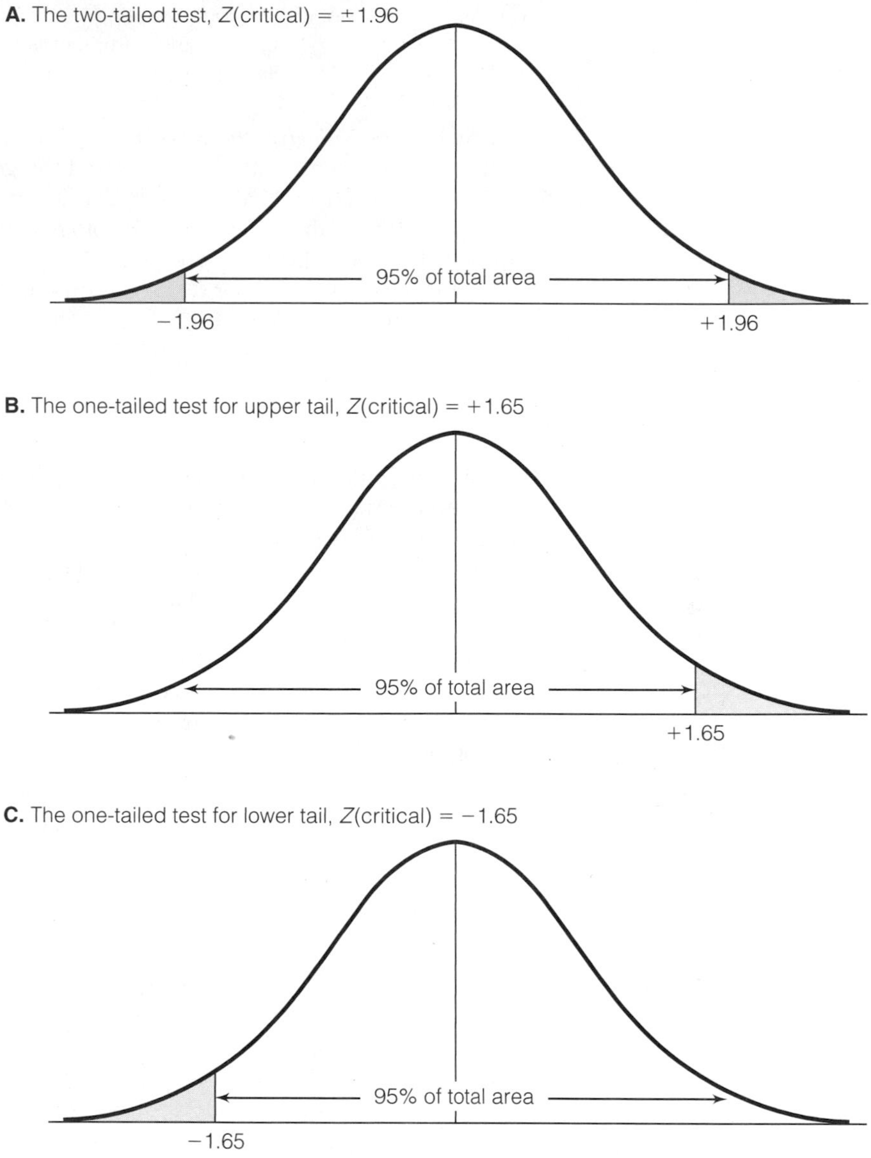

We will use the five-step model to test the H_0 of no difference between sociology majors and the general student body.

Step 1. Making Assumptions and Meeting Test Requirements. Since we are using a mean to summarize the sample outcome, we must assume that the Sophistication Scale generates interval-ratio-level data. With a sample size of 100, the Central Limit Theorem applies, and we can assume that the sampling distribution is normal in shape.

Model: Random sampling
Level of measurement is interval-ratio
Sampling distribution is normal

Step 2. Stating the Null Hypothesis (H_0). The null hypothesis states that there is no difference between sociology majors and the general student body. The research hypothesis (H_1) will also be stated at this point. The researcher has predicted a direction for the difference ("Sociology majors are *more* sophisticated"), so a one-tailed test is justified. The one-tailed research hypothesis asserts that sociology majors have a higher (>) score on the Sophistication Scale. The two hypotheses may be stated as

$$H_0: \mu = 17.3$$

$$(H_1: \mu > 17.3)$$

Step 3. Selecting the Sampling Distribution and Establishing the Critical Region. We will use the standardized normal distribution (Appendix A) to find areas under the sampling distribution. If alpha is set at 0.05, the critical region will begin at the Z score +1.65. That is, the researcher has predicted that sociology majors are *more* sophisticated and that this sample comes from a population that has a mean *greater than* 17.3, so he or she will be concerned only with sample outcomes in the upper tail of the sampling distribution. If sociology majors are *the same as* other students in terms of sophistication (if the H_0 is true) or if they are *less* sophisticated (and come from a population with a mean less than 17.3), the theory is disproved. These decisions may be summarized as

$$\text{Sampling distribution} = Z \text{ distribution}$$
$$\alpha = 0.05$$
$$Z(\text{critical}) = +1.65$$

Step 4. Computing the Test Statistic.

$$Z(\text{obtained}) = \frac{\bar{X} - \mu}{\sigma/\sqrt{N}}$$

$$Z(\text{obtained}) = \frac{19.2 - 17.3}{7.4/\sqrt{100}}$$

$$Z(\text{obtained}) = +2.57$$

Step 5. Making a Decision and Interpreting Test Results. Comparing the Z(obtained) with the Z(critical):

$$Z(\text{critical}) = +1.65$$
$$Z(\text{obtained}) = +2.57$$

We see that the test statistic falls into the critical region. This outcome is depicted graphically in Figure 8.6. We will reject the null hypothesis because, if the H_0 were true, a difference of this size would be very unlikely. There is a significant difference between sociology majors and the general student body in terms of sophistication. Since the null hypothesis has been rejected, the research hy-

FIGURE 8.6 Z(OBTAINED) VERSUS Z(CRITICAL) (α = 0.05, one tailed test)

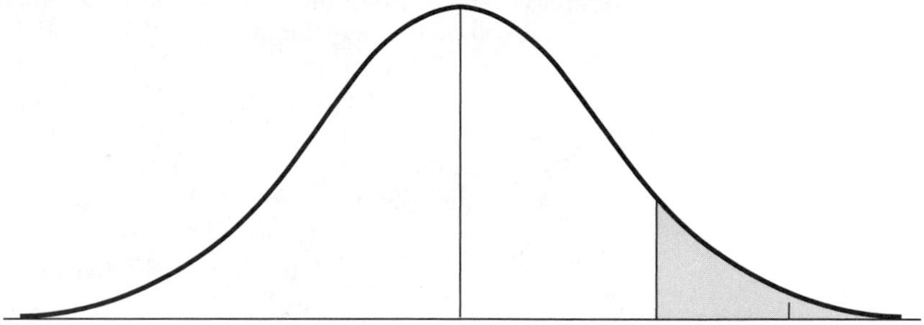

pothesis (sociology majors are more sophisticated) is supported. *(For practice in dealing with tests of significance for means that may call for one-tailed tests, see problems 8.2, 8.3, 8.6, 8.8, and 8.17).*

8.5 SELECTING AN ALPHA LEVEL

In addition to deciding between one-tailed and two-tailed tests, the researcher must select an alpha level. We have seen that the alpha level plays a crucial role in hypothesis testing. When we assign a value to alpha, we define what we mean by an "unlikely" sample outcome. If the probability of the observed sample outcome is lower than the alpha level (if the test statistic falls into the critical region), then we reject the null hypothesis as untrue. Thus, the alpha level will have important consequences for our decision in step 5.

How can reasonable decisions be made with respect to the value of alpha? Recall that, in addition to defining what will be meant by *unlikely,* the alpha level is the probability that the decision to reject the null hypothesis if the test statistic falls into the critical region will be incorrect. In hypothesis testing, the error of incorrectly rejecting the null hypothesis or rejecting a null hypothesis that is actually true is called **Type I error,** or **alpha error.** To minimize this type of error, use very small values for alpha.

To elaborate: When an alpha level is specified, the sampling distribution is divided into two sets of possible sample outcomes. The critical region includes all unlikely or rare sample outcomes. Outcomes in this region will cause us to reject the null hypothesis. The remainder of the area consists of all sample outcomes that are not rare. The lower the level of alpha, the smaller the critical region and the greater the distance between the mean of the sampling distribution and the beginnings of the critical region. Compare, for the sake of illustration, the following alpha levels and values for Z(critical) for two-tailed tests. As you may recall, Table 7.1 also presented this information.

If Alpha Equals	The Two-Tailed Critical Region Will Begin at Z(critical) Equal to
0.10	±1.65
0.05	±1.96
0.01	±2.58
0.001	±3.29

As alpha goes down, the critical region becomes smaller and moves farther away from the mean of the sampling distribution. The lower the alpha level, the harder it will be to reject the null hypothesis and, because a Type I error can occur only if our decision in step 5 is to reject the null hypothesis, the lower the probability of Type I error. To minimize the probability of rejecting a null hypothesis that is in fact true, use very low alpha levels.

However, there is a complication. As the critical region decreases in size (as alpha levels decrease), the noncritical region—the area between the two Z(critical) scores in a two-tailed test—becomes larger. All other things being equal, the lower the alpha level, the less likely that the sample outcome will fall into the critical region. This raises the possibility of a second type of incorrect decision, called **Type II error,** or **beta error:** failing to reject a null that is, in fact, false. The probability of Type I error decreases as the alpha level decreases, but the probability of Type II error increases. Thus, the two types of error are inversely related, and it is not possible to minimize both in the same test. As the probability of one type of error decreases, the other increases, and vice versa.

It may be helpful to clarify in table format the relationships between decision making and errors. Table 8.4 lists the two decisions we can make in step 5 of the five-step model: We either reject or fail to reject the null hypothesis. The other dimension of Table 8.4 lists the two possible conditions of the null hypothesis: It is either actually true or actually false. The table combines these possibilities into a total of four possible outcomes, two of which are desirable ("OK") and two of which indicate that an error has been made.

Let's consider the two desirable ("OK") outcomes first. We want to reject false null hypotheses and fail to reject true null hypotheses. The goal of any scientific investigation is to verify true statements and reject false statements.

The remaining two combinations are errors or situations we wish to avoid. If we reject a null hypothesis that is actually true, we are saying that a true statement is false. Likewise, if we fail to reject a null hypothesis that is actually false, we are saying that a false statement is true. Obviously, we would always prefer to wind up in one of the boxes labeled "OK" in Table 8.4—always to reject false statements and to accept the truth when we find it. Remember, however, that hypothesis testing always carries an element of risk and that it is not possible to minimize the chances of both Type I and Type II errors simultaneously.

What all of this means, finally, is that you must think of selecting an alpha level as an attempt to balance the two types of error. Higher alpha levels will minimize the probability of Type II error (saying that false statements are true), and lower alpha levels will minimize the probability of Type I error (saying that true statements are false). Normally, in social science research, we will want to

TABLE 8.4 DECISION MAKING AND THE NULL HYPOTHESIS

The H_0 Is Actually:	Decision: Reject	Decision: Fail to Reject
True	Type I, or α, error	OK
False	OK	Type II, or β, error

FIGURE 8.7 THE *t* DISTRIBUTION AND THE *Z* DISTRIBUTION

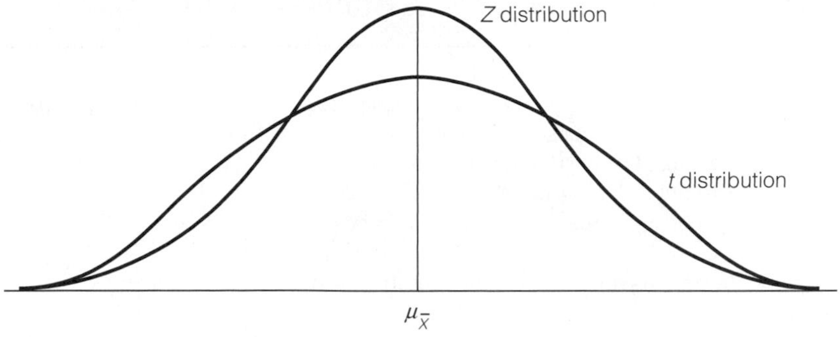

minimize Type I error, and lower alpha levels (.05, .01, .001 or lower) will be used. The 0.05 level in particular has emerged as a generally recognized indicator of a significant result. However, the widespread use of the 0.05 level is simply a convention, and there is no reason that alpha cannot be set at virtually any sensible level (such as 0.04, 0.027, 0.083). The researcher has the responsibility of selecting the alpha level that seems most reasonable in terms of the goals of the research project.

8.6 THE STUDENT'S *t* DISTRIBUTION

To this point, we have considered only one type of hypothesis test. Specifically, we have focused on situations involving single-sample means where the value of the population standard deviation (σ) was known. Needless to say, in most research situations, the value of σ will not be known. However, a value for σ is required in order to compute the standard error of the mean (σ/N), convert our sample outcome into a *Z* score, and place the *Z*(obtained) on the sampling distribution (step 4). How can we reasonably obtain a value for the population standard deviation?

It might seem sensible to estimate σ with *s*, the sample standard deviation. As we noted in Chapter 7, *s* is a biased estimator of σ, but the degree of bias decreases as sample size increases. For large samples (that is, samples with 100 or more cases), the sample standard deviation yields an adequate estimate of σ. Thus, for large samples, we simply substitute *s* for σ in the formula for *Z*(obtained) in step 4 and continue to use the standard normal curve to find areas under the sampling distribution.[1]

For smaller samples, however, when σ is unknown, an alternative distribution called the **Student's *t* distribution** must be used to find areas under the sampling distribution and establish the critical region. The shape of the *t* distribution varies as a function of sample size. The relative shapes of the *t* and *Z* distributions are depicted in Figure 8.7. For small samples, the *t* distribution is much flatter than the *Z* distribution, but, as sample size increases, the *t* distribution comes to resemble the *Z* distribution more and more until the two are

[1]Even though its effect will be minor and will decrease with sample size, we will always correct for the bias in *s* by using the term $N - 1$ rather than N in the computation for the standard deviation of the sampling distribution when σ is unknown.

Application 8.1

For a random sample of 152 felony cases tried in a local court, the average prison sentence was 27.3 months. Is this significantly different from the average prison term for felons nationally? We will use the five-step model to organize the decision-making process.

Step 1. Making Assumptions and Meeting Test Requirements.

Model: Random sampling
Level of measurement is interval-ratio
Sampling distribution is normal

From the information given (this is a large sample with $N > 100$, and length of sentence is an interval-ratio variable), we can conclude that the model assumptions are satisfied.

Step 2. Stating the Null Hypothesis (H_0). The null hypothesis would say that the average sentence locally (for *all* felony cases) is equal to the national average. In symbols:

$$H_0: \mu = 28.7$$

The research question does not specify a direction; it only asks if the local sentences are "different from" (not higher or lower than) national averages. This suggests a two-tailed test:

$$(H_1: \mu \neq 28.7)$$

Step 3. Selecting the Sampling Distribution and Establishing the Critical Region.

$$\text{Sampling distribution} = Z \text{ distribution}$$
$$\alpha = 0.05$$
$$Z(\text{critical}) = \pm 1.96$$

Step 4. Computing the Test Statistic. The necessary information for conducting a test of the null hypothesis is

$$\bar{X} = 27.3 \qquad \mu = 28.7$$
$$s = 3.7$$
$$N = 152$$

The test statistic, Z(obtained), would be

$$Z(\text{obtained}) = \frac{\bar{X} - \mu}{s/\sqrt{N-1}}$$

$$Z(\text{obtained}) = \frac{27.3 - 28.7}{3.7/\sqrt{152 - 1}}$$

$$Z(\text{obtained}) = \frac{-1.40}{3.7/\sqrt{151}}$$

$$Z(\text{obtained}) = \frac{-1.40}{0.30}$$

$$Z(\text{obtained}) = -4.67$$

Step 5. Making a Decision and Interpreting the Test Results. With alpha set at 0.05, the critical region begins at $Z(\text{critical}) = \pm 1.96$. With an obtained Z score of -4.67, the null would be rejected. This means that the difference between the prison sentences of felons convicted in the local court and felons convicted nationally is statistically significant. The difference is so large that we may conclude that it did not occur by random chance. The decision to reject the null hypothesis has a 0.05 probability of being wrong.

essentially identical when sample size is greater than 120. As N increases, the sample standard deviation (s) becomes a more and more adequate estimator of the population standard deviation (σ), and the t distribution becomes more and more like the Z distribution.

The Distribution of t: Using Appendix B. The t distribution is summarized in Appendix B. The t table differs from the Z table in several ways. First, there is a column at the left of the table labeled *df* for "degrees of free-

ONE STEP AT A TIME

Testing the Significance of the Difference Between a Sample Mean and a Population Mean Using Student's *t* Distribution: Computing *t*(obtained) and Interpreting Results

Use these procedures if the population standard deviation (σ) is unknown and sample size (N) is less than 100. See Section 8.3 for procedures when σ is known or N is more than 100.

Step 4: Computing *t*(obtained)

Use Formula 8.2 to compute the test statistic.

Step 1: Find the square root of $N - 1$

Step 2: Divide the quantity you found in step 1 into the sample standard deviation (s).

Step 3: Subtract the population mean (μ) from the sample mean ($\overline{X}$)

Step 4: Divide the quantity you found in step 3 by the quantity you found in step 2. This value is *t*(obtained).

Step 5: Making a Decision and Interpreting the Test Result

Step 5: Compare the *t*(obtained) you computed in step 4 to your *t*(critical). If *t*(obtained) is *in* the critical region, *reject* the null hypothesis. If *t*(obtained) is *not in* the critical region, *fail to reject* the null hypothesis.

Step 6: Interpret the decision to reject or fail to reject the null hypothesis in terms of the original question. For example, our conclusion for the example problem used in Section 8.6 was "There is no significant difference between the average GPAs of commuter students and the general student body."

dom."[2] As just mentioned, the exact shape of the *t* distribution and thus the exact location of the critical region for any alpha level varies as a function of sample size. Degrees of freedom, which are equal to $N - 1$ in the case of a single-sample mean, must first be computed before the critical region for any alpha can be located. Second, alpha levels are arrayed across the top of Appendix B in two rows, one row for the one-tailed tests and one for two-tailed tests. To use the table, begin by locating the selected alpha level in the appropriate row.

The third difference is that the entries in the table are the actual scores, called ***t*(critical),** that mark the beginnings of the critical regions and not areas under the sampling distribution. To illustrate the use of this table with single-sample means, find the critical region for alpha equal to 0.05, two-tailed test, for $N = 30$. The degrees of freedom will be $N - 1$, or 29; reading down the proper column, you should find a value of 2.045. Thus, the critical region for this test will begin at $t(\text{critical}) = \pm 2.045$.

Take a moment to notice some additional features of the *t* distribution. First, note that the *t*(critical) we found earlier is larger in value than the comparable *Z*(critical), which for a two-tailed test at an alpha of 0.05 would be ± 1.96. This is because the *t* distribution is flatter than the *Z* distribution (see Figure 8.6).

[2]Degrees of freedom refers to the number of values in a distribution that are free to vary. For a sample mean, a distribution has $N - 1$ degrees of freedom. This means that, for a specific value of the mean and of N, $N - 1$ scores are free to vary. For example, if the mean is 3 and $N = 5$, the distribution of five scores would have $5 - 1 = 4$ degrees of freedom. When the values of four of the scores are known, the value of the fifth is fixed. If four scores are 1, 2, 3, and 4, the fifth must be 5 and no other value.

When you use the t distribution, the critical regions will begin farther away from the mean of the sampling distribution; therefore, the null hypothesis will be harder to reject. Furthermore, the smaller the sample size (the lower the degrees of freedom), the larger the value of t(obtained) necessary to reject the H_0.

Second, scan the column for an alpha of 0.05, two-tailed test. Note that, for one degree of freedom, the t(critical) is ± 12.706 and that the value of t(critical) decreases as degrees of freedom increase. For degrees of freedom greater than 120, the value of t(critical) is the same as the comparable value of Z(critical), or ± 1.96. As sample size increases, the t distribution comes to resemble the Z distribution more and more until, with sample sizes greater than 120, the two distributions are essentially identical.[3]

A Test of Hypothesis Using Student's *t*. To demonstrate the uses of the t distribution in more detail, we will work through an example problem. Note that, in terms of the five-step model, the changes required by using t scores occur mostly in steps 3 and 4. In step 3, the sampling distribution will be the t distribution, and degrees of freedom (df) must be computed before locating the critical region as marked by t(critical). In step 4, a slightly different formula for computing the test statistic, ***t* (obtained),** will be used. As compared with the formula for Z(obtained), s will replace σ and $N - 1$ will replace N.

Specifically,

FORMULA 8.2

$$t(\text{obtained}) = \frac{\bar{X} - \mu}{s/\sqrt{N - 1}}$$

A researcher wonders if commuter students are different from the general student body in terms of academic achievement. She has gathered a random sample of 30 commuter students and has learned from the registrar that the mean grade-point average for all students is 2.50 ($\mu = 2.50$), but the standard deviation of the population (σ) has never been computed. Sample data are reported here. Is the sample from a population that has a mean of 2.50?

Student Body	Commuter Students
$\mu = 2.50\ (= \mu_{\bar{X}})$	$\bar{X} = 2.78$
$\sigma = ?$	$s = 1.23$
	$N = 30$

Step 1. Making Assumptions and Meeting Test Requirements.

Model: Random sampling
Level of measurement is interval-ratio
Sampling distribution is normal

[3]Appendix B abbreviates the t distribution by presenting a limited number of critical t scores for degrees of freedom between 31 and 120. If the degrees of freedom for a specific problem equal 77 and alpha equals 0.05, two-tailed, we have a choice between a t(critical) of ± 2.000 (df = 60) and a t(critical) of ± 1.980 (df = 120). In situations such as these, take the larger table value as t(critical). This will make rejection of H_0 less likely and is therefore the more conservative course of action.

Step 2. Stating the Null Hypothesis.

$$H_0: \mu = 2.50$$
$$(H_1: \mu \neq 2.50)$$

You can see from the research hypothesis that the researcher has not predicted a direction for the difference. This will be a two-tailed test.

Step 3. Selecting the Sampling Distribution and Establishing the Critical Region. Since σ is unknown and the sample size is small, the t distribution will be used to find the critical region. Alpha will be set at 0.01.

$$\begin{aligned} \text{Sampling distribution} &= t \text{ distribution} \\ \alpha &= 0.01, \text{ two-tailed test} \\ \text{df} &= (N-1) = 29 \\ t(\text{critical}) &= \pm 2.756 \end{aligned}$$

Step 4. Computing the Test Statistic.

$$\begin{aligned} t(\text{obtained}) &= \frac{\bar{X} - \mu}{s/\sqrt{N-1}} \\ t(\text{obtained}) &= \frac{2.78 - 2.50}{1.23/\sqrt{29}} \\ t(\text{obtained}) &= \frac{.28}{.23} \\ t(\text{obtained}) &= +1.22 \end{aligned}$$

Step 5. Making a Decision and Interpreting Test Results. The test statistic does not fall into the critical region. Therefore, the researcher fails to reject the H_0. The difference between the sample mean (2.78) and the population mean (2.50) is not statistically significant. The difference is no greater than what would be expected if only random chance were operating. The test statistic and critical regions are displayed in Figure 8.8.

(To summarize, when testing single-sample means, we must make a choice regarding the theoretical distribution we will use to establish the critical region.

FIGURE 8.8 SAMPLING DISTRIBUTION SHOWING t(OBTAINED) VERSUS t(CRITICAL) ($\alpha = 0.05$, two-tailed test, df = 29)

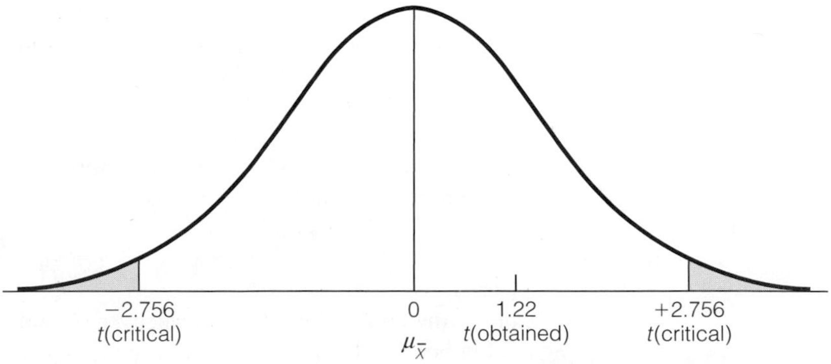

TABLE 8.5 CHOOSING A SAMPLING DISTRIBUTION WHEN TESTING SINGLE-SAMPLE MEANS FOR SIGNIFICANCE

If Population Standard Deviation (σ) is	Sampling Distribution
Known	Z distribution
Unknown and sample size (N) is large	Z distribution
Unknown and sample size (N) is small	t distribution

The choice is straightforward. If the population standard deviation (σ) is known or sample size is large, the Z distribution (summarized in Appendix A) will be used. If σ is unknown and the sample is small, the t distribution (summarized in Appendix B) will be used. These decisions are summarized in Table 8.5 *(For practice in using the t distribution in a test of hypothesis, see problems 8.8 to 8.10 and 8.17.)*

8.7 TESTS OF HYPOTHESES FOR SINGLE-SAMPLE PROPORTIONS (LARGE SAMPLES)

In many cases, the characteristic of interest in the sample will not be measured in a way that justifies the assumption of the interval-ratio level of measurement. One alternative in this situation would be to use a sample proportion (P_s), rather than a sample mean, as the test statistic. As we shall see, the overall procedures for testing single-sample proportions are the same as those for testing means. The central question is still "Does the population from which the sample was drawn have a certain characteristic?" We still conduct the test based on the assumption that the null hypothesis is true, and we still evaluate the probability of the obtained sample outcome against a sampling distribution of all possible sample outcomes. Our decision at the end of the test is also the same. If the obtained test statistic falls into the critical region (is unlikely, given the assumption that the H_0 is true), we reject the H_0.

Having stressed the continuity in procedures and logic, I must hastily point out the important differences as well. These differences are best related in terms of the five-step model for hypothesis testing. In step 1, when working with sample proportions, we assume that the variable is measured at the nominal level of measurement. In step 2, the symbols used to state the null hypothesis are different even though the null is still a statement of "no difference."

In step 3, we will use only the standardized normal curve (the Z distribution) to find areas under the sampling distribution and locate the critical region. This will be appropriate as long as sample size is large. We will not consider small-sample tests of hypothesis for proportions in this text.

In step 4, computing the test statistic, the form of the formula remains the same. That is, the test statistic, Z(obtained), equals the sample statistic minus the mean of the sampling distribution, divided by the standard deviation of the sampling distribution. However, the symbols will change because we are basing the tests on sample proportions. The formula can be stated as

FORMULA 8.3

$$Z(\text{obtained}) = \frac{P_s - P_u}{\sqrt{P_u(1 - P_u)/N}}$$

Step 5, making a decision, is exactly the same as before. If the test statistic, Z(obtained), falls into the critical region, reject the H_0.

Application 8.2

Seventy-six percent of the respondents in a random sample drawn from the most affluent neighborhood in a community reported that they had voted Republican in the most recent presidential election. For the community as a whole, 66% of the electorate voted Republican. Was the affluent neighborhood significantly more likely to have voted Republican?

Step 1. Making Assumptions and Meeting Test Requirements.

Model: Random sampling
Level of measurement is nominal
Sampling distribution is normal

This is a large sample, so we may assume a normal sampling distribution. The variable "percent Republican" is only nominal in level of measurement.

Step 2. Stating the Null Hypothesis (H_0). The null hypothesis says that the affluent neighborhood is not different from the community as a whole.

$$H_0: P_u = 0.66$$

The original question ("Was the affluent neighborhood *more* likely to vote Republican?") suggests a one-tailed research hypothesis:

$$(H_1: P_u > .66)$$

Step 3. Selecting the Sampling Distribution and Establishing the Critical Region.

$$\text{Sampling distribution} = Z \text{ distribution}$$
$$\alpha = 0.05$$
$$Z(\text{critical}) = +1.65$$

The research hypothesis says that we will be concerned only with outcomes in which the neighborhood is *more* likely to vote Republican or with sample outcomes in the upper tail of the sampling distribution.

Step 4. Computing the Test Statistic. The information necessary for a test of the null hypothesis, expressed in the form of proportions, is

Neighborhood	Community
$P_s = 0.76$	$P_u = 0.66$
$N = 103$	

The test statistic, Z(obtained), would be

$$Z(\text{obtained}) = \frac{P_s - P_u}{\sqrt{P_u(1 - P_u)/N}}$$

$$Z(\text{obtained}) = \frac{0.76 - 0.66}{\sqrt{(0.66)(1 - 0.66)/103}}$$

$$Z(\text{obtained}) = \frac{0.10}{\sqrt{(0.2244)/103}}$$

$$Z(\text{obtained}) = \frac{0.100}{0.047}$$

$$Z(\text{obtained}) = 2.13$$

Step 5. Making a Decision and Interpreting Test Results. With alpha set at 0.05, one-tailed, the critical region begins at Z(critical) $= +1.65$. With an obtained Z score of 2.13, the null hypothesis is rejected. The difference between the affluent neighborhood and the community as a whole is statistically significant and in the predicted direction. Residents of the affluent neighborhood were significantly more likely to have voted Republican in the last presidential election.

A Test of Hypothesis Using Sample Proportions. An example should clarify these procedures. A random sample of 122 households in a low-income neighborhood revealed that 53 (or a proportion of 0.43) of the households were headed by females. In the city as a whole, the proportion of female-headed households is 0.39. Are households in the lower-income neighborhood significantly different from the city as a whole in terms of this characteristic?

ONE STEP AT A TIME

Testing the Significance of the Difference Between a Sample Proportion and a Population Proportion: Computing *Z*(obtained) and Interpreting Results

Step 4: Computing *Z*(obtained)

Use Formula 8.3 to compute the test statistic.

Step 1: Start with the denominator of Formula 8.3 and substitute in the value for P_u. This value will be given in the statement of the problem.

Step 2: Find $(1 - P_u)$ by subtracting the value of P_u from 1.

Step 3: Multiply the value you found in step 2 by the value you found in step 1.

Step 4: Divide the quantity you found in step 3 by *N*.

Step 5: Take the square root of the value you found in step 4.

Step 6: Subtract P_u from P_s.

Step 7: Divide the quantity you found in step 6 by the quantity you found in step 5. This value is *Z*(obtained).

Step 5: Making a Decision and Interpreting the Test Result

Step 8: Compare the *Z*(obtained) you computed in step 7 to your *Z*(critical). If *Z*(obtained) is *in* the critical region, *reject* the null hypothesis. If *Z*(obtained) is *not in* the critical region, *fail to reject* the null hypothesis.

Step 9: Interpret the decision to reject or fail to reject the null hypothesis in terms of the original question. For example, our conclusion for the example problem used in Section 8.7 was "There is no significant difference between the low-income community and the city as a whole in the proportion of households that are headed by females."

Step 1. Making Assumptions and Meeting Test Requirements.

Model: Random sampling
Level of measurement is nominal
Sampling distribution is normal in shape

Step 2. Stating the Null Hypothesis. The research question, as stated earlier, asks only if the sample proportion is *different from* the population proportion. Because we have not predicted a direction for the difference, a two-tailed test will be used.

$$H_0: P_u = 0.39$$
$$(H_1: P_u \neq 0.39)$$

Step 3. Selecting the Sampling Distribution and Establishing the Critical Region.

$$\text{Sampling distribution} = Z \text{ distribution}$$
$$\alpha = 0.10, \text{ two-tailed test}$$
$$Z(\text{critical}) = \pm 1.65$$

Step 4. Computing the Test Statistic.

$$Z(\text{obtained}) = \frac{P_s - P_u}{\sqrt{P_u(1 - P_u)/N}}$$

$$Z(\text{obtained}) = \frac{0.43 - 0.39}{\sqrt{0.39(0.61)/122}}$$

$$Z(\text{obtained}) = +0.91$$

FIGURE 8.9 SAMPLING DISTRIBUTION SHOWING Z(OBTAINED) VERSUS Z(CRITICAL) ($\alpha = 0.10$, two-tailed test)

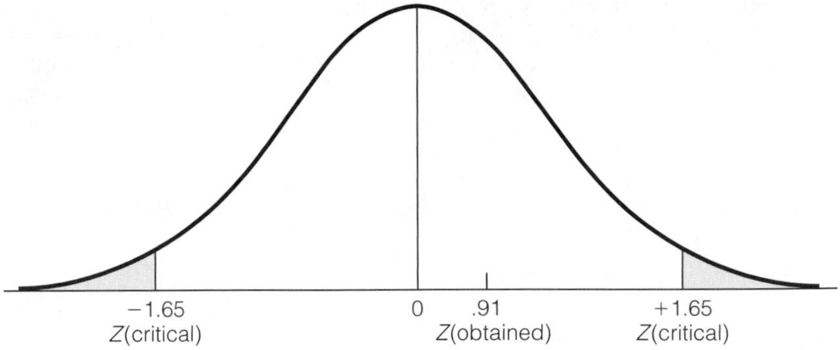

Step 5. Making a Decision and Interpreting Test Results. The test statistic, Z(obtained), does not fall into the critical region. Therefore, we fail to reject the H_0. There is no statistically significant difference between the low-income community and the city as a whole in terms of the proportion of households headed by females. Figure 8.9 displays the sampling distribution, the critical region, and the Z(obtained). *(For practice in tests of significance using sample proportions, see problems 8.1c, 8.11 to 8.14, 8.15a to d, and 8.16.)*

SUMMARY

1. All the basic concepts and techniques for testing hypotheses were presented in this chapter. We saw how to test the null hypothesis of "no difference" for single-sample means and proportions. In both cases, the central question is whether the population represented by the sample has a certain characteristic.
2. All tests of a hypothesis involve finding the probability of the observed sample outcome, given that the null hypothesis is true. If the outcomes have low probabilities, we reject the null hypothesis. In the usual research situation, we will wish to reject the null hypothesis and thereby support the research hypothesis.
3. The five-step model will be our framework for decision making throughout the hypothesis-testing chapters. We will always (1) make assumptions, (2) state the null hypothesis, (3) select a sampling distribution, specify alpha, and find the critical region, (4) compute a test statistic, and (5) make a decision. What we do during each step, however, will vary, depending on the specific test being conducted.
4. If we can predict a direction for the difference in stating the research hypothesis, a one-tailed test is called for. If no direction can be predicted, a two-tailed test is appropriate. There are two kinds of errors in hypothesis testing. Type I, or alpha, error is rejecting a true null; Type II, or beta, error is failing to reject a false null. The probabilities of committing these two types of error are inversely related and cannot be simultaneously minimized in the same test. By selecting an alpha level, we try to balance the probability of each of these two kinds of error.
5. When testing sample means, the t distribution must be used to find the critical region when the population standard deviation is unknown and sample size is small.
6. Sample proportions can also be tested for significance. Tests are conducted using the five-step model. Compared to the test for the sample mean, the major differences lie in the level-of-measurement assumption (step 1), the statement of the null hypothesis (step 2), and the computation of the test statistic (step 4).

7. If you are still confused about the uses of inferential statistics described in this chapter, don't be alarmed or discouraged. A sizeable volume of rather complex material has been presented, and only rarely will a beginning student fully comprehend the unique logic of hypothesis testing on the first exposure. After all, it is not every day that you learn how to test a statement you don't believe (the null hypothesis) against a distribution that doesn't exist (the sampling distribution)!

SUMMARY OF FORMULAS

Single-sample means, large samples:

8.1 $$Z(\text{obtained}) = \frac{\bar{X} - \mu}{\sigma/\sqrt{N}}$$

Single-sample means when samples are small and population standard deviation is unknown:

8.2 $$t(\text{obtained}) = \frac{\bar{X} - \mu}{s/\sqrt{N-1}}$$

Single-sample proportions, large samples:

8.3 $$Z(\text{obtained}) = \frac{P_s - P_u}{\sqrt{P_u(1 - P_u)/N}}$$

GLOSSARY

Alpha level (α). The proportion of area under the sampling distribution that contains unlikely sample outcomes, given that the null hypothesis is true. Also, the probability of Type I error.

Critical region (region of rejection). The area under the sampling distribution that, in advance of the test itself, is defined as including unlikely sample outcomes, given that the null hypothesis is true.

Five-step model. A step-by-step guideline for conducting tests of hypotheses. A framework that organizes decisions and computations for all tests of significance.

Hypothesis testing. Statistical tests that estimate the probability of sample outcomes if assumptions about the population (the null hypothesis) are true.

Null hypothesis (H_0). A statement of "no difference." In the context of single-sample tests of significance, the population from which the sample was drawn is assumed to have a certain characteristic or value.

One-tailed test. A type of hypothesis test used when (1) the direction of the difference can be predicted or (2) concern focuses on outcomes in only one tail of the sampling distribution.

Research hypothesis (H_1). A statement that contradicts the null hypothesis. In the context of single-sample tests of significance, the research hypothesis says that the population from which the sample was drawn does not have a certain characteristic or value.

Significance testing. See Hypothesis testing.

Student's *t* distribution. A distribution used to find the critical region for tests of sample means when σ is unknown and sample size is small.

***t* (critical).** The *t* score that marks the beginning of the critical region of a *t* distribution.

***t* (obtained).** The test statistic computed in step 4 of the file-step model. The sample outcome expressed as a *t* score.

Test statistic. The value computed in step 4 of the five-step model that converts the sample outcome into either a *t* score or a *Z* score.

Two-tailed test. A type of hypothesis test used when (1) the direction of the difference cannot be predicted or (2) concern focuses on outcomes in both tails of the sampling distribution.

Type I error (alpha error). The probability of rejecting a null hypothesis that is, in fact, true.

Type II error (beta error). The probability of failing to reject a null hypothesis that is, in fact, false.

***Z*(critical).** The *Z* score that marks the beginnings of the critical region on a *Z* distribution.

***Z*(obtained).** The test statistic computed in step 4 of the five-step model. The sample outcomes expressed as a *Z* score.

PROBLEMS

8.1 a. For each situation, find *Z*(critical).

Alpha	Form	*Z*(Critical)
.05	One-tailed	
.10	Two-tailed	
.06	Two-tailed	
.01	One-tailed	
.02	Two-tailed	

b. For each situation, find the critical *t* score.

Alpha	Form	*N*	*t*(Critical)
.10	Two-tailed	31	
.02	Two-tailed	24	
.01	Two-tailed	121	
.01	One-tailed	31	
.05	One-tailed	61	

c. Compute the appropriate test statistic (*Z* or *t*) for each situation:

1. $\mu = 2.40$	$\bar{X} = 2.20$
$\sigma = 0.75$	$N = 200$
2. $\mu = 17.1$	$\bar{X} = 16.8$
	$s = 0.9$
	$N = 45$
3. $\mu = 10.2$	$\bar{X} = 9.4$
	$s = 1.7$
	$N = 150$
4. $P_u = .57$	$P_s = 0.60$
	$N = 117$
5. $P_u = 0.32$	$P_s = 0.30$
	$N = 322$

8.2 SOC **a.** The student body at St. Algebra College attends an average of 3.3 parties per month. A random sample of 117 sociology majors averages 3.8 parties per month, with a standard deviation of 0.53. Are sociology majors significantly different from the student body as a whole? *(HINT: The wording of the research question suggests a two-tailed test. This means that the alternative, or research, hypothesis in step 2 will be stated as H_1: $\mu \neq 3.3$ and that the critical region will be split between the upper and lower tails of the sampling distribution. See Table 8.1 for values of Z(critical) for various alpha levels.)*

b. What if the research question were changed to "Do sociology majors attend a significantly *greater* number of parties"? How would the test conducted in problem 8.2a change? *(HINT: This wording implies a one-tailed test of significance. How would the research hypothesis change? For the alpha you used in problem 8.2a, what would the value of Z(critical) be?)*

8.3 SW **a.** Nationally, social workers average 10.2 years of experience. In a random sample, 203 social workers in greater metropolitan Shinbone average only 8.7 years, with a standard deviation of 0.52. Are social workers in Shinbone significantly less experienced? *(Note the wording of the research hypotheses. These situations may justify one-tailed tests of significance. If you chose a one-tailed test, what form would the research hypothesis take, and where would the critical region begin?)*

b. The same sample of social workers reports an average annual salary of $25,782, with a standard deviation of $622. Is this figure significantly higher than the national average of $24,509? *(The wording of the research hypotheses suggests a one-tailed test. What form would the research hypothesis take, and where would the critical region begin?)*

8.4 SOC Nationally, the average score on the college entrance exams (verbal test) is 453, with a standard deviation of 95. A random sample of 152 freshmen entering St. Algebra College shows a mean score of 502. Is there a significant difference?

8.5 SOC A random sample of 423 Chinese Americans has finished an average of 12.7 years of formal education, with a standard deviation of 1.7. Is this significantly different from the national average of 12.2 years?

8.6 SOC A sample of 105 workers in the Overkill Division of the Machismo Toy Factory earns an average of $24,375 per year. The average salary for all workers is $24,230, with a standard deviation of $523. Are workers in the Overkill Division overpaid? Conduct both one- and two-tailed tests.

8.7 GER **a.** Nationally, the population as a whole watches 6.2 hours of TV per day. A random sample of 1017 senior citizens report watching an average of 5.9 hours per day, with a standard deviation of 0.7. Is the difference significant?

b. The same sample of senior citizens reports that they belong to an average of 2.1 voluntary organizations and clubs, with a standard deviation of 0.5. Nationally, the average is 1.7. Is the difference significant?

8.8 SOC A school system has assigned several hundred "chronic and severe underachievers" to an alternative educational experience. To assess the program, a random sample of 35 has been selected for comparison with all students in the system.

a. In terms of GPA, did the program work?

Systemwide GPA	Program GPA
$\mu = 2.47$	$\bar{X} = 2.55$ $s = 0.70$ $N = 35$

b. In terms of absenteeism (number of days missed per year), what can be said about the success of the program?

Systemwide	Program
$\mu = 6.137$	$\bar{X} = 4.78$ $s = 1.11$ $N = 35$

c. In terms of standardized test scores in math and reading, was the program a success?

Math Test–Systemwide	Math Test–Program
$\mu = 103$	$\bar{X} = 106$ $s = 2.0$ $N = 35$

Reading Test–Systemwide	Reading Test–Program
$\mu = 110$	$\bar{X} = 113$ $s = 2.0$ $N = 35$

(HINT: Note the wording of the research questions. Is a one-tailed test justified? Is the program a success if the students in the program are no different from students systemwide? What if the program students were performing at lower levels? If a one-tailed test is used, what form should the research hypothesis take? Where will the critical region begin?)

8.9 SOC A random sample of 26 local sociology graduates scored an average of 458 on the GRE advanced sociology test, with a standard deviation of 20. Is this significantly different from the national average ($\mu = 440$)?

8.10 PA Nationally, the per capita property tax is \$130. A random sample of 36 southeastern cities average \$98, with a standard deviation of \$5. Is the difference significant? Summarize your conclusions in a sentence or two.

8.11 GER/CJ A survey shows that 10% of the population is victimized by property crime each year. A random sample of 527 older citizens (65 years or more of age) shows a victimization rate of 14%. Are older people more likely to be victimized? Conduct both one- and two-tailed tests of significance.

8.12 CJ A random sample of 113 convicted rapists in a state prison system completed a program designed to change their attitudes towards women, sex, and violence before being released on parole. Fifty-eight eventually became repeat sex offenders. Is this recidivism rate significantly different from the rate for all offenders in that state (57%)? Summarize your conclusions in a sentence or two. *(HINT: You must use the information given in the problem to compute a sample proportion. Remember to convert the population percentage to a proportion.)*

8.13 PS In a recent statewide election, 55% of the voters rejected a proposal to institute a state lottery. In a random sample of 150 urban precincts, 49% of the voters rejected the proposal. Is the difference significant? Summarize your conclusions in a sentence or two.

8.14 CJ Statewide, the police clear by arrest 35% of the robberies and 42% of the aggravated assaults reported to them. A researcher takes a random sample of all the robberies ($N = 207$) and aggravated assaults ($N = 178$) reported to a metropolitan police department in one year and finds that 83 of the robberies and 80 of the assaults were cleared by arrest. Are the local arrest rates significantly different from the statewide rate? Write a sentence or two interpreting your decision.

8.15 SOC/SW A researcher has compiled a file of information on a random sample of 317 families that have chronic, long-term patterns of child abuse. Reported here are some of the characteristics of the sample, along with values for the city as a whole. For each trait, test the null hypothesis of "no difference" and summarize your findings.

a. Mothers' educational level (proportion completing high school):

City	Sample
$P_u = 0.63$	$P_s = 0.61$

b. Family size (proportion of families with four or more children):

City	Sample
$P_u = 0.21$	$P_s = 0.26$

c. Mothers' work status (proportion of mothers with jobs outside the home):

City	Sample
$P_u = 0.51$	$P_s = 0.27$

d. Relations with kin (proportion of families that have contact with kin at least once a week):

City	Sample
$P_u = 0.82$	$P_s = 0.43$

e. Fathers' educational achievement (average years of formal schooling):

City	Sample
$\mu = 12.3$	$\bar{X} = 12.5$ $s = 1.7$

f. Fathers' occupational stability (average years in present job):

City	Sample
$\mu = 5.2$	$\bar{X} = 3.7$ $s = 0.5$

8.16 SW You are the head of an agency seeking funding for a program to reduce unemployment among teenage males. Nationally, the unemployment rate for this group is 18%. A random sample of 323 teenage males in your area reveals an unemployment rate of 21.7%. Is the difference significant? Can you demonstrate a need for the program? Should you use a one-tailed test in this situation? Why? Explain the result of your test of significance as you would to a funding agency.

8.17 PA The city manager of Shinbone has received a complaint from the local union of firefighters to the effect that they are underpaid. Not having much time, the city manager gathers the records of a random sample of 27 firefighters and finds that their average salary is $38,073, with a standard deviation of $575. If she knows that the average salary nationally is $38,202, how can she respond to the complaint? Should she use a one-tailed test in this situation? Why? What would she say in a memo to the union that would respond to the complaint?

8.18 The following essay questions review the basic principles and concepts of inferential statistics. The order of the questions roughly follows the five-step model.

a. Hypothesis testing or significance testing can be conducted only with a random sample. Why?

b. Under what specific conditions can it be assumed that the sampling distribution is normal in shape?

c. Explain the role of the sampling distribution in a test of hypothesis.

d. The null hypothesis is an assumption about reality that makes it possible to test sample outcomes for their significance. Explain.

e. What is the critical region? How is the size of the critical region determined?

f. Describe a research situation in which a one-tailed test of hypothesis would be appropriate.

g. Thinking about the shape of the sampling distribution, why does use of the *t* distribution (as opposed to the *Z* distribution) make it more difficult to reject the null hypothesis?

h. What exactly can be concluded in the one-sample case when the test statistic falls into the critical region?

9

Hypothesis Testing II
The Two-Sample Case

LEARNING OBJECTIVES

By the end of this chapter, you will be able to

1. Identify and cite examples of situations in which the two-sample test of hypothesis is afppropriate.
2. Explain the logic of hypothesis testing as applied to the two-sample case.
3. Explain what an independent random sample is.
4. Perform a test of hypothesis for two sample means or two sample proportions following the five-step model and correctly interpret the results.
5. List and explain each of the factors (especially sample size) that affect the probability of rejecting the null hypothesis. Explain the differences between statistical significance and importance.

9.1 INTRODUCTION

In Chapter 8, we dealt with hypothesis testing in the one-sample case. In that situation, we were concerned with the significance of the difference between a sample value and a population value. In this chapter, we consider a new research situation where we will be concerned with the significance of the difference between two separate populations. For example, do men and women in the United States vary in their support for gun control? Obviously, we cannot ask every male and female for their opinions on this issue. Instead, we must draw random samples of both groups and use the information gathered from these samples to infer population patterns.

The central question asked in hypothesis testing in the two-sample case is: Is the difference between the samples large enough to allow us to conclude (with a known probability of error) that the populations represented by the samples are different? Thus, if we find a large enough difference in support of gun control between random samples of men and women, we can argue that the difference between the samples did not occur by simple random chance but, rather, represents a real difference between men and women in the population.

In this chapter, we consider tests for the significance of the difference between sample means and sample proportions. In both tests, the five-step model will serve as a framework for organizing our decision making. The general flow of the hypothesis-testing process is very similar to the one followed in the one-sample case, but we also need to consider some important differences.

9.2 HYPOTHESIS TESTING WITH SAMPLE MEANS (LARGE SAMPLES)

The Five-Step Model. There are three important differences between the one-sample case considered in Chapter 8 and the two-sample case covered in this chapter. The first difference occurs in step 1 of the five-step model. The one-sample case (Chapter 8) requires that the sample be selected following the principle of EPSEM (each case in the population must have an equal chance of being selected for the sample). The two-sample situation requires that the samples be selected independently as well as randomly. This requirement is met when the selection of a case for one sample has no effect on the probability that any particular case will be included in the other sample. In our example concerning gender differences in support of gun control, this would mean that the selection of a specific male for the sample would have no effect on the probability of selecting any particular female. This new requirement will be stated as **independent random sampling** in step 1.

The requirement of independent random sampling can be satisfied by drawing EPSEM samples from separate lists (for example, one for females and one for males). It is usually more convenient, however, to draw a single EPSEM sample from a single list of the population and then to subdivide the cases into separate groups (males and females, for example). As long as the original sample is selected randomly, any subsamples created by the researcher will meet the assumption of independent random samples.

The second important difference is in the form of the null hypothesis stated in step 2 of the five-step model. The null is still a statement of "no difference." Now, however, instead of saying that the population from which the sample is drawn has a certain characteristic, it will say that the two populations are not different. ("There is no significant difference between men and women in their support of gun control.") If the test statistic falls in the critical region, the null hypothesis of no difference between the populations can be rejected, and the argument that the populations are different on the trait of interest will be supported.

A third important new element concerns the sampling distribution, or the distribution of all possible sample outcomes. In Chapter 8, the sample outcome was either a mean or a proportion. Now we are dealing with two samples (e.g., samples of men and women), and the sample outcome is the *difference between* the sample statistics. In terms of our example, the sampling distribution would include all possible differences in sample means for support of gun control between men and women. If the null hypothesis is true and men and women do *not* have different views about gun control, the difference between the population means would be zero, the mean of the sampling distribution will be zero, and the huge majority of differences between sample means would be zero or very close to zero. The greater the differences between the sample means, the further the sample outcome (the *difference* between the two sample means) will be from the mean of the sampling distribution (zero), and the more likely that the difference reflects a real difference between the populations represented by the samples.

Testing the Significance of the Difference Between Two Sample Means (Large Samples): An Example. To illustrate the procedure for testing sample means, assume that a researcher has access to a nationally representative random sample and that the individuals in the sample have responded to a scale that measures attitudes toward gun control. The sample is divided by sex, and sample statistics are computed for males and females. Assuming that the scale yields interval-ratio-level data, a test for the significance of the difference in sample means can be conducted.

As long as sample size is large (that is, as long as the combined number of cases in the two samples exceeds 100), the sampling distribution of the differences in sample means will be normal, and the normal curve (Appendix A) can be used to establish the critical regions. The test statistic, Z(obtained), will be computed by the usual formula: sample outcome (the difference between the sample means) minus the mean of the sampling distribution, divided by the standard deviation of the sampling distribution. The formula is presented as Formula 9.1. Note that numerical subscripts are used to identify the samples and the two populations they represent. The subscript attached to $\sigma(\sigma_{\bar{X}-\bar{X}})$ indicates that we are dealing with the sampling distribution of the *differences* in sample means.

FORMULA 9.1

$$Z(\text{obtained}) = \frac{(\bar{X}_1 - \bar{X}_2) - (\mu_1 - \mu_2)}{\sigma_{\bar{X}-\bar{X}}}$$

where $(\bar{X}_1 - \bar{X}_2)$ = the difference in the sample means
$(\mu_1 - \mu_2)$ = the difference in the population means
$\sigma_{\bar{X}-\bar{X}}$ = the standard deviation of the sampling distribution of the differences in sample means

The second term in the numerator, $\mu_1 - \mu_2$, reduces to zero because we assume that the null hypothesis (which will be stated as H_0: $\mu_1 = \mu_2$) is true. Recall that tests of significance are always based on the assumption that the null hypothesis is true. If the means of the two populations are equal, then the term $(\mu_1 - \mu_2)$ will be zero and can be dropped from the equation. In effect, then, the formula we will actually use to compute the test statistic in step 4 will be

FORMULA 9.2

$$Z(\text{obtained}) = \frac{(\bar{X}_1 - \bar{X}_2)}{\sigma_{\bar{X}-\bar{X}}}$$

For large samples, the standard deviation of the sampling distribution of the difference in sample means is defined as

FORMULA 9.3

$$\sigma_{\bar{X}-\bar{X}} = \sqrt{\frac{\sigma_1^2}{N_1} + \frac{\sigma_2^2}{N_2}}$$

Since we will rarely, if ever, be in a position to know the values of the population standard deviations (σ_1 and σ_2), we must use the sample standard deviations, suitably corrected for bias, to estimate them. Formula 9.4 displays the equation used to estimate the standard deviation of the sampling distribution in this situation. This is called a **pooled estimate** since it combines information from both samples.

FORMULA 9.4

$$\sigma_{\bar{X}-\bar{X}} = \sqrt{\frac{s_1^2}{N_1 - 1} + \frac{s_2^2}{N_2 - 1}}$$

Following are the sample outcomes for support of gun control; a test for the significance of the difference can now be conducted.

Sample 1 (Men)	Sample 2 (Women)
$\bar{X}_1 = 6.2$	$\bar{X}_2 = 6.5$
$s_1 = 1.3$	$s_2 = 1.4$
$N_1 = 324$	$N_2 = 317$

ONE STEP AT A TIME

Testing the Difference Between Sample Means for Significance (Large Samples): Computing *Z*(obtained) and Interpreting Results

Step 4: Computing *Z*(obtained)

Solve Formula 9.4 first and then solve Formula 9.2 to compute the test statistic.

Solving Formula 9.4

Step 1: Subtract 1 from N_1.

Step 2: Square the value of the standard deviation for the first sample (s_1^2).

Step 3: Divide the quantity you found in step 2 by the quantity you found in Step 1.

Step 4: Subtract 1 from N_2.

Step 5: Square the value of the standard deviation for the second sample (s_2^2).

Step 6: Divide the quantity you found in step 5 by the quantity you found in Step 4.

Step 7: Add the quantity you found in step 6 to the quantity you found in Step 3.

Step 8: Take the square root of the quantity you found in step 7.

Solving Formula 9.2

Step 9: Subtract $\bar{X}_2$ from $\bar{X}_1$.

Step 10: Divide the quantity you found in step 9 by the quantity you found in step 8.

Step 5: Making a Decision and Interpreting the Results of the Test

Step 11: Compare the *Z*(obtained) you computed in step 10 to your *Z*(critical). If *Z*(obtained) is *in* the critical region, *reject* the null hypothesis. If *Z*(obtained) is *not in* the critical region, *fail to reject* the null hypothesis

Step 12: Interpret the decision to reject or fail to reject the null hypothesis in terms of the original question. For example, our conclusion for the example problem used in Section 9.2 was "There is a significant difference between men and women in their support for gun control."

We see from the sample statistics that men have a lower average score on the Support for Gun Control Scale and are thus less supportive of gun control. The test of hypothesis will tell us if this difference is large enough to justify the conclusion that it did not occur by random chance alone but, rather, reflects an actual difference between the populations of men and women on this issue.

Step 1. Making Assumptions and Meeting Test Requirements. Note that, although we now assume that the random samples are independent, the rest of the model is the same as in the one-sample case.

Model: Independent random samples
Level of measurement is interval-ratio
Sampling distribution is normal

Step 2. Stating the Null Hypothesis. The null hypothesis states that the *populations* represented by the samples are not different on this variable. Since no direction for the difference has been predicted, a two-tailed test is called for, as reflected in the research hypothesis.

$$H_0: \mu_1 = \mu_2$$
$$(H_1: \mu_1 \neq \mu_2)$$

Application 9.1

An attitude scale measuring satisfaction with family life has been administered to a sample of married respondents. On this scale, higher scores indicate greater satisfaction. The sample has been divided into respondents with no children and respondents with at least one child, and means and standard deviations have been computed for both groups. Is there a significant difference in satisfaction with family life between these two groups? The sample information is:

Sample 1 (No Children)	Sample 2 (At Least One Child)
$\bar{X}_1 = 11.3$	$\bar{X}_2 = 10.8$
$s_1 = 0.6$	$s_2 = 0.5$
$N_1 = 78$	$N_2 = 93$

We can see from the sample results that respondents with no children are more satisfied. The significance of this difference will be tested following the five-step model.

Step 1. Making Assumptions and Meeting Test Requirements.

Model: Independent random samples
Level of measurement is interval-ratio
Sampling distribution is normal

Step 2. Stating the Null Hypothesis.

$$H_0: \mu_1 = \mu_2$$
$$(H_1: \mu_1 \neq \mu_2)$$

Step 3. Selecting the Sampling Distribution and Establishing the Critical Region.

$$\text{Sampling distribution} = Z \text{ distribution}$$
$$\text{Alpha} = 0.05, \text{ two-tailed}$$
$$Z(\text{critical}) = \pm 1.96$$

Step 4. Computing the Test Statistic.

$$\sigma_{\bar{X}-\bar{X}} = \sqrt{\frac{s_1^2}{N_1 - 1} + \frac{s_2^2}{N_2 - 1}}$$
$$\sigma_{\bar{X}-\bar{X}} = \sqrt{\frac{(0.6)^2}{78 - 1} + \frac{(0.5)^2}{93 - 1}}$$
$$\sigma_{\bar{X}-\bar{X}} = \sqrt{0.008}$$
$$\sigma_{\bar{X}-\bar{X}} = 0.09$$

$$Z(\text{obtained}) = \frac{(\bar{X}_1 - \bar{X}_2)}{\sigma_{\bar{X}-\bar{X}}}$$
$$Z(\text{obtained}) = \frac{11.3 - 10.8}{0.09}$$
$$Z(\text{obtained}) = \frac{0.50}{0.09}$$
$$Z(\text{obtained}) = 5.56$$

Step 5. Making a Decision and Interpreting the Results of the Test. Comparing the test statistic with the critical region,

$$Z(\text{obtained}) = 5.56$$
$$Z(\text{critical}) = \pm 1.96$$

we reject the null hypothesis. This test supports the conclusion that parents and childless couples are significantly different in their satisfaction with family life. Given the direction of the difference, we can also note that childless couples are significantly happier.

Step 3. Selecting the Sampling Distribution and Establishing the Critical Region. For large samples, the *Z* distribution can be used to find areas under the sampling distribution and establish the critical region. Alpha will be set at 0.05.

$$\text{Sampling distribution} = Z \text{ distribution}$$
$$\text{Alpha} = 0.05$$
$$Z(\text{critical}) = \pm 1.96$$

Step 4. Computing the Test Statistic. Since the population standard deviations are unknown, Formula 9.4 will be used to estimate the standard deviation of the sampling distribution. This value will then be substituted into Formula 9.2 and Z(obtained) will be computed.

$$\sigma_{\bar{X}-\bar{X}} = \sqrt{\frac{s_1^2}{N_1 - 1} + \frac{s_2^2}{N_2 - 1}}$$

$$\sigma_{\bar{X}-\bar{X}} = \sqrt{\frac{(1.3)^2}{324 - 1} + \frac{(1.4)^2}{317 - 1}}$$

$$\sigma_{\bar{X}-\bar{X}} = \sqrt{(0.0052) + (0.0062)}$$

$$\sigma_{\bar{X}-\bar{X}} = \sqrt{0.0114}$$

$$\sigma_{\bar{X}-\bar{X}} = 0.107$$

$$Z(\text{obtained}) = \frac{(\bar{X}_1 - \bar{X}_2)}{\sigma_{\bar{X}-\bar{X}}}$$

$$Z(\text{obtained}) = \frac{6.2 - 6.5}{0.107}$$

$$Z(\text{obtained}) = \frac{-0.300}{0.107}$$

$$Z(\text{obtained}) = -2.80$$

Step 5. Making a Decision and Interpreting the Results of the Test. Comparing the test statistic with the critical region:

$$Z(\text{obtained}) = -2.80$$
$$Z(\text{critical}) = \pm 1.96$$

We see that the Z score clearly falls into the critical region. This outcome indicates that a difference as large as -0.30 $(6.2 - 6.5)$ between the sample means is unlikely if the null hypothesis is true. The null hypothesis of no difference can be rejected, and the notion that men and women are different in terms of their support of gun control is supported. The decision to reject the null hypothesis has only a 0.05 probability (the alpha level) of being incorrect.

Note that the value for Z(obtained) is negative, indicating that men have significantly lower scores than women for support for gun control. The sign of the test statistics reflects our arbitrary decision to label men sample 1 and women sample 2. If we had reversed the labels and called women sample 1 and men sample 2, the sign of the Z(obtained) would have been positive, but its value (2.80) would have been exactly the same, as would our decision in step 5. *(For practice in testing the significance of the difference between sample means for large samples, see problems 9.1 to 9.6 and 9.15d to f.)*

9.3 HYPOTHESIS TESTING WITH SAMPLE MEANS (SMALL SAMPLES)

The Five-Step Model and the *t* distribution. As with single-sample means, when the population standard deviation is unknown and sample size is small (combined N's of less than 100), the Z distribution can no longer be used to find areas under the sampling distribution. Instead, we will use the t distribution to find the critical region and identify unlikely sample outcomes. To use the t distribution to test the significance of the difference between two sample means, we need to perform one additional calculation and make one additional assumption. The calculation is for degrees of freedom, a quantity required for proper use of the *t* table (Appendix B). In the two-sample case, degrees of freedom are equal to $N_1 + N_2 - 2$.

The additional assumption is a more complex matter. When samples are small, we must assume that the variances of the populations of interest are equal in order to justify the assumption of a normal sampling distribution and to form a pooled estimate of the standard deviation of the sampling distribution. The assumption of equal variance in the population can be tested by an inferential statistical technique known as the *analysis of variance*, or ANOVA (see Chapter 10). For our purposes here, however, we will simply assume equal population variances without formal testing. This assumption is safe as long as sample sizes are approximately equal.

A Test of Hypothesis Between Two Sample Means (Small Samples). To illustrate this procedure, assume that a researcher believes that center-city families have significantly more children than suburban families. Random samples from both areas are gathered and the following sample statistics computed.

Sample 1 (Suburban)	Sample 2 (Center-City)
$\bar{X}_1 = 2.37$	$\bar{X}_2 = 2.78$
$s_1 = 0.63$	$s_2 = 0.95$
$N_1 = 42$	$N_2 = 37$

The sample data reveal a difference in the predicted direction. The significance of this observed difference can be tested with the five-step model.

Step 1. Making Assumptions and Meeting Test Requirements. Sample size is small, and the population standard deviation is unknown. Hence, we must assume equal population variances in the model.

Model: Independent random samples
Level of measurement is interval-ratio
Population variances are equal ($\sigma_1^2 = \sigma_2^2$)
Sampling distribution is normal

Step 2. Stating the Null Hypothesis. Since a direction has been predicted (center-city families are larger), a one-tailed test will be used, and the research hypothesis is stated in accordance with this decision.

$$H_0: \mu_1 = \mu_2$$
$$(H_1: \mu_1 < \mu_2)$$

Step 3. Selecting the Sampling Distribution and Establishing the Critical Region. With small samples, the *t* distribution is used to establish the critical region. Alpha will be set at 0.05, and a one-tailed test will be used.

Sampling distribution = *t* distribution
Alpha = 0.05, one-tailed
Degrees of freedom = $N_1 + N_2 - 2 = 42 + 37 - 2 = 77$
$t(\text{critical}) = -1.671$

Note that the critical region is placed in the lower tail of the sampling distribution in accordance with the direction specified in H_1.

Step 4. Computing the Test Statistic. With small samples, a different formula (Formula 9.5) is used for the pooled estimate of the standard deviation of the sampling distribution. This value is then substituted directly into the denominator of the formula for t(obtained) given in Formula 9.6.

FORMULA 9.5

$$\sigma_{\bar{X}-\bar{X}} = \sqrt{\frac{N_1 s_1^2 + N_2 s_2^2}{N_1 + N_2 - 2}}\sqrt{\frac{N_1 + N_2}{N_1 N_2}}$$

$$\sigma_{\bar{X}-\bar{X}} = \sqrt{\frac{(42)(.63)^2 + (37)(.95)^2}{42 + 37 - 2}}\sqrt{\frac{42 + 37}{(42)(37)}}$$

$$\sigma_{\bar{X}-\bar{X}} = \sqrt{\frac{50.06}{77}}\sqrt{\frac{79}{1554}}$$

$$\sigma_{\bar{X}-\bar{X}} = (.81)(.23)$$

$$\sigma_{\bar{X}-\bar{X}} = 0.19$$

FORMULA 9.6

$$t(\text{obtained}) = \frac{(\bar{X}_1 - \bar{X}_2)}{\sigma_{\bar{X}-\bar{X}}}$$

$$t(\text{obtained}) = \frac{2.37 - 2.78}{0.19}$$

$$t(\text{obtained}) = \frac{-0.41}{0.19}$$

$$t(\text{obtained}) = -2.16$$

Step 5. Making a Decision and Interpreting the Results of the Test. Comparing the test statistic with the critical region,

$$t(\text{obtained}) = -2.16$$
$$t(\text{critical}) = -1.671$$

we can see that the test statistic falls into the critical region. If the null hypothesis ($\mu_1 = \mu_2$) were true, this would be a very unlikely outcome, so the null hypothesis can be rejected. There is a statistically significant difference (a difference so large that it is unlikely to be due to random chance) in the sizes of center-city and suburban families. Furthermore, center-city families are significantly larger in size. The test statistic and sampling distribution are depicted in Figure 9.1. *(For*

FIGURE 9.1 THE SAMPLING DISTRIBUTION, WITH CRITICAL REGION AND TEST STATISTIC DISPLAYED

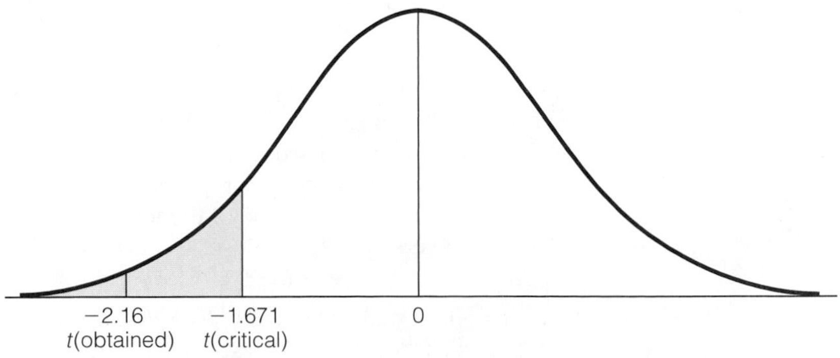

ONE STEP AT A TIME

Testing the Difference Between Sample Means for Significance (Small Samples): Computing t(obtained) and Interpreting Results

Step 4: Computing t(obtained)

Solve Formulas 9.5 first and then solve Formula 9.6 to compute the test statistic.

Solving Formula 9.5

Step 1: Add N_1 and N_2 and then subtract 2 from this total.

Step 2: Square the standard deviation for the first sample (s_1^2).

Step 3: Multiply the quantity you found in step 2 by N_1.

Step 4: Square the standard deviation for the second sample (s_2^2).

Step 5: Multiply the quantity you found in step 4 by N_2.

Step 6: Add the quantities you found in steps 3 and 5.

Step 7: Divide the quantity you found in step 6 by the quantity you found in step 1.

Step 8: Take the square root of the quantity you found in step 7.

Step 9: Multiply N_1 by N_2.

Step 10: Add N_1 and N_2

Step 11: Divide the quantity you found in step 10 by the quantity you found in step 9.

Step 12: Take the square root of the quantity you found in Step 11.

Step 13: Multiply the quantity you found in step 12 by the quantity you found in step 8.

Solving Formula 9.6

Step 14: Subtract from $\overline{X}_2$ from $\overline{X}_1$.

Step 15: Divide the quantity you found in step 14 by the quantity you found in step 13.

Step 5: Making a Decision and Interpreting the Results of the Test

Step 16: Compare the t(obtained) you computed in step 15 to your t(critical). If t(obtained) is *in* the critical region, *reject* the null hypothesis. If t(obtained) is *not in* the critical region, *fail to reject* the null hypothesis.

Step 17: Interpret the decision to reject or fail to reject the null hypothesis in terms of the original question. For example, our conclusion for the example problem used in this section was "There is a significant difference between the average size of center-city and suburban families."

practice in testing the significance of the difference between sample means for small samples, see problems 9.7 and 9.8.)

9.4 HYPOTHESIS TESTING WITH SAMPLE PROPORTIONS (LARGE SAMPLES)

The Five-Step Model. Testing for the significance of the difference between two sample proportions is analogous to testing sample means. The null hypothesis states that no difference exists between the populations from which the samples are drawn on the trait being tested. The sample proportions form the basis of the test statistic computed in step 4, which is then compared with the critical region. When sample sizes are large (combined N's of more than 100), the Z distribution may be used to find the critical region. In this text, we will not consider tests of significance for proportions based on small samples.

In order to find the value of the test statistics, several preliminary equations must be solved. Formula 9.7 uses the values of the two sample proportions (P_s) to give us an estimate of the population proportion (P_u), the proportion of cases in the population that have the trait under consideration assuming the null hypothesis is true.

ONE STEP AT A TIME

Testing the Difference Between Sample Proportions for Significance (Large Samples): Computing *Z*(obtained) and Interpreting Results

Step 4: Computing *Z*(obtained)

Solve Formulas 9.7, 9.8, and 9.10 to compute the test statistic.

Solving Formula 9.7

Step 1: Add N_1 and N_2.

Step 2: Multiply P_{s1} by N_1.

Step 3: Multiply P_{s2} by N_2.

Step 4: Add the quantity you found in step 3 to the quantity you found in step 2.

Step 5: Divide the quantity you found in step 4 by the quantity you found in step 1.

Solving Formula 9.8

Step 6: Multiply P_u (see step 5) by $(1 - P_u)$.

Step 7: Take the square root of the quantity you found in step 6.

Step 8: Multiply N_1 by N_2.

Step 9: Add N_1 and N_2. (See step 1.)

Step 10: Divide the quantity you found in step 9 by the quantity you found in step 8.

Step 11: Take the square root of the quantity you found in step 10.

Step 12: Multiply the quantity you found in step 11 by the quantity you found in step 7.

Solving Formula 9.10

Step 13: Subtract P_{s2} from P_{s1}.

Step 14: Divide the quantity you found in step 13 by the quantity you found in step 12.

Step 5: Making a Decision and Interpreting the Results of the Test

Step 15: Compare the *Z*(obtained) you computed in step 14 to your *Z*(critical). If *Z*(obtained) is *in* the critical region, *reject* the null hypothesis. If *Z*(obtained) is *not in* the critical region, *fail to reject* the null hypothesis.

Step 16: Interpret the decision to reject or fail to reject the null hypothesis in terms of the original question. For example, our conclusion for the example problem used in this section was "There is no significant difference between the participation patterns of black and white senior citizens."

FORMULA 9.7

$$P_u = \frac{N_1 P_{s1} + N_2 P_{s2}}{N_1 + N_2}$$

The estimated value of P_u is then used to determine a value for the standard deviation of the sampling distribution of the difference in sample proportions in Formula 9.8:

FORMULA 9.8

$$\sigma_{p-p} = \sqrt{P_u(1 - P_u)}\sqrt{\frac{N_1 + N_2}{N_1 N_2}}$$

This value is then substituted into the formula for computing the test statistic, presented as Formula 9.9:

FORMULA 9.9

$$Z(\text{obtained}) = \frac{(P_{s1} - P_{s2}) - (P_{u1} - P_{u2})}{\sigma_{p-p}}$$

where $(P_{s1} - P_{s2})$ = the difference between the sample proportions
$(P_{u1} - P_{u2})$ = the difference between the population proportions
σ_{p-p} = the standard deviation of the sampling distribution of the difference between sample proportions

Application 9.2

Do attitudes toward sex vary by gender? The respondents in a national survey have been asked if they think that premarital sex is "always wrong" or only "sometimes wrong." The proportion of each sex that feels that premarital sex is always wrong is

Females	Males
$P_{s1} = 0.35$	$P_{s2} = 0.32$
$N_1 = 450$	$N_2 = 417$

Females are more likely to say that premarital sex is always wrong. Is the difference significant? The table presents all the information we will need to conduct a test of the null hypothesis following the familiar five-step model with alpha set at .05, two-tailed test.

Step 1. Making Assumptions and Meeting Test Requirements.

Model: Independent random samples
Level of measurement is nominal
Sampling distribution is normal

Step 2. Stating the Null Hypothesis.

$$H_0: P_{u1} = P_{u2}$$
$$(H_1: P_{u1} \neq P_{u2})$$

Step 3. Selecting the Sampling Distribution and Establishing the Critical Region.

Sampling distribution = Z distribution
Alpha = 0.05, two-tailed
$Z(\text{critical}) = \pm 1.96$

Step 4. Computing the Test Statistic. Remember to start with Formula 9.7, substitute the value for P_u into Formula 9.8, and then substitute that value into Formula 9.10 to solve for Z(obtained).

$$P_u = \frac{N_1 P_{s1} + N_2 P_{s2}}{N_1 + N_2}$$
$$P_u = \frac{(450)(0.35) + (417)(0.32)}{450 + 417}$$
$$P_u = \frac{290.94}{867}$$
$$P_u = 0.34$$

$$\sigma_{p-p} = \sqrt{P_u(1 - P_u)}\sqrt{\frac{N_1 + N_2}{N_1 N_2}}$$
$$\sigma_{p-p} = \sqrt{(0.34)(0.66)}\sqrt{\frac{450 + 417}{(450)(417)}}$$
$$\sigma_{p-p} = \sqrt{0.2244}\sqrt{0.0046}$$
$$\sigma_{p-p} = (0.47)(0.068)$$
$$\sigma_{p-p} = 0.032$$

$$Z(\text{obtained}) = \frac{(P_{s1} - P_{s2})}{\sigma_{p-p}}$$
$$Z(\text{obtained}) = \frac{0.35 - 0.32}{0.032}$$
$$Z(\text{obtained}) = \frac{0.030}{0.032}$$
$$Z(\text{obtained}) = 0.94$$

Step 5. Making a Decision and Interpreting the Results of the Test. With an obtained Z score of 0.94, we would fail to reject the null hypothesis. Females are not significantly more likely to feel that premarital sex is always wrong.

As was the case with sample means, the second term in the numerator is assumed to be zero by the null hypothesis. Therefore, the formula reduces to

FORMULA 9.10

$$Z(\text{obtained}) = \frac{(P_{s1} - P_{s2})}{\sigma_{p-p}}$$

Remember to solve these equations in order, starting with Formula 9.7 (and skipping Formula 9.9).

A Test of Hypothesis Between Two Sample Proportions (Large Samples). An example will make these procedures clearer. Suppose we are researching social networks among senior citizens and wonder if there is a racial difference in the number of memberships in clubs and other organizations. Assume that random samples of black and white senior citizens have been selected and that each respondent has been classified as high or low in terms of the number of memberships he or she holds in voluntary associations. Is there a statistically significant difference in the participation patterns of black and white elderly? The proportion of each group classified as "high" in participation and sample size for both groups are as reported here:

Sample 1 (Black Senior Citizens)	Sample 2 (White Senior Citizens)
$P_{s1} = 0.34$	$P_{s2} = 0.25$
$N_1 = 83$	$N_2 = 103$

Step 1. Making Assumptions and Meeting Test Requirements.

Model: Independent random samples
Level of measurement is nominal
Sampling distribution is normal

Step 2. Stating the Null Hypothesis. Since no direction has been predicted, this will be a two-tailed test.

$$H_0: P_{u1} = P_{u2}$$
$$(H_1: P_{u1} \neq P_{u2})$$

Step 3. Selecting the Sampling Distribution and Establishing the Critical Region. Since sample size is large, the Z distribution will be used to establish the critical region. Setting alpha at 0.05, we have

Sampling distribution = Z distribution
Alpha = 0.05, two-tailed
Z(critical) = ±1.96

Step 4. Computing the Test Statistic. Begin with the formula for estimating P_u (Formula 9.7), substitute the resultant value into Formula 9.8, and then solve for Z(obtained) with Formula 9.10.

$$P_u = \frac{N_1P_{s1} + N_2P_{s2}}{N_1 + N_2}$$

$$P_u = \frac{(83)(0.34) + (103)(0.25)}{83 + 103}$$

$$P_u = 0.29$$

$$\sigma_{p-p} = \sqrt{P_u(1 - P_u)}\sqrt{\frac{N_1 + N_2}{N_1N_2}}$$

$$\sigma_{p-p} = \sqrt{(0.29)(0.71)}\sqrt{\frac{83 + 103}{(83)(103)}}$$

$$\sigma_{p-p} = (0.45)(0.15)$$

$$\sigma_{p-p} = 0.07$$

$$Z(\text{obtained}) = \frac{(P_{s1} - P_{s2})}{\sigma_{p-p}}$$

$$Z(\text{obtained}) = \frac{0.34 - 0.25}{0.07}$$

$$Z(\text{obtained}) = 1.29$$

Step 5. Making a Decision and Interpreting the Results of the Test. Since the test statistic, Z(obtained) = 1.29, does not fall into the critical region as marked by the Z(critical) of ±1.96, we fail to reject the null hypothesis. The difference between the sample proportions is no greater than what would be expected if the null hypothesis were true and only random chance were operating. Black and white senior citizens are not significantly different in terms of participation patterns as measured in this test. *(For practice in testing the significance of the difference between sample proportions, see problems 9.10 to 9.14 and 9.15a to c.)*

9.5 THE LIMITATIONS OF HYPOTHESIS TESTING: SIGNIFICANCE VERSUS IMPORTANCE

Given that we are usually interested in rejecting the null hypothesis, we should take a moment to consider systematically the factors that affect our decision in step 5. Generally speaking, the probability of rejecting the null hypothesis is a function of four independent factors:

1. The size of the observed difference(s)
2. The alpha level
3. The use of one- or two-tailed tests
4. The size of the sample

Only the first of these four is not under the direct control of the researcher. The size of the difference (either between the sample outcome and the population value or between two sample outcomes) is partly a function of the testing procedures (that is, how variables are measured) but should generally reflect the underlying realities we are trying to probe.

The relationship between alpha level and the probability of rejection is straightforward. The higher the alpha level, the larger the critical region, the higher the percentage of all possible sample outcomes that fall in the critical region, and the greater the probability of rejection. Thus, it is easier to reject the H_0 at the 0.05 level than at the 0.01 level, and easier still at the 0.10 level. The danger here, of course, is that higher alpha levels will lead to more frequent Type I errors, and we might find ourselves declaring small differences to be statistically significant. In similar fashion, using a one-tailed test will increase the probability of rejection (assuming that the proper direction has been predicted).

The final factor is sample size: With all other factors constant, the probability of rejecting H_0 increases with sample size. In other words, the larger the sample, the more likely we are to reject the null hypothesis and, with very large samples (say samples with thousands of cases), we may declare small, unimportant differences to be statistically significant.

This relationship may appear to be surprising, but the reasons for it can be appreciated with a brief consideration of the formulas used to compute test statistics in step 4. In all these formulas, for all tests of significance, sample size (N) is in the "denominator of the denominator." Algebraically, this is equivalent to being in the numerator of the formula and means that the value of the test statistic is directly proportional to N and that the two will increase together. To illustrate, consider Table 9.1, which shows the value of the test statistic for single-sample means from samples of various sizes. The value of the test statistic, Z(obtained), increases as N increases, even though none of the other terms in the formula changes. This pattern of higher probabilities for rejecting H_0 with larger samples holds for all tests of significance.

TABLE 9.1 TEST STATISTICS FOR SINGLE-SAMPLE MEANS COMPUTED FROM SAMPLES OF VARIOUS SIZES ($\bar{X} = 80$, $\mu = 79$, $s = 5$ throughout)

Sample Size	Test Statistic, Z (Obtained)
100	1.99
200	2.82
500	4.47

On one hand, the relationship between sample size and the probability of rejecting the null should not alarm us unduly. Larger samples are, after all, better approximations of the populations they represent. Thus, decisions based on larger samples can be trusted more than decisions based on small samples.

On the other hand, this relationship clearly underlines what is perhaps the most important limitation of hypothesis testing. Simply because a difference is statistically significant does not guarantee that it is important in any other sense. Particularly with very large samples, relatively small differences may be statistically significant. Even with small samples, of course, differences that are otherwise trivial or uninteresting may be statistically significant. The crucial point is that statistical significance and theoretical or practical importance can be two very different things. Statistical significance is a necessary but not sufficient condition for theoretical or practical importance. A difference that is not statistically significant is almost certainly unimportant. However, significance by itself does not guarantee importance. Even when it is clear that the research results were not produced by random chance, the researcher must still assess their importance. Do they firmly support a theory or hypothesis? Are they clearly consistent with a prediction or analysis? Do they strongly indicate a line of action in solving some problem? These are the kinds of questions a researcher must ask when assessing the importance of the results of a statistical test.

Also, we should note that researchers have access to some very powerful ways of analyzing the importance (vs. the statistical significance) of research results. These statistics include bivariate measures of association and multivariate statistical techniques, are introduced in Parts III and IV of this text.

9.6 INTERPRETING STATISTICS: ARE THERE SIGNIFICANT DIFFERENCES IN INCOME BETWEEN MEN AND WOMEN?

In the United States, as in many other nations around the globe, concerted efforts have been made to equalize working conditions for men and women. How successful have these efforts been? Do significant differences in the earnings of men and women persist? Is there a "gender gap" in income?

Note that we could answer these questions with absolute certainty only if we knew the earnings of every single man and woman in the United States. The U.S. Bureau of the Census does publish information on income and gender but usually in terms of medians, not means. Because income information for the population is not readily available, we will investigate the relationship between gender and income by using statistics calculated on randomly selected samples of men and women. If the difference between the sample means of men and women are large enough, we can infer that there is a difference in average salaries between men and women in the population. The General Social Survey, or GSS, is given to randomly selected samples of Americans and will be used as a basis for this test.

Before conducting the test, we need to deal with two issues. First and more importantly, the GSS measures personal income with a series of categories rather than in actual dollars. In other words, respondents were asked to choose from a list of salary ranges (e.g., $10,000 to $12,499, $12,500 to $14,999, and so forth). Thus, income is measured at the ordinal level, and we need an interval-ratio dependent variable to compute sample means and test for the significance of the difference between them. To deal with this problem, we can convert the variable to a form that more closely approximates interval-ratio data by substituting the midpoints for each of the salary intervals. For example, instead of trying to work

READING STATISTICS 6: Hypothesis Testing

Professional researchers use a vocabulary that is much terser than ours when presenting the results of tests of significance. This is partly because of space limitations in scientific journals and partly because professional researchers can assume a certain level of statistical literacy in their audiences. Thus, they omit many of the elements—such as the null hypothesis and the critical region—that we have been so careful to state.

Instead, researchers report only the sample values (for example, means or proportions), the value of the test statistic (for example, a *Z* or *t* score), the alpha level, the degrees of freedom (if applicable), and sample size. The results of the example problem in Section 9.3 might be reported in the professional literature as "The difference between the sample means of 2.37 (suburban families) and 2.78 (center-city families) was tested and found to be significant ($t = -2.16$, df = 77, $p < 0.05$)." Note that the alpha level is reported as "$p < 0.05$." This is shorthand for "the probability of a difference of this magnitude occurring by chance alone, if the null hypothesis of no difference is true, is less than 0.05" and is a good illustration of how researchers can convey a great deal of information in just a few symbols. In a similar fashion, our somewhat long-winded statement "The test statistic falls in the critical region and, therefore, the null hypothesis is rejected" is rendered tersely and simply: "The difference . . . was . . . found to be significant."

When researchers need to report the results of many tests of significance, they will often use a summary table to report the sample information and whether the difference is significant at a certain alpha level. If you read the researcher's description and analysis of such tables, you should have little difficulty interpreting and understanding them. As a final note, these comments about how significance tests are reported in the literature apply to all of the tests of hypotheses covered in Part II of this text.

Statistics in the Professional Literature

Sociologists Dana Haynie and Scott Smith used a representative national sample (the National Longitudinal Study of Adolescent Health) to study the relationship between residential mobility and violence for teenagers. The researchers examined variables that might affect the relationship between mobility and violence, including race, family type, psychological depression, and the characteristics of the adolescent's friendship networks. The following table presents some of their findings, using both means and percentages. All of the differences included in the table were statistically significant.

Violence, the dependent variable, was measured by asking the respondents about their involvement on six violent activities, including fighting and using a weapon to threaten someone. The scale measur-

	Movers (N = 1479)		Stayers (N = 6559)		Significant at
	$\bar{X}$ or %	s	$\bar{X}$ or %	s	p < 0.05?
Dependent Variable					
Violence	.67	1.15	.47	1.00	Yes
Background Variables					
Female	55.05%		52.09%		Yes
Two-parent family	62.98%		75.92%		Yes
Parent education	6.10	2.03	6.31	2.03	Yes
Distress					
Depression index	11.62	7.83	10.62	7.83	Yes
Network Behavior					
Peer deviance	3.12	2.62	2.88	2.62	Yes

(*continued next page*)

READING STATISTICS 6: *(continued)*

ing violence ranged from 0 to 3, so the means reported in the table indicate that violence was uncommon—or at least closer to zero than the maximum score of 3—for both groups. Still, residentially mobile adolescents were significantly more violent than "stayers." There were also significant differences between movers and stayers on a number of other variables that might impact the relationship between mobility and violence: gender (movers had a significantly higher proportion of females), family characteristics (the parents of movers were significantly less educated), level of depression (movers were, on the average, significantly more depressed than stayers), and the behavior of the network of friends to which they belonged (movers had significantly more deviant peers than stayers).

How did all these factors affect the relationship between mobility and violence for teens? You can follow up by consulting the actual article for yourself (see the following citation).

Haynie, Dana, and South, Smith. 2005. "Residential Mobility and Adolescent Violence." *Social Forces*: 84(1):361–374.

with the interval "12,500 to 14,999," we would use the midpoint ($13,750) as a score.

This technique makes income a more numerical variable, but it also creates possible inaccuracies. Estimates to the population parameters from this modified variable should be treated only as approximations. However, since we are concerned with the *difference* in income rather than actual income itself, we should still be able to come to some conclusions.

Second, we shouldn't compare the incomes of *all* men and *all* women because some difference could be the result of the fact that women are less likely to be in the paid labor market (i.e., they are more likely to be occupied as wives and mothers) and, thus, less likely to have an income. We will deal with this by restricting the comparison to respondents who work full time.

Once these adjustments have been made, we can test for the significance of the difference in income. The sample information calculated from the full 2006 GSS database (not the smaller version used for SPSS exercises in this text) is:

Males	Females
$\bar{X}_1 = 52{,}473.32$	$\bar{X}_2 = 35{,}928.85$
$s_1 = 42{,}138.47$	$s_2 = 29{,}431.02$
$N_1 = 1{,}068$	$N_2 = 889$

It appears that there is a large gender gap in income and that males average almost $17,000 a year more than females. Is this difference in sample means significant? Could it have occurred by random chance? Since we are dealing with large samples, we will use the test of significance described in Section 9.2. The null hypothesis is that males and females have the same average salary in the population ($\mu_1 = \mu_2$). We will skip the customary trip through the five-step model and simply report that the Z(obtained) calculated with Formula 9.2 (step 4) is 10.18, much greater than the customary Z(critical) score of ± 1.96 associated with an alpha level of 0.05. The difference in sample means is so large that we must reject the null hypothesis and conclude (with a probability of error of 0.05) that the population means are different: Males earn significantly more than females.

Is the gender gap in income due to differences in levels of education? If females were significantly less educated than males, this would account for at least

Application 9.3

The World Values Survey (WVS) has been administered to random samples in a variety of nations periodically since 1981. Like the General Social Survey used for computer applications in this text, the WVS tests opinions on a variety of issues and concerns. Some results of the surveys are available on the Internet at http://www.worldvaluessurvey.org/.

One item on the WVS asks about support for abortion. Are there significant differences in support between males and females in Canada and the United States? Support for abortion was measured with a ten-point scale, with 10 representing the most support and 1 indicating the most opposition. Thus, higher scores indicate more support. Results are shown for the most recent year available.

Canada (2000)		United States (1999)	
Males	Females	Males	Females
$\bar{X}_1 = 4.80$	$\bar{X}_2 = 4.48$	$\bar{X}_1 = 4.58$	$\bar{X}_2 = 4.21$
$s_1 = 2.99$	$s_2 = 3.07$	$s_1 = 2.96$	$s_2 = 2.97$
$N_1 = 927$	$N_2 = 958$	$N_1 = 596$	$N_2 = 595$

The sample results show that men are more supportive in both nations, but the differences seem small. Are they significant? We will test for the significance of these differences, but, to conserve space, only the most essential steps of the five-step model will be reported. Significance level will be set at 0.05 and the tests will be two-tailed. The null hypothesis will be that there is no difference in support for abortion between *all* males and *all* females in the respective nations (H_0: $\mu_1 = \mu_2$).

We will use Formulas 9.2 and 9.4 to compute the test statistics (step 4).

For Canada:

$$\sigma_{\bar{X}-\bar{X}} = \sqrt{\frac{s_1^2}{N_1 - 1} + \frac{s_2^2}{N_2 - 1}}$$

$$\sigma_{\bar{X}-\bar{X}} = \sqrt{\frac{(2.99)^2}{926} + \frac{(3.07)^2}{957}}$$

$$\sigma_{\bar{X}-\bar{X}} = \sqrt{\frac{8.94}{926} + \frac{9.43}{957}}$$

$$\sigma_{\bar{X}-\bar{X}} = \sqrt{.010 + .010}$$

$$\sigma_{\bar{X}-\bar{X}} = \sqrt{.02} = 0.14$$

$$Z(\text{obtained}) = \frac{(\bar{X}_1 - \bar{X}_2)}{\sigma_{\bar{X}-\bar{X}}} = \frac{4.80 - 4.48}{0.14}$$

$$Z(\text{obtained}) = \frac{0.32}{0.14} = 2.29$$

For the United States:

$$\sigma_{\bar{X}-\bar{X}} = \sqrt{\frac{s_1^2}{N_1 - 1} + \frac{s_2^2}{N_2 - 1}}$$

$$\sigma_{\bar{X}-\bar{X}} = \sqrt{\frac{(2.96)^2}{595} + \frac{(2.97)^2}{594}}$$

$$\sigma_{\bar{X}-\bar{X}} = \sqrt{\frac{8.76}{595} + \frac{8.82}{594}}$$

$$\sigma_{\bar{X}-\bar{X}} = \sqrt{.015 + .015}$$

$$\sigma_{\bar{X}-\bar{X}} = \sqrt{.03} = 0.17$$

$$Z(\text{obtained}) = \frac{(\bar{X}_1 - \bar{X}_2)}{\sigma_{\bar{X}-\bar{X}}} = \frac{4.58 - 4.21}{0.17}$$

$$Z(\text{obtained}) = \frac{0.37}{0.17} = 2.18$$

For both tests, we reject the null hypothesis. For both Canada and the United States, men are significantly more supportive of abortion than women.

some of the difference in income. Using the 2006 General Social Survey again to get information on years of education for random samples of males and females, we have:

Males	Females
$\bar{X}_1 = 13.35$	$\bar{X}_2 = 13.25$
$s_1 = 3.38$	$s_2 = 3.11$
$N_1 = 1998$	$N_2 = 2501$

We can see that both males and females average more than 13 years of schooling (one and a half years more than a high school education) and that there is almost no difference in the sample means. In fact, the test statistic computed in step 4 is a *Z*(obtained) of 1.02. The test statistic is not in the critical region, as marked by a *Z*(critical) score of ±1.96 at the 0.05 level, so we must fail to reject the null hypothesis of no difference. Males and females have essentially equal levels of schooling in the population, and education cannot account for the differences in income.

These results suggest that females are getting a lower return in income for their education. Although females and males are essentially equal in preparation for work and careers (at least as measured by level of schooling), a large and statistically significant gender gap in income persists. Why? To answer this question in detail would carry us far beyond the bounds of a statistics text. However, we can suggest that one important part of the answer lies in the tendency of men and women to pursue different kinds of careers and jobs. That is, men tend to dominate the more lucrative, higher-prestige occupations (lawyer, doctor), while women are concentrated in jobs that have lower levels of remuneration (elementary school teacher, nurse). For example, in 2005, women comprised 92% of all registered nurses and 86% of all paralegals and legal assistants vs. only 32% of all doctors and 30% of all lawyers.[1] Thus, the gender gap in income is sustained, in part, by the fact that women are more likely to be employed in occupations that command lower salaries.

[1] U.S. Bureau of the Census, 2007. *Statistical Abstract of the United States, 2007.* Washington, D.C.: Government Printing Office, pp. 388–389.

SUMMARY

1. A common research situation is to test for the significance of the difference between two populations. Sample statistics are calculated for random samples of each population, and then we test for the significance of the difference between the samples as a way of inferring differences between the specified populations.
2. When sample information is summarized in the form of sample means and *N* is large, the *Z* distribution is used to find the critical region. When *N* is small, the *t* distribution is used to establish the critical region. In the latter circumstance, we must also assume equal population variances before forming a pooled estimate of the standard deviation of the sampling distribution.
3. Differences in sample proportions may also be tested for significance. For large samples, the *Z* distribution is used to find the critical region.
4. In all tests of hypothesis, a number of factors affect the probability of rejecting the null hypothesis: the size of the difference, the alpha level, the use of one- versus two-tailed tests, and sample size. Statistical significance is not the same thing as theoretical or practical importance. Even after a difference is found to be statistically significant, the researcher must still demonstrate the relevance or importance of his or her findings. The statistics presented in Parts III and IV of this text will give us the tools we need to deal directly with issues beyond statistical significance.

SUMMARY OF FORMULAS

Test statistic for two sample means, large samples:

9.1 $$Z(\text{obtained}) = \frac{(\bar{X}_1 - \bar{X}_2) - (\mu_1 - \mu_2)}{\sigma_{\bar{X}-\bar{X}}}$$

Test statistic for two sample means, large samples (simplified formula):

9.2 $$Z(\text{obtained}) = \frac{(\bar{X}_1 - \bar{X}_2)}{\sigma_{\bar{X}-\bar{X}}}$$

Standard deviation of the sampling distribution of the difference in sample means, large samples:

9.3 $$\sigma_{\bar{X}-\bar{X}} = \sqrt{\frac{\sigma_1^2}{N_1} + \frac{\sigma_2^2}{N_2}}$$

Pooled estimate of the standard deviation of the sampling distribution of the difference in sample means, large samples:

9.4 $$\sigma_{\bar{X}-\bar{X}} = \sqrt{\frac{s_1^2}{N_1 - 1} + \frac{s_2^2}{N_2 - 1}}$$

Pooled estimate of the standard deviation of the sampling distribution of the difference in sample means, small samples:

9.5 $$\sigma_{\bar{X}-\bar{X}} = \sqrt{\frac{N_1 s_1^2 + N_2 s_2^2}{N_1 + N_2 - 2}}\sqrt{\frac{N_1 + N_2}{N_1 N_2}}$$

Test statistic for two sample means, small samples:

9.6 $$t(\text{obtained}) = \frac{(\bar{X}_1 - \bar{X}_2)}{\sigma_{\bar{X}-\bar{X}}}$$

Pooled estimate of population proportion, large samples:

9.7 $$P_u = \frac{N_1 P_{s1} + N_2 P_{s2}}{N_1 + N_2}$$

Standard deviation of the sampling distribution of the difference in sample proportions, large samples:

9.8 $$\sigma_{p-p} = \sqrt{P_u(1 - P_u)}\sqrt{\frac{N_1 + N_2}{N_1 N_2}}$$

Test statistic for two sample proportions, large samples:

9.9 $$Z(\text{obtained}) = \frac{(P_{s1} - P_{s2}) - (P_{u1} - P_{u2})}{\sigma_{p-p}}$$

Test statistic for two sample proportions, large samples (simplified formula):

9.10 $$Z(\text{obtained}) = \frac{(P_{s1} - P_{s2})}{\sigma_{p-p}}$$

GLOSSARY

Independent random samples. Random samples gathered in such a way that the selection of a particular case for one sample has no effect on the probability that any other particular case will be selected for the other samples.

Pooled estimate. An estimate of the standard deviation of the sampling distribution of the difference in sample means based on the standard deviations of both samples.

σ_{p-p}. symbol for the standard deviation of the sampling distribution of the differences in sample proportions.

$\sigma_{\bar{X}-\bar{X}}$. Symbol for the standard deviation of the sampling distribution of the differences in sample means.

PROBLEMS

9.1 For each of the following, test for the significance of the difference in sample statistics using the five-step model. *(HINT: Remember to solve Formula 9.4 before attempting to solve Formula 9.2. Also, in Formula 9.4, perform the mathematical operations in the proper sequence. First square each sample standard deviation, then divide by* $N - 1$*, add the resultant values, and then find the square root of the sum.)*

a.

Sample 1	Sample 2
$\bar{X}_1 = 72.5$	$\bar{X}_2 = 76.0$
$s_1 = 14.3$	$s_2 = 10.2$
$N_1 = 136$	$N_2 = 257$

b.

Sample 1	Sample 2
$\bar{X}_1 = 107$	$\bar{X}_2 = 103$
$s_1 = 14$	$s_2 = 17$
$N_1 = 175$	$N_2 = 200$

9.2 SOC Gessner and Healey administered questionnaires to samples of undergraduates. Among other things, the questionnaires contained a scale that measured attitudes toward interpersonal violence (higher scores indicate greater approval of interpersonal violence). Test the results as reported here for sexual, racial, and social-class differences.

a.

Sample 1 (Males)	Sample 2 (Females)
$\bar{X}_1 = 2.99$	$\bar{X}_2 = 2.29$
$s_1 = 0.88$	$s_2 = 0.91$
$N_1 = 122$	$N_2 = 251$

b.

Sample 1 (Blacks)	Sample 2 (Whites)
$\bar{X}_1 = 2.76$	$\bar{X}_2 = 2.49$
$s_1 = 0.68$	$s_2 = 0.91$
$N_1 = 43$	$N_2 = 304$

c.

Sample 1 (White Collar)	Sample 2 (Blue Collar)
$\bar{X}_1 = 2.46$	$\bar{X}_2 = 2.67$
$s_1 = 0.91$	$s_2 = 0.87$
$N_1 = 249$	$N_2 = 97$

d. Summarize your results in terms of the significance and the direction of the differences. Which of these three factors seems to make the biggest difference in attitudes toward interpersonal violence?

9.3 SOC Do athletes in different sports vary in terms of intelligence? Reported here are College Board scores of random samples of college basketball and football players. Is there a significant difference? Write a sentence or two explaining the difference.

a.

Sample 1 (Basketball Players)	Sample 2 (Football Players)
$\bar{X}_1 = 460$	$\bar{X}_2 = 442$
$s_1 = 92$	$s_2 = 57$
$N_1 = 102$	$N_2 = 117$

What about male and female college athletes?

b.

Sample 1 (Males)	Sample 2 (Females)
$\bar{X}_1 = 452$	$\bar{X}_2 = 480$
$s_1 = 88$	$s_2 = 75$
$N_1 = 107$	$N_2 = 105$

9.4 PA A number of years ago, the fire department in Shinbone, Kansas, began recruiting minority group members through an affirmative action program. In terms of efficiency ratings as compiled by their superiors, how do the affirmative action employees rate? The ratings of random samples of both groups were collected, and the results are reported here (higher ratings indicate greater efficiency).

Sample 1 (Affirmative Action)	Sample 2 (Regular)
$\bar{X}_1 = 15.2$	$\bar{X}_2 = 15.5$
$s_1 = 3.9$	$s_2 = 2.0$
$N_1 = 97$	$N_2 = 100$

Write a sentence or two of interpretation.

9.5 SOC Are middle-class families more likely than working-class families to maintain contact with kin? Write a paragraph summarizing the results of these tests.

a. A sample of middle-class families reported an average of 7.3 visits per year with close kin, while a sample of working-class families averaged 8.2 visits. Is the difference significant?

Visits

Sample 1 (Middle Class)	Sample 2 (Working Class)
$\bar{X}_1 = 7.3$	$\bar{X}_2 = 8.2$
$s_1 = 0.3$	$s_2 = 0.5$
$N_1 = 89$	$N_2 = 55$

b. The middle-class families averaged 2.3 phone calls and 8.7 e-mail messages per month with close kin. The working-class families averaged 2.7 calls and 5.7 e-mail messages per month. Are these differences significant?

Phone Calls

Sample 1 (Middle Class)	Sample 2 (Working Class)
$\bar{X}_1 = 2.3$	$\bar{X}_2 = 2.7$
$s_1 = 0.5$	$s_2 = 0.8$
$N_1 = 89$	$N_2 = 55$

E-mail Messages

Sample 1 (Middle Class)	Sample 2 (Working Class)
$\bar{X}_1 = 8.7$	$\bar{X}_2 = 5.7$
$s_1 = 0.3$	$s_2 = 1.1$
$N_1 = 89$	$N_2 = 55$

9.6 SOC Are college students who live in dormitories significantly more involved in campus life than students who commute to campus? The following data report the average number of hours per week students devote to extracurricular activities. Is the difference between these randomly selected samples of commuter and residential students significant?

Sample 1 (Residential)	Sample 2 (Commuter)
$\bar{X}_1 = 12.4$	$\bar{X}_2 = 10.2$
$s_1 = 2.0$	$s_2 = 1.9$
$N_1 = 158$	$N_2 = 173$

9.7 SOC Are senior citizens who live in retirement communities more socially active than those who live in age-integrated communities? Write a sentence or two explaining the results of these tests. *(HINT: Remember to use the proper formulas for small sample sizes.)*

a. A random sample of senior citizens living in a retirement village reported that they had an average of 1.42 face-to-face interactions per day with their neighbors. A random sample of those living in age-integrated communities reported 1.58 interactions. Is the difference significant?

Sample 1 (Retirement Community)	Sample 2 (Age-integrated Neighborhood)
$\bar{X}_1 = 1.42$	$\bar{X}_2 = 1.58$
$s_1 = 0.10$	$s_2 = 0.78$
$N_1 = 43$	$N_2 = 37$

b. Senior citizens living in the retirement village reported that they had an average of 7.43 telephone calls with friends and relatives each week, while those in the age-integrated communities reported an average of 5.50 calls. Is the difference significant?

Sample 1 (Retirement Community)	Sample 2 (Age-integrated Neighborhood)
$\bar{X}_1 = 7.43$	$\bar{X}_2 = 5.50$
$s_1 = 0.75$	$s_2 = 0.25$
$N_1 = 43$	$N_2 = 37$

9.8 SW As the director of the local Boys Club, you have claimed for years that membership in your club reduces juvenile delinquency. Now a cynical member of your funding agency has demanded proof of your claim. Fortunately, your local sociology department is on your side and springs to your aid with student assistants, computers, and hand calculators at the ready. Random samples of members and nonmembers are gathered and interviewed with respect to their involvement in delinquent activities. Each respondent is asked to enumerate the number of delinquent acts he has engaged in over the past year. The results are in and reported here (the average number of admitted acts of delinquency). What can you tell the funding agency?

Sample 1 (Members)	Sample 2 (Nonmembers)
$\bar{X}_1 = 10.3$	$\bar{X}_2 = 12.3$
$s_1 = 2.7$	$s_2 = 4.2$
$N_1 = 40$	$N_2 = 55$

9.9 SOC A survey has been administered to random samples of respondents in each of five nations. For each nation, are men and women significantly different in terms of their reported levels of satisfaction? Respondents were asked: "How satisfied are you with your life as a whole?" Responses varied from 1 (very dissatisfied) to 10 (very satisfied). Conduct a test for the significance of the difference in mean scores for each nation.

France

Males	Females
$\bar{X}_1 = 7.4$	$\bar{X}_2 = 7.7$
$s_1 = .20$	$s_2 = .25$
$N_1 = 1005$	$N_2 = 1234$

Nigeria

Males	Females
$\bar{X}_1 = 6.7$	$\bar{X}_2 = 7.8$
$s_1 = .16$	$s_2 = .23$
$N_1 = 1825$	$N_2 = 1256$

China

Males	Females
$\bar{X}_1 = 7.6$	$\bar{X}_2 = 7.1$
$s_1 = .21$	$s_2 = .11$
$N_1 = 1400$	$N_2 = 1200$

Mexico

Males	Females
$\bar{X}_1 = 8.3$	$\bar{X}_2 = 9.1$
$s_1 = .29$	$s_2 = .30$
$N_1 = 1645$	$N_2 = 1432$

Japan

Males	Females
$\bar{X}_1 = 8.8$	$\bar{X}_2 = 9.3$
$s_1 = .34$	$s_2 = .32$
$N_1 = 1621$	$N_2 = 1683$

9.10 For each problem, test the sample statistics for the significance of the difference.

a.

Sample 1	Sample 2
$P_{s1} = 0.17$	$P_{s2} = 0.20$
$N_1 = 101$	$N_2 = 114$

b.

Sample 1	Sample 2
$P_{s1} = 0.62$	$P_{s2} = 0.60$
$N_1 = 532$	$N_2 = 478$

9.11 CJ About half of the police officers in Shinbone, Kansas, have completed a special course in investigative procedures. Has the course increased their efficiency in clearing crimes by arrest? The proportions of cases cleared by arrest for samples of trained and untrained officers are reported here.

Sample 1 (Trained)	Sample 2 (Untrained)
$P_{s1} = 0.47$	$P_{s2} = 0.43$
$N_1 = 157$	$N_2 = 113$

9.12 SW A large counseling center needs to evaluate several experimental programs. Write a paragraph summarizing the results of these tests. Did the new programs work?

a. One program is designed for divorce counseling; the key feature of the program is its counselors, who are married couples working in teams. About half of all clients have been randomly assigned to this special program and half to the regular program, and the proportion of cases that eventually ended in divorce was recorded for both. The results for random samples of couples from both programs are reported here. In terms of preventing divorce, did the new program work?

Sample 1 (Special Program)	Sample 2 (Regular Program)
$P_{s1} = 0.53$	$P_{s2} = 0.59$
$N_1 = 78$	$N_2 = 82$

b. The agency is also experimenting with peer counseling for depressed children. About half of all clients were randomly assigned to peer counseling. After the program had run for a year, a random sample of children from the new program was compared with a random sample of children who did not receive peer counseling. In terms of the percentage who were judged to be "much improved," did the new program work?

Sample 1 (Peer Counseling)	Sample 2 (No Peer Counseling)
$P_{s1} = 0.10$	$P_{s2} = 0.15$
$N_1 = 52$	$N_2 = 56$

9.13 SOC At St. Algebra College, the sociology and psychology departments have been feuding for years about the respective quality of their programs. In an attempt to resolve the dispute, you have gathered data about the graduate school experience of random samples of both groups of majors. The results are presented here: the proportion of majors who applied to graduate schools, the proportion of majors accepted into their preferred programs, and the proportion of these who completed their programs. As measured by these data, is there a significant difference in program quality?

a. Proportion of majors who applied to graduate school:

Sample 1 Sociology	Sample 2 Psychology
$P_{s1} = 0.53$	$P_{s2} = 0.40$
$N_1 = 150$	$N_2 = 175$

b. Proportion accepted by program of first choice:

Sample 1 Sociology	Sample 2 Psychology
$P_{s1} = 0.75$	$P_{s2} = 0.85$
$N_1 = 80$	$N_2 = 70$

c. Proportion completing the programs:

Sample 1 Sociology	Sample 2 Psychology
$P_{s1} = 0.75$	$P_{s2} = 0.69$
$N_1 = 60$	$N_2 = 60$

9.14 CJ The local police chief started a "crimeline" program some years ago and wonders if it's really working. The program publicizes unsolved violent crimes in the local media and offers cash rewards for information leading to arrests. Are "featured" crimes more likely to be cleared by arrest than other violent crimes? Results from random samples of both types of crimes are reported as follows:

Sample 1 Crimeline Crimes Cleared by Arrest	Sample 2 Noncrimeline Crimes Cleared by Arrest
$P_{s1} = 0.35$	$P_{s2} = 0.25$
$N_1 = 178$	$N_2 = 212$

9.15 SOC Some results from a survey administered to a nationally representative sample are reported here in terms of differences by sex. Which of these differences, if any, are significant? Write a sentence or two of interpretation for each test.

a. Proportion favoring the legalization of marijuana:

Sample 1 (Males)	Sample 2 (Females)
$P_{s1} = 0.37$	$P_{s2} = 0.31$
$N_1 = 202$	$N_2 = 246$

b. Proportion strongly agreeing that "kids are life's greatest joy":

Sample 1 (Males)	Sample 2 (Females)
$P_{s1} = 0.47$	$P_{s2} = 0.58$
$N_1 = 251$	$N_2 = 351$

c. Proportion voting for President Bush in 2004:

Sample 1 (Males)	Sample 2 (Females)
$P_{s1} = 0.59$	$P_{s2} = 0.47$
$N_1 = 399$	$N_2 = 509$

d. Average hours spent with e-mail each week

Sample 1 (Males)	Sample 2 (Females)
$\bar{X}_1 = 4.18$	$\bar{X}_2 = 3.38$
$s_1 = 7.21$	$s_2 = 5.92$
$N_1 = 431$	$N_2 = 535$

e. Average rate of church attendance:

Sample 1 (Males)	Sample 2 (Females)
$\bar{X}_1 = 3.19$	$\bar{X}_2 = 3.99$
$s_1 = 2.60$	$s_2 = 2.72$
$N_1 = 641$	$N_2 = 808$

f. Number of children:

Sample 1 (Males)	Sample 2 (Females)
$\bar{X}_1 = 1.49$	$\bar{X}_2 = 1.93$
$s_1 = 1.50$	$s_2 = 1.50$
$N_1 = 635$	$N_2 = 803$

SPSS for Windows

Using *SPSS for Windows* to Test the Significance of the Difference Between Two Means

SPSS DEMONSTRATION 9.1 Do Men or Women Watch More TV?

SPSS for Windows includes several tests for the significance of the difference between means. In this demonstration, we'll use the **Independent-Samples T Test,** the test we covered in Section 9.2, to test for the significance of the difference between men and women in average hours of reported TV watching. If there is a statistically significant difference between the sample means for men and women, we can conclude that the populations (all U.S. adult men and women) are different on this variable.

Start *SPSS for Windows* and load the 2006 GSS database. From the main menu bar, click **Analyze,** then **Compare Means,** and then **Independent-Samples T Test.** The **Independent-Samples T Test** dialog box will open, with the usual list of variables on the left. Find and move the cursor over *tvhours* (the variable label is HOURS PER DAY WATCHING TV) and click the top arrow in the middle of the window to move *tvhours* to the **Test Variable(s)** box. Next, find and highlight *sex* (the label is RESPONDENTS SEX) and click the bottom arrow in the middle of the window to move *sex* to the **Grouping Variable** box. Two question marks will appear in the **Grouping Variable** box, and the **Define Groups** button will become active. SPSS needs to know which cases go in which groups, and, in the case at hand, the instructions we need to supply are straightforward. Males (indicated by a score of 1 on *sex*) go into group 1 and females (a score of 2) will go into group 2.

Click the **Define Groups** button, and the **Define Groups** window will appear. The cursor will be blinking in the box beside Group 1—SPSS is asking for the score that will determine which cases go into this group. Type a 1 in this box (for males) and then click the box next to Group 2 and type a 2 (for females). Click **Continue** to return to the **Independent-Samples T Test** window and click **OK,** and the following output will be produced. (The 95% confidence interval automatically produced by this program has been deleted to conserve space.)

Group Statistics

	RESPONDENTS SEX	N	Mean	Std. Deviation	Std. Error Mean
HOURS PER DAY WATCHING TV	MALE	293	2.86	2.233	.130
	FEMALE	325	3.14	2.545	.141

Independent Samples Test

		Levene's Test for Equality of Variances		t-test for Equality of Means				
		F	Sig.	t	df	Sig. (2-tailed)	Mean Difference	Std. Error Difference
HOURS PER DAY WATCHING TV	Equal variances assumed	.943	.332	-1.440	616	.150	-.279	.194
	Equal variances not assumed			-1.450	615.559	.148	-.279	.192

In the first block of output are some descriptive statistics. There were 293 males in the sample, and they watched TV an average of 2.86 hours a day, with a standard deviation of 2.233. The 325 females watched an average of 3.14 hours per day, with a standard deviation of 2.545. We can see from this output that the sample means are different and that, on the average, females watch more TV. Is the difference in sample means significant?

We will skip over the first columns of the next block of output (which reports the results of a test for equality of the population variances). The results of the test for significance are reported in this block. *SPSS for Windows* does a separate test for each assumption about the population variance (see Sections 9.2 and 9.3), but we will look only at the 'Equal variances assumed' reported in the top row. This is basically the same model used in this chapter.

SPSS for Windows reports the value of the obtained *t* score (-1.440), the degrees of freedom (df $= 616$), and the "Sig. (2-tailed)" (.150). This last piece of information is an alpha level (or a "*p*" level—see Reading Statistics 6) except that it is the *exact* probability of getting the observed difference in sample means if only chance is operating. Thus, there is no need to look up the test statistic in a *t* table. This value is greater than .05, our usual indicator of significance. We will fail to reject the null hypothesis and conclude that the difference is not statistically significant. On the average, men and women do not have significantly different TV-viewing habits.

SPSS DEMONSTRATION 9.2 Using the COMPUTE Command to Test for Gender Differences in Attitudes About Abortion

The **Compute** command was introduced in Demonstration 4.2. To refresh your memory, I used **Compute** to create a summary scale (*abscale*) for attitudes toward abortion by adding the scores on the two constituent items (*abrape* and *abany*). Remember that, once created, a computed variable is added to the active file and can be used like any of the variables actually recorded in the file. If you did not save the data file with *abscale* included, you can quickly recreate the variable by following the instructions in Demonstration 4.2.

Here, we will test *abscale* for the significance of the difference by gender. Our question is: Do men and women have different attitudes toward abortion? If the difference in the sample means is large enough, we can reject the null hypothesis and conclude that the populations are different. Before we conduct the test, I should point out that the abortion scale used in this test is only ordinal in level of measurement. Scales like this are often treated as interval-ratio variables, but we should still be cautious in interpreting our results.

Follow the instructions in Demonstration 9.1 for using the **Independent-Samples T TEST** command, and move *abscale* rather than *tvhours* to the **Test Variable(s)** box. If necessary, repeat the procedure for making *sex* the grouping variable. Your output will be as shown:

Group Statistics

	RESPONDENTS SEX	N	Mean	Std. Deviation	Std. Error Mean
ABSCALE	MALE	266	2.7105	.74911	.04593
	FEMALE	341	2.7918	.77120	.04176

Independent Samples Test

		Levene's Test for Equality of Variances		t-test for Equality of Means				
		F	Sig.	t	df	Sig. (2-tailed)	Mean Difference	Std. Error Difference
ABSCALE	Equal variances assumed	.044	.835	-1.304	605	.193	-.08126	.06230
	Equal variances not assumed			-1.309	576.945	.191	-.08126	.06208

The sample means are quite close in value (2.71 versus 2.79), and the test statistic ($t = -1.304$) is not significant, as indicated by the "Sig. (2-tailed)" value of .193. There is no significant difference in support for abortion by gender.

As a final point, let me direct your attention to the "Number of Cases" column. This test was based on 266 men and 341 women, for a total of 607 people. The original sample included almost 1500 people. What happened to all those cases?

Recall from Appendix F that relatively few items on the GSS are given to the entire sample. Furthermore, as I mentioned at the end of Chapter 4, the **Compute** command is designed to drop a case from the computations if it is missing a score on any of the constituent variables. So most of the "missing cases" were not asked about their attitudes about abortion, and the rest did not answer one or both of the constituent abortion items. This phenomenon of diminishing sample size is a common problem in survey research that, at some point, may jeopardize the integrity of the inquiry.

Exercises

9.1 Are men significantly different from women in occupational prestige or the number of hours they work each week? Using Demonstration 9.1 as a guide, substitute *prestg80* and *hrs1* for *tvhours*. Write a sentence or two summarizing the results of this test. (See Reading Statistics 6 for some ideas on writing up results.)

9.2 Using Demonstration 9.2 as a guide, construct a summary scale for attitudes about traditional gender roles using *fepresch* and *fefam* in place of *abrape* and *abany*. Test for the significance of the difference using *sex* as the independent variable. Write a sentence or two summarizing the results of this test.

9.3 What other independent variables might explain differences in opinions about abortion? Select several more independent variables besides *sex*, and conduct additional *t* tests with *abscale* as the dependent variable. Remember that the *t* test requires the independent variable to have *only* two categories, a description that fits only a few variables in the 2006 GSS data set. You can still use independent variables with more than two categories via one of two options:

a. Use the **Grouping Variable** box to select the specific categories of an independent variable that you would like to include. For example, you can compare Protestants with Catholics by choosing *relig* as the grouping variable and selecting 1 (Protestants) and 2 (Catholics) in the **Define Groups** window. To compare Protestants with "Nones," choose 1 and 4 in the **Define Groups** window.

b. You can collapse the scores of variables with more than two categories, such as *attend* or *income06*, by using the **Recode** command. Use Demonstration 2.3 as a guide and collapse your selected independent variables at their median scores or some other logical point. Don't forget to choose the "recode into different variable" option.

Write a sentence or two summarizing the results of these tests.

10 Hypothesis Testing III
The Analysis of Variance

LEARNING OBJECTIVES

By the end of this chapter, you will be able to

1. Identify and cite examples of situations in which ANOVA is appropriate.
2. Explain the logic of hypothesis testing as applied to ANOVA.
3. Perform the ANOVA test, using the five-step model as a guide, and correctly interpret the results.
4. Define and explain the concepts of population variance, total sum of squares, sum of squares between, sum of squares within, mean square estimates, and post hoc tests.
5. Explain the difference between the statistical significance and the importance of relationships between variables.

10.1 INTRODUCTION

In this chapter, we examine a very flexible and widely used test of significance called the **analysis of variance** (often abbreviated as **ANOVA**). This test can be used in a number of situations where previously discussed tests are less than optimum or entirely inappropriate. ANOVA is designed to be used with interval-ratio-level dependent variables and is a powerful tool for analyzing the most sophisticated and precise measurements you are likely to encounter.

It is perhaps easiest to think of ANOVA as an extension of the *t* test for the significance of the difference between two sample means, which was presented in Chapter 9. The *t* test can be used only in situations in which our independent variable has exactly two categories (e.g., Protestants and Catholics). The analysis of variance, on the other hand, is appropriate for independent variables with more than two categories (e.g., Protestants, Catholics, Jews, people with no religious affiliation, and so forth).

To illustrate, suppose we were interested in examining the social basis of support for capital punishment. Why does support for the death penalty vary from person to person? Could there be a relationship between religion (the independent variable) and support for capital punishment (the dependent variable)? Opinion about the death penalty has an obvious moral dimension and may well be affected by a person's religious background.

Suppose that we administered a scale that measures support for capital punishment at the interval-ratio level to a randomly selected sample that includes Protestants, Catholics, Jews, people with no religious affiliation ("Nones"), and people from other religions ("Others"). We will have five categories of subjects, and we will want to see if a particular attitude or opinion varies significantly by the category (religion) into which a person is classified. We will also want to answer other questions, such as: Which religion shows the least or most support for capital punishment? Are Protestants significantly more supportive than Catholics or Jews? How do people with no religious affiliation compare to people in the other categories? The analysis of variance provides a very useful statistical context in which the questions can be addressed.

10.2 THE LOGIC OF THE ANALYSIS OF VARIANCE

For ANOVA, the null hypothesis is that the populations from which the samples are drawn are equal on the characteristic of interest. As applied to our problem, the null hypothesis could be phrased as "People from different religious denominations do not vary in their support for the death penalty" or, symbolically, as $\mu_1 = \mu_2 = \mu_3 = \cdots = \mu_k$. (Note that this is an extended version of the null hypothesis for the two-sample *t* test). As usual, the researcher will normally be interested in rejecting the null and, in this case, showing that support is related to religion.

If the null hypothesis of "no difference" in the populations is true, then any means calculated from randomly selected samples should be roughly equal in value. The average score for the Protestant sample should be about the same as the average score for the Catholics and the Jews, and so forth. Note that the averages are unlikely to be exactly the same value even if the null hypothesis really is true, since we will always encounter some error or chance fluctuations in the measurement process. We are *not* asking: "Are there differences between the samples (or, in our example, the religions)?" Rather, we are asking: "Are the differences between the samples large enough to reject the null hypothesis and justify the conclusion that the populations represented by the samples are different?"

Now consider what kinds of outcomes we might encounter if we actually administered a "Support of Capital Punishment Scale" and organized the scores by religion. Of the infinite variety of possibilities, let's focus on the two extreme outcomes presented in Tables 10.1 and 10.2. In the first set of hypothetical results (Table 10.1), we see that the means and standard deviations of the groups are quite similar. The average scores are about the same for every religious group, and all five groups exhibit about the same dispersion. These results would be quite consistent with the null hypothesis of no difference between the populations from which the samples were selected on support for capital punishment. Neither the average score nor the dispersion of the scores varies in any important way by religion.

Now consider another set of fictitious results, as displayed in Table 10.2. Here we see substantial differences in average score from category to category, with Jews showing the lowest support and Protestants showing the highest. Also, the standard deviations are low and similar from category to category, indicating that there is not much variation within the religions. Table 10.2 shows marked differences *between* religions combined with homogeneity *within* religions, as indicated by the low values of the standard deviations. In other words, there are marked differences from religion to religion but little difference within

TABLE 10.1 SUPPORT FOR CAPITAL PUNISHMENT BY RELIGION (Fictitious Data)

	Protestant	Catholic	Jew	None	Other
Mean =	10.3	11.0	10.1	9.9	10.5
Standard deviation =	2.4	1.9	2.2	1.7	2.0

TABLE 10.2 SUPPORT FOR CAPITAL PUNISHMENT BY RELIGION (Fictitious Data)

	Protestant	Catholic	Jew	None	Other
Mean =	14.7	11.3	5.7	8.3	7.1
Standard deviation =	0.9	0.8	1.0	1.1	0.7

each religion. These results would contradict the null hypothesis and support the notion that support for the death penalty does vary by religion.

In principle, ANOVA proceeds by making the kinds of comparisons outlined above. The test compares the amount of variation between categories (for example, from Protestants to Catholics to Jews to "Nones" to "Others") with the amount of variation within categories (among Protestants, among Catholics, and so forth). The greater the difference *between* categories (as measured by the means) relative to the differences *within* categories (as measured by the standard deviations), the more likely that the null hypothesis of "no difference" is false and can be rejected. If support for capital punishment truly varies by religion, then the sample mean for each religion should be quite different from the others and dispersion within the categories should be relatively low.

10.3 THE COMPUTATION OF ANOVA

Even though we have been thinking of ANOVA as a test for the significance of the difference between sample means, the computational routine actually involves developing two separate estimates of the population variance, σ^2 (hence the name *analysis of variance*). Recall from Chapter 4 that the variance and standard deviation both measure dispersion and that the variance is simply the standard deviation squared. One estimate of the population variance is based on the amount of variation within each of the categories of the independent variable and the other is based on the amount of variation between categories.

Before constructing these estimates, we need to introduce some new concepts and statistics. The first new concept is the total variation of the scores, which is measured by a quantity called the **total sum of squares,** or **SST:**

FORMULA 10.1

$$\text{SST} = \sum X_i - N\bar{X}^2$$

To solve this formula, first find the sum of the squared scores (in other words, square each score and then add up the squared scores). Next, square the mean of all scores, multiply that value by the total number of cases in the sample (N), and subtract that quantity from the sum of the squared scores.

Formula 10.1 may seem vaguely familiar. A similar expression—$\sum(X_i - \bar{X})^2$—appears in the formula for the standard deviation and variance (see Chapter 4). All three statistics incorporate information about the variation of the scores (or, in the case of SST, the squared scores) around the mean (or, in the case of SST, the square of the mean multiplied by N). In other words, all three statistics are measures of the variation, or dispersion, of the scores.

To construct the two separate estimates of the population variance, the total variation (SST) is divided into two components. One component reflects the pattern of variation *within* each of the categories and is called the **sum of squares within (SSW).** In our example problem, SSW would measure the amount of variety in support for the death penalty within each of the religions.

The other component is based on the variation *between* categories and is called the **sum of squares between (SSB).** Again using our example to illustrate, SSB measures how different people in each religion are from each other in their support for capital punishment. SSW and SSB are components of SST, as reflected in Formula 10.2:

FORMULA 10.2

$$\text{SST} = \text{SSB} + \text{SSW}$$

Let's start with the computation of SSB, our measure of the variation in scores between categories. We use the category means as summary statistics to determine the size of the difference from category to category. In other words, we compare the average support for the death penalty for each religion with the average support for all other religions to determine SSB. The formula for the sum of squares between (SSB) is

FORMULA 10.3

$$\text{SSB} = \sum N_k(\overline{X}_k - \overline{X})^2$$

where SSB = the sum of squares between the categories
N_k = the number of cases in a category
$\overline{X}_k$ = the mean of a category

To find SSB, subtract the overall mean of all scores ($\overline{X}$) from each category mean ($\overline{X}_k$), square the difference, multiply by the number of cases in the category, and add the results across all the categories.

The second estimate of the population variance (SSW) is based on the amount of variation within the categories. Look at Formula 10.2 again and you will see that the total sum of squares (SST) is equal to the addition of SSW and SSB. This relationship provides us with an easy method for finding SSW by simple subtraction. Formula 10.4 rearranges the symbols in Formula 10.2:

FORMULA 10.4

$$\text{SSW} = \text{SST} - \text{SSB}$$

Let's pause for a second to remember what we are after here. If the null hypothesis is *true*, then there should not be much variation from category to category (see Table 10.1) relative to the variation within categories, and the two estimates of the population variance based on SSW and SSB should be roughly equal. If the null hypothesis is *not true*, there will be large differences between categories (see Table 10.2) relative to the differences within categories, and SSB should be much larger than SSW. SSB will increase as the differences *between* category means increases, especially when there is not much variation *within* the categories (SSW). The larger SSB is compared to SSW, the more likely that we will reject the null hypothesis.

The next step in the computational routine is to construct the estimates of the population variance. To do this, we divide each sum of squares by its respective degrees of freedom. To find the degrees of freedom associated with SSW, subtract the number of categories (k) from the number of cases (N). The degrees of freedom associated with SSB are the number of categories minus 1. In summary,

FORMULA 10.5

$$\text{dfw} = N - k$$

where dfw = degrees of freedom associated with SSW
N = total number of cases
k = number of categories

ONE STEP AT A TIME Computing ANOVA

It is strongly recommended that you use a computing table like Table 10.3 to organize these computations.

Step 1: Find SST by means of Formula 10.1:

$$\text{SST} = \sum X^2 - N\overline{X}^2$$

a. Find ΣX^2 by squaring each score and adding all of the squared scores together.

b. Find $N\overline{X}^2$ by squaring the value of the mean and then multiplying by N.

c. Subtract the quantity you found in step b from the quantity you found in step a.

Step 2: Find SSB by means of Formula 10.3:

$$\text{SSB} = \sum N_k(\overline{X}_k - \overline{X})^2$$

a. Subtract the mean of all scores ($\overline{X}$) from the mean of each category ($\overline{X}_k$) and then square each difference.

b. Multiply each of the squared differences you found in step a by the number of cases in the category (N_k).

c. Add together the quantities you found in step b.

Step 3: Find SSW by means of Formula 10.4:

$$\text{SSW} = \text{SST} - \text{SSB}$$

Subtract the value of SSB (see step 2) from the value of SST (see step 1).

Step 4: Calculate degrees of freedom:

a. For dfw, use Formula 10.5 (dfw $= N - k$): Subtract the number of categories (k) from the number of cases (N).

b. For dfb, use Formula 10.6 (dfb $= k - 1$): Subtract 1 from the number of categories (k).

Step 5: Construct the two mean square estimates of the population variance:

a. Find the mean square within (MSW) by using Formula 10.7 (MSW = SSW/dfw): Divide SSW (see step 3) by dfw (see step 4a).

b. Find the mean square between (MSB) by using Formula 10.8 (MSB = SSB/dfb): Divide SSB (see step 2) by dfb (see step 4b)

Step 6: Find the obtained F ratio by means of Formula 10.9 (F = MSB/MSW): Divide the MSB estimate (see step 5b) by the MSW estimate (see step 5a)

FORMULA 10.6

$$\text{dfb} = k - 1$$

where dfb = degrees of freedom associated with SSB
k = number of categories

The actual estimates of the population variance, called the **mean square estimates,** are calculated by dividing each sum of squares by its respective degrees of freedom:

FORMULA 10.7

$$\text{Mean square within} = \frac{\text{SSW}}{\text{dfw}}$$

FORMULA 10.8

$$\text{Mean square between} = \frac{\text{SSB}}{\text{dfb}}$$

The test statistic calculated in step 4 of the five-step model is called the *F ratio,* and its value is determined by the following formula:

FORMULA 10.9

$$F = \frac{\text{Mean square between}}{\text{Mean square within}}$$

As you can see, the value of the F ratio will be a function of the amount of variation between categories (based on SSB) to the amount of variation within the categories (based on SSW). The greater the variation between the categories relative to the variation within, the higher the value of the F ratio and the more likely we will reject the null hypothesis. These procedures are summarized in the One Step at a Time: Computing ANOVA box and illustrated in the next section.

10.4 A COMPUTATIONAL EXAMPLE

Assume that we have administered our "Support for Capital Punishment Scale" to a sample of 20 individuals who are equally divided into the five religions. (Obviously, this sample is much too small for any serious research and is intended solely for purposes of illustration.) All scores are reported in Table 10.3 along with the other quantities needed to complete the computations. The scores (X) are listed for each of the five categories (religions), and a column has been added for the squared scores (X^2). The sums of both X and X^2 are reported at the bottom of each column. The category means ($\bar{X}_k$) show that the four Protestants averaged 12.5 on the Support for Capital Punishment Scale, the four Catholics averaged 21.0, and so forth. Finally, the overall mean (sometimes called the *grand mean*) is reported in the bottom row of the table. This shows that all 20 respondents averaged 16.6 on the scale.

To organize our computations, we'll follow the routine summarized in the One Step at a Time box at the end of Section 10.3. We begin by finding SST by means of Formula 10.1:

$$\text{SST} = \sum X^2 - N\bar{X}^2$$
$$\text{SST} = (666 + 1898 + 1078 + 1794 + 712) - (20)(16.6)^2$$
$$\text{SST} = (6148) - (20)(275.56)$$
$$\text{SST} = 6148 - 5511.2$$
$$\text{SST} = 636.80$$

The sum of squares between (SSB) is found by means of Formula 10.3:

$$\text{SSB} = \sum N_k(\bar{X}_k - \bar{X})^2$$
$$\text{SSB} = 4(12.5 - 16.6)^2 + 4(21.0 - 16.6)^2 + 4(16.0 - 16.6)^2 + 4(20.5 - 16.6)^2 + 4(13.0 - 16.6)^2$$
$$\text{SSB} = 67.24 + 77.44 + 1.44 + 60.84 + 51.84$$
$$\text{SSB} = 258.80$$

Now SSW can be found by subtraction (Formula 10.4):

$$\text{SSW} = \text{SST} - \text{SSB}$$
$$\text{SSW} = 636.8 - 258.80$$
$$\text{SSW} = 378.00$$

TABLE 10.3 SUPPORT FOR CAPITAL PUNISHMENT BY RELIGION FOR 16 SUBJECTS (fictitious data)

Protestant		Catholic		Jew		None		Other	
X	X^2	X	X^2	X	X^2	X	X^2	X	X^2
8	64	12	144	12	144	15	225	10	100
12	144	20	400	13	169	16	256	18	324
13	169	25	625	18	324	23	529	12	144
17	289	27	729	21	441	28	784	12	144
50	666	84	1898	64	1078	82	1794	52	712
$\bar{X}_k =$	12.5	$\bar{X}_k = 21.0$		$\bar{X}_k = 16.0$		$\bar{X}_k = 20.5$		$\bar{X}_k = 13.0$	
				$\bar{X} = 16.6$					

To find the degrees of freedom for the two sums of squares, we use Formulas 10.5 and 10.6:

$$\text{dfw} = N - k = 20 - 5 = 15$$
$$\text{dfb} = k - 1 = 5 - 1 = 4$$

Finally, we are ready to construct the mean square estimates of the population variance. For the estimate based on SSW, we use Formula 10.7:

$$\text{Mean square within} = \frac{\text{SSW}}{\text{dfw}} = \frac{378.00}{15} = 25.20$$

For the between estimate, we use Formula 10.8:

$$\text{Mean square between} = \frac{\text{SSB}}{\text{dfb}} = \frac{258.80}{4} = 64.70$$

The test statistic, or F ratio, is found by means of Formula 10.9:

$$F = \frac{\text{Mean square between}}{\text{Mean square within}} = \frac{64.70}{25.20} = 2.57$$

This statistic must still be evaluated for its significance. *(Solve any of the end-of-chapter problems to practice computing these quantities and solving these formulas.)*

10.5 A TEST OF SIGNIFICANCE FOR ANOVA

In this section, we see how to test an F ratio for significance and also take a look at some of the assumptions underlying the ANOVA test. As usual, we will follow the five-step model as a convenient way of organizing the decision-making process.

Step 1. Making Assumptions and Meeting Test Requirements.

Model: Independent random samples
Level of measurement is interval-ratio
Populations are normally distributed
Population variances are equal

The model assumptions are quite stringent and underscore the fact that ANOVA should be used only with dependent variables that have been carefully and precisely measured. However, as long as the categories are roughly equal in size, ANOVA can tolerate some violation of the model assumptions. In situations where you are uncertain or have samples of very different size, it is probably advisable to use an alternative test. (Chi square in Chapter 11 is one option.)

Step 2. Stating the Null Hypothesis. For ANOVA, the null hypothesis always states that the means of the populations from which the samples were drawn are equal. For our example problem, we are concerned with five different populations, or categories, so our null hypothesis would be

$$H_0: \mu_1 = \mu_2 = \mu_3 = \mu_4 = \mu_5$$

Application 10.1

An experiment in teaching introductory biology was recently conducted at a large university. One section was taught by the traditional lecture-lab method, a second was taught by an all-lab/demonstration approach with no lectures, and a third was taught entirely by a series of videotaped lectures and demonstrations that the students were free to view at any time and as often as they wanted. Students were randomly assigned to each of the three sections, and, at the end of the semester, random samples of final exam scores were collected from each section. Is there a significant difference in student performance by teaching method?

Final Exam Scores by Teaching Method

	Lecture		Demonstration		Videotape	
	X	X^2	X	X^2	X	X^2
	55	3025	56	3136	50	2500
	57	3249	60	3600	52	2704
	60	3600	62	3844	60	3600
	63	3969	67	4489	61	3721
	72	5184	70	4900	63	3969
	73	5329	71	5041	69	4761
	79	6241	82	6724	71	5041
(continued)						
	85	7225	88	7744	80	6400
	92	8464	95	9025	82	6724
$\sum X =$	636		651		588	
$\sum X^2 =$		46286		48503		39420
$\bar{X}_k =$	70.67		72.33		65.33	

(continued next column)

$$\bar{X} = 1{,}875/27 = 69.44$$

We can see by inspection that the "Videotape" group had the lowest average score and that the "Demonstration" group had the highest average score. The ANOVA test will tell us if these differences are large enough to justify the conclusion that they did not occur by chance alone. We can organize the computations following the One Step at a Time box presented at the end of Section 10.3:

$$\text{SST} = \sum X^2 - N\bar{X}^2$$
$$\text{SST} = (46{,}286 + 48{,}503 + 39{,}420) - 27(69.44)^2$$
$$\text{SST} = 134{,}209 - 130{,}191.67$$
$$\text{SST} = 4017.33$$
$$\text{SSB} = \sum N_k(\bar{X}_k - \bar{X})^2$$

(continued next page)

where μ_1 represents the mean for Protestants, μ_2, the mean for Catholics, and so forth.

The alternative hypothesis states simply that at least one of the population means is different. The wording here is important. If we reject the null hypothesis, ANOVA does not identify which mean or means are significantly different. In the final section of the chapter, we briefly consider an advanced test that can help us identify which pairs of means are significantly different.

(H_1: At least one of the population means is different.)

Step 3. Selecting the Sampling Distribution and Establishing the Critical Region. The sampling distribution for ANOVA is the *F* distribution, which is summarized in Appendix D. Note that there are separate tables for alphas of .05 and .01. As with the *t* table, the value of the critical *F* score will vary by

Application 10.1: *(continued)*

$$SSB = (9)(70.67 - 69.44)^2 + (9)(72.33 - 69.44)^2 + (9)(65.33 - 69.44)^2$$
$$SSB = 13.62 + 75.17 + 152.03$$
$$SSB = 240.82$$
$$SSW = SST - SSB$$
$$SSW = 4017.33 - 240.82$$
$$SSW = 3776.51$$
$$dfw = N - k = 27 - 3 = 24$$
$$dfb = k - 1 = 3 - 1 = 2$$

$$\text{Mean square within} = \frac{SSW}{dfw} = \frac{3776.51}{24} = 157.36$$
$$\text{Mean square between} = \frac{SSB}{dfb} = \frac{240.82}{2} = 120.41$$

$$F = \frac{\text{Mean square between}}{\text{Mean square within}}$$
$$F = \frac{120.41}{157.36}$$
$$F = 0.77$$

We can now conduct the test of significance.

Step 1. Making Assumptions and Meeting Test Requirements.

Model: Independent random samples
Level of measurement is interval-ratio
Populations are normally distributed
Population variances are equal

Step 2. Stating the Null Hypothesis.

H_0: $\mu_1 = \mu_2 = \mu_3$
(H_1: At least one of the population means is different.)

Step 3. Selecting the Sampling Distribution and Establishing the Critical Region.

$$\begin{aligned} \text{Sampling distribution} &= F \text{ distribution} \\ \text{Alpha} &= .05 \\ \text{Degrees of freedom (within)} &= (N - k) \\ &= (27 - 3) = 24 \\ \text{Degrees of freedom (between)} &= (k - 1) \\ &= (3 - 1) = 2 \\ F(\text{critical}) &= 3.40 \end{aligned}$$

Step 4. Computing the Test Statistic. We found an obtained F ratio of 0.77.

Step 5. Making a Decision and Interpreting the Results of the Test. Compare the test statistic with the critical value:

$$F(\text{critical}) = 3.40$$
$$F(\text{obtained}) = 0.77$$

We would clearly fail to reject the null hypothesis ("The population means are equal") and would conclude that the observed differences among the category means were the results of random chance. Student performance in this course does not vary significantly by teaching method.

degrees of freedom. For ANOVA, there are two separate degrees of freedom, one for each estimate of the population variance. The numbers across the top of the table are the degrees of freedom associated with the between estimate (dfb), and the numbers down the side of the table are those associated with the within estimate (dfw). In our example, dfb is $(k - 1)$, or 4, and dfw is $(N - k)$, or 15 (see Formulas 10.5 and 10.6). So, if we set alpha at .05, our critical F score will be 3.06.

Summarizing these considerations:

$$\begin{aligned} \text{Sampling distribution} &= F \text{ distribution} \\ \text{Alpha} &= .05 \\ \text{Degrees of freedom (within)} &= (N - k) = 15 \\ \text{Degrees of freedom (between)} &= (k - 1) = 4 \\ F(\text{critical}) &= 3.06 \end{aligned}$$

Take a moment to inspect the two *F* tables and you will notice that all the values are greater than 1.00. This is because ANOVA is a one-tailed test and we are concerned only with outcomes in which there is more variance between categories than within categories. *F* values of less than 1.00 would indicate that the between estimate was lower in value than the within estimate, and, since we would always fail to reject the null in such cases, we simply ignore this class of outcomes.

Step 4. Computing the Test Statistic. This was done in the previous section, where we found an obtained *F* ratio of 2.57.

Step 5. Making a Decision and Interpreting the Results of the Test. Compare the test statistic with the critical value:

$$F(\text{critical}) = 3.06$$
$$F(\text{obtained}) = 2.57$$

Since the test statistic does not fall into the critical region, our decision would be to fail to reject the null. Support for capital punishment does not differ significantly by religion, and the variation we observed in the sample means is unimportant.

10.6 AN ADDITIONAL EXAMPLE FOR COMPUTING AND TESTING THE ANALYSIS OF VARIANCE

In this section, we work through an additional example of the computation and interpretation of the ANOVA test. We first review matters of computation, find the obtained *F* ratio, and then test the statistic for its significance. In the computational section, we follow the One Step at a Time box presented at the end of Section 10.3.

A researcher has been asked to evaluate the efficiency with which each of three social service agencies is administering a particular program. One area of concern is the speed of the agencies in processing paperwork and determining the eligibility of potential clients. The researcher has gathered information on the number of days required for processing a random sample of 10 cases in each agency. Is there a significant difference between the agencies? The data are reported in Table 10.4, which also includes some additional information we will need to complete our calculations.

To find SST by means of formula 10.1:

$$\text{SST} = \sum X^2 - N\overline{X}^2$$
$$\text{SST} = (524 + 1816 + 2462) - 30(11.67)^2$$
$$\text{SST} = 4802 - 30(136.19)$$
$$\text{SST} = 4802 - 4085.70$$
$$\text{SST} = 716.30$$

To find SSB by means of Formula 10.3:

$$\text{SSB} = \sum N_k(\overline{X}_k - \overline{X})^2$$
$$\text{SSB} = (10)(7.0 - 11.67)^2 + (10)(13.0 - 11.67)^2 + (10)(15.0 - 11.67)^2$$
$$\text{SSB} = (10)(21.81) + (10)(1.77) + (10)(11.09)$$
$$\text{SSB} = 218.10 + 17.70 + 110.90$$
$$\text{SSB} = 346.70$$

TABLE 10.4 NUMBER OF DAYS REQUIRED TO PROCESS CASES FOR THREE AGENCIES (fictitious data)

	Agency A		Agency B		Agency C	
Client	X	X^2	X	X^2	X	X^2
1	5	25	12	144	9	81
2	7	49	10	100	8	64
3	8	64	19	361	12	144
4	10	100	20	400	15	225
5	4	16	12	144	20	400
6	9	81	11	121	21	441
7	6	36	13	169	20	400
8	9	81	14	196	19	361
9	6	36	10	100	15	225
10	6	36	9	81	11	121
$\Sigma X =$	70		130		150	
$\Sigma X^2 =$		524		1816		2462
$\bar{X}_k =$	7.0		13.0		15.0	
			$\bar{X} = 350/30 = 11.67$			

Now we can find SSW using Formula 10.4:

$$\text{SSW} = \text{SST} - \text{SSB}$$
$$\text{SSW} = 716.30 - 346.70$$
$$\text{SSW} = 369.60$$

The degrees of freedom are found through Formulas 10.5 and 10.6:

$$\text{dfw} = N - k = 30 - 3 = 27$$
$$\text{dfb} = k - 1 = 3 - 1 = 2$$

The estimates of the population variances are found by means of Formulas 10.7 and 10.8:

$$\text{Mean square within} = \frac{\text{SSW}}{\text{dfw}} = \frac{369.60}{27} = 13.69$$

$$\text{Mean square between} = \frac{\text{SSB}}{\text{dfb}} = \frac{346.70}{2} = 173.35$$

The F ratio (Formula 10.9) is

$$F(\text{obtained}) = \frac{\text{Mean square between}}{\text{Mean square within}}$$

$$F(\text{obtained}) = \frac{173.35}{13.69}$$

$$F(\text{obtained}) = 12.66$$

And we can now test this value for its significance.

Step 1. Making Assumptions and Meeting Test Requirements.

Model: Independent random samples
Level of measurement is interval-ratio
Populations are normally distributed
Population variances are equal

Application 10.2

A random sample of 15 nations from three levels of development has been selected. "Least developed" nations are largely agricultural and have the lowest quality of life. Developed nations are industrial and the most affluent and modern. Developing nations are between these extremes. Are these general characteristics reflected in differences in life expectancy (the number of years the average citizen can expect to live at birth) between the three categories? The data for the 15 nations for 2005 are as follows:

Least Developed		Developing		Developed	
Nation	Life Expectancy	Nation	Life Expectancy	Nation	Life Expectancy
Cambodia	58.9	China	72.3	Australia	80.4
Malawi	41.6	Indonesia	69.6	Canada	80.1
Nepal	59.8	Pakistan	63.0	Japan	81.2
Niger	43.5	South Korea	76.9	Russia	65.8
Sudan	58.5	Turkey	72.4	United Kingdom	78.4

SOURCE: U.S. Bureau of the Census. 2006. *Statistical Abstract of the United States, 2006*. p. 837. Washington, DC: U.S. Government Printing Office.

To conduct the ANOVA test, the data will be organized into table format:

Life Expectancy by Level of Development

	Least Developed		Developing		Developed	
	X	X^2	X	X^2	X	X^2
	58.90	3469.21	72.30	5227.29	80.40	6464.16
	41.60	1730.56	69.60	4844.16	80.10	6416.01
	59.80	3576.04	63.00	3969.00	81.20	6593.44
	43.50	1892.25	76.90	5913.61	65.80	4329.64
	58.50	3422.25	72.40	5241.76	78.40	6146.56
$\sum X =$	262.30		354.20		385.90	
$\sum X^2 =$		14,090.31		25,195.82		29,949.81
$\bar{X}_k =$	52.46		70.84		77.18	

$$\bar{X} = \frac{1002.4}{15} = 66.83$$

The ANOVA test will tell us if these differences are large enough to justify the conclusion that they did not occur by chance alone. Following the computational routine established at the end of Section 10.3:

$$\text{SST} = \sum X^2 - N\bar{X}^2$$
$$\text{SST} = (14{,}090.31 + 25{,}195.82 + 29{,}949.81) - 15(66.83)^2$$
$$\text{SST} = 69{,}235.94 - 66{,}993.73$$
$$\text{SST} = 2242.21$$

(continued next page)

Application 10.2: *(continued)*

$SSB = \sum N_k (\bar{X}_k - \bar{X})^2$
$SSB = (5)(52.46 - 66.83)^2 + (5)(70.84 - 66.83)^2 + (5)(77.18 - 66.83)^2$
$SSB = 1032.49 + 80.40 + 535.61$
$SSB = 1648.50$

$SSW = SST - SSB$
$SSW = 2242.21 - 1648.50$
$SSW = 593.71$
$dfw = N - k = 15 - 3 = 12$
$dfb = k - 1 = 3 - 1 = 2$

$$\text{Mean square within} = \frac{SSW}{dfw} = \frac{593.71}{12} = 49.48$$

$$\text{Mean square between} = \frac{SSB}{dfb} = \frac{1648.50}{2} = 824.25$$

$$F = \frac{\text{Mean square between}}{\text{Mean square within}}$$

$$F = \frac{824.25}{49.48} = 16.66$$

We can now conduct the test of significance.

Step 1. Making Assumptions and Meeting Test Requirements.

Model: Independent random samples
Level of measurement is interval-ratio
Populations are normally distributed
Population variances are equal

Step 2. Stating the Null Hypothesis.

H_0: $\mu_1 = \mu_2 = \mu_3$
(H_1: At least one of the population means is different.)

Step 3. Selecting the Sampling Distribution and Establishing the Critical Region.

Sampling distribution = F distribution
Alpha = .05
Degrees of freedom (within) = $(N - k)$ = $(15 - 3) = 12$
Degrees of freedom (between) = $(k - 1)$ = $(3 - 1) = 2$
F(critical) = 3.88

Step 4. Computing the Test Statistic. We found an obtained F ratio of 16.66

Step 5. Making a Decision and Interpreting the Results of the Test. Compare the test statistic with the critical value:

F(critical) = 3.88
F(obtained) = 16.66

The null hypothesis ("The population means are equal") can be rejected. The differences in life expectancy between nations at different levels of economic development are statistically significant.

The researcher will always be in a position to judge the adequacy of the first two assumptions in the model. The second two assumptions are more problematical, but remember that ANOVA will tolerate some deviation from its assumptions as long as sample sizes are roughly equal.

Step 2. Stating the Null Hypothesis.

$$H_0: \mu_1 = \mu_2 = \mu_3$$
(H_1: At least one of the population means is different.)

Step 3. Selecting the Sampling Distribution and Establishing the Critical Region.

Sampling distribution = F distribution

$$\text{Alpha} = .05$$
$$\text{Degrees of freedom (within)} = (N - k) = (30 - 3) = 27$$
$$\text{Degrees of freedom (between)} = (k - 1) = (3 - 1) = 2$$
$$F(\text{critical}) = 3.35$$

Step 4. Computing the Test Statistic. We found an obtained F ratio of 12.66.

Step 5. Making a Decision and Interpreting the Results of the Test. Compare the test statistic with the critical value:

$$F(\text{critical}) = 3.35$$
$$F(\text{obtained}) = 12.66$$

The test statistic is in the critical region, and we would reject the null hypothesis of no difference. The differences between the three agencies are very unlikely to have occurred by chance alone. The agencies are significantly different in the speed with which they process paperwork and determine eligibility. *(For practice in conducting the ANOVA test, see problems 10.2 to 10.8. Begin with the lower-numbered problems since they have smaller data sets, fewer categories, and, therefore, the simplest calculations.)*

10.7 THE LIMITATIONS OF THE TEST

ANOVA is appropriate whenever you want to test differences between the means of an interval-ratio-level dependent variable across three or more categories of an independent variable. This application is called **one-way analysis of variance,** since it involves the effect of a single variable (for example, religion) on another (for example, support for capital punishment). This is the simplest application of ANOVA, and you should be aware that the technique has numerous more advanced and complex forms. For example, you may encounter research projects in which the effects of two separate variables (for example, religion and gender) on some third variable were observed.

One important limitation of ANOVA is that it requires interval-ratio measurement of the dependent variable and roughly equal numbers of cases in each of the categories of the independent variable. The former condition may be difficult to meet with complete confidence for many variables of interest to the social sciences. The latter condition may create problems when the research hypothesis calls for comparisons between groups that are, by their nature, unequal in numbers (for example, white versus black Americans) and may call for some unusual sampling schemes in the data-gathering phase of a research project. Neither of these limitations should be particularly crippling, since ANOVA can tolerate some deviation from its model assumptions, but you should be aware of these limitations in planning your own research as well as in judging the adequacy of research conducted by others.

A second limitation of ANOVA actually applies to all forms of significance testing and was introduced in Section 9.5. Tests of significance are designed to detect nonrandom differences or differences so large that they are very unlikely to be produced by random chance alone. The problem is that differences that are statistically significant are not necessarily important in any other sense. Parts III and IV of this text provide some statistical techniques that can assess the importance of results directly.

A final limitation of ANOVA relates to the research hypothesis. As you recall, when the null hypothesis is rejected, the alternative hypothesis is supported. The limitation is that the alternative hypothesis is not specific; it simply asserts that at least one of the population means is different from the others. Obviously, we would like to know which differences are significant. We can sometimes make this determination by simple inspection. In our problem involving social service agencies, for example, it is pretty clear from Table 10.4 that Agency A is the source of most of the differences. This informal, "eyeball" method can be misleading, however, and you should exercise caution in drawing conclusions about which means are significantly different. In the next section, we provide an extended example of the interpretation of ANOVA and introduce a technique (called **post hoc,** or "after the fact," analysis) that permits us to identify significant differences between the sample means reliably. The computational routines for post hoc analysis are beyond the scope of this text, but the tests are commonly available in computerized statistical packages such as SPSS.

10.8 INTERPRETING STATISTICS: DOES SEXUAL ACTIVITY VARY BY MARITAL STATUS?

Does a respondent's marital status make a difference in his or her sexual activity? Do married individuals have sex more often? Do unmarried individuals report more sexual partners than married individuals? We will use the data from the 2006 General Social Survey to answer these questions scientifically. The respondents to this survey are a representative sample (see Chapter 6) of the population of the United States.

This analysis involves three variables. The independent variable is marital status, and there will be two separate dependent variables: the number of different sexual partners in the past five years and the number of times the respondent had sex during the last 12 months. Marital status is a nominal-level variable with five categories. The two dependent variables have been measured at the ordinal level rather than the interval-ratio level required by ANOVA. For example, when asked for the number of times they had sex over the past year, the respondents were asked to select broad categories such as "2 or 3 times a month" (see Appendix G for a complete listing of the scores of the two variables that measure sexual activity). We encountered a similar problem in Section 9.6 when we assessed the gender gap in income. Then, we solved the problem by substituting midpoints for each of the intervals of the variable, and we will do the same here. For example, on number of sex partners, we will substitute 7.5 for the category "5 to 10," and, using the most conservative estimate, a value of 100 will be substituted for the "More than 100 partners" category. The variable measuring frequency of sexual activity has been transformed in a similar way: A value of 12 has been substituted for "about once a month," 52 for "about once a week," and so forth.

The summary statistics for the two sexual activity variables are presented in Table 10.5. As you can see, the sample averaged almost three sex partners (2.82) over the last five years and had sex at a rate of slightly less than once a week over the past year (45.44). Note that the standard deviation is high relative to the mean for both variables. This is caused by extreme positive skews: Most cases are grouped in the low range of the variable (e.g., they had only one or two sexual partners), but a few respondents have very high scores.

The means and standard deviations for number of partners over the last five years for each marital status are reported in Table 10.6. The F ratio for these differences is 26.43, which is significant at less than the .001 level. We would reject the null hypothesis of "no difference" and conclude that marital status does

TABLE 10.5 DESCRIPTIVE STATISTICS FOR SEXUAL ACTIVITY VARIABLES

	Number of Sex Partners, Last Five Years	Number of Times Respondent Had Sex During Last Year
Mean =	2.82	45.44
Standard deviation =	8.28	51.60
N =	2337	2333

TABLE 10.6 NUMBER OF SEXUAL PARTNERS IN THE LAST FIVE YEARS BY MARITAL STATUS

	Marital Status					ANOVA	
	Married	Widowed	Divorced	Separated	Never Married	*F* ratio	Alpha
Mean =	1.44	0.61	3.66	4.55	5.22	26.43	<.001
Standard deviation =	3.40	1.06	9.51	13.09	12.39		
N =	1050	197	397	78	578		

make a significant difference in the number of sex partners. Comparing the category means, the results are what one would expect. Widowed respondents had the fewest number of partners, and married individuals had the second lowest, averaging 1.44 partners. Divorced and separated individuals averaged a higher number of sex partners, and individuals who were never married had the highest average number.

Table 10.7 reports the means and standard deviations for the number of times the respondent reported having sex in the last 12 months by marital status. Once again, widowed individuals had by far the lowest rate. Divorced respondents had the second-lowest rate (although much higher than the widowed), followed by the never married and the separated. Married individuals, on the average, reported having sex more often than any other group (53.41 times a year). The *F* ratio of 35.68 is significant at less than .001, so we can conclude that the rate of sexual activity is significantly affected by marital status.

Tables 10.6 and 10.7 indicate significant relationships, but which differences in the tables are most important and contribute the most to the significant *F* ratios? We can see by inspection that the widowed group is dramatically different

TABLE 10.7 FREQUENCY OF SEX BY MARITAL STATUS

	Marital Status					ANOVA	
	Married	Widowed	Divorced	Separated	Never Married	*F* ratio	Alpha
Mean =	53.41	7.81	41.25	52.93	45.72	35.68	<.001
Standard deviation =	49.45	24.71	53.52	56.14	54.19		
N =	1050	198	401	78	606		

from the others, but what other differences might be important? A post hoc analysis of the differences in sample means can provide objective answers to these questions.

We will not present the computational routines for post hoc tests, but the interpretation of the results of these tests is straightforward. The tests essentially compare the means of all possible pairs of categories (i.e., married with divorced, widowed with separated, and so forth) and tell us exactly which combinations of means contribute the most to a significant *F* ratio. Comparing sample means in this way increases the probability of making an alpha error (falsely rejecting a true null hypothesis—see Chapter 8); to correct for this, post hoc tests use more stringent criteria to identify significant differences.

Table 10.8 presents the results of post hoc tests for both measures of sexual activity. The categories are listed at the left-hand side of the table and again across the top, along with the category means. The entries in the table are the differences in the means for the category listed at the left of the table and the category listed across the top. A positive difference indicates that the category listed to the left had a higher score, and a negative difference means that the category listed across the top had the higher score. For example, the average number of partners over the last five years was 1.44 for married respondents (category listed on the left) and 0.61 for widowed respondents (category listed across the top). The difference between the two category means is 1.44 − 0.61, or 0.83, the value reported at the intersection of the two categories. This value is positive, which indicates that married respondents (category to the left) had a higher score than widowed respondents (category across the top). An asterisk (*) next to the value means that the difference is significant at the .05 level.

TABLE 10.8 A POST HOC TEST FOR DIFFERENCES IN SEXUAL ACTIVITY BY MARITAL STATUS

	Mean Differences in Number of Partners Over Last Five Years				
Marital Status	Married (1.44)	Widowed (0.61)	Divorced (3.66)	Separated (4.55)	Never Married (5.22)
Married (1.44)		0.83	−2.22*	−3.11*	−3.78*
Widowed (0.61)			−3.05*	−3.94*	−4.61*
Divorced (3.66)				−0.89	−1.56*
Separated (4.55)					−0.67
Never Married (5.22)					

	Mean Difference in Frequency of Sex				
Marital Status	Married (53.41)	Widowed (7.81)	Divorced (41.25)	Separated (52.93)	Never Married (45.72)
Married (53.41)		45.60*	12.16*	0.48	7.69*
Widowed (7.81)			−33.44*	−45.12*	−37.91*
Divorced (41.25)				−11.68	−4.47*
Separated (52.93)					−7.21
Never Married (45.72)					

*Mean difference is significant at the 0.05 level, Modified Least Significant Differences Test.

READING STATISTICS 7: Delinquent Girls

Adolescent girls have become increasingly involved with the criminal justice system over the past several decades. What risk factors are associated with delinquency for females, and how are these factors associated with the treatment of delinquent girls? Researchers Ruffolo, Sarri, and Goodkind interviewed a small sample of girls ($N = 159$) that were involved with the juvenile justice system and documented their high levels of mental distress and negative life experiences. The girls had been assigned to one of three levels of treatment: home-based services where they continued to live with their families but were monitored by members of the system; an open residential facility (a group home with minimal restrictions of movement and freedom); or a closed residential facility (in which the girls were considered a security risk and locked down and controlled accordingly). The researchers found that the girls with higher levels of delinquency and mental stress and more negative life experiences were more likely to be in a more restrictive treatment program. The following table reports some of the significant differences between the groups of girls receiving different types of services.

	Home-Based Services		Community-Based, Open Residential		Closed Residential, Juvenile Justice		
	$\bar{X}$	s	$\bar{X}$	s	$\bar{X}$	s	F
Depression (possible score 0–60)	21.10	12.15	24.22	11.69	29.76	11.89	6.53**
Suicide attempts (1 = none, 5 = 6 or more)	1.32	0.72	1.52	0.98	1.85	1.37	3.57*
Use mental health services? (1 = yes, 0 = no)	0.20	0.40	0.29	0.46	0.43	0.50	3.45*
Sexual abuse? (possible score 1–5)	1.47	0.78	1.78	0.99	2.89	1.47	22.56**
Supportive family environment (possible score 1–5)	3.44	1.22	3.24	1.13	3.65	1.10	1.31
Negative peer involvement (possible score 1–5)	2.08	0.76	2.13	0.77	3.00	1.02	17.42***
Self-reported delinquency (possible score 16–80)	19.43	7.03	19.10	6.40	25.44	11.39	8.30***
Foster care (1 = yes, 0 = no)	0.17	0.38	0.27	0.45	0.57	0.50	11.75***
Age	15.28	2.11	17.28	2.13	15.50	1.33	15.85***

* = $p < .05$, ** = $p < .01$, *** = $p < .001$.

Group means and standard deviations are listed in the columns, and the significance of the differences from group to group is indicated by the number of asterisks associated with the *F* ratio in the right-hand column. All differences between category means are significant save one. Girls in the most restricted treatment setting were significantly more depressed, suicidal, abused, and delinquent than girls in the other settings. The only nonsignificant difference was in "supportive family environment," a variable on which all groups scored relatively low. The nature of these differences suggests that the girls were appropriately placed, but the researchers also point out that many of the risk factors associated with delinquency begin long before adolescence, and they stress the need for early intervention to divert the girls from delinquency.

Ruffalo, Mary, Sarri, Rosemary, and Goodkind, Sara. 2004. "Study of Delinquent, Diverted, and High-Risk Adolescent Girls: Implications for Mental Health Intervention." *Social Work Research* 28:237–245.

Looking first at the top portion of the table (differences for number of sexual partners over the last five years), we can see that the never married respondents had a significantly higher number of sex partners than all other categories except separated respondents. Divorced respondents had significantly more partners than every category except separated respondents, and the widowed had significantly fewer partners than every other category except married.

The differences for frequency of sexual activity (reported in the bottom portion of the table) show similar patterns. The widowed had sex significantly less often than every other group. However, the test also finds some less obvious differences: Divorced respondents had sex significantly less often than married respondents, and there was a significant difference between married and never married respondents. That the widowed group was the "most different" was obvious from Tables 10.6 and 10.7, but these additional significant differences might well have escaped our attention had a post hoc test not been conducted.

In conclusion, the analysis of variance test shows that marital status makes a significant difference in levels of sexual activity for adult Americans. Married individuals had sex significantly more often with significantly fewer partners. People who "never married" have significantly more sex partners, and widowed people rank the lowest on both measures of sexual activity.

SUMMARY

1. One-way analysis of variance is a powerful test of significance that is commonly used when comparisons across more than two categories or samples are of interest. It is perhaps easiest to conceptualize ANOVA as an extension of the test for the difference in sample means.
2. ANOVA compares the amount of variation within the categories to the amount of variation between categories. If the null hypothesis of no difference is false, there should be relatively great variation between categories and relatively little variation within categories. The greater the differences from category to category relative to the differences within the categories, the more likely we will be able to reject the null hypothesis.
3. The computational routine for even simple applications of ANOVA can quickly become quite complex. The basic process is to construct separate estimates of the population variance based on the variation within the categories and the variation between the categories. The test statistic is the *F* ratio, which is based on a comparison of these two estimates. The basic computational routine is summarized at the end of Section 10.4, and this is probably an appropriate time to mention the widespread availability of statistical packages such as SPSS, the purpose of which is to perform complex calculations such as these accurately and quickly. If you haven't yet learned how to use such programs, ANOVA may provide you with the necessary incentive.
4. The ANOVA test can be organized into the familiar five-step model for testing the significance of sample outcomes. Although the model assumptions (step 1) require high-quality data, the test can tolerate some deviation as long as sample sizes are roughly equal. The null hypothesis takes the familiar form of stating that there is no difference of any importance among the population values, while the alternative hypothesis asserts that at least one population mean is different. The sampling distribution is the *F* distribution, and the test is always one-tailed. The decision to reject or to fail to reject the null is based on a comparison of the obtained *F* ratio with the critical *F* ratio as determined for a given alpha level and degrees of freedom. The decision to reject the null hypothesis indicates only that one or more of the population means is different from the others. We can often determine which sample mean(s) account for the difference by inspecting the sample data, but this informal method should be used with caution, and post hoc tests are more reliable indicators of significant differences.

SUMMARY OF FORMULAS

Total sum of squares:

10.1 $$SST = \sum X^2 - N\bar{X}^2$$

The two components of the total sum of squares:

10.2 $SST = SSB + SSW$

Sum of squares between:

10.3 $$SSB = \sum N_k(\bar{X}_k - \bar{X})^2$$

Sum of squares within:

10.4 $SSW = SST - SSB$

Degrees of freedom for SSW:

10.5 $dfw = N - k$

Degrees of freedom for SSB:

10.6 $dfb = k - 1$

Mean square within:

10.7 $$\text{Mean square within} = \frac{SSW}{dfw}$$

Mean square between:

10.8 $$\text{Mean square between} = \frac{SSB}{dfb}$$

F ratio:

10.9 $$F = \frac{\text{Mean square between}}{\text{Mean square within}}$$

GLOSSARY

Analysis of variance. A test of significance appropriate for situations in which we are concerned with the differences among more than two sample means.

ANOVA. See Analysis of variance.

***F* ratio.** The test statistic computed in step 4 of the ANOVA test.

Mean square estimate. An estimate of the variance calculated by dividing the sum of squares within (SSW) or the sum of squares between (SSB) by the proper degrees of freedom.

One-Way analysis of variance. Applications of ANOVA in which the effect of a single independent variable on a dependent variable is observed.

Post hoc test. A technique for determining which pairs of means are significantly different.

Sum of squares between (SSB). The sum of the squared deviations of the sample means from the overall mean, weighted by sample size.

Sum of squares within (SSW). The sum of the squared deviations of scores from the category means.

Total sum of squares (SST). The sum of the squared deviations of the scores from the overall mean.

PROBLEMS

NOTE: The number of cases in these problems is very low—a fraction of the sample size necessary for any serious research—in order to simplify computations.

10.1 Conduct the ANOVA test for each of the following sets of scores. *(HINT: Follow the computational shortcut outlined in Section 10.4, and keep track of all sums and means by constructing computational tables like Table 10.3 or 10.4.)*

a. Category

A	B	C
5	10	12
7	12	16
8	14	18
9	15	20

b. Category

A	B	C
1	2	3
10	12	10
9	2	7
20	3	14
8	1	1

c. Category

A	B	C	D
13	45	23	10
15	40	78	20
10	47	80	25
11	50	34	27
10	45	30	20

10.2 SOC What type of person is most involved in the neighborhood and community? Who is more likely to volunteer for organizations such as PTA, scouts, or Little League? A random sample of 15 people have been asked for their number of memberships in community voluntary organizations and some other information. Which differences are significant?

a. Membership by education:

Less Than High School	High School	College
0	1	0
1	3	3
2	3	4
3	4	4
4	5	4

b. Membership by length of residence in present community:

Less Than 2 Years	2–5 Years	More Than 5 Years
0	0	1
1	2	3
3	3	3
4	4	4
4	5	4

c. Membership by extent of television watching:

Little or None	Moderate	High
0	3	4
0	3	4
1	3	4
1	3	4
2	4	5

d. Membership by number of children:

None	One Child	More Than One Child
0	2	0
1	3	3
1	4	4
3	4	4
3	4	5

10.3 SOC In a local community, a random sample of 18 couples has been assessed on a scale that measures the extent to which power and decision making are shared (lower scores) or monopolized by one party (higher scores) and on marital happiness (lower scores indicate lower levels of unhappiness). The couples were also classified by type of relationship: traditional (only the husband works outside the home), dual-career (both parties work), and cohabitational (parties living together but not legally married, regardless of work patterns). Does decision making or happiness vary significantly by type of relationship?

a. Decision Making

Traditional	Dual-career	Cohabitational
7	8	2
8	5	1
2	4	3
5	4	4
7	5	1
6	5	2

b. Happiness

Traditional	Dual-career	Cohabitational
10	12	12
14	12	14
20	12	15
22	14	17
23	15	18
24	20	22

10.4 CJ Two separate crime-reduction programs have been implemented in the city of Shinbone. One involves a neighborhood watch program, with citizens actively involved in crime prevention. The second involves officers patrolling the neighborhoods on foot rather than in patrol cars. In terms of the percentage reduction in crimes reported to the police over a one-year period, were the programs successful? The results are for random samples of 18 neighborhoods drawn from the entire city.

Neighborhood Watch	Foot Patrol	No Program
−10	−21	+30
−20	−15	−10
+10	−80	+14
+20	−10	+80
+70	−50	+50
+10	−10	−20

10.5 Are sexually active teenagers any better informed about AIDS and other potential health problems related to sex than teenagers who are sexually inactive? A 15-item test of general knowledge about sex and health was administered to random samples of teens who are sexually inactive, teens who are sexually active but with only a single partner ("going steady"), and teens who are sexually active with more than one partner. Is there any significant difference in the test scores?

Inactive	Active—One Partner	Active—More Than One Partner
10	11	12
12	11	12
8	6	10
10	5	4
8	15	3
5	10	15

10.6 SOC Does the rate of voter turnout vary significantly by the type of election? A random sample of voting precincts displays the following pattern of voter turnout by election type. Assess the results for significance.

Local Only	State	National
33	35	42
78	56	40
32	35	52
28	40	66
10	45	78
12	42	62
61	65	57
28	62	75
29	25	72
45	47	51
44	52	69
41	55	59

10.7 GER Do older citizens lose interest in politics and current affairs? A brief quiz on recent headline stories was administered to random samples of respondents from each of four different age groups. Is there a significant difference? The following data represent numbers of correct responses.

High School (15–18)	Young Adult (21–30)	Middle-aged (30–55)	Retired (65+)
0	0	2	5
1	0	3	6
1	2	3	6
2	2	4	6
2	4	4	7
2	4	5	7
3	4	6	8
5	6	7	10
5	7	7	10
7	7	8	10
7	7	8	10
9	10	10	10

10.8 SOC A small random sample of respondents has been selected from the General Social Survey database. Each respondent has been classified as either a city dweller, a suburbanite, or a rural dweller. Are there statistically significant differences by place of residence for any of the variables listed here? (See Appendix G for definitions and scores of the dependent variables)

a. Occupational Prestige (PRESTG80)

Urban	Suburban	Rural
32	40	30
45	48	40
42	50	40
47	55	45
48	55	45
50	60	50
51	65	52
55	70	55
60	75	55
65	75	60

b. Number of Children (CHILDS)

Urban	Suburban	Rural
1	0	1
1	1	4
0	0	2
2	0	3
1	2	3
0	2	2
2	3	5
2	2	0
1	2	4
0	1	6

c. Family Income (INCOME06)

Urban	Suburban	Rural
5	6	5
7	8	5
8	11	11
11	12	10
8	12	9
9	11	6
8	11	10
3	9	7
9	10	9
10	12	8

d. Church Attendance (ATTEND)

Urban	Suburban	Rural
0	0	1
7	0	5
0	2	4
4	5	4
5	8	0
8	5	4
7	8	8
5	7	8
7	2	8
4	6	5

e. Hours of TV Watching per Day (TVHOURS)

Urban	Suburban	Rural
5	5	3
3	7	7
12	10	5
2	2	0
0	3	1
2	0	8
3	1	5
4	3	10
5	4	3
9	1	1

10.9 SOC Does support for suicide ("death with dignity") vary by social class? Is this relationship different in different nations? Small samples in three nations were asked if it is ever justified for a person with an incurable disease to take his or her own life. Respondents answered in terms of a 10-point scale, on which 10 was "always justified" (the strongest support for "death with dignity") and 1 was "never justified" (the lowest level of support). Results are reported here.

Mexico

Lower Class	Working Class	Middle Class	Upper Class
5	2	1	2
2	2	1	4
4	1	3	5
5	1	4	7
4	6	1	8
2	5	2	10
3	7	1	10
1	2	5	9
1	3	1	8
3	1	1	8

Canada

Lower Class	Working Class	Middle Class	Upper Class
7	5	1	5
7	6	3	7
6	7	4	8
4	8	5	9
7	8	7	10
8	9	8	10
9	5	8	8
9	6	9	5
6	7	9	8
5	8	5	9

United States

Lower Class	Working Class	Middle Class	Upper Class
4	4	4	1
5	5	6	5
6	1	7	8
1	4	5	9
3	3	8	9
3	3	9	9
3	4	9	8
5	2	8	6
3	1	7	9
6	1	2	9

SSPS for Windows

Using *SPSS for Windows* to Conduct Analysis of Variance

SPSS DEMONSTRATION 10.1 Does Political Conservatism Increase with Age?

SPSS provides several different ways of conducting the analysis of variance test. The procedure summarized here is the most accessible of these, but it still incorporates options and capabilities that we have not covered in this chapter. If you wish to explore these possibilities, please consult the SPSS Help facility.

In designing examples to demonstrate the ANOVA procedure, my choices are constrained by the scarcity of interval-ratio variables in the GSS data set. Some variables which "should be" interval-ratio are actually measured at the ordinal level (e.g., income), while others are unsuitable dependent variables on logical grounds (e.g., age). To have something to discuss, I have taken some liberties with respect to level-of-measurement criteria in several of the examples that follow (a practice that is, in fact, common in social science research).

Let's begin by exploring the idea that political ideology is linked to age in U.S. society. People are often said to become more conservative about a wide range of issues as they age, and this might result in greater political conservatism in older age groups. For a measure of ideology, we will use *polviews*. Higher scores on this variable indicate higher levels of conservatism. We need to collapse *age* into a few categories to fit the ANOVA design. To find reasonable cutting points for age, I ran the **Frequencies** command to find the ages that divided the sample into three groups of roughly equal size. Use these scores to recode *age:*

Interval	% of Sample
18–37	34.2
38–53	31.7
54–89	34.1

Remember to select the "Recode into Different Variable" option. I named the recoded version of the variable *ager* (for "age recoded). As a convenience, I added labels to the new values to improve the readability of the output.

To use the ANOVA procedure, click **Analyze,Compare Means,** and then **One-Way Anova.** The **One-Way Anova** window appears. Find *polviews* (the label for this variable is "THINK OF SELF AS LIBERAL OR CONSERVATIVE") in the variable list on the left and click the arrow to move the variable name into the **Dependent List** box. Note that you can request more than one dependent variable at a time. Next, find the name of the recoded *age* variable (*ager?*) and click the arrow to move the variable name into the **Factor** box.

Click **Options** and then click the box next to **Descriptive** in the **Statistics** box to request means and standard deviations along with the analysis of variance. Click **Continue** and then click **OK,** and the following output will be produced. For the sake of clarity, several columns of the output have been deleted.

Descriptives: THINK OF SELF AS LIBERAL OR CONSERVATIVE

	N	Mean	Std. Deviation	Std. Error
18-37	466	3.93	1.368	.063
38-53	437	4.17	1.332	.064
54-89	464	4.33	1.419	.066
Total	1367	4.14	1.384	.037

ANOVA: THINK OF SELF AS LIBERAL OR CONSERVATIVE

	Sum of Squares	df	Mean Square	F	Sig.
Between Groups	38.218	2	19.109	10.112	.000
Within Groups	2577.679	1364	1.890		
Total	2615.898	1366			

The output box labeled ANOVA includes the various degrees of freedom, all of the sums of squares, the mean square estimates, the *F* ratio (10.112), and, at the far right, the exact probability ("Sig.") of getting these results if the null hypothesis is true.

This is reported as .000, which is lower than our usual alpha level of .05 (in fact, it's lower than .001). The differences in *polviews* for the various age groups are statistically significant.

The report also displays some summary statistics. An inspection of the means shows that the average score for the entire sample was 4.14. The oldest age group is the most conservative (remember that higher scores indicate greater conservatism). These results indicate that conservatism is significantly related to age and that older people are more conservative.

SPSS DEMONSTRATION 10.2 Does Political Ideology Vary by Social Class?

Let's continue our analysis of *polviews* to see if there are any significant differences in political ideology by social class. The GSS includes several variables that might be used as indicators of class, including *degree, income, class,* and *prestg80.* We should probably investigate all of these, but—to conserve space—we confine our attention to *class* (the label for this variable is SUBJECTIVE CLASS IDENTIFICATION). Follow the instructions for **One-Way ANOVA** in Demonstration 10.1, and specify *polviews* again as the dependent variable and *class* as the factor, or independent variable. The output is reproduced here. Again, I deleted several columns of output for the sake of clarity.

Descriptives: THINK OF SELF AS LIBERAL OR CONSERVATIVE

	N	Mean	Std. Deviation	Std. Error
LOWER CLASS	66	4.24	1.190	.147
WORKING CLASS	414	4.12	1.394	.069
MIDDLE CLASS	421	4.18	1.416	.069
UPPER CLASS	35	3.86	1.517	.256
Total	936	4.14	1.395	.046

ANOVA: THINK OF SELF AS LIBERAL OR CONSERVATIVE

	Sum of Squares	df	Mean Square	F	Sig.
Between Groups	4.407	3	1.469	.754	.520
Within Groups	1815.122	932	1.948		
Total	1819.529	935			

An inspection of the group means shows that upper-class respondents were most liberal (had the lowest average score on *polviews*). The scores for middle-, working-, and lower-class respondents were quite similar. The *F* ratio is quite low (.754), and the sig. (.520) reported at the far right of the ANOVA output box show that these differences are not statistically significant. According to these results, there is no significant difference in political ideology by social class

SPSS DEMONSTRATION 10.3 Another Test for Differences in Attitudes on Abortion

In Demonstration 9.2, we found that *sex* did not have a statistically significant impact on *abscale.* In this demonstration, we investigate the impact of religious affiliation (*relig*). Remember that higher scores on *abscale* indicate greater opposition to legalized abortion. Click **Analyze,Compare Means,** and **One Way Anova** and name *abscale* as the dependent variable and *relig* as the factor. If necessary, see Demonstration 9.2 for instructions on computing *abscale.* The output will look like this:

Descriptives: ABSCALE

					95% Confidence Interval for Mean			
	N	Mean	Std. Deviation	Std. Error	Lower Bound	Upper Bound	Minimum	Maximum
Protestant	235	3.3277	.90999	.05936	3.2107	3.4446	2.00	4.00
Catholic	121	3.2149	.90560	.08233	3.0519	3.3779	2.00	4.00
Jewish	7	2.4286	.78680	.29738	1.7009	3.1562	2.00	4.00
None	68	2.6176	.88147	.10689	2.4043	2.8310	2.00	4.00
OTHER (SPECIFY)	4	2.0000	.00000	.00000	2.0000	2.0000	2.00	2.00
Total	435	3.1586	.94070	.04510	3.0700	3.2473	2.00	4.00

ANOVA: ABSCALE

	Sum of Squares	df	Mean Square	F	Sig.
Between Groups	36.099	4	9.025	11.153	.000
Within Groups	347.957	430	.809		
Total	384.055	434			

The *F* ratio (11.153) and sig. (.000) indicate a significant difference at the .05 level. Protestant and Catholic respondents were the most opposed and "others" were the most supportive of the right to a legal abortion.

Post Hoc Analysis. To conduct a post hoc analysis to determine which differences are significant, you must select one of the many available tests from the **One Way Anova** window. To do this, rerun the **One Way Anova** command for *relig* and *abscale,* click the **Post Hoc** button on the **One Way Anova** window, and click the box next to **LSD** (the "least significant difference" test). Then click **Continue** and **OK,** and you will get new ANOVA output with the post hoc test included. The output is lengthy and so is not reproduced here, but the essential information is contained in the column labeled "Mean Difference." This column lists the difference between the mean scores of each category of the independent variable. For example, the top line compares Protestants with Catholics and reports a difference in mean scores of .1128. The next line compares Protestants with Jews (a difference of .8991), and differences that are significant at the .05 level are marked with an asterisk (*). Overall, the table shows that most of the significance in this relationship is accounted for by differences between Protestants and Catholics, on one hand, and Jews, "Nones," and "Others" on the other.

Exercises

10.1 As a follow-up on Demonstration 10.2, test *income91, prestg80,* and *papres80* as independent variables (factors) against *polviews.* Do these measures of social class display the same type of relationship with *polviews* as *class?* First, recode each of these independents into three categories. Run **Frequencies** for each, and find cutting points that divide the sample into three groups of roughly equal size.

10.2 What other variable might have a significant relationship with *abscale?* Pick three potential independent variables (factors) and test their relationships with *abscale.* Some possible independent variables would be *racecen1, attend,* and any of the measures of social class. If necessary, recode the independent variables into a few (three or four) categories.

11

Hypothesis Testing IV
Chi Square

LEARNING OBJECTIVES

By the end of this chapter, you will be able to

1. Identify and cite examples of situations in which the chi square test is appropriate.
2. Explain the structure of a bivariate table and the concept of independence as applied to expected and observed frequencies in a bivariate table.
3. Explain the logic of hypothesis testing as applied to a bivariate table.
4. Perform the chi square test using the five-step model and correctly interpret the results.
5. Explain the limitations of the chi square test and, especially, the difference between statistical significance and importance.

11.1 INTRODUCTION

The **chi square (χ^2) test** has probably been the most frequently used test of hypothesis in the social sciences, a popularity that is due largely to the fact that the assumptions and requirements in step 1 of the five-step model are easy to satisfy. Specifically, the test can be conducted with variables measured at the nominal level (the lowest level of measurement) and, because it is a **nonparametric,** or "distribution-free" test, chi square requires no assumption at all about the shape of the population or sampling distribution.

Why is it an advantage to have assumptions and requirements that are easy to satisfy? The decision to reject the null hypothesis (step 5) is not specific: It means only that one statement in the model (step 1) *or* the null hypothesis (step 2) is wrong. Usually, of course, we single out the null hypothesis for rejection. The more certain we are of the model, the greater our confidence that the null hypothesis is the faulty assumption. A "weak" or easily satisfied model means that our decision to reject the null hypothesis can be made with even greater certainty.

Chi square has also been popular for its flexibility. Not only can it be used with variables at any level of measurement, it can be used with variables that have many categories or scores. For example, in Chapter 9, we tested the significance of the difference in the proportions of black and white citizens who were "highly participatory" in voluntary associations. What if the researcher wished to expand the test to include Americans of Hispanic and Asian descent? The two-sample test would no longer be applicable, but chi square handles the more complex variable easily. Also, unlike the ANOVA test presented in Chapter 10, the chi square test can be conducted with variables at any level of measurement.

11.2 BIVARIATE TABLES

Chi square is computed from **bivariate tables,** so called because they display the scores of cases on two different variables at the same time. Bivariate tables are used to ascertain if there is a significant relationship between the variables and for other purposes that we will investigate in later chapters. In fact, these tables are very commonly used in research, and a detailed examination of them is in order.

First of all, bivariate tables have (of course) two dimensions. The horizontal (across) dimension is referred to in terms of **rows,** and the vertical dimension (up and down) is referred to in terms of **columns.** Each column or row represents a score on a variable, and the intersections of the row and columns **(cells)** represent the various combined scores on both variables.

Let's use an example to clarify. Suppose a researcher is interested in the relationship between racial group membership and participation in voluntary groups, community-service organizations, and so forth. Is there a difference in the level of involvement in volunteer groups between the races? We have two variables here (race and number of memberships) and, for the sake of simplicity, assume that both are simple dichotomies. That is, people have been classified as either black or white and as either high or low in their level of involvement in voluntary associations.

By convention, the independent variable (the variable that is taken to be the cause) is placed in the columns and the dependent variable in the rows. In the example at hand, race is the causal variable (the question was "Is membership *affected by* race?"), and each column will represent a score on this variable. Each row, on the other hand, will represent a score on level of membership (high or low). Table 11.1 displays the outline of the bivariate table for a sample of 100 people.

Note some further details of the table. First, subtotals have been added to each column and row. These are called the row or column **marginals,** and in this case they tell us that 50 members of the sample were black and 50 were white (the column marginals) and 50 were rated as high in participation and 50 were rated low (the row marginals). Second, the total number of cases in the sample ($N = 100$) is reported at the intersection of the row and column marginals. Finally, take careful note of the labeling of the table. Each row and column is identified, and the table has a descriptive title that includes the names of the variables, with the dependent variable listed first. Clear, complete labels and concise titles must be included in *all* tables, graphs, and charts.

As you have noticed, Table 11.1 lacks one piece of crucial information: the numbers of each racial group that rated high or low on the dependent variable. To finish the table, we need to classify each member of the sample in terms of both their race and their level of participation, keep count of how often each combination of scores occurs, and record these numbers in the appropriate cell of the table. Since each of our variables (race and participation rates) has two scores, four combinations of scores are possible, each corresponding to a cell

TABLE 11.1 RATES OF PARTICIPATION IN VOLUNTARY ASSOCIATIONS BY RACIAL GROUP FOR 100 SENIOR CITIZENS

	Racial Group		
Participation Rates	Black	White	
High			50
Low			50
	50	50	100

in the table. For example, blacks with high levels of participation would be counted in the upper-left-hand cell, whites with low levels of participation would be counted in the lower-right-hand cell, and so forth. When we are finished counting, each cell will display the number of times each *combination* of scores occurred.

Finally, note how we could expand the bivariate table to accommodate variables with more scores. If we wished to include more groups in the test (e.g., Asian Americans and Hispanic Americans), we would simply add additional columns to the table. More elaborate dependent variables could also be easily accommodated. If we had measured participation rates with three categories (e.g., high, moderate, and low) rather than two, we would simply add an additional row to the table.

11.3 THE LOGIC OF CHI SQUARE

The chi square test has several different uses, and most of this chapter deals with an application called the *chi square test for independence*. We have encountered the term *independence* in connection with the requirements for the two-sample case (Chapter 9) and for the ANOVA test (Chapter 10). In those situations, we noted that independent random samples are gathered such that the selection of a particular case for one sample has no effect on the probability that any particular case will be selected for the other sample (see Section 9.2).

In the context of chi square, the concept of **independence** takes on a slightly different meaning because it refers to the relationship between the variables, not between the samples. Two variables are independent if the classification of a case into a particular category of one variable has no effect on the probability that the case will fall into any particular category of the second variable. For example, race and participation in voluntary association would be independent of each other if the classification of a case as black or white has no effect on the classification of the case as high or low on participation. In other words, the variables are independent if level of participation and race were completely unrelated to each other.

Consider Table 11.1 again. If these two variables are truly independent, the cell frequencies will be determined solely by random chance and we would find that, just as an honest coin will show heads about 50% of the time when flipped, about half of the black respondents will rank high on participation and half will rank low. The same pattern would hold for the 50 white respondents and, therefore, each of the four cells would have about 25 cases in it, as illustrated in Table 11.2. This pattern of cell frequencies indicates that the racial classification of the subjects has no effect on the probability that they would be either high or low

TABLE 11.2 THE CELL FREQUENCIES THAT WOULD BE EXPECTED IF RATES OF PARTICIPATION AND RACIAL GROUP WERE INDEPENDENT

	Racial Group		
Participation Rates	Black	White	
High	25	25	50
Low	25	25	50
	50	50	100

in participation. The probability of being classified as high or low would be .50 for both blacks and whites, and the variables would therefore be independent.

The null hypothesis for chi square is that the variables are independent. Under the assumption that the null hypothesis is true, the cell frequencies we would expect to find if only random chance were operating are computed. These frequencies, called **expected frequencies** (symbolized f_e), are then compared, cell by cell, with the frequencies actually observed in the table (**observed frequencies,** symbolized f_o). If the null hypothesis is true and the variables are independent, then there should be little difference between the expected and observed frequencies. If the null hypothesis is false, however, there should be large differences between the two. The greater the differences between expected (f_e) and observed (f_o) frequencies, the less likely that the variables are independent and the more likely that we will be able to reject the null hypothesis.

11.4 THE COMPUTATION OF CHI SQUARE

As with all tests of hypothesis, with chi square we compute a test statistic, **χ^2(obtained),** from the sample data and then place that value on the sampling distribution of all possible sample outcomes. Specifically, the χ^2(obtained) will be compared with the value of **χ^2(critical)** that will be determined by consulting a chi square table (Appendix C) for a particular alpha level and degrees of freedom. Prior to conducting the formal test of hypothesis, let us take a moment to consider the calculation of chi square, as defined by Formula 11.1:

FORMULA 11.1

$$\chi^2(\text{obtained}) = \sum \frac{(f_o - f_e)^2}{f_e}$$

where f_o = the cell frequencies observed in the bivariate table
f_e = the cell frequencies that would be expected if the variables were independent

We must work on a cell-by-cell basis to solve this formula. To compute chi square, subtract the expected frequency from the observed frequency for each cell, square the result, divide by the expected frequency for that cell, and then sum the resultant values for all cells.

This formula requires an expected frequency for each cell in the table. In Table 11.2, the marginals are the same value for all rows and columns, and the expected frequencies are obvious by intuition: $f_e = 25$ for all four cells. In the more usual case, the expected frequencies will not be obvious, marginals will be unequal, and we must use Formula 11.2 to find the expected frequency for each cell:

FORMULA 11.2

$$f_e = \frac{\text{Row marginal} \times \text{Column marginal}}{N}$$

That is, the expected frequency for any cell is equal to the total number of cases in the row (the row marginal) times the total number of cases in the column (the column marginal) divided by the total number of cases in the table (N).

An example using Table 11.3 should clarify these procedures. A random sample of 100 social work majors have been classified in terms of whether the Council on Social Work Education has accredited their undergraduate programs (the column, or independent, variable) and whether they were hired in social work positions within three months of graduation (the row, or dependent, variable).

TABLE 11.3 EMPLOYMENT OF 100 SOCIAL WORK MAJORS BY ACCREDITATION STATUS OF UNDERGRADUATE PROGRAM

	Accreditation Status		
Employment Status	Accredited	Not Accredited	Totals
Working as a social worker	30	10	40
Not working as a social worker	25	35	60
Totals	55	45	100

TABLE 11.4 EXPECTED FREQUENCIES FOR TABLE 11.3

	Accreditation Status		
Employment Status	Accredited	Not Accredited	Totals
Working as a social worker	22	18	40
Not working as a social worker	33	27	60
Totals	55	45	100

TABLE 11.5 COMPUTATIONAL TABLE FOR TABLE 11.3

(1)	(2)	(3)	(4)	(5)
f_o	f_e	$f_o - f_e$	$(f_o - f_e)^2$	$(f_o - f_e)^2/f_e$
30	22	8	64	2.91
10	18	−8	64	3.56
25	33	−8	64	1.94
35	27	8	64	2.37
$N = 100$	$N = 100$	0	χ^2(obtained) =	10.78

Beginning with the upper-left-hand cell (graduates of accredited programs who are working as social workers), the expected frequency for this cell, using Formula 11.2, is (40 × 55)/100, or 22. For the other cell in this row (graduates of nonaccredited programs who are working as social workers), the expected frequency is (40 × 45)/100, or 18. For the two cells in the bottom row, the expected frequencies are (60 × 55)/100, or 33, and (60 × 45)/100, or 27, respectively. The expected frequencies for all four cells are displayed in Table 11.4.

Note that the row and column marginals as well as the total number of cases in Table 11.4 are exactly the same as those in Table 11.3. The row and column marginals for the expected frequencies must *always* equal those of the observed frequencies, a relationship that provides a convenient way of checking your arithmetic to this point.

The value for chi square for these data can now be found by solving Formula 11.1. It will be helpful to use a computing table, such as Table 11.5, to organize the several steps required to compute chi square. The table lists the observed frequencies (f_o) in column 1 in order from the upper-left-hand cell to the

ONE STEP AT A TIME Computing Chi Square

Step 1: Prepare a computing table similar to Table 11.5 to organize your computations. List the observed frequencies (f_0) in column 1. The total for column 1 is the number of cases (N).

Find the Expected Frequencies (f_e) Using Formula 11.2

Step 2: Start with the upper-left-hand cell and multiply the row marginal by the column marginal for that cell.

Step 3: Divide the quantity you found in Step 2 by N. The result is the expected frequency (f_e) for that cell. Record this value in the second column of the computing table. Make sure you place the value of f_e in the same row as the observed frequency for that cell.

Step 4: Repeat steps 2 and 3 for each cell in the table. Double-check to make sure that you are using the correct row and column marginals. Record each f_e in the second column of the computational table.

Step 5: Find the total of the expected frequencies column. This total *must* equal the total of the observed frequencies column (which is the same as N). If the two totals do not match (within rounding error), recompute the expected frequencies.

Find Chi Square Using Formula 11.1

Step 6: For each cell, subtract the expected frequency (f_e) from the observed frequency (f_o), and list these values in the third column of the computational table ($f_o - f_e$). Find the total for this column. If this total is not zero, you have made a mistake and need to check your computations.

Step 7: Square each of the values in the third column of the table, and record the result in the fourth column, labeled $(f_o - f_e)^2$.

Step 8: Divide each value in column 4 by the expected frequency for that cell, and record the result in the fifth column, labeled $(f_o - f_e)^2/f_e$.

Step 9: Find the total for the fifth column. This value is χ^2(obtained).

lower-right-hand cell, moving left to right across the table and top to bottom. Column 2 lists the expected frequencies (f_e) in exactly the same order. Double-check to make sure you have listed the cell frequencies in the same order for both of these columns.

The next step is to subtract the expected frequency from the observed frequency for each cell and list these values in column 3. To complete column 4, square the value in column 3 and then, in column 5, divide the column 4 value by the expected frequency for that cell. Finally, add up column 5. The sum of this column is χ^2(obtained):

$$\chi^2(\text{obtained}) = 10.78$$

Note that the totals for columns 1 and 2 (f_o and f_e) are exactly the same. This will always be the case, and if the totals do not match, you have made a computational error, probably in the calculation of the expected frequencies. Also note that the sum of column 3 will always be zero, another convenient way to check your math to this point.

This sample value for chi square must still be tested for its significance. *(For practice in computing chi square, see problem 11.1.)*

11.5 THE CHI SQUARE TEST FOR INDEPENDENCE

As always, the five-step model for significance testing will provide the framework for organizing our decision making. The data presented in Table 11.3 will serve as our example.

Application 11.1

Do members of different groups have different levels of "narrow-mindedness"? A random sample of 47 white and black Americans have been rated as high or low on a scale that measures intolerance of viewpoints or belief systems different from their own. The results are

	Group		
Intolerance	White	Black	Totals
High	15	5	20
Low	10	17	27
Totals	25	22	47

The frequencies we would expect to find if the null hypothesis (H_0: the variables are independent) were true are

	Group		
Intolerance	White	Black	Totals
High	10.64	9.36	20.00
Low	14.36	12.64	27.00
Totals	25.00	22.00	47.00

Expected frequencies are found on a cell-by-cell basis by means of the formula

$$f_e = \frac{\text{Row marginal} \times \text{Column marginal}}{N}$$

and the calculation of chi square will be organized into a computational table.

(1)	(2)	(3)	(4)	(5)
f_o	f_e	$f_o - f_e$	$(f_o - f_e)^2$	$(f_o - f_e)^2/f_e$
15	10.64	4.36	19.01	1.79
5	9.36	−4.36	19.01	2.03
10	14.36	−4.36	19.01	1.32
17	12.64	4.36	19.01	1.50
$N = 47$	$N = 47.00$	0.00	χ^2(obtained) =	6.64

$$\chi^2(\text{obtained}) = 6.64$$

Step 1. Making Assumptions and Meeting Test Requirements.

Model: Independent random samples
Level of measurement is nominal

Step 2. Stating the Null Hypothesis.

H_0: The two variables are independent.
(H_1: The two variables are dependent.)

Step 3. Selecting the Sampling Distribution and Establishing the Critical Region.

Sampling distribution = χ^2 distribution
Alpha = .05
Degrees of freedom = 1
χ^2(critical) = 3.841

Step 4. Computing the Test Statistic.

$$\chi^2(\text{obtained}) = \sum \frac{(f_o - f_e)^2}{f_e}$$

$$\chi^2(\text{obtained}) = 6.64$$

Step 5. Making a Decision and Interpreting the Results of the Test. With an obtained χ^2 of 6.64, we would reject the null hypothesis of independence. For this sample there is a statistically significant relationship between group membership and intolerance.

To complete the analysis, it would be useful to know exactly how the two variables are related. We can determine this by computing and analyzing column percents:

	Group		
Intolerance	White	Black	Totals
High	60.00%	22.73%	43.00%
Low	40.00%	77.27%	57.00%
Totals	100.00%	100.00%	100.00%

The column percents show that 60% of whites in this sample are high on intolerance vs. only 23% of blacks. We have already concluded that the relationship is significant, and now we know the pattern of the relationship: The white respondents were more likely to be high on intolerance.

Step 1. Making Assumptions and Meeting Test Requirements. Note that we make no assumptions at all about the shape of the sampling distribution.

Model: Independent random samples
Level of measurement is nominal

Step 2. Stating the Null Hypothesis. As stated previously, the null hypothesis in the case of chi square states that the two variables are independent. If the null hypothesis is true, the differences between the observed and expected frequencies will be small. As usual, the research hypothesis directly contradicts the null hypothesis. Thus, if we reject H_0, the research hypothesis will be supported.

H_0: The two variables are independent
(H_1: The two variables are dependent)

Step 3. Selecting the Sampling Distribution and Establishing the Critical Region. The sampling distribution of sample chi squares, unlike the Z and t distributions, is positively skewed, with higher values of sample chi squares in the upper tail of the distribution (to the right). Thus, with the chi square test, the critical region is established in the upper tail of the sampling distribution.

Values for χ^2(critical) are given in Appendix C. This table is similar to the t table, with alpha levels arrayed across the top and degrees of freedom down the side. A major difference, however, is that degrees of freedom (df) for chi square are found by the following formula:

FORMULA 11.3

$$\text{df} = (r - 1)(c - 1)$$

A table with two rows and two columns (a 2 × 2 table) has one degree of freedom regardless of the number of cases in the sample.[1] A table with two rows and three columns would have (2 − 1)(3 − 1), or two degrees of freedom. Our sample problem involves a 2 × 2 table with df = 1, so if we set alpha at 0.05, the critical chi square score would be 3.841. Summarizing these decisions, we have

$$\begin{aligned} \text{Sampling distribution} &= \chi^2 \text{ distribution} \\ \text{Alpha} &= .05 \\ \text{Degrees of freedom} &= 1 \\ \chi^2(\text{critical}) &= 3.841 \end{aligned}$$

[1] Degrees of freedom are the number of values in a distribution that are free to vary for any particular statistic. A 2 × 2 table has one degree of freedom because, for a given set of marginals, once one cell frequency is determined, all other cell frequencies are fixed (that is, they are no longer free to vary). In Table 11.3, for example, if any cell frequency is known, all others are determined. If the upper-left-hand cell is known to be 30, the remaining cell in that row must be 10, since there are 40 cases total in the row and 40 − 30 = 10. Once the frequencies of the cells in the top row are established, cell frequencies for the bottom row are determined by subtraction from the column marginals. Incidentally, this relationship can be used to good advantage when computing expected frequencies. For example, in a 2 × 2 table, only one expected frequency needs to be computed. The f_e's for all other cells can then be found by subtraction.

Step 4. Computing the Test Statistic. The mechanics of these computations were introduced in Section 11.4. As you recall, we had

$$\chi^2(\text{obtained}) = \sum \frac{(f_o - f_e)^2}{f_e}$$
$$\chi^2(\text{obtained}) = 10.78$$

Step 5. Making a Decision and Interpreting the Results of the Test. Comparing the test statistic with the critical region,

$$\chi^2(\text{obtained}) = 10.78$$
$$\chi^2(\text{critical}) = 3.841$$

we see that the test statistic falls into the critical region and, therefore, we reject the null hypothesis of independence. The pattern of cell frequencies observed in Table 11.3 is unlikely to have occurred by chance alone. The variables are dependent. Specifically, based on these sample data, the probability of securing employment in the field of social work is dependent on the accreditation status of the program. *(For practice in conducting and interpreting the chi square test for independence, see problems 11.2 to 11.15.)*

Let us stress exactly what the chi square test does and does not tell us. A significant chi square means that the variables are (probably) dependent on each other in the population: Accreditation status makes a difference in whether or not a person is working as a social worker. What chi square does *not* tell us is the exact nature of the relationship. In our example, it does not tell us which type of graduate—those from accredited or nonaccredited programs—is more likely to be working as social workers. To make this determination, we must perform some additional calculations. We can figure out how the independent variable (accreditation status) is affecting the dependent variable (employment as a social worker) by computing "column percentages" or by calculating percentages within each column of the bivariate table. This procedure is analogous to calculating percentages for frequency distributions (see Chapter 2).

To calculate column percentages, divide each cell frequency by the total number of cases in the column (the column marginal) and multiply the result by 100. For Table 11.3, starting in the upper-left-hand cell, we see that there are 30 cases in this cell and 55 cases in the column. In other words, 30 of the 55 graduates of accredited programs are working as social workers. The column percentage for this cell is therefore $(30/55) \times 100 = 54.55\%$. For the lower-left-hand cell, the column percent is $(25/55) \times 100 = 45.45\%$. For the two cells in the right-hand column (graduates of nonaccredited programs), the column percents are $(10/45) \times 100 = 22.22$ and $(35/45) \times 100 = 77.78$. Table 11.6 displays all column percents for Table 11.3.

Column percents help to make the relationship between the two variables more obvious, and we can see easily from Table 11.6 that it's the students from accredited programs that are more likely to be working as social workers. Nearly 55% of these students are working as social workers vs. less than 30% of the students from nonaccredited programs. We already know that this relationship is significant (unlikely to be caused by random chance), and now, with the aid of column percents, we know how the two variables are related. According to these results, graduating from an accredited program would be a decided advantage for people seeking to enter the social work profession.

ONE STEP AT A TIME Computing Column Percentages

Step 1: Start with the upper-left-hand cell. Divide the cell frequency (the number of cases in the cell) by the total number of cases in that column (or the column marginal). Multiply the result by 100 to convert to a percentage.

Step 2: Move down one cell and repeat step 1. Continue moving down the column, cell by cell, until you have converted all cell frequencies to percentages.

Step 3: Move to the next column. Start with the cell in the top row and repeat step 1 (making sure that you use the correct column total in the denominator of the fraction).

Step 4: Continue moving down the second column until you have converted all cell frequencies to percentages.

Step 5: Continue these operations, moving from column to column one at a time, until you have converted all cell frequencies to percentages.

TABLE 11.6 COLUMN PERCENTS FOR TABLE 11.3

Employment Status	Accreditation Status: Accredited	Accreditation Status: Not Accredited	Totals
Working as a social worker	54.55%	22.22%	40.00%
Not working as a social worker	45.45%	77.78%	60.00%
Totals	100.00% (55)	100.00% (45)	100.00%

Let's highlight two points in summary:

1. Chi square is a test of statistical significance. It tests the null hypothesis that the variables are independent in the population. If we reject the null hypothesis, we are concluding, with a known probability of error (determined by the alpha level), that the variables are dependent on each other in the population. In the terms of our example, this means that accreditation status makes a difference in the likelihood of finding work as a social worker. By itself, however, chi square does not tell us the exact nature of the relationship.
2. Computing column percents allows us to examine the bivariate relationship in more detail. By comparing the column percents for the various scores of the independent variable, we can see exactly how the independent variable affects the dependent variable. In this case, the column percents reveal that graduates of accredited programs are more likely to find work as social workers. We explore column percentages more extensively when we discuss bivariate association in Chapters 12–14.

11.6 THE CHI SQUARE TEST: AN ADDITIONAL EXAMPLE

To this point, we have confined our attention to 2 × 2 tables, that is, tables with two rows and two columns. For purposes of illustration, we will work through the computational routines and decision-making process for a larger table. As you will see, larger tables require more computations (because they have more cells); but in all other essentials they are dealt with in the same way as the 2 × 2 table.

A researcher is concerned with the possible effects of marital status on the academic progress of college students. Do married students, with their extra burden of family responsibilities, suffer academically as compared to unmarried students? Is academic performance dependent on marital status? A random sample of 453 students is gathered, and each student is classified as either married or unmarried and—using grade-point average (GPA) as a measure—as a good, average, or poor student. Results are presented in Table 11.7.

For the top-left-hand cell (married students with good GPAs) the expected frequency would be (160 × 175)/453, or 61.8. For the other cell in this row, expected frequency is (160 × 278)/453, or 98.2. In similar fashion, all expected frequencies are computed (being very careful to use the correct row and column marginals) and displayed in Table 11.8.

The next step is to solve the formula for χ^2(obtained), being very careful to be certain that we are using the proper f_o's and f_e's for each cell. Once again, we will use a computational table (Table 11.9) to organize the calculations and then test the obtained chi square for its statistical significance. Remember that obtained chi square is equal to the total of column 5.

The value of the obtained chi square (2.79) can now be tested for its significance.

Step 1. Making Assumptions and Meeting Test Requirements.

Model: Independent random samples
Level of measurement is nominal

Step 2. Stating the Null Hypothesis.

H_0: The two variables are independent
(H_1: The two variables are dependent)

TABLE 11.7 GRADE-POINT AVERAGE (GPA) BY MARITAL STATUS FOR 453 COLLEGE STUDENTS

	Marital Status		
GPA	Married	Not Married	Totals
Good	70	90	160
Average	60	110	170
Poor	45	78	123
Totals	175	278	453

TABLE 11.8 EXPECTED FREQUENCIES FOR TABLE 11.7

	Marital Status		
GPA	Married	Not Married	Totals
Good	61.8	98.2	160
Average	65.7	104.3	170
Poor	47.5	75.5	123
Totals	175.0	278.0	453

TABLE 11.9 COMPUTATIONAL TABLE FOR TABLE 11.7

(1)	(2)	(3)	(4)	(5)
f_o	f_e	$f_o - f_e$	$(f_o - f_e)^2$	$(f_o - f_e)^2/f_e$
70	61.8	8.2	67.24	1.09
90	98.2	−8.2	67.24	0.69
60	65.7	−5.7	32.49	0.49
110	104.3	5.7	32.49	0.31
45	47.5	−2.5	6.25	0.13
78	75.5	2.5	6.25	0.08
$N = 453$	$N = 453.0$	0.0		χ^2(obtained) = 2.79

Step 3. Selecting the Sampling Distribution and Establishing the Critical Region.

$$\text{Sampling distribution} = \chi^2 \text{ distribution}$$
$$\text{Alpha} = .05$$
$$\text{Degrees of freedom} = (r - 1)(c - 1) = (3 - 1)(2 - 1) = 2$$
$$\chi^2(\text{critical}) = 5.991$$

Step 4. Computing the Test Statistic.

$$\chi^2(\text{obtained}) = \sum \frac{(f_o - f_e)^2}{f_e}$$
$$\chi^2(\text{obtained}) = 2.79$$

Step 5. Making a Decision and Interpreting the Results of the Test. The test statistic, χ^2(obtained) = 2.79, does not fall into the critical region, which, for alpha = 0.05, df = 2, begins at χ^2(critical) of 5.991. Therefore, we fail to reject the null hypothesis. The observed frequencies are not significantly different from the frequencies we would expect to find if the variables were independent and only random chance were operating. Based on these sample results, we can conclude that the academic performance of college students is not dependent on their marital status. Since we failed to reject the null hypothesis, we will not examine column percentages as we did for Table 11.3

11.7 AN ADDITIONAL APPLICATION OF THE CHI SQUARE TEST: THE GOODNESS-OF-FIT TEST[2]

To this point we have dealt with the chi square test for independence for situations involving two variables, each of which has two or more categories. Another situation in which the chi square statistic is useful, called the **goodness-of-fit test,** is one in which the distribution of scores on a single variable must be tested for significance. The logic underlying this second research situation is quite similar to that of the test for independence. The chi square statistic will be computed by comparing the actual distribution of a variable against a set of expected frequencies. The greater the difference between the observed distribution of scores and the expected distribution, the more likely that the observed pattern did not

[2]This section is optional.

occur by random chance alone. If the observed and expected frequencies are similar, it is said that there is a "good fit" (hence the name of this application), and we would conclude that the two distributions were not significantly different.

A major difference in this new application lies in the way in which the expected frequencies are ascertained. Instead of computing these scores, the null hypothesis is used to figure out what the expected frequencies should be. An example may make this process clear. Suppose you were gambling on coin tosses and suspected that a particular coin was biased toward heads. Over a series of tosses, what percentage of heads would you expect to observe from an unbiased coin? In an actual test of significance for this problem, the null hypothesis would be that the coin was unbiased and that half of all tosses should be heads and half tails. Notice what we have just done: We figured out the expected frequencies (half of the flips should be heads) from the null hypothesis (the coin is unbiased) rather than by computing them.

Let's consider another example of a situation in which the goodness-of-fit test is appropriate. Is there a seasonal rhythm to the crime rate? If we gathered crime statistics on a monthly basis for a given jurisdiction, what would we expect to find? Given the way in which the problem has been stated, there is only one variable (crime rate), and the focus is on the distribution of that variable over a given set of categories (by month). If the crime rate does *not* vary by month, we would expect that about 1/12 of all crimes committed in a year would be committed in each and every month. Our null hypothesis would be that the crime rate does not vary across time, and our expected frequencies would be calculated by dividing the actual (observed) number of crimes equally by the number of months. (To conserve time and space, we will ignore the slight complexities that would be introduced by taking account of the varying number of days per month.)

In a particular jurisdiction, a total of 2172 crimes were committed last year. The actual distribution of crimes by month is displayed in Table 11.10. The expected distribution of crimes per month would be found by dividing the total number of crimes by 12: 2172/12 = 181. Note that the expected frequency will be the same for every month. With these values determined, we can calculate chi square by using Formula 11.1.

TABLE 11.10 NUMBER OF CRIMES PER MONTH

Month	Number of Crimes
January	190
February	152
March	121
April	110
May	147
June	199
July	250
August	247
September	201
October	150
November	193
December	212
	2172

$$\chi^2(\text{obtained}) = \sum \frac{(f_o - f_e)^2}{f_e}$$

$$\chi^2(\text{obtained}) = \frac{(190-181)^2}{181} + \frac{(152-181)^2}{181} + \frac{(121-181)^2}{181} + \frac{(110-181)^2}{181}$$
$$+ \frac{(147-181)^2}{181} + \frac{(199-181)^2}{181} + \frac{(250-181)^2}{181} + \frac{(247-181)^2}{181}$$
$$+ \frac{(201-181)^2}{181} + \frac{(150-181)^2}{181} + \frac{(193-181)^2}{181} + \frac{(212-181)^2}{181}$$

$$\chi^2(\text{obtained}) = \frac{81}{181} + \frac{841}{181} + \frac{3600}{181} + \frac{5041}{181}$$
$$+ \frac{1156}{181} + \frac{324}{181} + \frac{4761}{181} + \frac{4356}{181}$$
$$+ \frac{400}{181} + \frac{961}{181} + \frac{144}{181} + \frac{961}{181}$$

$$\chi^2(\text{obtained}) = 0.45 + 4.65 + 19.89 + 27.85 + 6.39 + 1.79 + 26.30 + 24.07 + 2.21 + 5.31 + 0.80 + 5.31$$

$$\chi^2(\text{obtained}) = 125.02$$

The χ^2 of 125.02 can now be tested for significance in the usual manner.

Step 1. Making Assumptions and Meeting Test Requirements.

Model: Random sampling
Level of measurement is nominal

In this application of significance testing, we are assuming that the observed frequencies represent a random sample of all possible frequency distributions.

Step 2. Stating the Null Hypothesis.

H_0: There is no difference in the crime rate by month.
(H_1: There is a difference in the crime rate by month.)

Step 3. Selecting the Sampling Distribution and Establishing the Critical Region. In the goodness-of-fit test, degrees of freedom are equal to the number of categories minus 1, or df $= k - 1$. In the problem under consideration, there are 12 months or categories and, therefore, 11 degrees of freedom.

Sampling distribution = χ^2 distribution
Alpha = .05
Degrees of freedom = $k - 1 = 12 - 1 = 11$
$\chi^2(\text{critical}) = 19.675$

Step 4. Computing the Test Statistic.

$$\chi^2(\text{obtained}) = 125.02$$

Step 5. Making a Decision and Interpreting the Results of the Test. The test statistic (125.02) clearly falls into the critical region (which begins at 19.675), so we may reject the null hypothesis. These data suggest that, at least for this jurisdiction, crime rate does vary by month in a nonrandom fashion. *(For* *practice in conducting and interpreting the chi square goodness-of-fit test, see problems 11.16 and 11.17.)*

11.8 THE LIMITATIONS OF THE CHI SQUARE TEST

Like any other test, chi square has limits, and you should be aware of several potential difficulties. First, even though chi square is very flexible and handles many different types of variables, it becomes difficult to interpret when the variables have many categories. For example, two variables with five categories each would generate a 5 × 5 table with 25 cells—far too many combinations of scores to be easily absorbed or understood. As a very rough rule of thumb, the chi square test is easiest to interpret and understand when both variables have four or fewer scores.

Two further limitations of the test are related to sample size. When sample size is small, we can no longer assume that the sampling distribution of all possible sample outcomes is accurately described by the chi square distribution. For chi square, a small sample is defined as one where a high percentage of the cells have expected frequencies (f_e) of 5 or less. Various rules of thumb have been developed to help the researcher decide what constitutes a "high percentage of cells." Probably the safest course is to take corrective action whenever *any* of the cells have expected frequencies of 5 or less.

In the case of 2 × 2 tables, the value of χ^2(obtained) can be adjusted by applying Yates' correction for continuity, the formula for which is

FORMULA 11.4

$$\chi^2_c = \sum \frac{(|f_o - f_e| - 0.5)^2}{f_e}$$

where χ^2_c = corrected chi square
$|f_o - f_e|$ = the absolute values of the difference between observed and expected frequency for each cell

The correction factor is applied by reducing the absolute value[3] of the term ($f_o - f_e$) by 0.5 before squaring the difference and dividing by the expected frequency for the cell.

For tables larger than 2 × 2, there is no correction formula for computing χ^2(obtained) for small samples. It may be possible to combine some of the categories of the variables and thereby increase cell sizes. Obviously, however, this course of action should be taken only when it is sensible to do so. In other words, distinctions that have clear theoretical justifications should not be erased merely to conform to the requirements of a statistical test. When you feel that categories cannot be combined to build up cell frequencies and the percentage of cells with expected frequencies of 5 or less is small, it is probably justifiable to continue with the uncorrected chi square test as long as the results are regarded with a suitable amount of caution.

A second potential problem related to sample size occurs with large samples. I pointed out in Chapter 9 that all tests of hypothesis are sensitive to sample size. That is, the probability of rejecting the null hypothesis increases as the number of cases increases, regardless of any other factor. It turns out that chi square is especially sensitive to sample size and that larger samples may lead

[3]Absolute values ignore plus and minus signs.

to the decision to reject the null when the actual relationship is trivial. In fact, chi square is more responsive to changes in sample size than other test statistics, since the value of χ^2(obtained) will increase at the same rate as sample size. That is, if sample size is doubled, the value of χ^2(obtained) will be doubled. *(For an illustration of this principle, see problem 11.14.)*

You should be aware of this relationship between sample size and the value of chi square because it once again raises the distinction between statistical significance and theoretical importance. On one hand, tests of significance play a crucial role in research. When we are working with random samples, we must know if our research results could have been produced by mere random chance.

On the other hand, like any other statistical technique, tests of hypothesis are limited in the range of questions they can answer. Specifically, these tests will tell us whether our results are statistically significant or not. They will not necessarily tell us if the results are important in any other sense. To deal more directly with questions of importance, we must use an additional set of statistical techniques called *measures of association.* We previewed these techniques in this chapter when we used column percents, and measures of association are the subject of Part III of this text.

11.9 INTERPRETING STATISTICS: FAMILY VALUES AND SOCIAL CLASS

When sociologists examine the patterns of family life in the United States, they commonly find that parents of different social classes raise their children in distinctly different ways. These differences include disciplinary techniques, expectations regarding success in school, and a host of other variables. In this installment of Interpreting Statistics, we examine differences in the goals parents of different social classes might have for their children. Specifically, we look at an item from the 2006 General Social Science survey that asks parents: "If you had to choose, which thing on this list would you pick as the most important for a child to learn to prepare him or her for life"? Respondents were offered a list of traits from which to choose; we focus on two of these: "To obey" and "To think for himself or herself." Respondents could rank each trait from one ("most important") to five ("least important"), but, to simplify the analysis, I have collapsed the scale to three scores:

1. Most important
2. Second or third most important
3. Fourth or fifth most important

Why would parents of different social classes vary in their socialization values? One possible causal factor is the nature of the work associated with each of the social classes. For people in the higher social classes, work tends to stress individual initiative and creativity and be less supervised. Lower-class and working-class occupations, in contrast, are more likely to be highly routinized, regimented, and closely supervised.

Thus, in a reflection of their work situations, we can hypothesize that parents in the higher social classes will be more likely to stress independent thinking, while parents in the lower social classes will be more likely to endorse obedience as the most important socialization goal. In the language of chi square, we can hypothesize that socialization goal will be dependent on social class.

Tables 11.11 and 11.12 show the results of two chi square tests, one for the relationship between social class and "think for self" and one for the relationship

TABLE 11.11 SOCIALIZATION GOAL ("THINK FOR SELF") BY SOCIAL CLASS

Importance of Thinking for Self	Social Class				
	Lower Class	Working Class	Middle Class	Upper Class	Totals
Highest	27	195	217	17	456
Moderate	25	154	155	11	345
Lowest	19	95	67	7	188
Totals	71	444	439	35	989

Chi square = 9.55

TABLE 11.12 SOCIALIZATION GOAL ("OBEY") BY SOCIAL CLASS

Importance of Obedience	Social Class				
	Lower Class	Working Class	Middle Class	Upper Class	Total
Highest	17	88	72	6	183
Moderate	23	137	121	7	288
Lowest	31	219	246	22	518
Totals	71	444	439	35	989

Chi square = 8.40

between social class and "obey." Before analyzing the results, remember that the General Social Survey is administered to random samples of adult Americans, selected in accordance with the rules of EPSEM (see Chapter 6). Thus, we can assume that the conclusions of this analysis will apply to U.S. society in general.

Both tables have 6 degrees of freedom ($4-1 \times 3-1$), so, at the .05 alpha level, the critical chi square value is 12.592. Neither of the obtained chi square approaches significance, and we may conclude that, contrary to our hypothesis, these socialization goals are independent of social class.

Since we failed to reject the null hypothesis, it would be quite sensible to end the investigation at this point. However, since the tables are already constructed, it might be interesting to look at the direction of the relationship to see if we can gather any insights.

Recall that we hypothesized that obedience would be more strongly endorsed by lower- and working-class respondents and that independent thinking would be more endorsed by the middle and upper classes. Do the patterns in Tables 11.13 and 11.14 conform to these hypotheses?

There is lot of detail in the tables, but for our purposes we can focus on the top row, or the percentage of respondents who selected the goals as most important. Starting with Table 11.13, independent thinking had a lot of support in all classes, and there is little difference between the column percentages. Lower-class respondents were the least likely to rate the goal as the highest, but the

TABLE 11.13 SOCIALIZATION GOAL ("THINK FOR SELF") BY SOCIAL CLASS (PERCENTAGES)

Importance of Thinking for Self	Social Class				
	Lower Class	Working Class	Middle Class	Upper Class	Totals
Highest	38.0%	43.9%	49.4%	48.6%	46.1%
Moderate	35.2%	34.7%	35.3%	31.4%	34.9%
Lowest	26.8%	21.4%	15.3%	20.0%	19.0%
Totals	100.0%	100.0%	100.0%	100.0%	100.0%

TABLE 11.14 SOCIALIZATION GOAL ("OBEY") BY SOCIAL CLASS

Importance of Obedience	Social Class				
	Lower Class	Working Class	Middle Class	Upper Class	Totals
Highest	23.9%	19.8%	16.4%	17.1%	18.5%
Moderate	32.4%	30.9%	27.6%	20.0%	29.1%
Lowest	43.7%	49.3%	56.0%	62.9%	52.4%
Totals	100.0%	100.0%	100.0%	100.0%	100.0%

differences from column to column are small. Note also that there is virtually no difference between middle-class and upper-class respondents. Overall, these slight (statistically insignificant) differences in column percentages are not consistent with the hypothesis that support for independent thinking increases with social class.

Turning to Table 11.4, and again focusing on the top row, we see that small minorities of the respondents from each class support obedience as a socialization goal. There are only slight differences from column to column, but, consistent with our hypothesis lower-class respondents were the most likely to rate obedience as most important (but remember that the relationship between the variables is not significant).

What can we conclude? These results show little support for our hypotheses. The relationships are not statistically significant even though our expectations about the pattern of column percentages were confirmed in both tables. Perhaps the problem with our hypothesis is that it is more appropriate for an industrial society, in which a large portion of the workforce is engaged in manual labor, blue-collar, manufacturing jobs. The United States is now a postindustrial society, in which good jobs increasingly demand high levels of education and well-developed critical thinking abilities. If people of all class backgrounds have come to recognize this economic reality, we would expect to find little difference in socialization goals by social class, exactly the pattern displayed in these tables.

Application 11.2

The World Values Survey, which is administered periodically to randomly selected samples of citizens from around the globe, allows us to test theories in a variety of settings. In Section 11.9, we examined the relationship between socialization goals and social class for a random sample of adults in the United States. In this section, we test the significance of the relationship between support for obedience as a socialization goal and gender using a randomly selected sample of Canadians and U.S. citizens. Do males place more emphasis on this trait than females? Does the relationship between these two variables change from nation to nation?

Respondents to the World Values Survey were given a list of possible socialization goals and asked to select the goals they thought were most important. Those who included obedience in their list were coded as "Yes" in the tables that follow. The U.S. sample was tested in 1999 and the Canadian data were collected in 2000.

The relationship for the U.S. sample is as follows:

United States (frequenceies)

Obedience Mentioned?	Gender: Male	Gender: Female	Totals
Yes	410	406	816
No	190	194	384
Totals	600	600	1,200

And chi square is computed with the following table:

(1) f_o	(2) f_e	(3) $f_o - f_e$	(4) $(f_o - f_e)^2$	(5) $(f_o - f_e)^2/f_e$
410	408	2	4	0.01
406	408	−2	4	0.01
190	192	−2	4	0.02
194	192	2	4	0.02
$N = 1200$	$N = 1200$	.0		χ^2(obtained) = 0.06

This is a 2 × 2 table, so there is one degree of freedom (df = 1 × 1 = 1) and the critical chi square is 3.841. The relationship is not significant, but we will still look at the pattern of the column percentages:

United States (percentages)

Obedience Mentioned?	Gender: Male	Gender: Female	Totals
Yes	68.3%	67.7	68.0
No	31.7%	32.3	32.0
Totals	100.00	100.00	100.00

Not only is the relationship not significant, but there is virtually no difference between U.S. males and females in their support for obedience.

Are the variables related in the Canadian sample? The bivariate table is as follows:

Canada (frequenceies)

Obedience Mentioned?	Gender: Male	Gender: Female	Totals
Yes	662	694	1,356
No	287	298	585
Totals	949	992	1,941

And we will use a computing table to get the value of chi square:

(1) f_o	(2) f_e	(3) $f_o - f_e$	(4) $(f_o - f_e)^2$	(5) $(f_o - f_e)^2/f_e$
662	662.98	−0.98	0.96	0.00
694	693.02	0.98	0.96	0.00
287	286.02	0.98	0.96	0.00
298	298.98	−0.98	0.96	0.00
$N = 1941$	$N = 1941.00$	0.00		χ^2(obtained) = 0.00

A 2 × 2 table has one degree of freedom and the critical chi square is 3.841. As was the case for the U.S. sample, there is no significant relationship between gender and support for this socialization goal in the Canadian sample.

Looking at the column percentages, we find virtually no difference between Canadian males and females in their support for obedience:

Canada (percentages)

Obedience Mentioned?	Gender: Male	Gender: Female	Totals
Yes	69.8	70.0	69.9
No	30.2	30.0	30.1
Totals	100.00	100.00	100.00

In summary, there is no relationship between gender and support for obedience as a socialization goal in either nation. Furthermore, the two nations are very similar in their support for obedience: Virtually identical percentages of the two samples (68% for the U.S., 70% for Canada) mentioned obedience as an important goal of child rearing.

READING STATISTICS 8: Does Treatment for Drug Abusers Work?

A team of researchers evaluated the effectiveness of a treatment program for inmates with a history of substance abuse. They gathered the records of 569 offenders, 495 of whom were paroled to a treatment facility in the community and 74 of whom were released without treatment. Criminal records were collected for all subjects for the two-year period following release. About a third of the nontreatment offenders had been convicted of a new crime by the end of the two-year period, versus only 22% of the treated offenders. The difference was statistically significant.

	Parole Condition	
Criminal Record	Treatment	No Treatment
New Conviction	22%	34%
No conviction	78%	66%
Totals	100%	100%
	$\chi^2 = 4.57$ (df = 1, $p = .03$)	

Furthermore, of the 178 offenders who not only were paroled to the treatment facility but also completed the program, only 12% were convicted of a new crime, compared to 29% of the offenders who did not complete treatment. This difference was also significant ($\chi^2 = 11.50$, df = 1, $p < 0.01$)

The researchers conducted a number of other, more sophisticated statistical tests on the sample and concluded that their study "supports other research demonstrating that substance abuse treatment may reduce criminal recidivism" (p. 231).

Zanis, David, Mulvaney, Frank, Coviello, Donna, Alterman, Arthur, et al. 2003. "The Effectiveness of Early Parole to Substance Abuse Treatment Facilities on 24-Month Criminal Recidivism." *Journal of Drug Issues, 33*: 223–235.

SUMMARY

1. The chi square test for independence is appropriate for situations in which the variables of interest have been organized into table format. The null hypothesis is that the variables are independent, or that the classification of a case into a particular category on one variable has no effect on the probability that the case will be classified into any particular category of the second variable.

2. Since chi square is nonparametric and requires only nominally measured variables, its model assumptions are easily satisfied. Furthermore, since it is computed from bivariate tables, in which the number of rows and columns can be easily expanded, the chi square test can be used in many situations in which other tests are inapplicable.

3. In the chi square test, we first find the frequencies that would appear in the cells if the variables were independent (f_e) and then compare those frequencies, cell by cell, with the frequencies actually observed in the cells (f_o). If the null is true, expected and observed frequencies should be quite close in value. The greater the difference between the observed and expected frequencies, the greater the possibility of rejecting the null.

4. The chi square test has several important limitations. It is often difficult to interpret when tables have many (more than four or five) dimensions. Also, as sample size (N) decreases, the chi square test becomes less trustworthy, and corrective action may be required. Finally, with very large samples, we may declare relatively trivial relationships to be statistically significant. As is the case with all tests of hypothesis, statistical significance is not the same thing as "importance" in any other sense. As a general rule, statistical significance is a necessary but not sufficient condition for theoretical or practical importance.

SUMMARY OF FORMULAS

Chi square (obtained):

$$11.1 \qquad \chi^2(\text{obtained}) = \sum \frac{(f_o - f_e)^2}{f_e}$$

Expected frequencies:

$$11.2 \qquad f_e = \frac{\text{Row marginal} \times \text{Column marginal}}{N}$$

Degrees of freedom, bivariate tables:

11.3 $\text{df} = (r - 1)(c - 1)$

Yates' correction for continuity:

11.4 $$\chi_c^2 = \sum \frac{(|f_o - f_e| - 0.5)^2}{f_e}$$

GLOSSARY

Bivariate table. A table that displays the joint frequency distributions of two variables.

Cells. The cross-classification categories of the variables in a bivariate table.

χ^2(critical). The score on the sampling distribution of all possible sample chi squares that marks the beginning of the critical region.

χ^2(obtained). The test statistic as computed from sample results.

Chi square test. A nonparametric test of hypothesis for variables that have been organized into a bivariate table.

Column. The vertical dimension of a bivariate table. By convention, each column represents a score on the independent variable.

Expected frequency (f_e). The cell frequencies that would be expected in a bivariate table if the variables were independent.

Goodness-of-fit test. An additional use for chi square that tests the significance of the distribution of a single variable.

Independence. The null hypothesis in the chi square test. Two variables are independent if, for all cases, the classification of a case on one variable has no effect on the probability that the case will be classified in any particular category of the second variable.

Marginals. The row and column subtotals in a bivariate table.

Nonparametric. A "distribution-free" test. These tests do not assume a normal sampling distribution.

Observed frequency (f_o). The cell frequencies actually observed in a bivariate table.

Row. The horizontal dimension of a bivariate table, conventionally representing a score on the dependent variable.

PROBLEMS

11.1 For each of the following tables, calculate the obtained chi square. *(HINT: Calculate the expected frequencies for each cell with Formula 11.2. Double-check to make sure you are using the correct row and column marginals for each cell. It may be helpful to record the expected frequencies in table format as well—see Tables 11.2, 11.4, and 11.7. Next, use a computational table to organize the calculation for Formula 11.1—see Tables 11.5 and 11.8. For each cell, subtract expected frequency from observed frequency and record the result in column 3. Square the value in column 3 and record the result in column 4, and then divide the value in column 4 by the expected frequency for that cell and record the result in column 5. Remember that the sum of column 5 in the computational table is obtained chi square. As you proceed, double-check to make sure you are using the correct values for each cell.)*

a.

20	25	45
25	20	45
45	45	90

b.

10	15	25
20	30	50
30	45	75

c.

25	15	40
30	30	60
55	45	100

d.

20	45	65
15	20	35
35	65	100

11.2 SOC A sample of 25 cities have been classified as high or low on their homicide rates and on the number of handguns sold within the city limits. Is there a relationship between these two variables? Do cities with higher homicide rates have significantly higher gun sales? Explain your results in a sentence or two.

Volume of Gun Sales	Homicide Rate: Low	Homicide Rate: High	Totals
High	8	5	13
Low	4	8	12
Totals	12	13	25

11.3 SW A local politician is concerned that a program for the homeless in her city is discriminating against blacks and other minorities. The following data were taken from a random sample of black and white homeless people.

Received Services?	Race: Black	White	Totals
Yes	6	7	13
No	4	9	13
Totals	10	16	26

a. Is there a statistically significant relationship between race and whether or not the person has received services from the program?
b. Compute column percents for the table to determine the pattern of the relationship. Which group was more likely to get services?

11.4 PS Many analysts have noted a "gender gap" in elections for the U.S. presidency, with women more likely to vote for the Democratic candidate. A sample of university faculty has been asked about their political party preference. Do their responses indicate a significant relationship between gender and party preference?

Party Preference	Gender: Male	Female	Totals
Democrats	10	15	25
Republican	15	10	25
Totals	25	25	50

a. Is there a statistically significant relationship between gender and party preference?
b. Compute column percents for the table to determine the pattern of the relationship. Which gender is more likely to prefer the Democrats?

11.5 PA Is there a relationship between salary levels and unionization for public employees? The following data represent this relationship for fire departments in a random sample of 100 cities of roughly the same size. Salary data have been dichotomized at the median. Summarize your findings.

Salary	Status: Union	Nonunion	Totals
High	21	29	50
Low	14	36	50
Totals	35	65	100

a. Is there a statistically significant relationship between these variables?
b. Compute column percents for the table to determine the pattern of the relationship. Which group was more likely to get high salaries?

11.6 SOC A program of pet therapy has been running at a local nursing home. Are the participants in the program more alert and responsive than nonparticipants? The results, drawn from a random sample of residents, are reported here.

Alertness	Status: Participants	Non-participants	Totals
High	23	15	38
Low	11	18	29
Totals	34	33	67

a. Is there a statistically significant relationship between participation and alertness?
b. Compute column percents for the table to determine the pattern of the relationship. Which group was more likely to be alert?

11.7 SOC The state Department of Education has rated a sample of local school systems for compliance with state-mandated guidelines for quality. Is the quality of a school system significantly related to the affluence of the community as measured by per capita income?

Quality	Per Capita Income: Low	High	Totals
Low	16	8	24
High	9	17	26
Totals	25	25	50

a. Is there a statistically significant relationship between these variables?
b. Compute column percents for the table to determine the pattern of the relationship. Are high- or low-income communities more likely to have high-quality schools?

11.8 CJ A local judge has been allowing some individuals convicted of "driving under the influence" to work in a hospital emergency room as an alternative to fines, suspensions, and other penalties. A random sample of offenders has been drawn. Do participants in this program have lower rates of recidivism for this offense?

	Status		
Recidivist?	Participants	Non-participants	Totals
Yes	60	123	183
No	55	108	163
Totals	115	231	346

a. Is there a statistically significant relationship between these variables?

b. Compute column percents for the table to determine the pattern of the relationship. Which group is more likely to be rearrested for driving under the influence?

11.9 SOC Is there a relationship between length of marriage and satisfaction with marriage? The necessary information has been collected from a random sample of 100 respondents drawn from a local community. Write a sentence or two explaining your decision.

	Length of Marriage (in years)			
Satisfaction	Less than 5	5–10	More than 10	Totals
Low	10	20	20	50
High	20	20	10	50
Totals	30	40	30	100

a. Is there a statistically significant relationship between these variables?

b. Compute column percents for the table to determine the pattern of the relationship. Which group is more likely to be highly satisfied?

11.10 PS Is there a relationship between political ideology and social class standing? Are upperclass students significantly different from underclass students on this variable? The following table reports the relationship between these two variables for a random sample of 267 college students.

	Class Standing		
Ideology	Underclass	Upperclass	Totals
Liberal	43	40	83
Moderate	50	50	100
Conservative	40	44	84
Totals	133	134	267

a. Is there a statistically significant relationship between these variables?

b. Compute column percents for the table to determine the pattern of the relationship. Which group is more likely to be conservative?

11.11 SOC At a large urban college, about half of the students live off campus in various arrangements, and the other half live in dormitories on campus. Is academic performance dependent on living arrangements? The results based on a random sample of 300 students are presented here.

	Residential Status			
GPA	Off Campus with Roommates	Off Campus with Parent	On Campus	Totals
Low	22	20	48	90
Moderate	36	40	54	130
High	32	10	38	80
Totals	90	70	140	300

a. Is there a statistically significant relationship between these variables?

b. Compute column percents for the table to determine the pattern of the relationship. Which group is more likely to have a high GPA?

11.12 SOC An urban sociologist has built up a database describing a sample of the neighborhoods in her city and has developed a scale by which each area can be rated for the "quality of life" (this includes measures of pollution, noise, open space, services available, and so on). She has also asked samples of residents of these areas about their level of satisfaction with their neighborhoods. Is there significant agreement between the sociologist's objective ratings of quality and the respondents' self-reports of satisfaction?

	Quality of Life			
Satisfaction	Low	Moderate	High	Totals
Low	21	15	6	42
Moderate	12	25	21	58
High	8	17	32	57
Totals	41	57	59	157

a. Is there a statistically significant relationship between these variables?

b. Compute column percents for the table to determine the pattern of the relationship? Which group is most likely to say that their satisfaction is high?

11.13 SOC Does support for the legalization of marijuana vary by region of the country? The table displays the relationship between the two variables for a random sample of 1020 adult citizens. Is the relationship significant?

	Region				
Legalize?	North	Midwest	South	West	Totals
Yes	60	65	42	78	245
No	245	200	180	150	775
Totals	305	265	222	228	1020

a. Is there a statistically significant relationship between these variables?

b. Compute column percents for the table to determine the pattern of the relationship. Which region is most likely to favor the legalization of marijuana?

11.14 SOC A researcher is concerned with the relationship between attitudes toward violence and violent behavior. If attitudes "cause" behavior (a very debatable proposition), then people who have positive attitudes toward violence should have high rates of violent behavior. A pretest was conducted on 70 respondents, and, among other things, the respondents were asked "Have you been involved in a violent incident of any kind over the past six months?" The researcher established the following relationship:

	Attitude Toward Violence		
Involvement	Favorable	Unfavorable	Totals
Yes	16	19	35
No	14	21	35
Totals	30	40	70

The chi square calculated on these data is .23, which is not significant at the .05 level (confirm this conclusion with your own calculations). Undeterred by this result, the researcher proceeded with the project and gathered a random sample of 7000. In terms of percentage distributions, the results for the full sample were exactly the same as for the pretest:

	Attitude Toward Violence		
Involvement	Favorable	Unfavorable	Totals
Yes	1600	1900	3500
No	1400	2100	3500
Totals	3000	4000	7000

However, the chi square obtained is a very healthy 23.4 (confirm with your own calculations). Why is the full-sample chi square significant when the pretest was not? What happened? Do you think that the second result is important?

11.15 SOC Some results from a survey administered to a nationally representative sample are presented here. For each table, conduct the chi square test of significance and compute column percents. Write a sentence or two of interpretation for each test.

a. Support for the legal right to an abortion by age:

	Age			
Support?	Younger than 30	30–49	50 and Older	Totals
Yes	154	360	213	727
No	179	441	429	1049
Totals	333	801	642	1776

b. Support for the death penalty by age:

	Age			
Support?	Younger than 30	30–49	50 and Older	Totals
Favor	361	867	675	1903
Oppose	144	297	252	693
Totals	505	1164	927	2596

c. Fear of walking alone at night by age:

	Age			
Fear?	Younger than 30	30–49	50 and Older	Totals
Yes	147	325	300	772
No	202	507	368	1077
Totals	349	832	668	1849

d. Support for legalizing marijuana by age:

	Age			
Legalize?	Younger than 30	30–49	50 and Older	Totals
Should	128	254	142	524
Should not	224	534	504	1262
Totals	352	788	646	1786

e. Support for suicide when a person has an incurable disease by age:

Support?	Age: Younger than 30	30–49	50 and Older	Totals
Yes	225	537	367	1129
No	107	270	266	643
Totals	332	807	633	1772

11.16[4] SOC The director of athletics at the local high school wonders if the sports program is getting a proportional amount of support from each of the four classes. If there are roughly equal numbers of students in each of the classes, what does the following breakdown of attendance figures from a random sample of students in attendance at a recent basketball game suggest?

Class	Frequencies
Freshman	200
Sophomore	150
Junior	120
Senior	110
Totals	580

11.17[5] SOC A small western town has roughly equal numbers of Hispanic, Asian, and Anglo-American residents. Are the three groups equally represented at town meetings? The attendance figures for a random sample drawn from those attending a meeting were

Group	Frequencies
Hispanic	74
Asian	55
Anglo	53
Totals	182

Is there a statistically significant pattern here?

[4]This problem is optional.

[5]This problem is optional.

SPSS for Windows

Using *SPSS for Windows* to Conduct the Chi Square Test

SPSS DEMONSTRATION 11.1 Do Childcare Practices Vary by Social Class?

The **Crosstabs** procedure in SPSS produces bivariate tables and a wide variety of statistics. This procedure is very commonly used in social science research at all levels, and you will see many references to **Crosstabs** in chapters to come. I introduce the command here, and we will return to it often in later sessions.

Do people in different social classes raise their children differently? We examined this issue in Section 11.9 and we'll continue the analysis by examining the relationship between social class and approval of spanking as a disciplinary technique. We will have SPSS construct a bivariate table, with a chi square test and column percentages, to display the relationship between *class* and *spanking.*

Begin by clicking **Analyze,** then click **Descriptive Statistics** and **Crosstabs.** The **Crosstabs** dialog box will appear, with the variables listed in a box on the left. Highlight *spanking* (the label for this variable is FAVOR SPANKING TO DISCIPLINE CHILD) and click the arrow to move the variable name into the **Rows** box, and then highlight *class* (the label for this variable is SUBJECTIVE CLASS IDENTIFICATION) and move it into the **Columns** box. Click the **Statistics** button at the bottom of the window and click the box next to Chi-square. Then click the **Cells** button and select "column" in the **Percentages** box. This will generate column percents for the table. Click **Continue** and **OK,** and the following output will be produced (*NOTE: This output has been slightly edited for clarity. It will not exactly match the output on your screen*).

FAVOR SPANKING TO DISCIPLINE CHILD * SUBJECTIVE CLASS IDENTIFICATION Crosstabulation

		SUBJECTIVE CLASS IDENTIFICATION				
FAVOR SPANKING TO DISCIPLINE CHILD		LOWER CLASS	WORKING CLASS	MIDDLE CLASS	UPPER CLASS	Total
STRONGLY AGREE	Count	14	91	71	3	179
	%	29.8%	34.0%	26.4%	15.0%	29.6%
AGREE	Count	19	118	124	8	269
	%	40.4%	44.0%	46.1%	40.0%	44.5%
DISAGREE	Count	12	44	56	5	117
	%	25.5%	16.4%	20.8%	25.0%	19.4%
STRONGLY DISAGREE	Count	2	15	18	4	39
	%	4.3%	5.6%	6.7%	20.0%	6.5%
Total	Count	47	268	269	20	604
	%	100.0%	100.0%	100.0%	100.0%	100.0%

Chi-Square Tests

	Value	df	Asymp. Sig. (2-sided)
Pearson Chi-Square	13.629[a]	9	.136
Likelihood Ratio	11.877	9	.220
Linear-by-Linear Association	4.861	1	.027
N of Valid Cases	604		

[a] 3 cells (18.8%) have expected count less than 5. The minimum expected count is 1.29.

The crosstab table is quite large (4 × 4), and it will probably take a bit of effort to absorb the information. Begin with the cells. Each cell displays the number of cases in the cell and the column percent for that cell. For example, there were 14 respondents who were lower class and who "strongly agreed" with spanking, and these were 29.8% of all lower-class respondents. If you focus on the top row, you can see that the percentage of each class that strongly support spanking increases slightly for working-class respondents and then decreases for middle- and upper-class respondents. The differences are not large, however, and this relationship seems to be weak.

The results of the chi square test are reported in the output block that follows the table. The value of chi square (obtained) is 13.629, the degrees of freedom are 9, and the exact significance of the chi square is .136. This is greater than the standard indicator of a significant result (alpha = .05), so we may conclude that there is no statistically significant relationship between *class* and *spanking.* Support for spanking is independent of social class. This conclusion is consistent with our analysis in Section 11.9: the 2006 GSS shows little support for the idea that there is a relationship between social class and socialization practices.

SSPS DEMONSTRATION 11.2 Do Attitudes About Immigration Vary By Social Class?

Is there a relationship between social class and attitudes about immigration? Which class is most likely to support reductions in the number of immigrants permitted to enter the nation each year? One hypothesis would be that the lower and working classes would feel more threatened by job competition from immigrants and be more likely to feel that the numbers should be reduced. Is this hypothesis supported?

Run the **Crosstabs** procedure again, with *letin1* ("The number of immigrants should be increased, remain the same, or decreased") as the row variable (in place of *spanking*) and class as the column variable. Don't forget to request chi square and column percents. The output is reproduced here. *(NOTE: This output has been slightly edited for clarity. It will not exactly match the output on your screen.)*

NUMBER OF IMMIGRANTS BY SUBJECTIVE CLASS IDENTIFICATION Crosstabulation

		SUBJECTIVE CLASS IDENTIFICATION				
NUMBER OF IMMIGRANTS TO AMERICA NOWADAYS SHOULD BE		LOWER CLASS	WORKING CLASS	MIDDLE CLASS	UPPER CLASS	Total
INCREASED A LOT	Count	6	12	11	3	32
	%	12.8%	4.5%	4.2%	15.8%	5.4%
INCREASED A LITTLE	Count	2	20	21	6	49
	%	4.3%	7.5%	8.0%	31.6%	8.2%
REMAIN THE SAME AS IT IS	Count	11	96	109	6	222
	%	23.4%	36.1%	41.4%	31.6%	37.3%
REDUCED A LITTLE	Count	8	52	57	2	119
	%	17.0%	19.5%	21.7%	10.5%	20.0%
REDUCED A LOT	Count	20	86	65	2	173
	%	42.6%	32.3%	24.7%	10.5%	29.1%
TOTAL	Count	47	266	263	19	595
	%	100.0%	100.0%	100.0%	100.0%	100.0%

Chi-Square Tests

	Value	df	Asymp. Sig. (2-sided)
Pearson Chi-Square	36.577(a)	12	.000
Likelihood Ratio	29.829	12	.003
Linear-by-Linear Association	7.497	1	.006
N of Valid Cases	595		

[a]5 cells (25.0%) have expected count less than 5. The minimum expected count is 1.02.

Chi square is 36.577, degrees of freedom are 12, and the exact probability of getting this pattern of cell frequencies by random chance alone is less than .001. There is a significant relationship between the variables. Inspect the column percents in the bottom row of the table ("Immigration Should Be Reduced a Lot") and you will see that opposition to immigration decreases as class increases. About 43% of the lower-class respondents supported a reduction in immigration vs. about 11% of the upper-class respondents.

What other variables might be significantly related to attitudes about immigration? See exercise 11.2 for an opportunity to investigate further.

SPSS DEMONSTRATION 11.3 Is Ignorance Bliss?

We can test the accuracy of the ancient folk wisdom that "ignorance is bliss" with the 2006 GSS. Are more-educated people unhappier? To test the relationship, we will use *degree* as our measure of education; to measure the dependent variable, we will use an item from the GSS (*happy*), which asked respondents to rate their overall level of happiness. As originally coded, *degree* has five categories (see Appendix G), but we will simplify the analysis and distinguish only between people with a high school education or less and people with at least some education beyond high school. When you recode *degree,* remember to choose "Recode into Different Variable," and give the recoded variable a new name (say, *rdeg*). The recode instruction in the **Old → New** box of the **Recode into Different Variable** window should look like this:

0 thru 1 → 1
2 thru 4 → 2

Follow the instructions in Demonstration 11.1 for the **Crosstabs** command. Specify *happy* as the row variable and recoded *degree* as the column variable. Don't forget to request chi square and column percents. The output should look like this *(NOTE: This output has been slightly edited for clarity. It will not exactly match the output on your screen)*:

GENERAL HAPPINESS * Recoded Degree Crosstabulation

GENERAL HAPPINESS		Recoded Degree		Total
		1.00	2.00	
VERY HAPPY	Count	183	141	324
	%	29.7%	41.0%	33.7%
PRETTY HAPPY	Count	334	178	512
	%	54.1%	51.7%	53.3%
NOT TOO HAPPY	Count	100	25	125
	%	16.2%	7.3%	13.0%
Total	Count	617	344	961
		100.0%	100.0%	100.0%

Chi-Square Tests

	Value	df	Asymp. Sig. (2-sided)
Pearson Chi-Square	22.215(a)	2	.000
Likelihood Ratio	23.310	2	.000
Linear-by-Linear Association	21.360	1	.000
N of Valid Cases	961		

[a]0 cells (.0%) have expected count less than 5. The minimum expected count is 44.75.

The table displays the observed frequencies for the cells along with column percents. The value of chi square (22.215), the degrees of freedom (2), and the exact probability that this pattern of cell frequencies occurred by chance (.000) are reported in the output block below the table. The reported significance is less than .001, so we would reject the null hypothesis of independence. There is a relationship between level of education and degree of happiness.

Remember that chi square tells us only that the overall relationship between the variables is significant. To assess the specific idea that "ignorance is bliss," we need to analyze the column percents. If the idea is true, a higher percentage of the less educated (category 1 of *rdeg*) should be happy. Unfortunately for the old adage, the bivariate table shows the reverse pattern. While 29.7% of the less educated respondents are "very happy," 41.0% of the more educated place themselves in this category.

Exercises

11.1 For a follow-up on Demonstration 11.1, pick two variables that measure social class. If necessary, recode these variables so that they have only two or three categories. Run the **Crosstabs** procedure with *spanking* as the row variable and your new measures of social class as the column variables. How do these relationships compare with the table generated in Demonstration 11.1? Are these tables consistent with the idea that approval of spanking is dependent on social class?

11.2 Find three more variables that might be related to *letin1*. Use the **Crosstabs** procedure to see if any of your variables have a significant relationship with *letin1*. Which of the variables had the most significant relationship?

11.3 Since we already recoded *degree* for Demonstration 11.3, let's see if education has a significant relationship with *grass* and *cappun*. Run the **Crosstabs** procedure with the dependent variables in the **Rows** box and recoded *degree* in the **Columns** box. Write a paragraph summarizing the results of these three tests. Which relationships are significant? At what **levels**?

PART II CUMULATIVE EXERCISES

1. Conduct the appropriate test of significance for each research situation. Problems are stated in no particular order and include research situations from each chapter in Part II.

a. Is there a gender gap in use of the Internet? Random samples of men and women have been questioned about the average number of minutes they spend each week on the Internet for any purpose. Is the difference significant?

Women	Men
$\bar{X}_1 = 55$	$\bar{X}_2 = 60$
$s_1 = 2.5$	$s_2 = 2.0$
$N_1 = 520$	$N_2 = 515$

b. For high school students, is there a relationship between social class and involvement in activities such as clubs and sports? Data have been gathered for a random sample of students. Is the relationship significant?

	Social Class		
Involvement	Middle	Lower	Totals
High	11	19	30
Moderate	19	21	40
Low	8	12	20
Totals	38	52	90

c. The General Social Survey asks respondents how many total sex partners they have had over their lifetimes. For a subsample of 23 respondents, does the number vary significantly by educational level?

Less Than High School	High School	At Least Some College
1	1	2
1	2	1
2	3	5
1	8	9
9	10	11
2	5	2
9	4	1
	3	1

d. A random sample of the U.S. population was asked how many times they had moved since they were 18 years of age. The results are presented here. On the average, how many times do adult Americans move?

$$\bar{X} = 3.5$$
$$s = 0.4$$
$$N = 1450$$

e. On the average, school districts in a state receive budget support from the state government of $623 per student. A random sample of 107 rural schools reports they received an average of $605 per pupil. Is the difference significant?

$$\mu = 623 \qquad \bar{X} = 605$$
$$s = 74$$
$$N = 107$$

2. Following are a number of research questions that can be answered by the techniques presented in Chapters 7–11. For each question, select the most appropriate test, compute the necessary statistics, and state your conclusions.

In order to complete some problems, you must first calculate the sample statistics (for example, means or proportions) that are used in the test of hypothesis. Use alpha = 0.05 throughout. The questions are presented in random order. There is at least one problem for each chapter, but I have not included a research situation for every statistical procedure covered in the chapters.

In selecting tests and procedures, you need to consider the question, the number of samples or categories being compared, and the level of measurement of the variables. The database for this problem is based on the General Social Survey (GSS). The actual questions asked and the complete response codes are presented in Appendix G.

Abbreviated codes are listed under the following "Survey Items" head. The scores of some variables have been simplified or collapsed for this exercise. These problems are based on a small sample, and you may have to violate some assumptions about sample size in order to complete this exercise.

a. Is there a statistically significant difference in average hours of TV watching by income level? By race?

b. Is there a statistically significant relationship between age and happiness?

c. Estimate the average number of hours spent watching TV for the entire population.

d. If Americans currently average 2.3 children, is this sample representative of the population?

e. Are the educational levels of Catholics and Protestants significantly different?

f. Does average hours of TV watching vary by level of happiness?

g. Based on the sample data, estimate the proportion of black Americans in the population.

SURVEY ITEMS

1. How many children have you ever had? (CHILDS) Scores are actual numbers.

2. Respondent's educational level (DEGREE)
   0. Less than HS
   1. HS
   2. At least some college

3. Race (RACE)
   1. White
   2. Black

4. Age (AGE)
   1. Younger than 35
   2. 35 and older
5. Number of hours of TV watched per day (TVHOURS). Values are actual number of hours.
6. What is your religious preference? (RELIG)
   1. Protestant
   2. Catholic
7. Respondent's income (INCOME06)
   1. $24,999 or less
   2. $25,000 or more
8. Respondent's overall level of happiness (HAPPY)
   1. Very happy
   2. Pretty happy
   3. Not too happy

Case	No. of Children	Educational Level	Race	Age	TV Hours	Religious Preference	Income	Happiness
1	3	1	1	1	3	1	1	2
2	2	0	1	1	1	1	2	3
3	4	2	1	2	3	1	1	1
4	0	2	1	1	2	1	1	1
5	5	1	1	1	2	1	2	2
6	1	1	1	2	3	1	2	1
7	9	0	1	1	6	1	1	1
8	6	1	1	2	4	1	1	2
9	4	2	1	1	2	2	2	2
10	2	1	1	1	1	1	2	3
11	2	0	1	2	4	1	1	3
12	4	1	2	1	5	2	1	2
13	0	1	1	2	2	2	1	2
14	2	1	1	2	2	1	1	2
15	3	1	2	2	4	1	1	1
16	2	0	1	2	2	1	2	1
17	2	1	1	2	2	1	2	3
18	0	2	1	2	2	1	2	1
19	3	0	1	2	5	2	1	1
20	2	1	2	1	10	1	1	3
21	2	1	1	2	4	1	2	1
22	1	0	1	2	5	1	2	1
23	0	2	1	1	2	2	1	1
24	0	1	1	2	0	2	2	1
25	2	2	1	1	1	2	2	1

Part III Bivariate Measures of Association

The four chapters in this section cover the computation and analysis of a class of statistics known as *measures of association*. These statistics are extremely useful in scientific research and commonly reported in the professional literature. They provide, in a single number, an indication of the strength and—if applicable—the direction of a bivariate association.

It is important to remember the difference between statistical significance, covered in Part II, and association, the topic of this part. Tests for statistical significance answer certain questions: Were the differences or relationships observed in the sample caused by mere random chance? What is the probability that the sample results reflect patterns in the population(s) from which the sample(s) were selected? Measures of association address a different set of questions: How strong is the relationship between the variables? What is the direction or pattern of the relationship?

Thus, measures of association provide information complementary to tests of significance. Association and significance are two different things. While the most satisfying results are those that are *both* statistically significant and strong, it is common to find mixed or ambiguous results: relationships that are statistically significant but weak, not statistically significant but strong, and so forth.

Chapter 12 introduces the basic ideas behind the analysis of association in terms of bivariate tables and column percentages. The remaining three chapters are organized by level of measurement. Chapter 13 presents measures of association for use with nominal-level variables, Chapter 14 covers measures for ordinal-level variables, and Chapter 15 presents Pearson's r, the most important measure of association and the only one designed for interval-ratio-level variables.

12 Bivariate Association

Introduction and Basic Concepts

LEARNING OBJECTIVES

By the end of this chapter, you will be able to

1. Explain how we can use measures of association to describe and analyze the importance of relationships (vs. their statistical significance).
2. Define *association* in the context of bivariate tables and in terms of changing conditional distributions.
3. List and explain the three characteristics of a bivariate relationship: existence, strength, and pattern or direction.
4. Investigate a bivariate association by properly calculating percentages for a bivariate table and interpreting the results.

12.1 STATISTICAL SIGNIFICANCE AND THEORETICAL IMPORTANCE

As we have seen over the past several chapters, tests of statistical significance are extremely important in social science research. As long as social scientists must work with random samples rather than populations, these tests are indispensable for dealing with the possibility that our research results are the products of mere random chance. However, tests of significance are often merely the first step in the analysis of research results. These tests do have limitations, and statistical significance is not necessarily the same thing as relevance or importance. Furthermore, all tests of significance are affected by sample size: Tests performed on large samples may result in decisions to reject the null hypothesis when, in fact, the observed differences are quite minor.

Beginning with this chapter, we will be working with **measures of association.** Whereas tests of significance detect nonrandom relationships, measures of association provide information about the strength and direction of relationships, information that is more directly relevant for assessing the importance of relationships and testing the power and validity of our theories. The theories that guide scientific research are almost always stated in cause-and-effect terms (for example: "variable *X* causes variable *Y*"). As an example, recall our discussion of the contact hypothesis in Chapter 1. In that theory, the causal (or independent) variable was equal status contacts between groups, and the effect (or dependent) variable was level of individual prejudice. The theory asserts that involvement in equal-status-contact situations *causes* prejudice to decline. Measures of association help us trace causal relationships among variables, and they are our most important and powerful statistical tools for documenting, measuring, and analyzing cause-and-effect relationships.

As useful as they are, measures of association, like any class of statistics, do have their limitations. Most importantly, these statistics cannot *prove* that two variables are causally related. Even if there is a strong (and significant) statistical association between two variables, we cannot necessarily conclude that one variable is a cause of the other. We will explore causation in more detail in Part

IV, but for now you should keep in mind that causation and association are two different things. We can use a statistical association between variables as evidence for a causal relationship, but association by itself is not proof that a causal relationship exists.

Another important use for measures of association is prediction. If two variables are associated, we can predict the score of a case on one variable from the score of that case on the other variable. For example, if equal status contacts and prejudice are associated, we can predict that people who have experienced many such contacts will be less prejudiced than those who have had few or no contacts. Note that prediction and causation can be two separate matters. If variables are associated, we can predict from one to the other even if the variables are not causally related.

This chapter introduces the concept of **association** between variables in the context of bivariate tables and stresses the use of percentages to analyze associations between variables. In the chapters that follow, we will concentrate on the logic, calculation, and interpretation of the various measures of association. Finally, in Part IV, we will extend some of these ideas to the multivariate (more than two variables) case.

12.2 ASSOCIATION BETWEEN VARIABLES AND THE BIVARIATE TABLE

Most generally, two variables are said to be associated if the distribution of one of them changes under the various categories or scores of the other. For example, suppose that an industrial sociologist was concerned with the relationship between job satisfaction and productivity for assembly-line workers. If these two variables are associated, then scores on productivity will change under the different conditions of satisfaction. Highly satisfied workers will have different scores on productivity than workers who are low on satisfaction, and levels of productivity will vary by levels of satisfaction.

This relationship will become clearer with the use of bivariate tables. As you recall (see Chapter 11), bivariate tables display the scores of cases on two different variables. By convention, the **independent,** or ***X*, variable** (that is, the variable taken as causal) is arrayed in the columns, and the **dependent,** or ***Y*, variable** in the rows.[1] That is, each column of the table (the vertical dimension) represents a score or category of the independent variable (X), and each row (the horizontal dimension) represents a score or category of the dependent variable (Y).

Table 12.1 displays a relationship between productivity and job satisfaction for a fictitious sample of 173 factory workers. We focus on the columns to detect the presence of an association between variables displayed in table format. Each column shows the pattern of scores on the dependent variable for each score on the independent variable. For example, the left-hand column indicates that 30 of the 60 workers who were low on job satisfaction were low on productivity, 20 were moderately productive, and 10 were high on productivity. The middle column shows that 21 of the 61 moderately satisfied workers were low on productivity, 25 were moderately productive, and 15 were high on productivity. Of the 52 workers who are highly satisfied (the right-hand column), 7 were low on productivity, 18 were moderate, and 27 were high.

[1]In the material that follows, we will often, for the sake of brevity, refer to the independent variable as X and the dependent variable as Y.

TABLE 12.1 PRODUCTIVITY BY JOB SATISFACTION (frequencies)

	Job Satisfaction (X)			
Productivity (Y)	Low	Moderate	High	Totals
Low	30	21	7	58
Moderate	20	25	18	63
High	10	15	27	52
Totals	60	61	52	173

By inspecting the table from column to column we can observe the effects of the independent variable on the dependent variable (provided, of course, that the table is constructed with the independent variable in the columns). These "within-column" frequency distributions are called the **conditional distributions of *Y*,** since they display the distribution of scores on the dependent variable for each condition (or score) of the independent variable.

Table 12.1 indicates that productivity and satisfaction are associated: The distribution of scores on Y (productivity) changes across the various conditions of X (satisfaction). For example, half of the workers who were low on satisfaction were also low on productivity (30 out of 60). On the other hand, over half of the workers who were high on satisfaction were high on productivity (27 out of 52).

Although it is intended to be a test of significance, the chi square statistic provides another way to detect the existence of an association between two variables that have been organized into table format. Any nonzero value for obtained chi square indicates that the variables are associated. For example, the obtained chi square for Table 12.1 is 24.2, a value that affirms our previous conclusion, based on the conditional distributions of Y, that an association of some sort exists between job satisfaction and productivity.

Often, the researcher will have already conducted a chi square test before considering matters of association. In such cases, it will not be necessary to inspect the conditional distributions of Y to ascertain whether or not the two variables are associated. If the obtained chi square is zero, the two variables are independent and not associated. Any value other than zero indicates some association between the variables. Remember, however, that statistical significance and association are two different things. It is perfectly possible for two variables to be associated (as indicated by a nonzero chi square) but still independent (if we fail to reject the null hypothesis).

In this section, we have defined, in a general way, the concept of association between two variables. We have also shown two different ways to detect the presence of an association. In the next section, we extend the analysis beyond questions of the mere presence or absence of an association and, in a systematic way, show how additional very useful information about the relationship between two variables can be developed.

12.3 THREE CHARACTERISTICS OF BIVARIATE ASSOCIATIONS

Bivariate associations possess three different characteristics, each of which must be analyzed for a full investigation of the relationship. Investigating these characteristics may be thought of as a process of finding answers to three questions:

1. Does an association exist?
2. If an association does exist, how strong is it?
3. What is the pattern and/or the direction of the association?

We will consider each of these questions separately.

Does an Association Exist? We have already discussed the general definition of association, and we have seen that we can detect an association by observing the conditional distributions of Y in a table or by using chi square. In Table 12.1, we know that the two variables are associated to some extent because the conditional distributions of productivity (Y) are different across the various categories of satisfaction (X) and because the chi square statistic is a nonzero value.

Comparisons from column to column in Table 12.1 are relatively easy to make because the column totals are roughly equal. This will not usually be the case, and it is helpful to compute percentages to control for varying column totals. These **column percentages,** introduced in Chapter 11, are computed within each column separately and make the pattern of association more visible.

The general procedure for detecting association with bivariate tables is to compute percentages within the columns (vertically, or down each column) and then compare column to column across the table (horizontally, or across the rows). Table 12.2 presents column percentages calculated from the data in Table 12.1. Note that this table reports the row and column marginals, but in parentheses. Besides controlling for any differences in column totals, tables in percentage form are usually easier to read because changes in the conditional distributions of Y are easier to detect.

In Table 12.2, we can see that the largest cell changes position from column to column. For workers who are low on satisfaction, the single largest cell is in the top row (low on productivity). For the middle column (moderate on satisfaction), the largest cell is in the middle row (moderate on productivity), and, for the right-hand column (high on satisfaction), it is in the bottom row (high on productivity). Even a cursory glance at the conditional distributions of Y in Table 12.2 reinforces our conclusion that an association does exist between these two variables.

If two variables are not associated, then the conditional distributions of Y will not change across the columns. The distribution of Y would be the same for each condition of X. Table 12.3 illustrates a "perfect nonassociation" between height and productivity. Table 12.3 is only one of many patterns that indicate "no association." The important point is that the conditional distributions of Y

TABLE 12.2 PRODUCTIVITY BY JOB SATISFACTION (percentages)

	Job Satisfaction (X)			
Productivity (Y)	Low	Moderate	High	Totals
Low	50.00%	34.43%	13.46%	33.53% (58)
Moderate	33.33%	40.98%	34.62%	36.42% (63)
High	16.67%	24.59%	51.92%	30.06% (52)
Totals	100.00% (60)	100.00% (61)	100.00% (52)	100.00% (173)

Application 12.1

Why are many Americans attracted to movies that emphasize graphic displays of violence? One idea is that "slash" movie fans feel threatened by violence in their daily lives and use these movies as a means of coping with their fears. In the safety of the theater, violence can be vicariously experienced, and feelings and fears can be expressed privately. Also, highly violent movies almost always, as a necessary plot element, provide a role model of one character who does deal with violence successfully (usually, of course, with more violence).

Is fear of violence associated with frequent attendance at high-violence movies? The following table reports the joint frequency distributions of "fear" and "attendance" in percentages for a fictitious sample of 600.

	Fear		
Attendance	Low	Moderate	High
Rare	50%	20%	30%
Occasional	30%	60%	30%
Frequent	20%	20%	40%
Totals	100%	100%	100%
	(200)	(200)	(200)

The conditional distributions of attendance (Y) do change across the values of fear (X), so these variables are associated. The clustering of cases in the diagonal from upper left to lower right suggests a substantial relationship in the predicted direction. People who are low on fear attend violent movies infrequently, and people who are high on fear are frequent attendees. Since the maximum difference in column percentages in the table is 30 (in both the top and middle rows), the relationship can be characterized as moderate to strong

These results do suggest an important relationship between fear and attendance. Notice, however, that these results pose an interesting causal problem. The table supports the idea that fearful and threatened people attend violent movies as a coping mechanism (X causes Y) but is also consistent with the reverse causal argument: Attendance at violent movies increases fears for one's personal safety (Y causes X). The results support *both* causal arguments and remind us that association is not the same thing as causation.

are the same. Levels of productivity do not change at all for the various heights; therefore, no association exists between these variables. Also, the obtained chi square computed from this table would have a value of zero, again indicating no association.[2]

How Strong Is the Association? Once we know there is an association between two variables, we need to know how strong it is. This is essentially a matter of determining the amount of change in the conditional distributions of Y. At one extreme, of course, is the case of "no association," where the conditional distributions of Y do not change at all (see Table 12.3). At the other extreme is a perfect association, the strongest possible relationship. A perfect association exists between two variables if each value of the dependent variable is associated with one and only one value of the independent variable.[3] In a bivariate

[2] See Section 11.5 for detailed instructions on computing column percentages.

[3] Each measure of association to be introduced in the following chapters incorporates its own definition of a "perfect association," and these definitions vary somewhat, depending on the specific logic and mathematics of the statistic. That is, for different measures computed from the same table, some measures will possibly indicate perfect relationships when others will not. We will note these variations in the mathematical definitions of a perfect association at the appropriate times.

TABLE 12.3 PRODUCTIVITY BY HEIGHT (an illustration of no association)

	Height (X)		
Productivity (Y)	Short	Medium	Tall
Low	33.33%	33.33%	33.33%
Moderate	33.33%	33.33%	33.33%
High	33.33%	33.33%	33.33%
Totals	100.00%	100.00%	100.00%

TABLE 12.4 PRODUCTIVITY BY HEIGHT (an illustration of perfect association)

	Height (X)		
Productivity (Y)	Short	Medium	Tall
Low	0%	0%	100%
Moderate	0%	100%	0%
High	100%	0%	0%
Totals	100%	100%	100%

table, the variables would have a perfect relationship if all cases in each column are located in a single cell and there is no variation in Y for a given value of X (see Table 12.4).

A perfect relationship would be taken as very strong evidence of a causal relationship between the variables, at least for the sample at hand. In fact, the results presented in Table 12.4 would indicate that, for this sample, height is the sole cause of productivity. Also, in the case of a perfect relationship, predictions from one variable to the other could be made without error. If we know that a particular worker is short, for example, we could be sure that he or she is highly productive.

Of course, the huge majority of relationships will fall somewhere between the two extremes of no association and perfect association, and we need to develop some way of describing these intermediate relationships consistently and meaningfully. For example, Tables 12.1 and 12.2 show an association between productivity and job satisfaction. How could this relationship be described in terms of strength? How close is the relationship to perfect? How far away from no association?

Researchers rely on the measures of association to be presented in Chapters 13–15 to provide precise, objective indicators of the strength of a relationship. Virtually all of these statistics are designed so that they have a lower limit of 0.00 and an upper limit of 1.00 ($\pm$1.00 for ordinal and interval-ratio measures of association). A measure that equals 0.00 indicates no association between the variables (the conditional distributions of Y do not vary), and a measure of 1.00 ($\pm$1.00 in the case of ordinal and interval-ratio measures) indicates a perfect relationship. The exact meaning of values between 0.00 and 1.00 varies from measure to measure, but, for all measures, the closer the value is to 1.00, the stronger the relationship (the greater the change in the conditional distributions of Y).

Application 12.2

It is very common for societies to be male dominated, but the status of women relative to men is also highly variable, ranging from abject oppression to relative equality (and, on a few indicators for a few nations, higher status than men). Is the relative status of women related to a nation's religiosity? Since many religions sanction the lower status of women, are women in the most religious nations at a comparative disadvantage?

To examine this question, a study of 47 nations, selected from all levels of development and parts of the world, has been completed. Each nation has been rated on women's status relative to men in schooling, occupational prestige, politics, health care, and several areas of social life. Nations have been scored as high (women have high status relative to men) or low (women have low status relative to men). Nations have also been characterized as high or low on religiosity. The following table presents the results of the comparison in both frequencies and column percentages. Analyzing this table step by step, we see that the column percentages vary, so the variables are related. The size of the difference from column to column is substantial, and the maximum difference is 68.00% − 36.36% = 31.64%, so this is a moderate-to-strong relationship. (Note that, in a 2 × 2 table, the maximum difference will be the same for either row.) Looking at the pattern of the relationship, we see that the majority of women (68.00%) have low status in nations that are high on religiosity and, also, that the majority of women (64%) in less religious nations have high status. This is a negative relationship: As religiosity increases, the status of women relative to men decreases.

STATUS OF WOMEN BY RELIGIOSITY FOR 47 NATIONS: Frequencies and (*Percentages*)

Women's Status	Religiosity: Low	Religiosity: High	Totals
Low	8 (*36.36%*)	17 (*68.00%*)	25
High	14 (*63.64%*)	8 (*32.00%*)	22
Totals	22 (*100.00%*)	25 (*100.00%*)	47

What other factors might be related to the status of women? Another possibility is that more industrialized nations will raise the status of women as they upgrade the quality of the workforce and the literacy of the population. More traditional agricultural societies can function without an educated or literate workforce, but industrial, "high-tech" societies cannot. The same 47 nations have been rated as "LDCs" (least developed countries, or nations that are largely agricultural), developed (fully industrialized), or developing (nations between the more agricultural LDCs and the fully industrialized nations). The following table shows the relationship between the two variables.

STATUS OF WOMEN BY LEVEL OF DEVELOPMENT FOR 47 NATIONS: Frequencies and (*Percentages*)

Women's Status	Level of Development: LDCs	Developing	Developed	Totals
Low	13 (*81.25%*)	8 (*53.33%*)	4 (*25.00%*)	25 (*53.19%*)
High	3 (*18.75%*)	7 (*46.67%*)	12 (*75.00%*)	22 (*46.81%*)
Totals	16 (*100.00%*)	15 (*100.00%*)	16 (*100.00%*)	47 (*100.00%*)

Analyzing the table step by step, we see that these variables are related, because there is a very substantial change in the column percentages across the table. The maximum difference is 81.25 − 25.00 = 56.25, indicating a strong relationship. This is a positive relationship: Women's status increases as level of development increases. Only 18.75% of the LDCs are high on women's status, versus 75% of the developed nations. In sum, there is a strong, positive relationship between the status of women and the development level of a nation.

We begin our consideration of measures of association in the next chapter. For now, we can consider a more informal way of assessing the strength of a relationship based on comparing column percentages across the rows and called the **maximum difference.** This technique is best regarded as a "quick and easy" method for assessing the strength of a relationship: easy to apply but limited in its usefulness. To use this technique, compute the column percentages as usual and then skim the table across each of the rows to find the largest difference—in any row—between column percentages. For example, the largest difference in column percentages in Table 12.2 is in the top row between the "Low" column and the "High" column: 50.00% − 13.46% = 36.54%. The maximum difference in the middle row is between "moderates" and "highs" (40.98% − 33.33% = 7.65%), and in the bottom row it is between "highs" and "lows" (51.92% − 16.67% = 35.25%). Both of these values are less than the maximum difference in the top row.

Once you have found the maximum difference in the table, the scale presented in Table 12.5 can be used to describe the strength of the relationship. For instance, it can help us describe the relationship between productivity and job satisfaction in Table 12.2 as strong.

You should be aware that the relationships between the size of the maximum difference and the descriptive terms (weak, moderate, and strong) in Table 12.5 are arbitrary and approximate. We will get more precise and useful information when we compute and analyze the measures of association that will be presented in Chapters 13 through 15. Also, maximum differences are easiest to find and most useful for smaller tables. In large tables, with many (say, more than three) columns and rows, it can be cumbersome to find the high and low percentages, and it is advisable to consider only measures of association as indicators of the strength for these tables. Finally, note that the maximum difference is based on only two values (the high and low column percentages within any row). Like the range (see Chapter 4), this statistic can give a misleading impression of the overall strength of the relationship

As a final caution, do not mistake chi square as an indicator of the strength of a relationship. Even very large values for chi square do not necessarily mean that the relationship is strong. Remember that significance and association are two separate matters and that chi square, by itself, is not a measure of association. While a nonzero value indicates some association between the variables, the magnitude of chi square bears no particular relationship to the strength of the association. Chapter 13 will introduce some ways to transform chi square into other statistics that do measure the strength of the association between two variables. *(For practice in computing percentages and judging the existence and strength of an association, see any of the problems at the end of this chapter.)*

TABLE 12.5 THE RELATIONSHIP BETWEEN THE MAXIMUM DIFFERENCE AND THE STRENGTH OF THE RELATIONSHIP

Maximum Difference	Strength
If the maximum difference is:	*The strength of the relationship is:*
between 0 and 10 percentage points	Weak
between 10 and 30 percentage points	Moderate
more than 30 percentage points	Strong

TABLE 12.6 LIBRARY USE BY EDUCATION (an illustration of a positive relationship)

Library Use	Education		
	Low	Moderate	High
Low	60%	20%	10%
Moderate	30	60	30
High	10	20	60
Total	100%	100%	100%

TABLE 12.7 AMOUNT OF TELEVISION VIEWING BY EDUCATION (an illustration of a negative relationship)

Television Viewing	Education		
	Low	Moderate	High
Low	10%	20%	60%
Moderate	30%	60%	30%
High	60%	20%	10%
Total	100%	100%	100%

What Is the Pattern and/or the Direction of the Association? Investigating the pattern of the association requires that we ascertain which values or categories of one variable are associated with which values or categories of the other. We have already remarked on the pattern of the relationship between productivity and satisfaction. Table 12.2 indicates that low scores on satisfaction are associated with low scores on productivity, moderate satisfaction with moderate productivity, and high satisfaction with high productivity.

When working with nominal-level variables, we can discuss the pattern of the relationship only.[4] However, when both variables are at least ordinal in level of measurement, the association between the variables may also be described in terms of direction. The direction of the association can be either positive or negative. An association is positive if the variables vary in the same direction. That is, in a **positive association,** high scores on one variable are associated with high scores on the other variable, and low scores on one variable are associated with low scores on the other. In a positive association, as one variable increases in value, the other also increases; and as one variable decreases, the other also decreases. Table 12.6 displays, with fictitious data, a positive relationship between education and use of public libraries. As education increases (as you move from left to right across the table), library use also increases (the percentage of "high" users increases). The association between job satisfaction and productivity, as displayed in Tables 12.1 and 12.2, is also a positive association.

In a **negative association,** the variables vary in opposite directions. High scores on one variable are associated with low scores on the other, and increases in one variable are accompanied by decreases in the other. Table 12.7 displays a negative relationship, again with fictitious data, between education and television viewing. The amount of television viewing decreases as education increases. In other words, as you move from left to right across the top of the table (as education increases), the percentage of heavy viewers decreases.

Measures of association for ordinal and interval-ratio variables are designed so that they will take on positive values for positive associations and negative values for negative associations. Thus, a measure of association preceded by a plus sign indicates a positive relationship between the two variables, with the

[4]Variables measured at the nominal level have no numerical order to them (by definition). Therefore, associations including nominal-level variables, while they may have a pattern, cannot be said to have a direction.

READING STATISTICS 9: Bivariate Tables

The conventions for constructing and interpreting bivariate tables presented in this text are commonly but not universally followed in the professional literature. Tables will usually be constructed with the independent variable in the columns, the dependent variable in the rows, and percentages calculated in the columns. However, you should be careful to check the format of every table you attempt to read to see if these conventions have been observed. If the table is presented with the independent variable in the rows, for example, you will have to reorient your analysis (or redraw the table) to account for this. Above all, you should convince yourself that the percentages have been calculated in the correct direction. Even skilled professionals occasionally calculate percentages incorrectly and misinterpret the data (see the note on the upcoming table and Reading Statistics 10).

Once you have assured yourself that the table is properly presented, you can apply the analytical techniques developed in this chapter. By comparing the conditional distributions of the dependent variable, you can ascertain for yourself if the variables are associated and check the strength, pattern and (for tables with ordinal-level variables) the direction of the association. You may then compare your conclusions with those of the researchers. As an aid in the interpretation of bivariate tables, researchers will usually compute and report other statistics in addition to percentages. We talk about these statistics in Reading Statistics 11 in Chapter 14.

STATISTICS IN THE PROFESSIONAL LITERATURE

Researchers Dorothy Seals and Jerry Young used a survey to measure the extent of school bullying among 7th and 8th graders at several Mississippi public schools. They wanted to measure the prevalence of the behavior and explore its relationship to a variety of sociological variables including gender and ethnicity. About half of the 7th and 8th graders in five separate public school districts were included in the study, and 24% of them reported some involvement with the behavior: 10% reported that they bullied others at least one or more times per week and 13% reported that they were victims of bullying at the same rate.

Their table, shown here, indicates the relationship that was found between gender and perceptions of the frequency of various types of bullying.

FREQUENCY OF TYPE OF BULLYING BEHAVIOR REPORTED AS OCCURRING "OFTEN" BY GENDER OF RESPONDENT: Frequency and (*Percentages*)

	Gender	
Type of Bullying	Male	Female
Physical	25 (*21.9%*)	26 (*23.2%*)
Threats of harm	26 (*22.8%*)	14 (*12.5%*)
Name calling	27 (*23.6%*)	34 (*30.4%*)
Mean teasing	22 (*19.3%*)	19 (*17.0%*)
Exclusion	14 (*12.3%*)	19 (*17.0%*)
	114 (*100.0%*)	112 (*100.0%*)

Redrawn from Table 4, p. 743. The original table calculated percentages within each row. This showed gender by type of reported bullying—for example, 49% of the people that saw physical bullying as frequent were male and 51% were female. For this relationship, this is a valid point. But it seemed more appropriate to follow the usual convention of this text and treat gender as the independent variable. Either way, the conclusion is the same: There is, at best, a weak relationship between these variables.

An inspection of the column percentages shows that this relationship is weak. Males report more threats, and females are a little more likely to report name calling and exclusion, but the differences for the remaining forms of bullying are quite small. The maximum difference is 10.3% (for "Threats of harm"), a value that verifies the characterization of the relationship as weak. Male and female students report essentially the same types of bullying.

The researchers also found that males were more involved in bullying than females and that bullies tended to choose victims of their own gender. They also found no significant differences in bullying between African American and Caucasian students and no significant differences in self-esteem between bullies and victims. They did find, however, that victims of bullying had higher levels of depression than either bullies or students not involved in bullying. Want to learn more? See the following citation.

Seals, Dorothy, and Young, Jerry. 2003. "Bullying and Victimization: Prevalence and Relationship to Gender, Grade Level, Ethnicity, Self-Esteem, and Depression." *Adolescence*, 38:735–747.

READING STATISTICS 10: The Importance of Percentages

Starting with the next chapter, we will be primarily concerned with measures of association for bivariate tables. These statistics are extremely useful for summarizing the strength and, for relationships in which both variables are at least ordinal in level of measurement, the direction of relationships. Nonetheless, the first step in analyzing bivariate tables should always be to apply the techniques introduced in this chapter. The column percentages and conditional distributions will give you more detail about the relationship than measures of association. As useful as they are, the latter should be regarded as summary statements rather than analysis in depth.

Percentages may be humble statistics, but they are not necessarily simple; and they can be miscalculated and misunderstood. One type of error can occur when the researcher misunderstands which variable is cause (the independent variable) and which is effect (the dependent variable). A closely related error can happen when the researcher asks questions about the relationship incorrectly.

To illustrate these errors, let's review the proper method for analyzing bivariate relationships with tables. Recall that, by convention, we array the independent variable in the columns and the dependent variable in the rows and compute percentages within each column. When we follow this procedure, we are asking: "Does *Y* (the dependent variable) vary by *X* (the independent variable)?" or "Is *Y* caused by *X*?" We conclude that there is evidence for a causal relationship if the values of *Y* change under the different values of *X*.

To illustrate further, consider Table 1, which shows the relationship between race and support for affirmative action from the 2006 General Social Survey, a representative national sample. Race must be the independent, or causal, variable in this relationship. A person's race may cause or shape his or her attitudes and opinions, but the reverse cannot be true: A person's opinion cannot cause or shape his or her race. The percentages in Table 1 are computed in the proper direction and show that support for affirmative action does vary by race: There may be a causal relationship between these variables. The maximum difference between the columns is about 31 percentage points, indicating that the relationship is moderate to strong.

What if we had misunderstood this causal relationship or had asked the wrong question about it? If

TABLE 1 SUPPORT FOR AFFIRMATIVE ACTION BY RACIAL GROUP : Frequencies and (*Percentages*)

Support Affirmative Action?	Racial Group		
	White	Black	
Yes	158 (*11.5%*)	113 (*43.0%*)	271
No	1221 (*88.5%*)	150 (*57.0%*)	1,371
	1379 (*100.0%*)	263 (*100.0%*)	1,642

TABLE 2 ROW PERCENTAGES FOR TABLE 1

Support Affirmative Action?	Racial Group		
	White	Black	
Yes	58.3	41.7	100.0%
No	89.0	11.0	100.0%

we had computed percentages within each row, for example, we would be asking "Does support for affirmative action have a causal impact on race?" or "Does race vary by support for affirmative action?" Table 2 shows the results of asking these incorrect questions. A casual glance at the table might give the impression that there is a causal relationship, since nearly 58% of the supporters of affirmative action are white and only about 42% are black. If we looked *only* at this row (as people sometimes do), we would conclude that whites are more supportive of affirmative action than blacks. But the second row shows that whites are also the huge majority (almost 90%) of those who *oppose* the policy. How can this be? The row percentages in this table simply reflect the fact that whites vastly outnumber blacks in the sample: Whites outnumber blacks in both rows because there are five times as many whites in the sample. Computing percentages within the rows would make sense only if race could vary by attitude or opinion, and Table 2 could easily lead to false conclusions about this relationship.

Professional researchers sometimes compute percentages in the wrong direction or ask a question about the relationship incorrectly, and you should always check bivariate tables to make sure the analy-

(*continued*)

TABLE 3 RACE AND POSITION, NATIONAL FOOTBALL LEAGUE, 2005 (Offensive Positions Only)

	Position						
Race	Quarter-back	Running Back	Wide Receiver	Tight End	Tackle	Guard	Center
White	82%	9%	9%	57%	44%	54%	69%
Black	16%	89%	91%	40%	55%	39%	24%
Other	2%	2%	1%	3%	1%		
	100%	100%	100%	100%	100%	100%	100%

Source: Lapchick, Richard. 2004. 2005 Racial and Gender Report Card. Available at http://www.bus.ucf.edu/sport/cgi-bin/site/sitew.cgi?page=/ides/index.htx. p 56.

sis agrees with the patterns in the table. To illustrate this error, consider the phenomenon of racial stacking in sports. This is the practice of reserving certain positions for white players, usually the positions that require leadership skills or decision making (e.g., the quarterback in football or the catcher in baseball). This practice was quite common in the past, although most agree that it is less prevalent now. Racial stacking was widely documented but sometimes misinterpreted because of incorrectly calculated percentages (e.g., see Leonard and Phillips, 1997)[5].

It has been common to report racial breakdown by position, as in Table 3. Note that this table does not follow the conventions introduced in this chapter. Race must be the independent variable in this relationship, but its values are placed in the rows, not in the columns. Position, the dependent variable, is placed in the columns. This arrangement is not unusual, and the real problem is that the percentages are incorrectly calculated within the columns (or by position): The table shows how race varies by position. To assess accurately the persistence of racial stacking we need to know how positions are distributed by race. By computing percentages within the races, we would control for the fact that the number of professional football players varies by race (in 2005, blacks outnumbered whites almost 2 to 1) and get a more accurate picture of the extent of racial stacking.

Unfortunately, we cannot reconstruct this table with the proper percentages because detailed information on the race and position of players is not readily available. However, if we mix data from two different seasons, the proper calculation can be demonstrated. A quick search of the Internet reveals that 24 of the 32 starting quarterbacks on the last weekend of the 2006–2007 NFL season were white and 8 were black. In 2005, there were 537 white players in the league and 1116 black players. Mixing the two years, this means that about 5% (24/537) of all white professional football players were starting quarterbacks, while less than 1% (8/1116, or 0.7%) of all black players held the same position. These results are much more consistent with the racial stacking hypothesis and dramatically illustrate how rare it is for blacks, in proportion to their numbers in the league, to occupy the most important decision-making position on the team. (On the other hand, whites comprise the majority of the general population, and we would get a very different picture if we used the total population numbers as the basis for the comparison. Also, we should note again that there is much less segregation of positions in professional sports now than in the past).

The differences between our calculation and Table 3 might seem slight. Both show racial stacking for the quarterback position. Table 3 shows that white quarterbacks outnumbered blacks by about 5 to 1, and our calculation showed approximately the same ratio. However, our calculation demonstrated how rare it is for blacks to occupy this elite position in proportion to their numbers in the league. Remember that the simple error of computing percentages in the wrong direction can lead to huge errors in conclusions and serious misinterpretations of results.

[5]W. M. Leonard II and J. Phillips, 1997, "The Cause and Effect Rule for Percentaging Tables: An Overdue Statistical Correction for 'Stacking' Studies." *Sociology of Sport Journal,* 14(5): 283–289.

value +1.00 indicating a perfect positive relationship. A negative sign indicates a negative relationship, with −1.00 indicating a perfect negative relationship. *(For practice in determining the pattern of an association, see any of the end-of-chapter problems. For practice in determining the direction of a relationship, see problems 12.1, 12.6, 12.7, 12.8, 12.9, 12.10, and 12.12.)*

SUMMARY

1. Analyzing the association between variables provides information that is complementary to tests of significance. The latter are designed to detect nonrandom relationships, whereas measures of association are designed to quantify the importance or strength of a relationship.
2. Relationships between variables have three characteristics: the existence of an association, the strength of the association, and the direction or pattern of the association. These three characteristics can be investigated by calculating percentages for a bivariate table in the direction of the independent variable (vertically) and then comparing in the opposite direction (horizontally). It is often useful (as well as quick and easy) to assess the strength of a relationship by finding the maximum difference in column percentages in any row of the table.
3. Tables 12.1 and 12.2 can be analyzed in terms of these three characteristics. Clearly, a relationship does exist between job satisfaction and productivity, since the conditional distributions of the dependent variable (productivity) are different for the three different conditions of the independent variable (job satisfaction). Even without a measure of association, we can see that the association is substantial in that the change in Y (productivity) across the three categories of X (satisfaction) is marked. The maximum difference of 36.54% confirms that the relationship is substantial (moderate to strong). Furthermore, the relationship is positive in direction. Productivity increases as job satisfaction rises, and workers who report high job satisfaction tend also to be high on productivity. Workers with little job satisfaction tend to be low on productivity.
4. Given the nature and strength of the relationship, it could be predicted with fair accuracy that highly satisfied workers tend to be highly productive ("Happy workers are busy workers"). These results might be taken as evidence of a causal relationship between these two variables, but they cannot, by themselves, prove that a causal relationship exists: Association is not the same thing as causation. In fact, although we have presumed that job satisfaction is the independent variable, we could have argued the reverse causal sequence ("Busy workers are happy workers"). The results presented in Tables 12.1 and 12.2 are consistent with both causal arguments.
5. The analysis of the association between variables produces systematic evidence for (or against) a causal relationship between two variables. Ultimate proof that variables have a causal relationship depends less on statistics and more on logic, theory, and methodology (actually proving causation is a rather difficult task). As we shall see in Part IV, some of the multivariate techniques are quite useful for probing possible causal relationships, and we will return to some of these concerns at that point.

GLOSSARY

Association. The relationship between two (or more) variables. Two variables are said to be associated if the distribution of one variable changes for the various categories or scores of the other variable.

Column percentages. Percentages computed with each column of a bivariate table.

Conditional distribution of *Y*. The distribution of scores on the dependent variable for a specific score or category of the independent variable when the variables have been organized into table format.

Dependent variable. In a bivariate relationship, the variable that is taken as the effect.

Independent variable. In a bivariate relationship, the variable that is taken as the cause.

Maximum difference. A way to assess the strength of an association between variables that have been organized into a bivariate table. The maximum difference is the largest difference between column percentages for any row of the table.

Measures of association. Statistics that quantify the strength of the association between variables.

Negative association. A bivariate relationship where the variables vary in opposite directions. As one variable increases, the other decreases, and high scores on one variable are associated with low scores on the other.

Positive association. A bivariate relationship where the variables vary in the same direction. As one variable increases, the other also increases, and high scores on one variable are associated with high scores on the other.

X. Symbol used for any independent variable.

Y. Symbol used for any dependent variable.

PROBLEMS

12.1 PA Various supervisors in the city government of Shinbone, Kansas, have been rated on the extent to which they practice authoritarian styles of leadership and decision making. The efficiency of each department has also been rated, and the results are summarized here. Calculate percentages for the table so that it shows the effect of leadership style on efficiency. Is there an association between these two variables? What is the strength and direction of the relationship?

	Authoritarianism		
Efficiency	Low	High	Totals
Low	10	12	22
High	17	5	22
Totals	27	17	44

12.2 SOC The administration of a local college campus has proposed an increase in the mandatory student fee in order to finance an upgrading of the intercollegiate football program. A sample of the faculty has completed a survey on the issues. Is there any association between support for raising fees and the gender, discipline, or tenured status of the faculty? Describe the strength and direction of the association.

a. Support for raising fees by gender:

	Gender		
Support	Males	Females	Totals
For	12	8	20
Against	15	12	27
Totals	27	20	47

b. Support for raising fees by discipline:

	Discipline		
Support	Liberal Arts	Science & Business	Totals
For	6	13	19
Against	14	14	28
Totals	20	27	47

c. Support for raising fees by tenured status:

	Status		
Support	Tenured	Nontenured	Totals
For	15	4	19
Against	18	10	28
Totals	33	14	47

12.3 PS How consistent are people in their voting habits? Do people vote for the same party from election to election? Given here are the results of a poll in which people were asked if they had voted Democrat or Republican in each of the last two presidential elections. Assess the strength of this relationship.

	2000 Election		
2004 Election	Democrat	Republican	Totals
Democrat	117	23	140
Republican	17	178	195
Totals	134	201	335

12.4 SOC A needs assessment survey has been distributed in a large retirement community. Residents were asked to check off the services or programs they thought should be added. Is there any association between gender and the perception of a need for more social occasions? Write a few sentences describing the relationship in terms of pattern and strength of the association.

	Gender		
More Parties?	Males	Females	Totals
Yes	321	426	747
No	175	251	426
Totals	496	677	1173

12.5 Reproduced here are problems 11.3–11.5. In Chapter 11, you tested these relationships for their significance (the chi square test), and now you will test the relationships to determine the existence, strength, and pattern or direction of the relationship. Find column percentages and the maximum difference for each table, and write a short paragraph summarizing the relationship.

a. Services by race for a sample of the homeless:

Received Services?	Race: Black	White	Totals
Yes	6	7	13
No	4	9	13
Totals	10	16	26

b. Party preference by gender for a sample of college faculty:

Party Preference	Gender: Male	Female	Totals
Democrats	10	15	25
Republicans	15	10	25
Totals	25	25	50

c. Salary level by unionization status for 100 workplaces:

Salary	Status: Union	Nonunion	Totals
High	21	29	50
Low	14	36	50
Totals	35	65	100

12.6 SW As the state director of mental health programs, you note that some local mental health facilities have very high rates of staff turnover. You believe that part of this problem is a result of the fact that some of the local directors have very little training in administration and poorly developed leadership skills. Before implementing a program to address this problem, you collect some data to make sure your beliefs are supported by the facts. Is there a relationship between staff turnover and the administrative experience of the directors? Describe the relationship in terms of pattern and strength of the association.

Turnover	Director Experienced?: No	Yes	Totals
Low	4	9	13
Moderate	9	8	17
High	15	5	20
Totals	28	22	50

12.7 CJ About half the neighborhoods in a large city have instituted programs to increase citizen involvement in crime prevention. Do these areas experience less crime? Write a few sentences describing the relationship in terms of pattern and strength of the association.

Crime Rate	Program: No	Yes	Totals
Low	29	15	44
Moderate	33	27	60
High	52	45	97
Totals	114	87	201

12.8 SOC What types of people are most concerned about the future of the environment? The World Values Survey includes an item asking people if they would agree to an increase in taxes if the extra money was used to prevent environmental damage. The following tables show the relationship between this variable and level of education for Canada, the United States, and Mexico. Education has been collapsed into three levels. Compute column percentages and find the maximum difference for each table. Is there a relationship between and concern for the environment? Describe the strength and direction of the relationship for each nation. Does the relationship between these variables change from nation to nation? How?

a. Canada:

Support Higher Tax to Help the Environment?	Education: Low	Moderate	High	Totals
Yes	252	502	344	1098
No	203	394	190	787
	455	896	534	1885

b. United States:

Support Higher Tax to Help the Environment?	Education: Low	Moderate	High	Totals
Yes	142	204	374	720
No	90	148	225	463
	232	352	599	1183

c. Mexico:

Support Higher Tax to Help the Environment?	Education: Low	Moderate	High	Totals
Yes	407	308	96	811
No	344	206	58	608
	751	514	154	1419

12.9 SOC The latest fad to sweep college campuses is streaking to panty raids while swallowing live goldfish. A researcher is interested in how closely the spread of this bizarre behavior is linked to the amount of coverage and publicity provided by local campus newspapers. For a sample of 25 universities, the researcher has rated the amount of press coverage (as extensive, moderate, or no coverage) and how much the student body was involved in this new fad. The data for each campus are reported here. Organize the data into a properly labeled table in percentage form. Does the table indicate an association between press coverage and fad behavior?

Campus	Press Coverage	Student Involvement
1	Extensive	Extensive
2	Extensive	Some
3	Moderate	Some
4	Moderate	Some
5	Moderate	Extensive
6	Extensive	Some
7	Extensive	Extensive
8	Moderate	Some
9	None	Some
10	Moderate	None
11	None	None
12	Extensive	Extensive
13	None	Some
14	Extensive	Extensive
15	Moderate	None
16	Moderate	Some
17	Moderate	Extensive
18	Moderate	None
19	None	None
20	Extensive	Some
21	None	Extensive
22	Moderate	None
23	None	None
24	Extensive	Extensive
25	Moderate	Extensive

12.10 In any social science journal, find an article that includes a bivariate table. Inspect the table and the related text carefully and answer the following questions.

a. Identify the variables in the table. What values (categories) does each possess? What is the level of measurement for each variable?

b. Is the table in percentage form? In what direction are the percentages calculated? Are comparisons made between columns or rows?

c. Is one of the variables identified by the author as independent? Are the percentages in the direction of the independent variable?

d. How is the relationship characterized by the author in terms of the strength of the association? In terms of the direction (if any) of the association?

e. Find the measure of association (if any) calculated for the table. What is the numerical value of the measure? What is the sign (if any) of the measure?

12.11 SOC If a person's political ideology (liberal, moderate, or conservative) is known, can we predict that person's position on issues? If liberals are generally progressive and conservatives are generally traditional (with moderates in between), what relationships would you expect to find between political ideology and these issues?

a. Support for the legal right to an abortion

b. The death penalty

c. The legal right to commit suicide for people with incurable disease

d. Sex education in schools

e. Support for traditional gender roles

The following tables show the results of a recent public-opinion survey. For each table, compute column percentages and the maximum difference. Summarize the strength and direction of each relationship in a brief paragraph. Were your expectations confirmed?

a. Support for the legal right to an abortion by political ideology:

Supports Legal Abortion?	Political Ideology			
	Liberal	Moderate	Conservative	Totals
Yes	309	234	154	697
No	211	360	419	990
Totals	520	594	573	1687

b. Support for capital punishment by political ideology:

Supports Capital Punishment?	Political Ideology			
	Liberal	Moderate	Conservative	Totals
Yes	440	693	693	1826
No	265	214	186	665
Totals	705	907	879	2491

c. Support for the right of people with an incurable disease to commit suicide by political ideology:

Supports the Right to Suicide?	Political Ideology: Liberal	Moderate	Conservative	Totals
Yes	381	394	319	1094
No	120	229	261	610
Totals	501	623	580	1704

d. Support for sex education in public schools by political ideology:

Supports Sex Education?	Political Ideology: Liberal	Moderate	Conservative	Totals
Yes	481	572	476	1529
No	28	63	123	214
Totals	509	635	599	1743

e. Support for traditional gender roles by political ideology:

Supports Traditional Gender Roles?	Political Ideology: Liberal	Moderate	Conservative	Totals
Yes	59	90	108	257
No	454	548	484	1486
Totals	513	638	592	1743

f. Support for legalizing marijuana:

Should Marijuana Be Legalized?	Ideology: Liberals	Moderates	Conservatives	Totals
Yes	132	78	52	262
No	101	87	109	297
Totals	233	165	161	559

SSPS for Windows

Using *SPSS for Windows* to Analyze Bivariate Association

SPSS DEMONSTRATION 12.1 Does Support for Gay Marriage Vary by Social Class?

In Demonstrations 11.1 and 11.2, we conducted tests using social class as an independent variable and found nonsignificant relationships with attitudes about spanking but significant relationships with attitude about immigration. In this demonstration, we continue to assess the importance of social class as an independent variable, this time focusing on its relationship with support for gay marriage.

Click **Analyze, Descriptive Statistics,** and **Crosstabs** and name *marhomo* as the row variable and *class* as the column variable. Click the **Cells** button and request column percentages by clicking the box next to **Column** in the **Percentages** box. Also, request chi square by clicking the **Statistics** button. Click **Continue** and **OK,** and the following output will be produced *(NOTE: the output has been modified slightly to improve readability)*:

HOMOSEXUALS SHOULD HAVE RIGHT TO MARRY * SUBJECTIVE CLASS IDENTIFICATION Crosstabulation

HOMOSEXUALS SHOULD HAVE RIGHT TO MARRY		SUBJECTIVE CLASS IDENTIFICATION: LOWER CLASS	WORKING CLASS	MIDDLE CLASS	UPPER CLASS	Total
STRONGLY AGREE	Count	10	43	48	6	107
	%	25.0%	15.0%	16.6%	24.0%	16.7%
AGREE	Count	9	57	59	3	128
	%	22.5%	19.9%	20.4%	12.0%	20.0%

(continued next page)

(*continued*)

HOMOSEXUALS SHOULD HAVE RIGHT TO MARY		SUBJECTIVE CLASS IDENTIFICATION				Total
		LOWER CLASS	WORKING CLASS	MIDDLE CLASS	UPPER CLASS	
NEITHER AGREE	Count	3	28	42	3	76
NOR DISAGREE	%	7.5%	9.8%	14.5%	12.0%	11.9%
DISAGREE	Count	5	60	53	4	122
	%	12.5%	21.0%	18.3%	16.0%	19.1%
STRONGLY	Count	13	98	87	9	207
DISAGREE	%	32.5%	34.3%	30.1%	36.0%	32.3%
TOTAL	Count	40	286	289	25	640
	%	100.0%	100.0%	100.0%	100.0%	100.0%

Chi-Square Tests

	Value	df	Asymp. Sig. (2-sided)
Pearson Chi-Square	9.818(a)	12	.632
Likelihood Ratio	9.819	12	.632
Linear-by-Linear Association	.067	1	.795
N of Valid Cases	640		

[a]4 cells (20.0%) have expected count less than 5. The minimum expected count is 2.97.

The sample is variable on this issue, but the majority of respondents either disagree (about 19%) or disagree strongly (about 32%) with marriage for homosexuals. There is an association between the variables (the conditional distribution of *Y*, or *marhomo*, changes from column to column), but the maximum difference is 10.5 (in the "Agree" row), so the relationship is weak. As for the direction of the relationship, the picture is complex. The highest levels of support ("Agree" and "Strongly Agree") for gay marriage are found in the lower and upper classes, and it is difficult to detect a general pattern in the relationship.

Finally, the chi square of 9.818 is not significant. The relationship between these variables is weak and not statistically significant, a disappointing result if we wanted to argue that support for gay marriage was associated with social class.

SPSS DEMONSTRATION 12.2 Does Support for Capital Punishment Vary by Political Ideology?

Capital punishment is a highly politicized issue in our society, and, given the way in which this issue is usually debated, we would expect liberals to oppose the death penalty, conservatives to support it, and moderates to be intermediate. In the 2006 GSS, attitude toward capital punishment is measured by the variable *cappun* and political ideology is measured by *polviews*. The latter has too many categories (7) to be used in a bivariate table, so we must first recode the variable. Click **Transform** from the main menu, and remember to choose "Recode into Different Variable." In this demonstration, I named the recoded version of the variable *polR*. The original scores for this variable are given in Appendix G, and the recoding instructions that should appear in the **Old → New** box of the **Recode into Different Variable** window should be:

1 thru 3 → 1
4 → 2
5 thru 7 → 3

This scheme groups all liberals (a score of 1 on *polR*) and conservatives (a score of 3 on *polR*) together and changes the score associated with "moderates" (a score of 2 on *polR*) so that they remain intermediate between the other two scores.

The **Crosstabs** procedure is accessed by following the steps explained in Demonstration 12.1. Click **Analyze, Descriptive Statistics,** and **Crosstabs.** Specify *cappun* as the row variable, or the dependent variable, and *polR* (recoded *polviews*) as the column, or independent, variable. In the **Crosstabs** window, click **Cells,** and, under **Percentages,** click the button next to **Columns.** With the dependent variable in the rows and the independent variable in the columns and with percentages calculated within columns, we will be able to read the table by following the rules developed in this chapter. Also, click the **Statistics** button and request chi square. The bivariate table for *cappun* and *polR* is shown here *(NOTE: the appearance of the table has been slightly changed to improve readability).*

FAVOR OR OPPOSE DEATH PENALTY FOR MURDER * Recoded polviews Crosstabulation

FAVOR OR OPPOSE DEATH PENALTY FOR MURDER		Recoded polviews			
		1.00	2.00	3.00	Total
FAVOR	Count	111	239	239	589
	%	49.1%	67.7%	77.9%	66.5%
OPPOSE	Count	115	114	68	297
	%	50.9%	32.3%	22.1%	33.5%
Total	Count	226	353	307	886
	%	100.0%	100.0%	100.0%	100.0%

Chi-Square Tests

	Value	df	Asymp. Sig. (2-sided)
Pearson Chi-Square	48.629(a)	2	.000
Likelihood Ratio	48.162	2	.000
Linear-by-Linear Association	46.893	1	.000
N of Valid Cases	886		

[a]0 cells (.0%) have expected count less than 5. The minimum expected count is 75.76.

In each cell of the table, the count is reported first. For example, there are 111 cases in the upper-left-hand cell. These are liberals (1 on *polR*) who favor the death penalty. Below the count is the percentage of all cases in that column (all liberals) in favor of capital punishment. Reading across the columns, 49.1% of the liberals favor the death penalty, as compared with 67.7% of the moderates (2 on *polR*) and 77.9% of the conservatives (3 on *polR*). Support for capital punishment increases as political conservatism increases. The maximum difference is 28.8%, between conservatives and liberals, indicating that this is a moderate-to-strong relationship.

These results are similar to what we expected and suggest that there is an important relationship between these variables. This conclusion is reinforced by the fact that chi square is significant at less than .05 ("Significance" is reported as .000, to be exact).

SPSS DEMONSTRATION 12.3 The Direction of Relationships: Do Attitudes Toward Sex Vary by Age?

Let's take a look at a relationship between two ordinal-level variables so that you can develop some experience in describing the direction as well as the strength of bivariate relationships. Both of our variables have to be at least ordinal in level of measurement, so let's start with *age* as an independent variable and **Recode** it into three categories, as in Demonstration 10.1. The recoding instructions were

18 thru 37 → 1
38 thru 53 → 2
54 thru 89 → 3

I used these particular cutting points to divide the sample into three categories of roughly equal size.

Without looking at the data, I'm willing to bet that *age* will have a negative relationship with approval of premarital sex (*premarsx*). Run the **Crosstabs** procedure (see Demonstration 12.1) with *premarsx* as the row variable and *ager* (recoded *age*) as the column variable. Click the **Cells** button and make sure that the button next to **Columns** under **Percentages** is checked. If you wish, click the **Statistics** button and request a chi square test. The bivariate table for these two variables is shown *(NOTE: The appearance of the table has been slightly changed to improve readability. Numerical scores have been added to the row variable).*

SEX BEFORE MARRIAGE * Recorded age Crosstabulation

SEX BEFORE MARRIAGE		RECODED AGE			
		1.00	2.00	3.00	Total
ALWAYS WRONG (1)	Count	47	47	56	150
	%	23.0%	25.3%	26.4%	24.9%
ALMOST ALWAYS WRONG (2)	Count	9	12	25	46
	%	4.4%	6.5%	11.8%	7.6%
SOMETIMES WRONG (3)	Count	32	32	41	105
	%	15.7%	17.2%	19.3%	17.4%
NOT WRONG AT ALL (4)	Count	116	95	90	301
	%	56.9%	51.1%	42.5%	50.0%
Total	Count	204	186	212	602
	%	100.0%	100.0%	100.0%	100.0%

Is there a relationship between these two variables? Do the conditional distributions change? Inspect the table column by column, and I think you will agree that there is a relationship. The maximum difference is on the bottom row ("Not wrong at all") between the youngest (1) and oldest (3) age groups: 56.9 − 42.5 = 14.4. This is a moderate relationship.

Is this relationship positive or negative? Remember that in a positive association, high scores on one variable will be associated with high scores on the other and low scores will be associated with low. In a negative relationship, high scores on one variable are associated with low scores on the other.

Now go back and look at the table again. Find the single largest cell in each column and see if you can detect the pattern. The lowest score (youngest age group = 1) on *ager* is in the left-hand column. For this age group, the most common score (56.9% of this age group) on *premarsx* is 4 (not wrong at all). (To view the numerical codes associated with each response on a variable, see Appendix G or click **Utilities** and **Variables.**) In other words, a slight majority of the youngest respondents supports the most permissive position on premarital sex. This score is also the most common response for the middle group on age, but the older respondents are less likely to endorse this view. About 43% of the oldest age group felt than premarital sex was "Not wrong at all." So, we could say that permissive attitudes tend to *decline* as age *increases*—the relationship is *negative* in direction.

Now look at the top two rows of the table. These are the least permissive positions on premarital sex ("Always wrong" and "Almost Always Wrong") and the percentage of cases in these rows is about the same for the youngest and middle groups but then increases for the oldest group. So we could say that opposition to premarital sex *increases* as age *increases*—the variables change in the *same* direction, so the relationship is *positive*.

Which characterization of the direction of the relationship is correct? Both. The relationship is negative if you think of *premarsx* as measuring *approval* and positive if you think of the variable as measuring *opposition* to premarital sex. The point, of course, is to call your attention to the fact that the direction of a relationship can be ambiguous and confusing when we are dealing with ordinal-level variables like *premarsx*. In a positive association, the numerical *scores* always vary in the same direction; in a negative relationship, they always vary in opposite directions. However, because the coding for ordinal-level variables is arbitrary, higher scores don't necessarily mean that the quantity being measured is increasing, and lower scores don't always mean that the quantity is decreasing. Always pay careful attention to the coding scheme for the variable, and exercise caution when analyzing the direction of relationships that involve ordinal-level variables.

Exercises

12.1 As long as *polviews* has already been recoded, pick a few more "social issues" (such as *grass* and *abany*) and, with recoded *polviews* as the independent variable, see if the patterns of association conform to expectations. Are "liberals" really more liberal on these issues? For each table, be sure to request column percentage in the cells and the chi square test. For each table, write a sentence or two of interpretation.

12.2 See if you can assess the strongest determinant of attitudes on an issue such as *cappun, grass,* or *abany*. You already have results for recoded *polviews* as an independent variable. Run the same dependent, or row, variables against *sex, relig, class,* and *racecen1* and compare the strength and direction or pattern of the relationships with each other and with *polviews*. Which independent variable has the strongest effect on the "social issue" you choose to analyze? Describe the relationships in terms of strength and direction or pattern.

13 Association Between Variables Measured at the Nominal Level

LEARNING OBJECTIVES

By the end of this chapter, you will be able to

1. Calculate and interpret phi, Cramer's V, and lambda.
2. Explain the logic of proportional reduction in error in terms of lambda.
3. Use any of the three measures of association to analyze and describe a bivariate relationship in terms of the three questions introduced in Chapter 12.

13.1 INTRODUCTION

Measures of association are descriptive statistics that summarize the overall strength of the association between two variables. Because they represent that relationship in a single number, these statistics are more efficient methods of expressing an association than conditional distributions, column percentages, and the maximum difference. If the researcher considers only the measure of association, however, a certain amount of information (detail and nuance) about the relationship will be lost, as is the case with any summarizing technique. Always inspect the patterns of cell frequencies or percentages in the table along with the summary measure of association in order to maximize the amount of information you have about the relationship. You should do this regardless of the level of measurement of the data or the specific measure that has been calculated.

As we shall see, there are many measures of association. In this text, these statistics have been organized according to the level of measurement for which they are most appropriate. In this chapter, we consider measures appropriate for nominally measured variables. You will note that several of the research situations used as examples involve variables measured at different levels (for example, one nominal-level variable and one ordinal-level variable). The general procedure in the situation of "mixed levels" is to select measures of association appropriate for the lower of the two levels of measurement.

13.2 CHI SQUARE–BASED MEASURES OF ASSOCIATION

Over the years, social science researchers have relied heavily on measures of association based on the value of chi square. When the value of chi square is already known, these measures are easy to calculate. To illustrate, let us reconsider Table 11.3, which displayed, with fictitious data, a relationship between accreditation and employment for social work majors. For the sake of convenience, this table is reproduced here as Table 13.1.

We saw in Chapter 11 that this relationship is statistically significant ($\chi^2 = 10.78$, which is significant at $\alpha = 0.05$), but the question now concerns the *strength* of the association. A brief glance at Table 13.1 shows that the conditional distributions of employment status do change, so the variables are associated. To emphasize this point, it is always helpful to calculate column percentages, as in Table 13.2.

TABLE 13.1 EMPLOYMENT OF 100 SOCIAL WORK MAJORS BY ACCREDITATION STATUS OF UNDERGRADUATE PROGRAM (fictitious data)

	Accreditation Status		
Employment Status	Accredited	Not Accredited	Totals
Working as a social worker	30	10	40
Not working as a social worker	25	35	60
Totals	55	45	100

TABLE 13.2 EMPLOYMENT BY ACCREDITATION STATUS (percentages)

	Accreditation Status		
Employment Status	Accredited	Not Accredited	Totals
Working as a social worker	54.55%	22.22%	40.00%
Not working as a social worker	45.45%	77.78%	60.00%
Totals	100.00%	100.00%	100.00%

TABLE 13.3 THE RELATIONSHIP BETWEEN THE VALUE OF NOMINAL-LEVEL MEASURES OF ASSOCIATION AND THE STRENGTH OF THE RELATIONSHIP

Value	Strength
If the value is	*The strength of the relationship is*
between 0.00 and 0.10	weak
between 0.11 and 0.30	moderate
greater than 0.30	strong

So far, we know that the relationship between these two variables is statistically significant and that there is an association of some kind between accreditation and employment. To assess the strength of the association, we will compute a **phi (ϕ).** This statistic is a frequently used chi square–based measure of association appropriate for 2 × 2 tables (that is, tables with two rows and two columns).

Before considering the computation of phi, it will be helpful to establish some general guidelines for interpreting the value of measures of association for nominally measured variables, similar to the guidelines we used for interpreting the maximum difference in column percentages in Chapter 12. For phi and the other measures introduced in this chapter, the general relationship between the value of the statistic and the strength of the relationship is presented in Table 13.3.

As was the case for Table 12.5, the relationships between the numerical values and the descriptive terms in Table 13.3 are arbitrary and meant as general guidelines only. Measures of association generally have mathematical definitions that yield interpretations more meaningful and exact than these.

One of the attractions of phi is that it is easy to calculate. Simply divide the value of the obtained chi square by N and take the square root of the result. Expressed in symbols, the formula for phi is

ONE STEP AT A TIME **Computing and Interpreting Phi**

To calculate phi, solve Formula 13.1:

$$\phi = \sqrt{\frac{\chi^2}{N}}$$

Step 1: Divide the value of chi square by *N*.

Step 2: Find the square root of the quantity you found in step 1. The resulting value is phi.

Step 3: Consult Table 13.3 to help interpret the value of phi.

FORMULA 13.1

$$\phi = \sqrt{\frac{\chi^2}{N}}$$

For the data displayed in Table 13.1, the chi square was 10.78. Therefore, phi is

$$\phi = \sqrt{\frac{\chi^2}{N}}$$

$$\phi = \sqrt{\frac{10.78}{100}}$$

$$\phi = 0.33$$

For a 2 × 2 table, phi ranges in value from 0 (no association) to 1.00 (perfect association). The closer to 1.00, the stronger the relationship; the closer to 0.00, the weaker the relationship. For Table 13.1, we already knew that the relationship was statistically significant at the 0.05 level. Phi, as a measure of association, adds information about the strength of the relationship: There is a moderate-to-strong relationship between these two variables. As for the pattern of the association, the column percentages in Table 13.2 shows that graduates of accredited programs were more often employed as social workers.

For tables larger than 2 × 2 (specifically, for tables with more than two columns and more than two rows), the upper limit of phi can exceed 1.00. This makes phi difficult to interpret, and a more general form of the statistic called **Cramer's *V*** must be used for larger tables. The formula for Cramer's *V* is

FORMULA 13.2

$$V = \sqrt{\frac{\chi^2}{(N)(\min r - 1, c - 1)}}$$

where (min $r - 1$, $c - 1$) = the minimum value of $r - 1$ (number of rows minus 1) or $c - 1$ (number of columns minus 1)

In words: To calculate *V*, find the lesser of the number of rows minus 1 ($r - 1$) or the number of columns minus 1 ($c - 1$), multiply this value by *N*, divide the result into the value of chi square, and then find the square root. Cramer's *V* has an upper limit of 1.00 for a table of any size and will be the same value as phi if the table has either two rows or two columns. Like phi, Cramer's *V* can be interpreted as an index that measures the strength of the association between two variables.

To illustrate the computation of *V*, suppose you had gathered the data displayed in Table 13.4, which shows the relationship between membership in

TABLE 13.4 ACADEMIC ACHIEVEMENT BY CLUB MEMBERSHIP

	Membership			
Academic Achievement	Fraternity or Sorority	Other Organization	No Memberships	Totals
Low	4	4	17	25
Moderate	15	6	4	25
High	4	16	5	25
Totals	23	26	26	75

TABLE 13.5 ACADEMIC ACHIEVEMENT BY CLUB MEMBERSHIP (percentages)

	Membership			
Academic Achievement	Fraternity or Sorority	Other Organization	No Memberships	Totals
Low	17.39	15.39	65.39	33.33%
Moderate	65.22	23.08	15.39	33.33%
High	17.39	61.54	19.23	33.33%
Totals	100.00	100.01	100.01	100.00%

student organizations and academic achievement for a sample of college students. The obtained chi square for this table is 31.5, a value that is significant at the 0.05 level. Cramer's V is

$$V = \sqrt{\frac{\chi^2}{(N)(\min r - 1, c - 1)}}$$

$$V = \sqrt{\frac{31.50}{(75)(2)}}$$

$$V = \sqrt{\frac{31.50}{150}}$$

$$V = \sqrt{0.21}$$

$$V = 0.46$$

Since Table 13.4 has the same number of rows and columns, we may use either $(r - 1)$ or $(c - 1)$ in the denominator. In either case, the value of the denominator is N multiplied by $(3 - 1)$, or 2. The computed value of V of 0.46 means there is a strong association between club membership and academic achievement. Column percentages are presented in Table 13.5 to help identify the pattern of this relationship. Fraternity and sorority members tend to be moderate, members of other organizations tend to be high, and nonmembers tend to be low in academic achievement.

One limitation of phi and Cramer's V is that they are only general indicators of the strength of the relationship. Of course, the closer these measures are to 0.00, the weaker the relationship and the closer to 1.00, the stronger the relationship. Values between 0.00 and 1.00 can be described as weak, moderate, or

ONE STEP AT A TIME **Computing and Interpreting Cramer's *V***

To calculate Cramer's *V*, solve Formula 13.2:

$$V = \sqrt{\frac{\chi^2}{(N)(\min r - 1, c - 1)}}$$

Step 1: Find the number of rows (*r*) and number of columns (*c*) in the table. Subtract 1 from the lesser of these two numbers to find (min *r* − 1, *c* − 1).

Step 2: Multiply the value you found in step 1 by *N*.

Step 3: Divide the value of chi square by the value you found in step 2.

Step 4: Take the square root of the quantity you found in step 3. The resulting value is *V*.

Step 5: Consult Table 13.3 to help interpret the value of *V*.

strong according to the general convention introduced earlier but have no direct or meaningful interpretation. On the other hand, phi and *V* are easy to calculate (once the value of chi square has been obtained) and are commonly used indicators of the importance of an association.[1] *(For practice in computing phi and Cramer's V, see any of the problems at the end of this chapter or at the end of Chapter 12. To minimize computations, however, use problems from the end of Chapter 11 for which you already have a value for chi square. Otherwise, use the problems based on 2 × 2 tables. Remember that for tables that have either two rows or two columns, phi and Cramer's V will have the same value.)*

13.3 PROPORTIONAL REDUCTION IN ERROR (PRE)

In recent years, a group of measures based on a logic known as **proportional reduction in error (PRE)** have been developed to complement the older chi square–based measures of association. Most generally stated, the logic of these measures requires us to make two different predictions about the scores of cases. In the first prediction, we ignore information about the independent variable and, therefore, make many errors in predicting the score on the dependent variable. In the second prediction, we take account of the score of the case on the independent variable to help predict the score on the dependent variable. If there is an association between the variables, we will make fewer errors when taking the independent variable into account. PRE measures of association express the proportional reduction in errors between the two predictions. Applying these general thoughts to the case of nominal-level variables will make the logic clearer.

For nominal-level variables, we first predict the category into which each case will fall on the dependent variable (*Y*) while ignoring the independent variable (*X*). Since we would, in effect, be predicting blindly in this case, we would make many errors (that is, we would often predict the value of a case on the dependent variable).

[1]Two other chi square–based measures of association, T^2 and *C* (the contingency coefficient), are sometimes reported in the literature. Both of these measures have serious limitations. T^2 has an upper limit of 1.00 only for tables with an equal number of rows and columns, and the upper limit of *C* varies, depending on the dimensions of the table. These characteristics make these measures more difficult to interpret and thus less useful than phi or Cramer's *V*.

The second prediction allows us to take the independent variable into account. If the two variables are associated, the additional information supplied by the independent variable will reduce our errors of prediction (that is, we should misclassify fewer cases). The stronger the association between the variables, the greater the reduction in errors. In the case of a perfect association, we would make no errors at all when predicting score on Y from score on X. When there is no association between the variables, on the other hand, knowledge of the independent variable will not improve the accuracy of our predictions. We would make just as many errors of prediction with knowledge of the independent variable as we did without knowledge of the independent variable.

An illustration should make these principles clearer. Suppose you were placed in the rather unusual position of having to predict whether each of the next 100 people you meet will be shorter or taller than 5 feet 9 inches in height under the condition that you would have no knowledge of these people at all. With absolutely no information about these people, your predictions will be wrong quite often (you will frequently misclassify a tall person as short, and vice versa).

Now assume that you must go through this ordeal twice; but that on the second round you know the sex of the person whose height you must predict. Since height is associated with sex and females are, on the average, shorter than males, the optimal strategy would be to predict that all females are short and all males are tall. Of course, you will still make errors on this second round; but, if the variables are associated, the number of errors on the second round will be less than the number of errors on the first. That is, using information about the independent variable will reduce the number of errors (if, of course, the two variables are related). How can these unusual thoughts be translated into a useful statistic?

13.4 A PRE MEASURE FOR NOMINAL-LEVEL VARIABLES: LAMBDA

One hundred individuals have been categorized by gender and height, and the data are displayed in Table 13.6. It is clear, even without percentages, that the two variables are associated. To measure the strength of this association, a PRE measure called **lambda** (symbolized by the Greek letter λ) will be calculated. Following the logic introduced in the previous section, we must find two quantities. First, the number of prediction errors made while ignoring the independent variable (gender) must be found. Then we will find the number of prediction errors made while taking gender into account. These two sums will then be compared to derive the statistic.

First, the information given by the independent variable (gender) can be ignored, in effect, by working only with the row marginals. Two different predictions can be made about height (the dependent variable) by using these

TABLE 13.6 HEIGHT BY GENDER

	Gender		
Height	Male	Female	Totals
Tall	44	8	52
Short	6	42	48
Totals	50	50	100

marginals. We can predict either that all subjects are tall or that all subjects are short.[2] For the first prediction (all subjects are tall), 48 errors will be made. That is, for this prediction, all 100 cases would be placed in the first row. Since only 52 of the cases actually belong in this row, this prediction would result in (100 − 52), or 48, errors. If we had predicted that all subjects were short, on the other hand, we would have made 52 errors (100 − 48 = 52). We will take the *lesser* of these two numbers and refer to this quantity as E_1, for the number of errors made while ignoring the independent variable. So $E_1 = 48$.

In the second step in the computation of lambda, we predict score on Y (height) again, but this time we take X (gender) into account. To do this, follow the same procedure as in the first step, but this time move from column to column. Since each column is a category of X, we thus take X into account in making our predictions. For the left-hand column (males), we predict that all 50 cases will be tall and make six errors (50 − 44 = 6). For the second column (females), our prediction is that all females are short, and eight errors will be made. By moving from column to column, we have taken X into account and have made a total of 14 errors of prediction, a quantity we will label E_2 ($E_2 = 6 + 8 = 14$).

If the variables are associated, we will make fewer errors under the second procedure than under the first. In other words, E_2 will be smaller than E_1. In this case, we made fewer errors of prediction while taking gender into account ($E_2 = 14$) than while ignoring gender ($E_1 = 48$), so gender and height are clearly associated. Our errors were reduced from 48 to only 14. To find the *proportional* reduction in error, use Formula 13.3:

FORMULA 13.3

$$\lambda = \frac{E_1 - E_2}{E_1}$$

For the sample problem, the value of lambda would be

$$\lambda = \frac{E_1 - E_2}{E_1}$$

$$\lambda = \frac{48 - 14}{48}$$

$$\lambda = \frac{34}{48}$$

$$\lambda = 0.71$$

The value of lambda ranges from 0.00 to 1.00. Of course, a value of 0.00 means that the variables are not associated at all (E_1 is the same as E_2), and a value of 1.00 means that the association is prefect (E_2 is zero and scores on the dependent variable can be predicted without error from the independent variable). Unlike phi or V, however, the numerical value of lambda between the extremes of 0.00 and 1.00 has a precise meaning: It is an index of the extent to which the independent variable (X) helps us to predict (or, more loosely, understand) the dependent variable (Y). When multiplied by 100, the value of lambda indicates the strength of the association in terms of the percentage reduction in error. Thus, the preceding lambda would be interpreted by concluding that knowledge of gender improves our ability to predict height by 71%. That is, we are 71% better off knowing gender when attempting to predict height.

[2] Other predictions are, of course, possible, but these are the only two permitted by lambda.

Application 13.1

A random sample of students at a large urban university have been classified as either "traditional" (18–23 years of age and unmarried) or "nontraditional" (24 or older or married). Subjects have also been classified as "vocational," if their primary motivation for college attendance is career or job oriented, or "academic," if their motivation is to pursue knowledge for its own sake. Are these two variables associated?

	Type		
Motivation	Traditional	Non-traditional	Totals
Vocational	25	60	85
Academic	75	15	90
Totals	100	75	175

Since this is a 2 × 2 table, we can compute phi as a measure of association. The chi square for the table is 51.89, so phi is

$$\phi = \sqrt{\frac{\chi^2}{N}}$$

$$\phi = \sqrt{\frac{51.89}{175}}$$

$$\phi = \sqrt{0.30}$$

$$\phi = 0.55$$

which indicates a strong relationship between the two variables.

A lambda can also be computed as an additional measure of association:

$$E_1 = 175 - 90 = 85$$

$$E_2 = (100 - 75) + (75 - 60)$$

$$= 25 + 15 = 40$$

$$\lambda = \frac{E_1 - E_2}{E_1}$$

$$\lambda = \frac{85 - 40}{85}$$

$$\lambda = \frac{45}{85}$$

$$\lambda = 0.53$$

A lambda of 0.53 indicates that we would make 53% fewer errors in predicting motivation (Y) from student type (X), as opposed to predicting motivation while ignoring student type. The association is strong, and, by inspection of the table, we can see that traditional students are more likely to have academic motivations (75%) and that nontraditional types are more likely to be vocational in motivation (80%).

13.5 THE COMPUTATION OF LAMBDA

In this section, we work through another example in order to state the computational routine for lambda in more general terms. Suppose a researcher was concerned with the relationship between religious denomination and attitude toward capital punishment and had collected the data presented in Table 13.7 from a sample of 130 respondents.

TABLE 13.7 ATTITUDE TOWARD CAPITAL PUNISHMENT BY RELIGIOUS DENOMINATION (fictitious data)

	Religion				
Attitude	Catholic	Protestant	Other	None	Totals
Favors	10	9	5	14	38
Neutral	14	12	10	6	42
Opposed	11	4	25	10	50
Totals	35	25	40	30	130

Application 13.2

In application 12.2, we examined the relationships between the status of women and two different independent variables—religiosity and level of development—for a sample of 47 nations. We found a moderate-to-strong relationship between religiosity and the status of women and a strong relationship between level of development and the status of women. In this application, we use the measures of association introduced in this chapter to verify these characterizations of the strength of the relationship.

STATUS OF WOMEN BY RELIGIOSITY FOR 47 NATIONS: Frequencies and (*Percentages*)

Women's Status	Religiosity: Low	Religiosity: High	Totals
Low	8 (*36.36%*)	17 (*68.00%*)	25
High	14 (*64.64%*)	8 (*32.00%*)	22
Totals	22 (*100.00%*)	25 (*100.00%*)	47

We have already examined the conditional distributions and the maximum difference for both tables, so we can proceed to the measures of association. For the relationship between the status of women and religiosity, chi square is 4.7, so phi is

$$\phi = \sqrt{\frac{\chi^2}{N}}$$
$$\phi = \sqrt{\frac{4.7}{47}}$$
$$\phi = \sqrt{0.10}$$
$$\phi = 0.32$$

Lambda is

$$E_1 = 47 - 25 = 22$$
$$E_2 = (22 - 14) + (25 - 17)$$
$$= (8 + 8) = 16$$
$$\lambda = \frac{E_1 - E_2}{E_1}$$
$$\lambda = \frac{22 - 16}{22}$$
$$\lambda = \frac{6}{22}$$
$$\lambda = 0.27$$

A lambda of 0.27 indicates that we would make 27% fewer errors using the independent variable (religiosity) to predict women's status. According to the guidelines suggested in Table 13.3, both measures of association indicate that the relationship is on the borderline between moderate and strong.

(*continued next page*)

Step 1. To find E_1, the number of errors made while ignoring X (religion, in this case), subtract the largest row total from N. For Table 13.7, E_1 will be

$$E_1 = N - (\text{Largest row total})$$
$$E_1 = 130 - 50$$
$$E_1 = 80$$

Thus, we will misclassify 80 cases on attitude toward capital punishment while ignoring religion.

Step 2. Next, E_2—the number of errors made when taking the independent variable into account—must be found. For each column, subtract the largest cell frequency from the column total and then add the subtotals together. For the data presented in Table 13.7:

For Catholics: $35 - 14 = 21$
For Protestants: $25 - 12 = 13$
For "Others": $40 - 25 = 15$
For "None": $30 - 14 = 16$
$E_2 = 65$

Application 13.2: *(continued)*

STATUS OF WOMEN BY LEVEL OF DEVELOPMENT FOR 47 NATIONS: Frequencies and (*Percentages*)

Women's	Level of Development			
Status	LDCs	Developing	Developed	Totals
Low	13 (*81.25%*)	8 (*53.33%*)	4 (*25.00%*)	25 (53.19%)
High	3 (*18.75%*)	7 (*46.67%*)	12 (*75.00%*)	22 (46.81%)
Totals	16 (*100.00%*)	15 (*100.00%*)	16 (*100.00%*)	47 (100.00%)

Turning to the relationship between level of development and the status of women, the chi square is 10.17, so phi is

$$\phi = \sqrt{\frac{\chi^2}{N}}$$

$$\phi = \sqrt{\frac{10.17}{47}}$$

$$\phi = \sqrt{0.22}$$

$$\phi = 0.47$$

The lambda is

$$E_1 = 47 - 25 = 22$$

$$E_2 = (16 - 13) + (15 - 8) + (16 - 12)$$
$$= (3 + 7 + 4) = 14$$

$$\lambda = \frac{E_1 - E_2}{E_1}$$

$$\lambda = \frac{22 - 14}{22}$$

$$\lambda = \frac{8}{22}$$

$$\lambda = 0.36$$

A lambda of 0.36 means that we will make 36% fewer errors of prediction using information from the independent variable (level of development) to predict score on the dependent variable (women's status). Using the guidelines suggested in Table 13.3, we see that both phi (0.47) and lambda (0.36) indicate that this is a strong relationship.

A total of 65 errors are made when predicting attitude on capital punishment while taking religion into account.

Step 3. In step 1, 80 errors of prediction were made, as compared to 65 errors in step 2. Since the number of errors has been reduced, the variables are associated. To find the proportional reduction in error, the values for E_1 and E_2 can be substituted directly into Formula 13.3:

$$\lambda = \frac{80 - 65}{80}$$

$$\lambda = \frac{15}{80}$$

$$\lambda = 0.19$$

ONE STEP AT A TIME **Computing and Interpreting Lambda**

To calculate lambda, solve Formula 13.3:

$$\lambda = \frac{E_1 - E_2}{E_1}$$

Step 1: To find E_1, subtract the largest row subtotal (marginal) from *N*.

Step 2: Starting with the far left-hand column, subtract the largest cell frequency in the column from the column total. Repeat this step for all columns in the table.

Step 3: Add up all the values you found in step 2. The result is E_2.

Step 4: Subtract E_2 from E_1.

Step 5: Divide the quantity you found in step 4 by E_1. The resultant quantity is lambda.

Step 6: To interpret lambda, multiply the value of lambda by 100. This percentage tells us the extent to which our predictions of the dependent variable are improved by taking the independent variable into account. Also, lambda may be interpreted using the descriptive terms in Table 13.3.

Using our conventional labels, we would call this a moderate relationship. Using PRE logic, we can add more detail to the characterization: When attempting to predict attitude toward capital punishment, we would make 19% fewer errors by taking religion into account. Knowledge of a respondent's religious denomination improves the accuracy of our predictions by a factor of 19%. The moderate strength of lambda indicates that factors other than religion are associated with the dependent variable.

13.6 THE LIMITATIONS OF LAMBDA

As a measure of association, lambda has two characteristics that should be stressed. First, lambda is asymmetric. This means that the value of the statistic will vary, depending on which variable is taken as independent. For example, in Table 13.7, the value of lambda would be .14 if attitude toward capital punishment had been taken as the independent variable (verify this with your own computation). Thus, you should exercise some caution in the designation of an independent variable. If you consistently follow the convention of arraying the independent variable in the columns and compute lambda as outlined earlier, the asymmetry of the statistic should not be confusing.

Second, when one of the row totals is much larger than the others, lambda can be misleading. It can be 0.00 even when other measures of association are greater than 0.00 and the conditional distributions for the table indicate that there is an association between the variables. This anomaly is a function of the way lambda is calculated and suggests that great caution should be exercised in the interpretation of lambda when the row marginals are very unequal. In fact, in the case of very unequal row marginals, a chi square–based measure of association would be the preferred measure of association *(For practice in computing lambda, see any of the problems at the end of this chapter, Chapter 12, or Chapter 11. As with phi and Cramer's V, it's probably a good idea to start with small samples and 2 × 2 tables.)*

SUMMARY

1. Three measures of association—phi, Cramer's *V*, and lambda—were introduced. Each is used to summarize the overall strength of the association between two variables that have been organized into a bivariate table.
2. Phi and Cramer's *V* are chi square–based measures of association and have the advantage of being easy to compute (once the value of chi square is found). Phi is used for 2 × 2 tables; Cramer's *V* can be used for tables of any size. Both indicate the strength of the relationship, but values between 0.00 and 1.00 have no direct interpretation.
3. Lambda is a PRE-based measure and provides a more direct interpretation for values between the extremes of 0.00 and 1.00. Lambda indicates the improvement in predicting the dependent variable with knowledge of the independent, compared to predicting the dependent without knowledge of the independent. Because of the meaningfulness of values between the extremes, lambda is often preferred over the more traditional chi square–based measures (except when row totals are very unequal).

SUMMARY OF FORMULAS

Phi 13.1 $$\phi = \sqrt{\frac{\chi^2}{N}}$$

Cramer's *V* 13.2 $$V = \sqrt{\frac{\chi^2}{(N)(\min r - 1, c - 1)}}$$

Lambda 13.3 $$\lambda = \frac{E_1 - E_2}{E_1}$$

GLOSSARY

Cramer's *V*. A chi square–based measure of association. Appropriate for nominally measured variables that have been organized into a bivariate table of any number of rows and columns.

E_1. For lambda, the number of errors of prediction made when predicting which category of the dependent variable cases will fall into while ignoring the independent variable.

E_2. For lambda, the number of errors of prediction made when predicting which category of the dependent variable cases will fall into while taking account of the independent variable.

Lambda (λ). A measure of association appropriate for nominally measured variables that have been organized into a bivariate table. Lambda is based on the logic of proportional reduction in error (PRE).

Phi (ϕ). A chi square–based measure of association. Appropriate for nominally measured variables that have been organized into a 2 × 2 bivariate table.

Proportional reduction in error (PRE). The logic that underlies the definition and computation of lambda. The statistic compares the number of errors made when predicting the dependent variable while ignoring the independent variable (E_1) with the number of errors made while taking the independent variable into account (E_2).

PROBLEMS

13.1 SOC Who is most likely to be victimized by crime? A small sample of city residents has been asked if they were the victims of burglary or robbery over the past year. The following tables report relationships between several variables and victimization. Compute phi and lambda for each table. Which relationship is strongest?

a. Victimization by sex of respondent:

	Sex		
Victimized?	Male	Female	Totals
Yes	10	12	22
No	15	18	33
Totals	25	30	55

b. Victimization by age of respondent:

	Age		
Victimized?	21 or Younger	22 or Older	Totals
Yes	12	10	22
No	15	18	33
Totals	27	28	55

c. Victimization by race of respondent:

	Race		
Victimized?	Black	White	Totals
Yes	9	13	22
No	6	27	33
Totals	15	40	55

13.2 Compute a phi and a lambda for problems 12.1 to 12.4. Compare the value of the measure of association with your impressions of the strength of the relationships based solely on the percentages you calculated in Chapter 12.

13.3 SOC There is concern that suicides are motivated, in part, by imitation. Especially among young people, it may be that "epidemics" of self-destructive behaviors follow publication of suicides in local media. A number of cities have been classified by rate of suicide and by whether or not they experienced a publicized suicide within the past year. Is there an association between these two variables? Summarize your conclusions in a sentence or two.

	Publicized Suicide?		
Suicide Rate	Yes	No	Totals
Low	15	20	35
High	15	10	25
Totals	30	30	60

13.4 SW The director of a shelter for battered women has noticed that many of the women who are referred to the shelter eventually return to their violent husbands, even when there is every indication that the husband will continue the pattern of abuse. The director suspects that the women who return to their husbands do so because they have no place else to go—for example, no close relatives in the area with whom the women could reside. Do the following data support the director's suspicion? (Data are from the case files of former clients.)

	Relatives Nearby?		
Return to Husband?	Yes	No	Totals
Yes	10	23	33
No	50	17	67
Totals	60	40	100

13.5 PA Traditionally, bus ridership in your town has been confined to lower-income and blue-collar patrons. As head of transportation planning for the city, you believe that ridership from white-collar, middle-income neighborhoods can be increased if bus routes linking these neighborhoods to the downtown area (where most people work) are increased. A survey is conducted, and the results are displayed here. Is willingness to ride the bus related to job location? What is the pattern of the relationship (if any)?

	Job Location		
Potential Ridership	Downtown	Other	Totals
Would use bus	55	20	75
Would not use bus	15	21	36
Totals	70	41	111

13.6 GER A survey of senior citizens who live in either a housing development specifically designed for retirees or an age-integrated neighborhood has been conducted. Is type of living arrangement related to sense of social isolation?

	Living Arrangement		
Sense of Isolation	Housing Development	Integrated Neighborhood	Totals
Low	80	30	110
High	20	120	140
Totals	100	150	250

13.7 SOC Is there an association between the gender of college instructors and the teaching effectiveness ratings they receive from students? Write a few sentences summarizing your findings.

Teaching Effectiveness	Gender: Female	Gender: Male	Totals
High	115	241	356
Low	54	113	167
Totals	169	354	523

13.8 SOC A researcher has conducted a survey on sexual attitudes for a sample of 317 teenagers. The respondents were asked whether they considered premarital sex to be "always wrong" or "OK under certain circumstances." The following tables summarize the relationship between responses to this item and several other variables. For each table, assess the strength and pattern of the relationship and write a paragraph interpreting these results.

a. Attitudes toward premarital sex by gender:

Premarital Sex	Gender: Female	Gender: Male	Totals
Always wrong	90	105	195
Not always wrong	65	57	122
Totals	155	162	317

b. Attitudes toward premarital sex by courtship status:

Premarital Sex	Ever "Gone Steady": No	Ever "Gone Steady": Yes	Totals
Always wrong	148	47	195
Not always wrong	42	80	122
Totals	190	127	317

c. Attitudes toward premarital sex by social class:

Premarital Sex	Social Class: Blue Collar	Social Class: White Collar	Totals
Always wrong	72	123	195
Not always wrong	47	75	122
Totals	119	198	317

13.9 SOC St. Algebra College has a problem with attrition. A sizeable number of nongraduating students do not return to classes each semester. Is attrition importantly related to race, status, or age? For each of the following tables, how strong is the association (if any) between each of the independent variables and attrition? Does the relationship have a pattern? (Calculating percentages for the table will help to answer the second question.) Write a paragraph summarizing the results presented in these three tables.

a. Attrition by race for 532 students enrolled in fall semester:

Attrition	Race: White	Race: Black	Totals
Returned spring semester	280	100	380
Did not return	105	47	152
Totals	385	147	532

b. Attrition by status for 532 students enrolled in fall semester:

Attrition	Status: Part-time	Status: Full-time	Totals
Returned spring semester	42	338	380
Did not return	87	65	152
Totals	129	403	532

c. Attrition by age for 532 students enrolled in fall semester:

Attrition	Age: 18–24	Age: 25 and Older	Totals
Returned spring semester	307	73	380
Did not return	72	80	152
Totals	379	153	532

13.10 SOC A sociologist is researching public attitudes toward crime and has asked a sample of residents of his city if they think that the crime rate in their neighborhoods is rising. Is there a relationship between sex and perception of the crime rate? Between race and perception of the crime rate? What is the pattern of the relationship? Write a paragraph summarizing the information presented in these tables.

a. Perception of crime rate by sex:

Crime Rate Is	Gender: Male	Gender: Female	Totals
Rising	200	225	425
Stable	175	150	325
Falling	125	125	250
Totals	500	500	1000

b. Perception of crime rate by race:

Crime Rate Is	Race: White	Black	Totals
Rising	300	150	425
Stable	230	85	325
Falling	170	65	250
Totals	700	300	1000

13.11 PS You are running for mayor of Shinbone, Kansas, and realize that, if you are to win, you must win the support of blue-collar voters. (You already have strong support in the white-collar neighborhoods.) You have a very limited advertising budget and wonder how best to reach your intended audience. An aide has found the following data, which show the relationship between social class and "main source of news" for a sample of Shinboneites. Will this information help you make a decision?

Main Source of News	Social Class: Blue Collar	White Collar	Totals
Television	140	200	340
Radio	25	40	65
Newspapers	85	100	185
Totals	250	340	590

13.12 Problem 12.12 analyzed some bivariate relationships taken from a recent public opinion survey. Political ideology was used as the independent variable for five different dependent variables, and, using only percentages, you were asked to characterize the relationships in terms of strength and direction. Now, with the aid of measures of association, these characterizations should be easier to develop. Compute a Cramer's *V* and a lambda for each table in problem 12.12. Compare the measures of association with your characterizations based on the percentages.

Following are the same five dependent variables cross-tabulated against gender as an independent variable. Compare the strength of these relationships with those for political ideology. Which independent variable has the stronger associations?

a. Support for the legal right to an abortion by gender:

Right to Abortion?	Gender: Male	Female	Totals
Yes	310	418	728
No	432	618	1050
Totals	742	1036	1778

b. Support for capital punishment by gender:

Capital Punishment?	Gender: Male	Female	Totals
Favor	908	998	1906
Oppose	246	447	693
Totals	1154	1445	2559

c. Approve of suicide for people with incurable disease by gender:

Right to Suicide?	Gender: Male	Female	Totals
Yes	524	608	1132
No	246	398	644
Totals	770	1006	1776

d. Support for sex education in public schools by gender:

Sex Education?	Gender: Male	Female	Totals
Favor	685	900	1585
Oppose	102	134	236
Totals	787	1034	1821

e. Support for traditional gender roles by gender:

Women Should Take Care of Running Their Homes and Leave Running the Country to Men	Gender: Male	Female	Totals
Agree	116	164	280
Disagree	669	865	1534
Totals	785	1029	1814

SPSS for Windows

Using *SPSS for Windows* to Produce Nominal-Level Measures of Association

SPSS DEMONSTRATION 13.1 Does Support for Gun Control Vary by Political Ideology? Another Look

In Demonstration 12.2, we used **Crosstabs** to look for an association between *cappun* and recoded *polviews*. We saw that these variables were associated and that liberals were the least supportive and conservatives were the most supportive of the death penalty. The differences in conditional distributions were not great, and the relationship seemed to be, at best, moderate in strength. In this demonstration, we will reexamine the relationship and have SPSS compute some measures of association. The commands should repeat those in Demonstration 12.2 (remember to recode *polviews* into three categories). Click **Analyze,** then **Descriptive Statistics,** and then **Crosstabs.** Move *cappun* into the Row(s) box and recoded *polviews* into the Column(s) box. Click the **Cells** button and request column percentages. Click the **Statistics** button and request chi square, phi, Cramer's *V*, and lambda. Click **Continue** and **OK,** and the following output will be produced. (NOTE: The output has been slightly edited to improve readability).

FAVOR OR OPPOSE DEATH PENALTY FOR MURDER * Recoded polviews Crosstabulation

FAVOR OR OPPOSE DEATH PENALTY FOR MURDER		RECODED POLVIEWS			Total
		1.00	2.00	3.00	
FAVOR	Count	111	239	239	589
	%	49.1%	67.7%	77.9%	66.5%
OPPOSE	Count	115	114	68	297
	%	50.9%	32.3%	22.1%	33.5%
Total	Count	226	353	307	886
	%	100.0%	100.0%	100.0%	100.0%

Chi-Square Tests

	Value	df	Asymp. Sig. (2-sided)
Pearson Chi-Square	48.629(a)	2	.000
Likelihood Ratio	48.162	2	.000
Linear-by-Linear Association	46.893	1	.000
N of Valid Cases	886		

[a] 0 cells (.0%) have expected count less than 5. The minimum expected count is 75.76.

Directional Measures

			Value	Asymp. Std. Error(a)	Approx. T(b)	Approx. Sig.
Nominal by Nominal	Lambda	Symmetric	.006	.031	.191	.848
		FAVOR OR OPPOSE DEATH PENALTY FOR MURDER Dependent	.013	.050	.266	.790
		Recoded polviews Dependent	.002	.028	.066	.947

(continued next page)

Directional Measures (*continued*)

		Value	Asymp. Std. Error (a)	Approx. T(b)	Approx. Sig.
Goodman and Kruskal tau	FAVOR OR OPPOSE DEATH PENALTY FOR MURDER Dependent	.055	.016		.000
	Recoded polviews Dependent	.024	.007		.000(c)

[a]Not assuming the null hypothesis.
[b]Using the asymptotic standard error assuming the null hypothesis.
[c]Based on chi-square approximation.

Symmetric Measures

		Value	Approx. Sig.
Nominal by Nominal	Phi	.234	.000
	Cramer's *V*	.234	.000
N of Valid Cases		886	

[a]Not assuming the null hypothesis.
[b]Using the asymptotic standard error assuming the null hypothesis.

The measures of association are reported below the output for the chi square tests. Three values for lambda are reported in the Directional Measures output block. Remember that lambda is asymmetric and will change value depending on which variable is taken as dependent. In this case, *cappun* (FAVOR OR OPPOSE DEATH PENALTY FOR MURDER) is the dependent variable, so lambda is .013, a value that indicates no relationship between the variables. We saw previously, however, that the conditional distributions in the table do change, indicating that there actually is a relationship. The problem here is that the row totals are very unequal, and lambda is misleading and should be disregarded (see Section 13.6).

Phi and Cramer's *V* are reported in the Symmetric Measures output block. The statistics are identical in value (.234), as they will be whenever the table has either two rows or two columns. These measures indicate, as we suspected, a moderate association between the variables.

SSPS DEMONSTRATION 13.2 What Variables Affect Support for Capital Punishment?

Let's see if we can do any better in "explaining" support for capital punishment with some other commonly used independent variables. As you are no doubt aware, attitudes are often associated with various measures of social location. In keeping with our focus on the nominal level of measurement, let's investigate relationships between *cappun* and *sex, marital* (marital status), *racecen1,* and religious preference, or *relig.* Incorporating a number of potential independent variables, although quite common in social science research, generates a large volume of output, and we will abbreviate the actual output from *SPSS* in this demonstration.

	Percent in Favor	Significance of Chi Square	Phi* or *V*	Lambda
SEX				
Male	70.5%			
Female	63.0%	.017	.08	.00
RELIGION				
Protestant	70.7%			
Catholic	64.9%			
Jew	57.1%			
None	56.8%			
Other	57.1%	.018	.115	.000
RACE				
White	71.5%			
Black	48.8%			
Native Am.	57.1%			
Asian Am.	70.3%			
Hispanic	52.4%	.000	.192	.010
MARITAL				
Married	72.9%			
Widowed	63.1%			
Divorced	64.7%			
Separated	45.8%			
Never married	58.0%	.000	.150	.007

*Phi may be displayed as a minus value. If so, disregard the sign.

Basically, these results indicate that *cappun* is not strongly related to any of these demographic variables. Using phi or *V* as a guide, we can see that all four relationships are weak or, at best, moderate, with race having the strongest relationship. However, note the number of cases are small for some cells in this table (e.g., for Native Americans), so we should be cautious in interpreting these results. Since the rows are very unequal, lambda should be disregarded.

Exercises

13.1 See if you can find a variable with a strong association with *cappun.* If necessary, use the **Recode** command to reduce the number of categories in the column variable. As a suggestion, try *degree* and *income06* as independent variables.

13.2 Run some of the tables you did for Exercises 12.1 and 12.2 again, with a request for phi, Cramer's *V,* and lambda. Do the measures of association help you interpret the relationships?

14 Association Between Variables Measured at the Ordinal Level

LEARNING OBJECTIVES

By the end of this chapter, you will be able to

1. Calculate and interpret gamma and Spearman's rho.
2. Explain the logic of proportional reduction in error in terms of gamma.
3. Use gamma and Spearman's rho to analyze and describe a bivariate relationship in terms of the three questions introduced in Chapter 12.
4. Test gamma and Spearman's rho for significance.

14.1 INTRODUCTION

There are two common types of ordinal-level variables. Some variables have many possible scores and look, at least at first glance, like interval-ratio-level variables. We will call these *continuous ordinal variables.* An attitude scale that incorporated many different items and, therefore, had many possible values would produce this type of variable.

The second type, which we will call a *collapsed ordinal variable*, has only a few (no more than five or six) values or scores and can be created either by collecting data in collapsed form or by collapsing a continuous ordinal scale. For example, we would produce collapsed ordinal variables by measuring social class as upper, middle, or lower or by reducing the scores on an attitude scale into just a few categories (such as high, moderate, and low).

A number of measures of association have been invented for use with collapsed ordinal-level variables. Rather than attempt a comprehensive coverage of all of these statistics, we will concentrate on **gamma (*G*).** Other measures suitable for collapsed ordinal-level data (Somer's *d* and Kendall's tau-*b*) are covered at the Web site for this text. For "continuous" ordinal variables, a statistic called **Spearman's rho (r_s)** is commonly used, and we cover this measure of association toward the end of this chapter.

This chapter will expand your understanding of how bivariate associations can be described and analyzed, but it is important to remember that we are still trying to answer the three questions raised in Chapter 12: Are the variables associated? How strong is the association? What is the direction of the association?

14.2 PROPORTIONAL REDUCTION IN ERROR (PRE)

For nominal-level variables, the logic of PRE was based on two different "predictions" of the scores of cases on the dependent variable (Y): one that ignored the independent variable (X) and a second that took the independent variable into account. The value of lambda showed the extent to which taking the independent variable into account improved the accuracy when predicting the score of the dependent variable (see Section 13.4). The PRE logic for variables measured at the ordinal level is similar, and gamma, like lambda, measures the proportional reduction in error gained by predicting one variable while taking the other into account. The major difference lies in the way predictions are made.

In the case of gamma, we predict the order of pairs of cases rather than a score on the dependent variable. That is, we predict whether one case will have a higher or lower score than the other. First, we predict the order of a pair of cases on one variable while ignoring their order on the other. Second, we predict the order on one variable while taking order on the other variable into account.

As an illustration, assume that a researcher is concerned about the causes of "burnout" (that is, demoralization and loss of commitment) among elementary school teachers and wonders about the relationship between levels of burnout and years of service. One way to state the research question would be to ask if teachers with more years of service have higher levels of burnout. Another way to ask the same question is: Do teachers who *rank higher* on years of service also *rank higher* on burnout? If we knew that teacher A had more years of service than teacher B, would we be able to predict that teacher A is also more "burned out" than teacher B? That is, would knowledge of the order of this pair of cases on one variable help us predict their order on the other?

If the two variables are associated, we will reduce our errors when our predictions about one of the variables are based on knowledge of the other. Furthermore, the stronger the association, the fewer the errors we will make. When there is no association between the variables, gamma will be 0.00, and knowledge of the order of a pair of cases on one variable will not improve our ability to predict their order on the other. A gamma of ± 1.00 denotes a perfect relationship: The order of all pairs of cases on one variable would be predictable without error from their order on the other variable

With nominal-level variables, we analyzed the pattern of the relationship between the variables. That it, we looked to see which value on one variable (e.g., "male" on the variable gender) was associated with which value on the other variable (e.g., "tall" on the variable height). Recall that a defining characteristic of variables measured at the ordinal level is that the scores or values can be rank ordered from high to low or from more to less (see Chapter 1). This means that relationships between ordinal-level variables can have a direction as well as a pattern. In terms of the logic of gamma, the overall relationship between the variables is *positive* if cases tend to be ranked in the same order on both variables. For example, if a relationship is positive and Case A ranks above Case B on one variable, Case A would also be ranked above Case B on the second variable. The relationship suggested earlier between years of service and burnout is positive. In a *negative* relationship, the order of the cases would be reversed between the two variables. If Case A ranked above Case B on one variable, it would tend to rank below Case B on the second variable. If there is a negative relationship between prejudice and education and Case A was more educated than Case B (that is, ranked above Case B on education), then Case A would be less prejudiced (that is, would rank below Case B on prejudice).

14.3 THE COMPUTATION OF GAMMA

Table 14.1 summarizes the relationship between "length of service" and "burnout" for a fictitious sample of 100 teachers. To compute gamma, two sums are needed. First, we must find the number of pairs of cases that are ranked the same on both variables (we will label this N_s) and then the number of pairs of

Application 14.1

A group of 40 nations have been rated as high or low on religiosity (based on the percentage of a random sample of citizens that described themselves as "a religious person) and as high or low in their support for single mothers (based on the percentage of a random sample of citizens who said they would approve of a woman's choosing to be a single parent). Are more religious nations less approving of single mothers?

	Religiosity		
Approval	Low	High	Totals
Low	4	9	13
High	11	16	27
Totals	15	25	40

The number of pairs of cases ranked in the same order on both variables (N_s) would be

$$N_s = 4(16) = 64$$

The number of pairs of cases ranked in different order on both variables (N_d) would be

$$N_d = 9(11) = 99$$

Gamma is

$$G = \frac{N_s - N_d}{N_s + N_d} = \frac{64 - 99}{64 + 99} = \frac{-35}{163} = -0.22$$

A gamma of – 0.22 means that, when predicting the order of pairs of cases on the dependent variable (approval of single mothers), we would make 22% fewer errors by taking the independent variable (religiosity) into account. There is a moderate-to-weak, negative association between these two variables. As religiosity increases, approval decreases (that is, more religious nations are less approving of single mothers).

cases ranked differently on the variables **(N_d).** We find these sums by working with the cell frequencies.

To find the number of pairs of cases ranked the same (N_s), begin with the cell containing the cases that were ranked the lowest on both variables. In Table 14.1, this would be the upper-left-hand cell. *(NOTE: Not all tables are constructed with values increasing from left to right across the columns and from top to bottom across the rows. When using other tables, always be certain that you have located the proper cell.)* The 20 cases in the upper-left-hand cell all rank low on both burnout and length of service, and we will refer to these cases as "low-lows," or "LL's."

Now form a pair of cases by selecting one case from this cell and one from any other cell—for example, the middle cell in the table. All 15 cases in this cell are moderate on both variables and, following our earlier practice, can be labeled moderate-moderates, or MM's. Any pair of cases formed between these two cells will be ranked the same on both variables. That is, all LL's are lower

TABLE 14.1 BURNOUT BY LENGTH OF SERVICE (fictitious data)

	Length of Service			
Burnout	Low	Moderate	High	Totals
Low	20	6	4	30
Moderate	10	15	5	30
High	8	11	21	40
Totals	38	32	30	100

than all MM's on both variables (on X, low is less than moderate; on Y, low is less than moderate). The total number of pairs of cases is given by multiplying the cell frequencies. So the contribution of these two cells to the total N_s is (20)(15), or 300.

Gamma ignores all pairs of cases that are tied on either variable. For example, any pair of cases formed between the LL's and any other cell in the top row (low on burnout) or the left-hand column (low on length of service) will be tied on one variable. Also, any pair of cases formed within any cell will be tied on both X and Y. Gamma ignores all pairs of cases formed within the same row, column, or cell. Practically, this means that, in computing N_s, we will work with only the pairs of cases that can be formed between each cell and the cells below and to the right of it.

In summary: To find the total number of pairs of cases ranked the same on both variables (N_s), multiply the frequency in each cell by the total of all frequencies below and to the right of that cell. Repeat this procedure for each cell and add the resultant products. The total of these products is N_s. This procedure is displayed here for each cell in Table 14.1:

	Contribution to N_s
For LL's, 20(15 + 5 + 11 + 21)	= 1040
For ML's, 6(21 + 5)	= 156
For HL's, 4(0)	= 0
For LM's, 10(11 + 21)	= 320
For MM's, 15(21)	= 315
For HM's, 5(0)	= 0
For LH's, 8(0)	= 0
For MH's, 11(0)	= 0
For HH's, 21(0)	= 0
	N_s = 1831

FIGURE 14.1 COMPUTING N_s IN A 3 × 3 TABLE

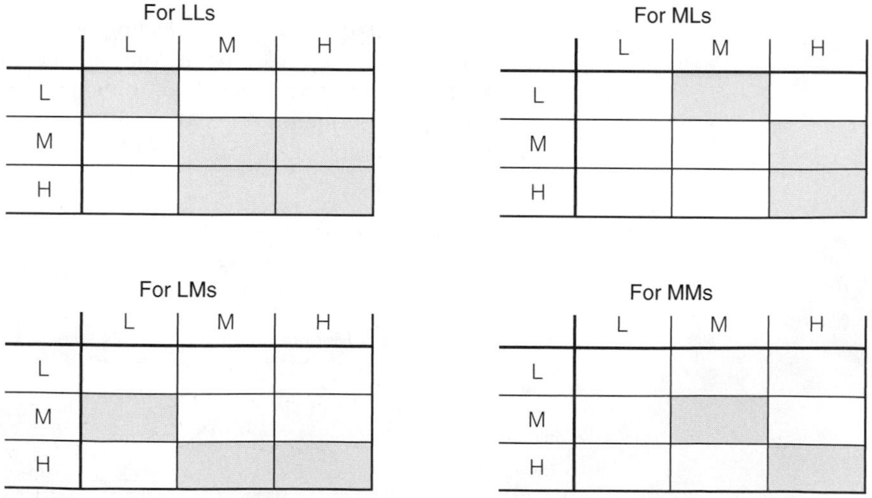

Note that none of the cells in the bottom row or the right-hand column of Table 14.1 can contribute to N_s because they have no cells below and to the right of them. Figure 14.1 shows the direction of multiplication for each of the four cells that, in a 3 × 3 table, can contribute to N_s. Computing N_s for Table 14.1, we find that a total of 1831 pairs of cases are ranked the same on both variables.

Our next step is to find the number of pairs of cases ranked differently (N_d) on both variables. To find the total number of pairs of cases ranked in different order on the variables, multiply the frequency in each cell by the total of all frequencies below and to the left of that cell. Note that the pattern for computing N_d is the reverse of the pattern for N_s. This time, we begin with the upper-right-hand cell (high-lows, or HL's) and multiply the number of cases in the cell by the total frequency of cases below and to the left. The four cases in the upper-right-hand cell of Table 14.1 are low on Y and high on X. If a pair of cases is formed with any case from this cell and any cell below and to the left, the cases will be ranked differently on the two variables. For example, if a pair is formed between any HL case and any case from the middle cell (moderate-moderates, or MM's), the HL case would be less than the MM case on Y ("low" is less than "moderate") but more than the MM case on X ("high" is greater than "moderate"). The computation of N_d is detailed here and shown graphically in Figure 14.2. In the computations, we have omitted cells that cannot contribute to N_d because they have no cells below and to the left of them.

FIGURE 14.2 COMPUTING N_d IN A 3 × 3 TABLE

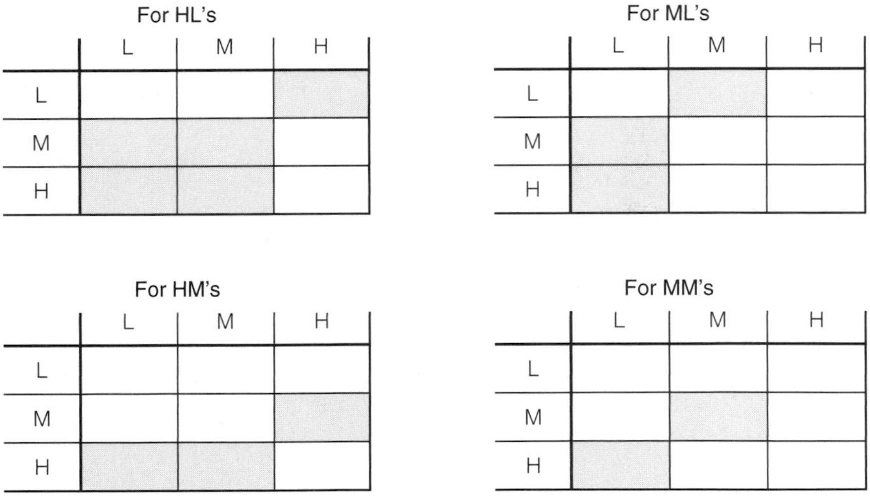

	Contribution to N_d
For HL's, 4(10 + 15 + 8 + 11)	= 176
For ML's, 6(10 + 8)	= 108
For HM's, 5(8 + 11)	= 95
For MM's, 15(8)	= 120
	N_d = 499

Application 14.2

For local political elections, do voters turn out at higher rates when the candidates for office spend more money on media advertising? Each of 177 localities has been rated as high or low on voter turnout. Also, the total advertising budgets for all candidates in each locality have been classified as high, moderate, or low. Are these variables associated?

Voter Turnout	Expenditure: Low	Moderate	High	Totals
Low	35	32	17	84
High	23	27	43	93
Totals	58	59	60	177

Because both variables are ordinal in level of measurement, we will compute a gamma to summarize the strength and direction of the association. The number of pairs of cases ranked in the same order on both variables (N_s) would be

$$N_s = 35(27 + 43) + 32(43) = 2450 + 1376 = 3826$$

The number of pairs of cases ranked in different order on both variables (N_d) would be

$$N_d = 17(27 + 23) + 32(23) = 850 + 736 = 1586$$

Gamma is

$$G = \frac{N_s - N_d}{N_s + N_d}$$

$$G = \frac{3826 - 1586}{3826 + 1586}$$

$$G = \frac{2240}{5412}$$

$$G = 0.41$$

A gamma of 0.41 means that, when predicting the order of pairs of cases on voter turnout, we would make 41% fewer errors by taking candidates' advertising expenditures into account, as opposed to ignoring the latter variable. There is a moderate, positive association between these two variables.

Table 14.1 has 499 pairs of cases ranked in different order and 1831 pairs of cases ranked in the same order. The formula for computing gamma is

FORMULA 14.1

$$G = \frac{N_s - N_d}{N_s + N_d}$$

where N_s = the number of pairs of cases ranked the same as both variables
N_d = the number of pairs of cases ranked differently on the two variables

For Table 14.1, the value of gamma would be

$$G = \frac{N_s - N_d}{N_s + N_d}$$

$$G = \frac{1831 - 499}{1831 + 499}$$

$$G = \frac{1332}{2330}$$

$$G = 0.57$$

A gamma of 0.57 indicates that we would make 57% fewer errors if we predicted the order of pairs of cases on one variable from the order of pairs of cases on the other (as opposed to predicting order while ignoring the other variable.) Length of service is associated with degree of burnout, and the relationship is positive. Knowing the respective rankings of two teachers on length of service

ONE STEP AT A TIME **Computing and Interpreting Gamma**

Computation

Step 1: Make sure the table is arranged with the column variable increasing from left to right and the row variable increasing from top to bottom.

Step 2: To compute N_s, start with the upper-left-hand cell. Multiply the number of cases in this cell by the total number of cases in all cells below and to the right. Repeat this process for each cell in the table. Add up these subtotals to find N_s.

Step 3: To compute N_d, start with the upper-right-hand cell. Multiply the number of cases in this cell by the total number of cases in all cells below and to the left. Repeat this process for each cell in the table. Add up these subtotals to find N_d.

Step 4: Subtract the value of N_d from N_s.

Step 5: Add the value of N_d to N_s.

Step 6: Divide the value you found in step 4 by the value you found in step 5. The resultant value is gamma.

Interpretation

Step 7: To interpret the *strength* of the relationship, always begin with the column percentages: The bigger the change in column percentages, the stronger the relationship.

Step 8: Next, you can use gamma to interpret the strength of the relationship in either or both of two ways:

- **a.** Use Table 14.2 to describe strength in general terms.
- **b.** To use the logic of proportional reduction in error, multiply gamma by 100. This value represents the percentage by which we improve our prediction of the dependent variable by taking the independent variable into account.

Step 9: To interpret the *direction* of the relationship, always begin by looking at the pattern of the column percentages. If the cases tend to fall in a diagonal from upper-left to lower-right, the relationship is positive. If the cases tend to fall in a diagonal from lower-left to upper-right, the relationship is negative.

Step 10: The sign of the gamma also tells the direction of the relationship. However, be very careful when interpreting direction with ordinal-level variables. Remember that coding schemes for these variables are arbitrary and that a positive gamma may mean that the actual relationship is negative, and vice versa.

(Case A is higher on length of service than Case B) will help us predict their ranking on burnout (we would predict that Case A will also be higher than Case B on burnout).

Table 14.2 provides some additional assistance for interpreting gamma in a format similar to Tables 12.5 and 13.3. As before, the relationship between the values and the descriptive terms are arbitrary and intended as general guidelines only. Note, in particular, that the strength of the relationship is independent of its direction. That is, a gamma of −0.35 is exactly as strong as a gamma of +0.35 but is opposite in direction.

TABLE 14.2 RELATIONSHIP BETWEEN THE VALUE OF GAMMA AND THE STRENGTH OF THE RELATIONSHIP

Value	Strength
If the value is	*The strength of the relationship is*
between 0.00 and 0.30	Weak
between 0.31 and 0.60	Moderate
greater than 0.60	Strong

To use the computational routine for gamma presented earlier, you must arrange the table in the manner of Table 14.1, with the column variable increasing in value as you move from left to right and the row variable increasing from top to bottom. Be careful to construct your tables according to this format; if you are working with data already in table format, you may have to rearrange the table or rethink the direction of patterns. Gamma is a symmetrical measure of association; that is, the value of gamma will be the same regardless of which variable is taken as independent. *(To practice computing and interpreting gamma, see problems 14.1 to 14.10 and 14.15. Begin with some of the smaller, 2 × 2 tables until you are comfortable with these procedures.)*

14.4 DETERMINING THE DIRECTION OF RELATIONSHIPS

Nominal measures of association like phi and lambda measure only the strength of a bivariate association. Ordinal measures of association, like gamma, are more sophisticated and add information about the overall direction of the relationship (positive or negative). In one way, it is easy to determine direction: If gamma is a plus value, the relationship is positive. A minus sign for gamma indicates a negative relationship. Often, however, direction is confusing when working with ordinal-level variables, so it will be helpful if we focus on the matter specifically. We discuss positive relationships first and then relationships in the negative direction.

With gamma, a positive relationship means that the *scores* of cases tend to be ranked in the same order on both variables. In more general terms, a positive relationship means that the variables change in the same direction. That is, as scores on one variable increase (or decrease), scores on the other variable also increase (or decrease). Cases tend to have scores in the same range on both variables (i.e., low scores go with low scores, moderate with moderate, and so forth). Table 14.3 illustrates the general shape of a positive relationship. In a positive relationship, cases tend to fall along a diagonal from upper left to lower right (assuming, of course, that tables have been constructed with the column variable increasing from left to right and the row variable from top to bottom).

Tables 14.4 and 14.5 present an example of a positive relationship with actual data. The sample consists of 186 preindustrial societies from around the globe. Each has been rated in terms of its degree of stratification or inequality and the type of political institution it has. In the societies that are the lowest in stratification, there are virtually no differences between people in terms of wealth and power, and the degree of inequality in the society increases from left

TABLE 14.3 A GENERALIZED POSITIVE RELATIONSHIP

	Variable *X*		
Variable *Y*	Low	Moderate	High
Low	X		
Moderate		X	
High			X

TABLE 14.4 STATE STRUCTURE BY DEGREE OF STRATIFICATION* (frequencies)

	Degree of Stratification		
Type of State	Low	Medium	High
Stateless	77	5	0
Semi-state	28	15	4
State	12	19	26
Totals	117	39	30

*Data are from the Human Relation Area File, standard cross-cultural sample.

TABLE 14.5 STATE STRUCTURE BY DEGREE OF STRATIFICATION* (percentages)

Type of State	Degree of Stratification		
	Low	Medium	High
Stateless	66%	13%	0%
Semi-state	24%	38%	13%
State	10%	49%	87%
Totals	100%	100%	100%

TABLE 14.6 A GENERALIZED NEGATIVE RELATIONSHIP

	Variable X		
Variable *Y*	Low	Moderate	High
Low			X
Moderate		X	
High	X		

to right across the columns of Table 14.4. In "stateless" societies there is no formal political institution and the political institution becomes more elaborate and stronger as you move down the rows from top to bottom. The gamma for this table is 0.86, so the relationship is strong and positive. By inspection, we can see that most cases fall in the diagonal from upper left to lower right.

The percentages in Table 14.5 make it even clearer that societies with little inequality tend to be stateless and that the political institution becomes more elaborate as inequality increases. The great majority of the least stratified societies had no political institution, and none of the highly stratified societies were stateless.

Negative relationships are the opposite of positive relationships. Low scores on one variable are associated with high scores on the other and high scores with low scores. This pattern means that the cases will tend to fall along a diagonal from lower left to upper right (at least for all tables in this text). Table 14.6 illustrates a generalized negative relationship. The cases with higher scores on variable *X* tend to have lower scores on variable *Y*, and score on *Y* decreases as score on *X* increases.

Tables 14.7 and 14.8 present an example of a negative relationship, with data taken from a recent public-opinion poll. The independent variable is church attendance, and the dependent variable is approval of cohabitation ("Is it alright for a couple to live together without intending to get married?"). Note that rates of attendance increase from left to right, and approval of cohabitation increases from top to bottom of the table.

Once again, the percentages in Table 14.8 make the pattern more obvious. The great majority of people who were low on attendance ("never") were high on approval of cohabitation, and most people who were high on attendance

TABLE 14.7 APPROVAL OF COHABITATION BY CHURCH ATTENDANCE (frequencies)

	Attendance		
Approval	Never	Monthly or Yearly	Weekly
Low	37	186	195
Moderate	25	126	46
High	156	324	52
Totals	218	636	293

TABLE 14.8 APPROVAL OF COHABITATION BY CHURCH ATTENDANCE (percentages)

	Attendance		
Approval	Never	Monthly or Yearly	Weekly
Low	17.0	29.2	66.6
Moderate	11.5	19.8	15.7
High	71.6	50.9	17.8
Totals	100.1%	99.9%	100.1%

were low on approval. As attendance increases, approval of cohabitation tends to decrease. The gamma for this table is −0.57, indicating a strong, negative relationship between attendance and approval of this living arrangement.

You should be aware of an additional complication. The coding for ordinal-level variables is arbitrary, and a higher score may mean "more" or "less" of the variable being measured. For example, if we measured social class as upper, middle, or lower, we could assign scores to the categories in either of two ways:

A	B
(1) Upper	(3) Upper
(2) Middle	(2) Middle
(3) Lower	(1) Lower

While coding scheme B might seem preferable (because higher scores go with higher class position), *both* schemes are perfectly legitimate, and the direction of a relationship will change depending on which scheme is selected. Using scheme B, we would find positive relationships between social class and education: As education increased, so would class.

With scheme A, however, the same relationship would appear to be negative because the numerical scores (1, 2, 3) are coded in reverse order: The highest social class is assigned the lowest score, and so forth. If you didn't check the coding scheme, you might conclude that the negative gamma means that class decreases as education increases, when, actually, the opposite is true.

Unfortunately, this source of confusion cannot be avoided when working with ordinal-level variables. Coding schemes will always be arbitrary for these variables, so you need to exercise additional caution when interpreting the direction of ordinal-level variables.

14.5 INTERPRETING ASSOCIATION WITH BIVARIATE TABLES: WHAT ARE THE SOURCES OF CIVIC ENGAGEMENT IN U.S. SOCIETY?

Up to this point, we have discussed the association between two variables using tables, column percentages, and both nominal and ordinal measures of association. In this section, we summarize this material by using these techniques to address an issue in U.S. social life about which many people are concerned. Specifically, social scientists, politicians, ministers, newspaper editorialists, and others have complained that Americans are "dropping out" of community life and minimizing their involvement with other people.[1] They worry that the bonds of friendship and neighborliness are withering away and that Americans are deserting the groups and associations (like PTAs, Little League, and Girl Scouts) that kept our communities functioning in the past. This is a complex, multifaceted issue, and we cannot hope to resolve it in these few paragraphs. We can, however, use data from the 2006 General Social Survey to examine the correlates of involvement with others in the community. What kind of person is most heavily engaged in social life?

To measure involvement in social life (the dependent variable), I used a series of variables that asked people about how often they spent an evening with friends, relatives, or neighbors or at a bar. I created a composite variable that rated people as high, moderate, or low on their involvement across all these different social activities. This simple variable does not measure all possible forms

[1] See, for example, Putnam, Robert D. 2000. *Bowling Alone*. New York: Simon & Schuster.

of social involvement, so we will not be able to come to any final conclusions on the issues. Like almost all variables of interest to the social sciences, involvement in social life is complex and subtle and cannot be adequately captured with a single measurement. Nonetheless, we may be able to develop some insight into the issue, even if the variable is not perfectly valid. Researchers typically have to settle for partial or incomplete measurement of their concepts.

What should we use as independent variables? In other words, what factors might have causal relationships with participation in social life? Here, we will investigate three possible relationships. First, some have suggested that married people are more likely to be involved with others socially. Second, it may be that the quality of social life of a community depends on people who are not fully engaged in the paid workforce (e.g., housewives or "stay-at-home" moms). We'll use work status (working full time vs. not working full time) as an independent variable for this argument. Finally, some have blamed television for a decline in sociality: Do people choose to amuse themselves with the fictional lives of televised characters rather than get involved with the real lives around them? Which of these arguments has support? How strong are the relationships between these variables and social involvement, and what is the pattern or direction of the relationships?

Take a moment to consider the level of measurement of each of these variables. The dependent variable (social involvement) is ordinal, since the categories can be distinguished in terms of "more or less." Television viewing, measured as low, medium, or high for this exercise, is also an ordinal-level variable. Marital status, on the other hand, is clearly nominal, and work status might be considered either nominal (if the categories are just different) or ordinal (if "full time" is regarded as working *more than* not "full time"). Ideally, we should use gamma for ordinal variables and lambda or Cramer's V for the nominal variables, but this creates problems of comparability: If we used different measures, how could we tell which relationship was strongest? We can resolve this problem in two ways. First, we can use column percentages to make comparisons from table to table. Second, we can ignore level of measurement and compute both ordinal and nominal measures of association for each table.

Table 14.9 shows that the relationship between marital status and social involvement is significant ($p < .05$) but not particularly strong. Married people are actually *less* likely to be highly involved than unmarried people, and the maximum difference (about 15%) and Cramer's V indicate a moderate relationship. These results are contrary to the idea that married people are more likely to be highly involved socially.

TABLE 14.9 VOLUNTEERING BY MARITAL STATUS

	Marital Status		
Involvement	Married	Not Married	Totals
Low	385 (*40.1%*)	298 (*29.0%*)	683 (*34.4%*)
Moderate	287 (*29.9%*)	263 (*25.6%*)	550 (*27.7%*)
High	287 (*29.9%*)	468 (*45.5%*)	755 (*38.0%*)
Totals	959 (*99.9%*)	1,029 (*100.1%*)	1,988 (*100.1%*)

$\chi^2 = 53.12$ (df = 2, $p < 0.000$); $V = 0.16$; gamma = 0.26

The relationship between involvement and work status (Table 14.10) is also statistically significant, but the differences in the column percentages are small, and the measures of association are low in value. The pattern of the relationship is that people employed full time are more involved socially.

Table 14.11 shows the relationship between social involvement and extent of television watching. The relationship is significant, but once again the relationship is weak. Viewers who watch little TV are more likely to be highly involved (42%) than viewers who watch a lot (35%). As indicated by gamma, this is a negative relationship: As TV viewing increases, involvement decreases. The relationship is weak, which means that factors other than TV viewing are important causes of involvement. (Note also that these results cannot show which variable is cause and which is effect. Do viewers who watch a lot of TV avoid involvement with others? Or do people who choose not to have an active social life become viewers who watch a lot? Remember that correlation is not the same thing as causation.)

In summary, we can say that unmarried people, full-time workers, and viewers who watch little TV tend to be more actively involved in the social life of their communities (but not by much). Altogether, the relative weakness of these relationships might send us on a search for other variables that explain better the various levels of involvement in community social life.

14.6 SPEARMAN'S RHO (r_s)

To this point, we have considered ordinal variables that have a limited number of categories (possible values) and are presented in tables. However, many ordinal-level variables have a broad range of scores and many distinct values. Such

TABLE 14.10 VOLUNTEERING BY WORK STATUS

	Working Full Time?		
Involvement	Yes	No	Totals
Low	301 (*30.6%*)	383 (*36.3%*)	684 (*34.4%*)
Moderate	285 (*29.0%*)	265 (*25.1%*)	550 (*27.7%*)
High	398 (*40.4%*)	357 (*33.8%*)	755 (*38.0%*)
Totals	984 (*100.0%*)	1,055 (*100.0%*)	1,989 (*100.1%*)

$\chi^2 = 12.56$ (df = 2, $p < 0.002$); $V = 0.08$; gamma = −0.12

TABLE 14.11 INVOLVEMENT BY LEVEL OF TV WATCHING

	TV Watching per Day			
Involvement	Low (0–1 hour)	Moderate (2–3 hours)	High (4 or more hours)	Totals
Low	160 (*32.1%*)	283 (*31.0%*)	241 (*41.8%*)	684 (*34.4%*)
Moderate	128 (*25.7%*)	285 (*31.2%*)	137 (*23.7%*)	550 (*27.7%*)
High	211 (*42.3%*)	345 (*37.8%*)	199 (*34.5%*)	755 (*38.0%*)
Totals	499 (*100.0%*)	913 (*100.0%*)	577 (*100.0%*)	1,989 (*100.00%*)

$\chi^2 = 25.38$ (df = 4, $p < 0.000$); $V = 0.08$; gamma = −0.11

data may be collapsed into a few broad categories (such as high, moderate, and low), organized into a bivariate table, and analyzed with gamma. Collapsing scores in this manner may be beneficial and desirable in many instances, but some important distinctions between cases may be obscured or lost as a consequence.

For example, suppose a researcher wished to test the claim that jogging is beneficial, not only physically but also psychologically. Do joggers have an enhanced sense of self-esteem? To deal with this issue, 10 female joggers are evaluated on two scales, the first measuring involvement in jogging and the other measuring self-esteem. Scores are reported in Table 14.12.

These data could be collapsed and a bivariate table produced. We could, for example, dichotomize both variables to create only two values (high and low) for both variables. Although collapsing scores in this way is certainly legitimate and often necessary,[2] two difficulties with this practice must be noted. First, the scores seem continuous, and there are no obvious or natural division points in the distribution that would allow us to distinguish, in a nonarbitrary fashion, between high scores and low ones. Second, and more important, grouping these cases into broader categories will lose information. That is, if both Wendy and Debbie are placed in the category "high" on involvement, the fact that they had different scores on the variable would be obscured. If differences like this are important and meaningful, then we should opt for a measure of association that permits the retention of as much detail and precision in the scores as possible.

Spearman's rho (r_s) is a measure of association for ordinal-level variables that have a broad range of many different scores and few ties between cases on either variable. Scores on ordinal-level variables cannot, of course, be manipulated mathematically except for judgments of "greater than" or "less than." To compute Spearman's rho, first the cases are ranked from high to low on each

TABLE 14.12 THE SCORES OF 10 SUBJECTS ON INVOLVEMENT IN JOGGING AND A MEASURE OF SELF-ESTEEM

Jogger	Involvement in Jogging (X)	Self-esteem (Y)
Wendy	18	15
Debbie	17	18
Phyllis	15	12
Stacey	12	16
Evelyn	10	6
Tricia	9	10
Christy	8	8
Patsy	8	7
Marsha	5	5
Lynn	1	2

[2]For example, collapsing scores may be advisable when the researcher is not sure that fine distinctions between scores are meaningful.

Application 14.3

Five cities have been rated on an index that measures the quality of life. Also, the percentage of the population that has moved into each city over the past year has been determined. Have cities with higher quality-of-life scores attracted more new residents? The following table below summarizes the scores, ranks, and differences in ranks for each of the five cities.

City	Quality of Life	Rank	% New Residents	Rank	D	D^2
A	30	1	17	1	0	0
B	25	2	14	3	−1	1
C	20	3	15	2	−1	1
D	10	4	3	5	−1	1
E	2	5	5	4	−1	1
					$\sum D = 0$	$\sum D^2 = 4$

Spearman's rho for these variables is

$$r_s = 1 - \frac{6\sum D^2}{N(N^2 - 1)}$$

$$r_s = 1 - \frac{(6)(4)}{5(25 - 1)}$$

$$r_s = 1 - \left(\frac{24}{120}\right)$$

$$r_s = 1 - 0.20$$

$$r_s = 0.80$$

These variables have a strong, positive association. The higher the quality-of-life score, the greater the percentage of new residents. The value of r_s^2 is 0.64 ($0.80^2 = 0.64$), which indicates that we will make 64% fewer errors when predicting rank on one variable from rank on the other, as opposed to ignoring rank on the other variable.

TABLE 14.13 COMPUTING SPEARMAN'S RHO

	Involvement (X)	Rank	Self-Image (Y)	Rank	D	D^2
Wendy	18	1	15	3	−2	4
Debbie	17	2	18	1	1	1
Phyllis	15	3	12	4	−1	1
Stacey	12	4	16	2	2	4
Evelyn	10	5	6	8	−3	9
Tricia	9	6	10	5	1	1
Christy	8	7.5	8	6	1.5	2.25
Patsy	8	7.5	7	7	0.5	.25
Marsha	5	9	5	9	0	0
Lynn	1	10	2	10	0	0
					$\sum D = 0$	$\sum D^2 = 22.50$

variable and then the ranks (not the scores) are manipulated to produce the final measure. Table 14.13 displays the original scores and the rankings of the cases on both variables.

To rank the cases, first find the highest score on each variable and assign it rank 1. Wendy has the high score on X (18) and is thus ranked number 1. Debbie, on the other hand, is highest on *Y* and is ranked first on that variable. All other cases are then ranked in descending order of scores. If any cases have the same score on a variable, assign them the average of the ranks they would have used up had they not been tied. Christy and Patsy have identical scores of 8 on involvement. Had they not been tied, they would have used up ranks 7 and 8. The average of these two ranks is 7.5, and this average of used ranks is assigned to all tied cases. (For example, if Marsha had also had a score of 8, three ranks—7, 8, and 9—would have been used, and all three tied cases would have been ranked eighth.)

The formula for Spearman's rho is

FORMULA 14.2

$$r_s = 1 - \frac{6\sum D^2}{N(N^2 - 1)}$$

where $\sum D^2$ = the sum of the differences in ranks, the quantity squared

To compute $\sum D^2$, the rank of each case on Y is subtracted from its rank on X (*D* is the difference between rank on *Y* and rank on *X*). A column has been provided in Table 14.10 so that these differences may be recorded on a case-by-case basis. Note that the sum of this column ($\sum D$) is 0. That is, the negative differences in rank are equal to the positive differences, as will always be the case, and you should find the total of this column as a check on your computations to this point. If $\sum D$ is not equal to 0, you have made a mistake either in ranking the cases or in subtracting the differences.

In the column headed D^2, each difference is squared to eliminate negative signs. The sum of this column is $\sum D^2$, and this quantity is entered directly into the formula. For our sample problem:

$$r_s = 1 - \frac{6\sum D^2}{N(N^2 - 1)}$$

$$r_s = 1 - \frac{6(22.5)}{10(100 - 1)}$$

$$r_s = 1 - \frac{135}{990}$$

$$r_s = 1 - 0.14$$

$$r_s = 0.86$$

Spearman's rho is an index of the strength of association between the variables; it ranges from 0 (no association) to ±1.00 (perfect association). A perfect positive association ($r_s = +1.00$) would exist if there were no disagreements in ranks between the two variables (if cases were ranked in exactly the same order on both variables). A perfect negative relationship ($r_s = -1.00$) would exist if the ranks were in perfect disagreement (if the case ranked highest on one variable were lowest on the other, and so forth). A Spearman's rho of 0.86 indicates a strong, positive relationship between these two variables. The respondents who were highly involved in jogging also ranked high on self-image. These results are supportive of claims regarding the psychological benefits of jogging.

Since Spearman's rho is an index of the relative strength of a relationship, values between 0 and ±1.00 have no direct interpretation. However, if the value

ONE STEP AT A TIME Computing and Interpreting Spearman's Rho

Computation

Step 1: Set up a computing table like Table 14.13 to help organize the computations. In the far left-hand column, list the cases in order, with the case with the highest score on the independent variable (X) stated first.

Step 2: In the next column, list the scores on X.

Step 3: In the third column, list the rank of each case on X, beginning with rank 1 for the highest score. If any cases have the same score, assign them the average of the ranks they would have used up had they not been tied.

Step 4: In the fourth column, list the score of each case on Y, and then, in the fifth column, rank the cases on Y from high to low. Start by assigning the rank of 1 to the case with the highest score on Y, and assign any tied cases the average of the ranks they would have used up had they not been tied.

Step 5: For each case, subtract the rank on Y from the rank on X and write the difference (D) in the sixth column. Add up this column. If the sum is not zero, you have made a mistake and need to recompute.

Step 6: Square the value of each D and record the result in the seventh column. Add this column up to find ΣD^2, and substitute this value into the numerator of Formula 14.2:

$$r_s = 1 - \frac{6\Sigma D^2}{N(N^2 - 1)}$$

Step 7: Multiply ΣD^2 (the total of column 7 in the computing table) by 6.

Step 8: Square N and subtract 1 from the result.

Step 9: Multiply the quantity you found in step 8 by N.

Step 10: Divide the quantity you found in step 7 by the quantity you found in step 9.

Step 11: Subtract the quantity you found in step 10 from 1. The result is r_s.

Interpretation

Step 12: To interpret the *strength* of Spearman's rho, you can do either of the following:

- **a.** Use Table 14.2 to describe strength in general terms.
- **b.** Square the value of rho and multiply by 100. This value represents the percentage by which we improve our prediction of the dependent variable by taking the independent variable into account.

Step 13: To interpret the *direction* of the relationship, look at the sign of r_s. However, be careful when interpreting direction with ordinal-level variables. Remember that coding schemes for these variables are arbitrary and that a positive r_s may mean that the actual relationship is negative, and vice versa.

of rho is squared, a PRE interpretation is possible. Rho squared (r_s^2) represents the proportional reduction in errors of prediction when predicting rank on one variable from rank on the other variable, as compared to predicting rank while ignoring the other variable. In our example, r_s was 0.86 and r_s^2 would be 0.74. Thus, our errors of prediction would be reduced by 74% if we used the rank on jogging to help predict rank on self-image. *(For practice in computing and interpreting Spearman's rho, see problems 14.11 to 14.14. Problem 14.11 has the fewest number of cases and is probably a good choice for a first attempt at these procedures.)*

14.7 TESTING THE NULL HYPOTHESIS OF "NO ASSOCIATION" WITH GAMMA AND SPEARMAN'S RHO

Whenever a researcher is working with EPSEM, or random, samples, he or she will need to ascertain if the sample findings can be generalized to the population. In Part II of this text, we considered various ways that information taken from samples—for example, the difference between two sample means—could be generalized to the populations from which the samples were drawn. A test of the null hypothesis, regardless of the form or specific test used, asks essentially if the patterns (or differences or relationships) that have been observed in the samples can be assumed to exist in the population. Measures of association can also be tested for significance. When data have been collected from a random sample, we will not only need to measure the existence, strength, and direction of the association, but we will also want to know if we can assume that the variables are related in the population.

For nominal-level variables, the statistical significance of a relationship is usually judged by the chi square test. Chi square tests could also be conducted on tables displaying the relationship between ordinal-level variables. However, chi square tests deal with the probability that the observed cell frequencies occurred by chance alone and are therefore not a direct test of the significance of the measure of association (gamma or Spearman's rho) itself.

When testing gamma and Spearman's rho for statistical significance, the null hypothesis will state that there is no association between the variables in the population and that, therefore, the population value for the measure is 0.00. Population values will be denoted by the Greek letters gamma (γ) and rho (ρ_s). For both measures, the test procedures will be organized around the familiar five-step model (see Chapter 8).

Testing Gamma for Significance. To illustrate the test of significance for gamma, we will use Table 14.1, where gamma was 0.57.

Step 1. Making Assumptions and Meeting Test Requirements. When sample size is greater than 10, the sampling distribution of all possible sample gammas can be assumed to be normal in shape.

Model: Random sampling
Level of measurement is ordinal
Sampling distribution is normal

Step 2. Stating the Null Hypothesis.

$$H_0: \gamma = 0.0$$
$$(H_1: \gamma \pm 0.0)$$

Step 3. Selecting the Sampling Distribution and Establishing the Critical Region. For samples of 10 or more, the Z distribution (Appendix A) can be used to find areas under the sampling distribution:

Sampling distribution = Z distribution
Alpha = .05
Z(critical) = ±1.96

Step 4. Computing the Test Statistic.

FORMULA 14.3

$$Z(\text{obtained}) = G\sqrt{\frac{N_s + N_d}{N(1 - G^2)}}$$

$$Z(\text{obtained}) = G\sqrt{\frac{N_s + N_d}{N(1 - G^2)}} = 0.57\sqrt{\frac{1831 - 499}{100(1 - 0.33)}} = 0.57\sqrt{\frac{2330}{100(0.67)}}$$

$$Z(\text{obtained}) = 0.57\sqrt{34.78} = 3.36$$

Step 5. Making a Decision and Interpreting the Results of the Test. Comparing the Z (obtained) with the Z (critical), we get:

$$Z(\text{obtained}) = 3.36$$
$$Z(\text{critical}) = \pm 1.96$$

We see that the null hypothesis can be rejected. The sample gamma is unlikely to have occurred by chance alone, and we may conclude that these variables are related in the population from which the sample was drawn. *(For practice in conducting and interpreting the test of significance for gamma, see problems 14.2, 14.4, 14.7, 14.10, and 14.15.)*

Testing Spearman's Rho for Significance. When testing Spearman's rho, the null hypothesis states that the population value (ρ_s) is actually 0 and, therefore, the value of the sample Spearman's rho (r_s) is the result of mere random chance. When the number of cases in the sample is 10 or more, the sampling distribution of Spearman's rho approximates the t distribution, and we will use this distribution to conduct the test. To illustrate, the Spearman's rho computed in Section 14.6 will be used.

Step 1. Making Assumptions and Meeting Test Requirements.

Model: Random sampling
Level of measurement is ordinal
Sampling distribution is normal

Step 2. Stating the Null Hypothesis.

$$H_0: \rho_s = 0.0$$
$$(H_1: \rho_s \pm 0.0)$$

Step 3. Selecting the Sampling Distribution and Establishing the Critical Region.

Sampling distribution = t distribution
Alpha = .05
Degrees of freedom = $N - 2 = 8$
t(critical) = ± 2.306

READING STATISTICS 11: Bivariate Tables and Associated Statistics

The statistics associated with any bivariate table will usually be reported directly below the table itself. This information would include the name of the measure of association, its value, and (if relevant) its sign. Also, if the research involves a random sample, the results of the chi square test or a test for the significance of the measure itself (see Section 14.7) will be reported. So the information might look something like this:

$$\lambda = 0.47$$
$$\chi^2 = 13.23 \text{ (2 df, } p < 0.05)$$

Note that the alpha level is reported in the "$p <$" format, which we discussed in Reading Statistics 5.

Besides simply reporting the value of these statistics, the researcher will interpret them in the text of the article. The preceding statistics might be characterized by the statement "The association between the variables is statistically significant and strong." Again we see that researchers will avoid our rather wordy (but more careful) style of stating results. Where we might say, "A lambda of 0.47 indicates that we will reduce our errors of prediction by 47% when predicting the dependent variable from the independent variable, as opposed to predicting the dependent while ignoring the independent," the researcher will simply report the value of lambda and characterize that value in a word or two (e.g., "The association is strong"). The researcher assumes that his or her audience is statistically literate and can make the more detailed interpretations for themselves.

Statistics in the Professional Literature

While gender inequality seems to be a nearly universal characteristic of human society, the extent of male advantage is highly variable. Why are some societies more unequal than others? Sociologists Stephen Sanderson, Alex Heckert, and Joshua Dubrow sought some answers to this question by examining the correlates of gender inequality in a sample of preindustrial societies. They tested a number of hypotheses but found the most support for what they called "materialist" theories, which argued that "the greater the extent to which women are involved in economic production, the higher their status tends to be" (p. 1426). Table 1 reports their findings for the hypotheses derived from this theory. The entries in the table are gammas, and the asterisks indicate the statistical significance of the relationship based on a chi square test.

To read this table, note that the three variables in the columns measure the dependent variable, or the status of women. The four row variables measure the independent variable, or the importance of women for various economic activities. The entries in the table are gammas. The first three row variables are self-explanatory, and the higher the score of the society on these variables, the greater the economic importance of women. The variable in the bottom row measures the extent to which the society depends on the plow as an instrument of production, and its meaning may not be so obvious. Since it requires a good deal of upper-body strength, plow agriculture is generally "men's work," and the greater

TABLE 1

Measures Related to Materialist Theories (Independent Variables)	**Measures of Women's Status (Dependent Variables)**		
	Domestic Authority	Control of Sexuality	Female Solidarity
Female contribution to gathering	0.23	0.38*	0.45***
Female contribution to subsistence	0.05	0.13	−0.01
Female contribution to agriculture	0.46***	0.12	0.57***
Use of the plow	−0.51**	−0.63**	−0.58***

N's range from 67 to 93.
* $p < .05$; ** $p <$ >01; *** $p < .001$.

(*continued next page*)

READING STATISTICS 11: *(continued)*

the extent to which the society uses plows, the lower the economic importance of women.

Is the materialist theory supported? The relationships between the first three measures of women's economic importance and women's status (the three column variables) are positive in direction (with one exception) and weak to moderate in strength. The positive relationships indicate that women's status increases as their economic importance increases, and the strength and significance of the relationships provide moderate (but not overwhelming) support for the theory being tested. Looking at the bottom row, we see somewhat stronger support for the materialist theory. All three relationships are statistically significant at less than the .05 level, moderate to strong, and negative—the direction predicted by the materialist theory ("The greater the reliance on the plow, the lower the status of women").

Although I have not reproduced them here, the tests of competing theories showed weaker and less significant relationships, and the authors conclude that the materialist perspective is most consistent with the empirical relationships they were able to examine. Although these results cannot be considered proof (remember that correlation is not the same thing as causation), they are very consistent with the predictions of materialist theory.

Sanderson, Stephen, Heckert, Alex, and Dubrow, Joshua. 2005. "Militarist, Marxian, and Non-Marxian Theories of Gender Inequality: A Cross-Cultural Test." *Social Forces,* 83: 1425–1441.

Step 4. Computing the Test Statistic.

FORMULA 14.4

$$t(\text{obtained}) = r_s\sqrt{\frac{N-2}{1-r_s^2}}$$

$$t(\text{obtained}) = r_s\sqrt{\frac{N-2}{1-r_s^2}} = 0.86\sqrt{\frac{8}{1-0.74}} = 0.86\sqrt{\frac{8}{0.26}}$$

$$t(\text{obtained}) = 0.86\sqrt{30.77} = (0.86)(5.55) = 4.77$$

Step 5. Making a Decision and Interpreting the Results of the Test. Comparing the test statistic with the critical region:

$$t(\text{obtained}) = 4.77$$

$$t(\text{critical}) = \pm 2.306$$

We see that the null hypothesis can be rejected. We may conclude, with a .05 chance of making an error, that the variables are related in the population from which the samples were drawn. *(For practice in conducting and interpreting the test of significance for Spearman's rho, see problems 14.11 to 14.14.)*

SUMMARY

1. A measure of association for variables with collapsed ordinal scales (gamma) was covered along with a measure (Spearman's rho) appropriate for "continuous" ordinal variables. Both measures summarize the overall strength and direction of the association between the variables.
2. Gamma is a PRE-based measure that shows the improvement in our ability to predict the order of pairs of cases on one variable from the order of pairs of cases on the other variable, as opposed to ignoring the order of the pairs of cases on the other variable.

3. Spearman's rho is computed from the ranks of the scores of the cases on two "continuous" ordinal variables and, when squared, can be interpreted by the logic of PRE.
4. Both gamma and Spearman's rho should be tested for their statistical significance when computed for a random sample drawn from a defined population. The null hypothesis is that the variables are not related in the population, and the test can be organized by using the familiar five-step model.

SUMMARY OF FORMULAS

Gamma	14.1	$G = \frac{N_s - N_d}{N_s + N_d}$
Spearman's rho	14.2	$r_s = 1 - \frac{6\sum D^2}{N(N^2 - 1)}$
Z(obtained) for gamma	14.3	$Z(\text{obtained}) = G\sqrt{\frac{N_s + N_d}{N(1 - G^2)}}$
t(obtained) for Spearman's rho	14.4	$t(\text{obtained}) = r_s\sqrt{\frac{N - 2}{1 - r_s^2}}$

GLOSSARY

Gamma (G). A measure of association appropriate for variables measured with "collapsed" ordinal scales that have been organized into table format; G is the symbol for any sample gamma, γ is the symbol for any population gamma.

N_d. The number of pairs of cases ranked in different order on two variables.

N_s. The number of pairs of cases ranked in the same order on two variables.

Spearman's rho (r_s). A measure of association appropriate for ordinally measured variables that are "continuous" in form; r_s is the symbol for any sample Spearman's rho; ρ_s is the symbol for any population Spearman's rho.

PROBLEMS

For problems 14.1 to 14.10 and 14.15 calculate percentages for the bivariate tables as described in Chapter 12. Use the percentages to help analyze the strength and direction of the association.

14.1 SOC A small sample of non-English-speaking immigrants to the United States has been interviewed about their level of assimilation. Is the pattern of adjustment affected by length of residence in the United States? For each table compute gamma and summarize the relationship in terms of strength and direction. *(HINT: In 2 × 2 tables, only two cells can contribute to N_s or N_d. To compute N_s, multiply the number of cases in the upper-left-hand cell by the number of cases in the lower-right-hand cell. For N_d, multiply the number of cases in the upper-right-hand cell by the number of cases in the lower-left-hand cell.)*

a. Facility in English:

	Length of Residence		
English Facility	Less Than Five Years (Low)	More Than Five Years (High)	Totals
Low	20	10	30
High	5	15	20
Totals	25	25	50

b. Total family income:

Income	Length of Residence: Less Than Five Years (Low)	More Than Five Years (High)	Totals
Below national average (1)	18	8	26
Above national average (2)	7	17	24
Totals	25	25	50

c. Extent of contact with country of origin:

Contact	Length of Residence: Less Than Five Years (Low)	More Than Five Years (High)	Totals
Rare (1)	5	20	25
Frequent (2)	20	5	25
Totals	25	25	50

14.2 Compute gamma for the tables presented in problems 11.2, 11.6, 11.7, 11.9, 11.10, and 11.12. Since these tables are based on random samples, test the gammas you computed for significance.

14.3 Compute gamma for the table presented in problem 12.1. If you computed a nominal measure of association for this table in problem 13.2, compare the measures of association. Are they similar in value? Do they characterize the strength for the association in the same way? What information about the relationship does gamma provide that is not available from nominal measures of association?

14.4 CJ A random sample of 150 cities has been classified as small, medium, or large by population and as high or low on crime rate. Is there a relationship between city size and crime rate?

Crime Rate	City Size: Small	Medium	Large	Totals
Low	21	17	8	46
High	29	33	42	104
Totals	50	50	50	150

a. Describe the strength and direction of the relationship.

b. Is the relationship significant?

14.5 SOC Some research has shown that families vary by how they socialize their children to sports, games, and other leisure-time activities. In middle-class families, such activities are carefully monitored by parents and are, in general, dominated by adults (for example, Little League baseball). In working-class families, children more often organize and initiate such activities themselves, and parents are much less involved (for example, sandlot or playground baseball games). Are the following data consistent with these findings? Summarize your conclusions in a few sentences.

As a Child, Did You Play Mostly Organized or Sandlot Sports?	Social Class: White-Collar	Blue-Collar	Totals
Organized	155	123	278
Sandlot	101	138	239
Totals	256	261	517

14.6 Is there a relationship between education and support for women in the paid labor force? Is the relationship between the variables different for different nations? The World Values Survey has been administered to random samples drawn from Canada, the United States, and Mexico. Respondents were asked if they agree or disagree that both husbands and wives should contribute to the family income. Compute column percentages and gamma for each table. Is there a relationship? Describe the strength and direction of the relationship. Which educational level is most supportive of women being in the paid labor force? How does the relationship change from nation to nation?

a. Canada

"Husband and Wives Should Contribute to Income"	Education: Low	Moderate	High	Totals
Agree	352	682	365	1399
Disagree	98	204	154	456
Totals	450	886	519	1855

b. United States

"Husband and Wives Should Contribute to Income"	Education: Low	Moderate	High	Totals
Agree	177	239	380	796
Disagree	54	107	203	364
Totals	231	346	583	1160

c. Mexico

"Husband and Wives Should Contribute to Income"	Education			
	Low	Moderate	High	Totals
Agree	718	471	140	1329
Disagree	105	48	14	167
Totals	823	519	154	1496

14.7 PA All applicants for municipal jobs in Shinbone, Kansas, are given an aptitude test, but the test has never been evaluated to see if test scores are in any way related to job performance. The following table reports aptitude test scores and job performance ratings for a random sample of 75 city employees.

Efficiency Ratings	Test Scores			
	Low	Moderate	High	Totals
Low	11	6	7	24
Moderate	9	10	9	28
High	5	9	9	23
Totals	25	25	25	75

a. Are these two variables associated? Describe the strength and direction of the relationship in a sentence or two.
b. Is gamma statistically significant?
c. Should the aptitude test continue to be administered? Why or why not?

14.8 SW A sample of children has been observed and rated for symptoms of depression. Their parents have been rated for authoritarianism. Is there any relationship between these variables? Write a few sentences stating your conclusions.

Symptoms of Depression	Authoritarianism			
	Low	Moderate	High	Totals
Few	7	8	9	24
Some	15	10	18	43
Many	8	12	3	23
Totals	30	30	30	90

14.9 SOC Are prejudice and level of education related? State your conclusion in a few sentences.

Prejudice	Level of Education				
	Elementary School	High School	Some College	College Graduate	Totals
Low	48	50	61	42	201
High	45	43	33	27	148
Totals	93	93	94	69	349

14.10 SOC In a recent survey, a random sample of respondents was asked to indicate how happy they were with their situations in life. Are their responses related to income level?

Happiness	Income			
	Low	Moderate	High	Totals
Not happy	101	82	36	219
Pretty happy	40	227	100	367
Very happy	216	198	203	617
Totals	357	507	339	1203

a. Describe the strength and direction of the relationship.
b. Is the relationship significant?

14.11 SOC A random sample of 11 neighborhoods in Shinbone, Kansas, have been rated by an urban sociologist on a "quality-of-life" scale (which includes measures of affluence, availability of medical care, and recreational facilities) and a social cohesion scale. The results are presented here in scores. Higher scores indicate higher "quality of life" and greater social cohesion.

Neighborhood	Quality of Life	Social Cohesion
Queens Lake	17	8.8
North End	40	3.9
Brentwood	47	4.0
Denbigh Plantation	90	3.1
Phoebus	35	7.5
Kingswood	52	3.5
Chesapeake Shores	23	6.3
Windsor Forest	67	1.7
College Park	65	9.2
Beaconsdale	63	3.0
Riverview	100	5.3

a. Are the two variables associated? What is the strength and direction of the association? Summarize the relationship in a sentence or two. *(HINT: Don't forget to square the value of Spearman's rho for a PRE interpretation.)*
b. Conduct a test of significance for this relationship. Summarize your findings.

14.12 SW Several years ago, a job-training program began, and a team of social workers screened the candidates for suitability for employment. Now the screening process is being evaluated, and the actual work performance of a sample of hired candidates has been rated. Did the screening process work? Is there a relationship between the original scores and performance evaluation on the job?

Case	Original Score	Performance Evaluation
A	17	78
B	17	85
C	15	82
D	13	92
E	13	75
F	13	72
G	11	70
H	10	75
I	10	92
J	10	70
K	9	32
L	8	55
M	7	21
N	5	45
O	2	25

14.13 SOC Following are the scores of a sample of 15 nations on a measure of ethnic diversity (the higher the number, the greater the diversity) and a measure of economic inequality (the higher the score, the greater the inequality). Are these variables related? Are ethnically diverse nations more economically unequal?

Nation	Diversity	Inequality
India	91	29.7
South Africa	87	58.4
Kenya	83	57.5
Canada	75	31.5
Malaysia	72	48.4
Kazakhstan	69	32.7
Egypt	65	32.0
United States	63	41.0
Sri Lanka	57	30.1
Mexico	50	50.3
Spain	44	32.5
Australia	31	33.7
Finland	16	25.6
Ireland	4	35.9
Poland	3	27.2

14.14 Twenty ethnic, racial, or national groups were rated by a random sample of white and black students on a Social Distance Scale. Lower scores represent less social distance and less prejudice. How similar are these rankings? Is the relationship statistically significant?

	Average Social Distance Scale Score	
Group	White Students	Black Students
1 White Americans	1.2	2.6
2 English	1.4	2.9
3 Canadians	1.5	3.6
4 Irish	1.6	3.6
5 Germans	1.8	3.9
6 Italians	1.9	3.3
7 Norwegians	2.0	3.8
8 American Indians	2.1	2.7
9 Spanish	2.2	3.0
10 Jews	2.3	3.3
11 Poles	2.4	4.2
12 Black Americans	2.4	1.3
13 Japanese	2.8	3.5
14 Mexicans	2.9	3.4
15 Koreans	3.4	3.7
16 Russians	3.7	5.1
17 Arabs	3.9	3.9
18 Vietnamese	3.9	4.1
19 Turks	4.2	4.4
20 Iranians	5.3	5.4

14.15 In problems 12.12 and 13.12, we looked at the relationships between five dependent variables and, respectively, political ideology and sex. In this exercise, we use income as an independent variable and assess its relationship with this set of variables. For each table, calculate percentages and gamma. Describe the strength, direction, and statistical significance of each relationship in a few sentences. *Be careful in interpreting direction.*

a. Support for the legal right to an abortion by income:

Right to an Abortion?	Income			
	Low	Moderate	High	Totals
Yes	220	218	226	664
No	366	299	250	915
Totals	586	517	476	1579

b. Support for capital punishment by income:

Capital Punishment?	Income: Low	Moderate	High	Totals
Favor	567	574	552	1693
Oppose	270	183	160	613
Totals	837	757	712	2306

c. Approval of suicide for people with an incurable disease by income:

Right to Suicide?	Income: Low	Moderate	High	Totals
Approve	343	341	338	1022
Oppose	227	194	147	568
Totals	570	535	485	1590

d. Support for sex education in public schools by income:

Sex Education?	Income: Low	Moderate	High	Totals
For	492	478	451	1421
Against	85	68	53	206
Totals	577	546	504	1627

e. Support for traditional gender roles by income:

Women Should Take Care of Running Their Homes and Leave Running the Country to Men	Income: Low	Moderate	High	Totals
Agree	130	71	39	240
Disagree	448	479	461	1388
Totals	578	550	500	1628

SPSS for Windows

Using *SPSS for Windows* to Produce Ordinal-Level Measures of Association

SPSS DEMONSTRATION 14.1 Do Sexual Attitudes Vary by Age? Another Look

In Demonstration 12.3, we used percentages to look at the relationships between *premarsx* (attitude toward premarital sex) and recoded *age.* Let's reexamine this relationship and see if gamma can add any new information. We will use the **Crosstabs** program, with *ager* (recoded age) as the independent (column) variable. If you no longer have access to the recoded version of this variable, follow the directions in Demonstration 12.3. The dependent variable *premarsx* will go in the rows. On the Crosstabs dialog window, click the Statistics button and request gamma. Don't forget to click the Cells button and request column percentages. Click OK, and the output should looks as reproduced here. (This table has been edited for clarity and will not match your output exactly)

Crosstab

SEX BEFORE MARRIAGE		Recoded age 1.00	2.00	3.00	Total
ALWAYS WRONG	Count	47	47	56	150
	%	23.0%	25.3%	26.4%	24.9%
ALMOST ALWAYS WRONG	Count	9	12	25	46
	%	4.4%	6.5%	11.8%	7.6%
SOMETIMES WRONG	Count	32	32	41	105
	%	15.7%	17.2%	19.3%	17.4%
NOT WRONG AT ALL	Count	116	95	90	301
	%	56.9%	51.1%	42.5%	50.0%
Total	Count	204	186	212	602
	%	100.0%	100.0%	100.0%	100.0%

Symmetric Measures

		Value	Asymp. Std. Error(a)	Approx. T(b)	Approx. Sig.
Ordinal by Ordinal	Gamma	-.140	.054	-2.611	.009
N of Valid Cases		602			

[a]Not assuming the null hypothesis.
[b]Using the asymptotic standard error assuming the null hypothesis.

A gamma of −.14 indicates a weak, negative relationship. Older respondents (a score of 3 on *ager*) were most opposed to premarital sex (they had the highest percentage who felt it was "always wrong"), and younger respondents were least opposed (they had the highest percentage who felt it was "not wrong at all"). As age increases, support of premarital sex (the percentage who say it's "not wrong at all") decreases.

SPSS DEMONSTRATION 14.2 A Note about the Direction of Ordinal Relationships

Let's pause once again to consider the direction of relationships between ordinal-level variables. Direction for ordinal variables is a surprisingly tricky matter, mostly because of the arbitrary nature of coding at the ordinal level. We usually think of *higher* scores as indicating *more* of the quantity being measured, and, for every interval-ratio-level variable I can think of, this pattern will be true. For ordinal variables, however, higher scores may indicate *less* of the quantity because the codes are arbitrary. Depending on how you code the values, a high score on a scale measuring, say, prejudice might indicate great prejudice or its complete absence.

Looking at the table for *premarsx* and *ager* in Demonstration 14.1, remember that a negative gamma means that *high* numerical scores on one variable are associated with *low* numerical scores on the other. This means, for example, that those who scored a 1 (the lowest possible value) on *ager* tended to score a 4 (the highest possible value) on *premarsx*. You may think of a 4 on *premarsx* as representing "high support for premarital sex" or "low opposition to premarital sex." A negative gamma always means that the numerical scores of the variables are inversely related, but this does not necessarily mean that the underlying relationship is truly negative. Always inspect tables carefully to make sure you are interpreting the direction of the relationship properly.

SPSS DEMONSTRATION 14.3 Interpreting the Direction of Relationships: Are Prestigious Jobs More Satisfying?

What's the relationship between the prestige of an occupation and job satisfaction? Some would argue that high-prestige jobs are rewarding in a variety of ways (besides simply income) and that people in such jobs would express greater satisfaction with their work. Others might argue an opposing point of view: Because people in lower-prestige jobs have fewer responsibilities and less pressure, they will be more satisfied.

In an attempt to resolve the debate, we'll run a Crosstabs on *prestg80* and *satjob*. The former variable is a "continuous" ordinal scale whose scores range from a low of 17 to a high of 86. We need to collapse *prestg80* into a form suitable for use in bivariate tables. I did this by first finding the median on *prestg80* with the Frequencies command and then recoding the variable into a dichotomy. I also recoded *satjob* into two approximately equal categories. Don't forget to recode "into different variable" and give the recoded variables new names (I used *rprest* and *rsat*). The recoding instructions for *prestg80* are

0 thru 43 = 1
44 thru 86 = 2

For *satjob,* I left the score of 1 alone and collapsed scores 2–4:

1 = 1
2 thru 4 = 2

Use Crosstabs once again, with recoded *prestg80* as the independent (column) variable and recoded *satjob* as the dependent variable. Request gamma, chi square, and column percentages. The output should look like this

Recoded satjob * Recoded prestg80 Crosstabulation

			Recorded prestg80		
			1.00	2.00	Total
Recoded satjob	1.00	Count	154	195	349
		% within Recoded prestg80	42.2%	58.0%	49.8%
	2.00	Count	211	141	352
		% within Recoded prestg80	57.8%	42.0%	50.2%
Total		Count	365	336	701
		% within Recoded prestg80	100.0%	100.0%	100.0%

Chi-Square Tests

	Value	df	Asymp. Sig. (2-sided)	Exact Sig. (2-sided)	Exact Sig. (1-sided)
Pearson Chi-Square	17.567(b)	1	.000		
Continuity Correction(a)	16.939	1	.000		
Likelihood Ratio	17.641	1	.000		
Fisher's Exact Test				.000	.000
Linear-by-Linear Association	17.542	1	.000		
N of Valid Cases	701				

[a]Computed only for a 2 × 2 table.
[b]0 cells (.0%) have expected count less than 5. The minimum expected count is 167.28.

Symmetric Measures

		Value	Asymp. Std. Error(a)	Approx. T(b)	Approx. Sig.
Ordinal by Ordinal	Gamma	-.309	.069	-4.245	.000
N of Valid Cases		701			

[a]Not assuming the null hypothesis.
[b]Using the asymptotic standard error assuming the null hypothesis.

This relationship is statistically significant (the significance of the chi square is less than .05), moderate to weak in strength, and negative in direction (gamma = −.309). The higher the prestige, the lower the job satisfaction. Right? Wrong. Look at the codes for *satjob.* A *higher* score indicates a *lower* level of satisfaction. So "high" (or a score of 2) on *prestg80* is associated with "high" (or a score of 1) on *satjob.* In spite of the negative sign for gamma, this is a *positive* relationship, and job satisfaction increases with prestige.

Exercises

14.1 Follow up on Demonstration 14.1 with some new ordinal-level independent variables. What other factors might affect attitudes toward premarital sex? Among other possibilities, consider *degree* and *class* as independent variables. Do sexual attitudes vary by education or social class? Other possibilities include *income06, prestg80,* and *educ* (these would have to be recoded), *polviews,* and *attend.*

14.2 Following the procedure in Demonstration 14.3, use other measures of social class besides prestige (*class* and *degree*) as independent variables in the relationship with recoded *satjob.* Summarize the strength and direction of the relationships in a few sentences. Are these relationships stronger or weaker than those with recoded *prestg80?* Be careful in interpreting the direction of the relationships.

15

Association Between Variables Measured at the Interval-Ratio Level

LEARNING OBJECTIVES

By the end of this chapter, you will be able to

1. Interpret a scattergram.
2. Calculate and interpret slope (b), Y intercept (a), and Pearson's r and r^2.
3. Find and explain the least-squares regression line and use it to predict values of Y.
4. Explain the concepts of total, explained, and unexplained variance.
5. Use regression and correlation techniques to analyze and describe a bivariate relationship in terms of the three questions introduced in Chapter 12.
6. Test Pearson's r for significance.

15.1 INTRODUCTION

This chapter presents a set of statistical techniques for analyzing the association, or correlation, between variables measured at the interval-ratio level.[1] As I have repeatedly noted, such variables are relatively rare in social science research, and the techniques presented in this chapter are commonly used on variables measured at the ordinal level, especially with ordinal variables that are "continuous" (see Chapter 14). In fact, as we shall see in Section 15.8, there is a way to include variables measured at the nominal level. Nevertheless, the statistical techniques of correlation and regression are most appropriately used with high-quality, precisely measured variables at the interval-ratio level.

As we shall see, the techniques presented in this chapter are rather different in their logic and computation from those covered in Chapters 13 and 14. Let me stress at the outset, therefore, that we are still asking the same three questions: Is there a relationship between the variables? How strong is the relationship? What is the direction of the relationship? You might become preoccupied with some of the technical details and computational routines in this chapter, so remind yourself occasionally that our ultimate goals are unchanged: We are trying to understand bivariate relationships, explore possible causal ties between variables, and improve our ability to predict scores.

15.2 SCATTERGRAMS

Introduction. As we have seen over the past several chapters, properly percentaged tables provide important information about bivariate associations between nominal- and ordinal-level variables. In addition to measures of association like phi and gamma, the conditional distributions and patterns of cell frequency almost always provide useful information and a better understanding of the relationship between variables.

By the same token, the usual first step in analyzing a relationship between interval-ratio variables is to construct and examine a **scattergram.** Like bivari-

[1] The term *correlation* is commonly used instead of *association* when discussing the relationship between interval-ratio variables. We will use the two terms interchangeably.

ate tables, these graphs allow us to identify quickly several important features of the relationship. An example will illustrate the construction and use of scattergrams. Suppose a researcher is interested in analyzing how dual-wage-earner families (that is, families where both husband and wife have jobs outside the home) cope with housework. Specifically, the researcher wonders if the number of children in the family is related to the amount of time the husband contributes to housekeeping chores. The relevant data for a sample of 12 dual-wage-earner families are displayed in Table 15.1.

Constructing Scattergrams. A scattergram, like a bivariate table, has two dimensions. The scores of the independent (X) variable are arrayed along the horizontal axis, and the scores of the dependent (Y) variable along the vertical axis. Each dot on the scattergram represents a case in the sample and is located at a point determined by the scores of the case. Figure 15.1 shows a scattergram displaying the relationship between "number of children" and "husband's housework" for the sample of 12 families presented in Table 15.1. Family A has a score of 1 on the X variable (number of children) and 1 on the Y variable (husband's housework) and is represented by the dot above the score of 1 on the X axis and directly to the right of the score of 1 on the Y axis. All 12 cases are similarly represented by dots on Figure 15.1. Also note that, as always, the scattergram is clearly titled and both axes are labeled.

Interpreting Scattergrams. The overall pattern of the dots or cases summarizes the nature of the relationship between the two variables. The clarity of the pattern can be enhanced by drawing a straight line through the cluster of dots such that the line touches every dot or comes as close to doing so as possible. In Section 15.3, a precise technique for fitting this line to the pattern of the dots is explained. For now, an "eyeball" approximation will suffice. This summarizing line, called the **regression line,** has already been added to the scattergram.

Scattergrams, even when they are crudely drawn, can be used for a variety of purposes. They provide at least impressionistic information about the exis-

TABLE 15.1 NUMBER OF CHILDREN AND HUSBAND'S CONTRIBUTION TO HOUSEWORK (fictitious data)

Family	Number of Children	Hours per Week Husband Spends on Housework
A	1	1
B	1	2
C	1	3
D	1	5
E	2	3
F	2	1
G	3	5
H	3	0
I	4	6
J	4	3
K	5	7
L	5	4

FIGURE 15.1 HUSBAND'S HOUSEWORK BY NUMBER OF CHILDREN

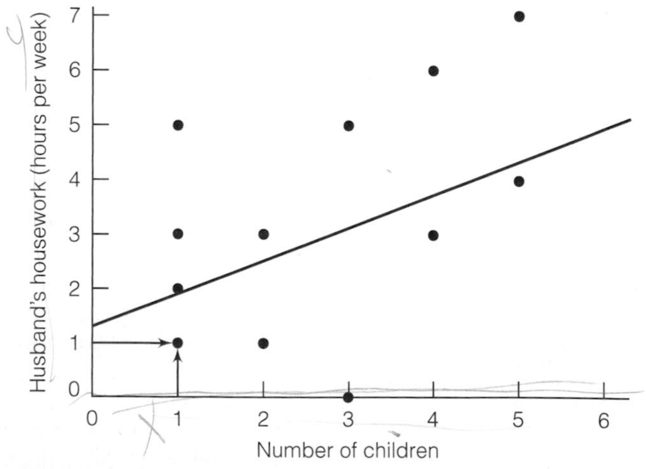

tence, strength, and direction of the relationship and can also be used to check the relationship for linearity (that is, how well the pattern of dots can be approximated with a straight line). Finally, the scattergram can be used to predict the score of a case on one variable from the score of that case on the other variable. We now briefly examine each of these uses in terms of the three questions we first asked in Chapter 12.

- **Does a relationship exist?** To ascertain the existence of a relationship, we can return to the basic definition of an association stated in Chapter 12. Two variables are associated if the distributions of *Y* (the dependent variable) change for the various conditions of *X* (the independent variable). In Figure 15.1, scores on *X* (number of children) are arrayed along the horizontal axis. The dots above each score on *X* are the scores (or conditional distributions) of *Y*. That is, the dots represent scores on *Y* for each value of *X*. Figure 15.1 shows that there is a relationship between these variables, because the conditional distributions of *Y* (the dots above each score on *X*) change as *X* changes. The existence of an association is further reinforced by the fact that the regression line lies at an angle to the *X* axis. If these two variables had not been associated, the conditional distributions of *Y* would not have changed, and the regression line would have been parallel to the horizontal axis.
- **How strong is the relationship?** The strength of the bivariate association can be judged by observing the spread of the dots around the regression line. In a perfect association, all dots would lie on the regression line. The more the dots are clustered around the regression line, the stronger the association.
- **What is the direction of the relationship?** The direction of the relationship can be detected by observing the angle of the regression line. Figure 15.1 shows a positive relationship: As *X* (number of children) increases, husband's housework (*Y*) also increases. Husbands in families with more children tend to do more housework. If the relationship had been negative, the regression line would have sloped in the opposite direction to indicate that high scores on one variable were associated with low scores on the other.

To summarize these points, Figure 15.2 shows a perfect positive and a perfect negative relationship and a "zero relationship," or "nonrelationship," between two variables.

Linearity. One key assumption underlying the statistical techniques to be introduced later in this chapter is that the two variables have an essentially **linear relationship.** In other words, the observation points or dots in the scattergram must form a pattern that can be approximated with a straight line. Significant departures from linearity would require the use of statistical techniques beyond the scope of this text. Examples of some common curvilinear relationships are presented in Figure 15.3. If the scattergram shows that the variables have a nonlinear relationship, the techniques described in this chapter should be used with great caution or not at all. Checking for the linearity of the relationship is perhaps the most important reason for constructing at least a crude, hand-drawn scattergram before proceeding with the statistical analysis. If the relationship is nonlinear, you might need to treat the variables as if they were ordinal rather than interval-ratio in level of measurement. *(For practice in constructing and interpreting scattergrams, see problems 15.1 to 15.4.)*

FIGURE 15.2 POSITIVE (A), NEGATIVE (B), AND ZERO (C) RELATIONSHIPS

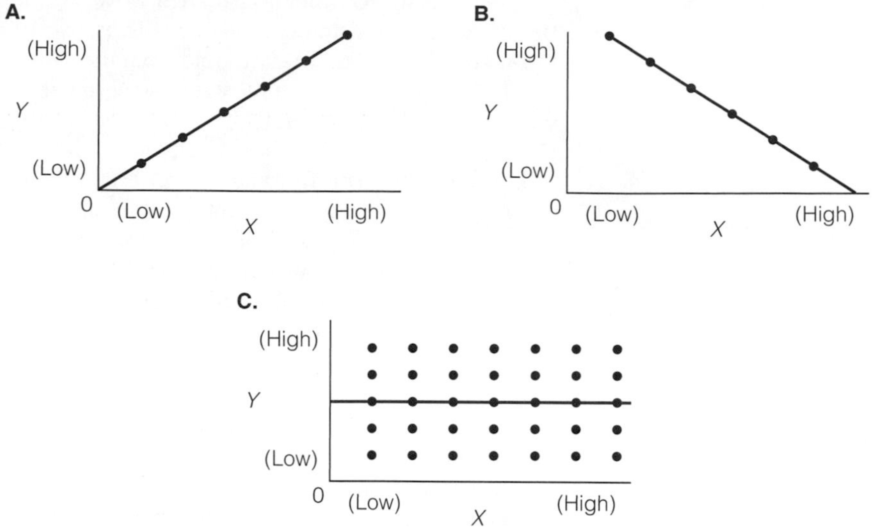

FIGURE 15.3 SOME NONLINEAR RELATIONSHIPS

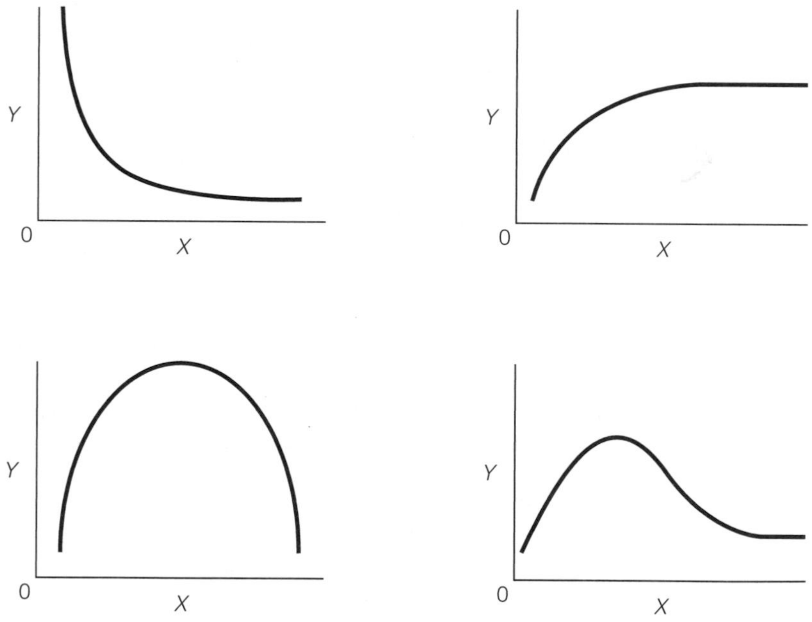

15.3 REGRESSION AND PREDICTION

Prediction. A final use of the scattergram is to predict scores of cases on one variable from their score on the other. To illustrate, suppose that, based on the relationship between number of children and husband's housework displayed in Figure 15.1, we wish to predict the number of hours of housework a husband with a family of six children would do each week. The sample has no families with six children, but if we extend the axes and regression line in Figure 15.1 to

FIGURE 15.4 PREDICTING HUSBAND'S HOUSEWORK

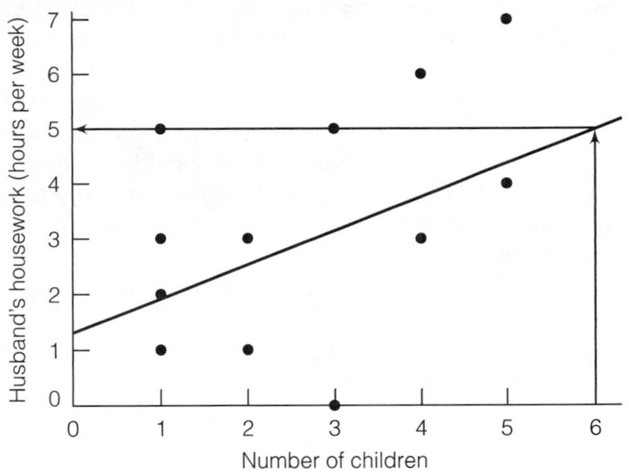

incorporate this score, a prediction is possible. Figure 15.4 reproduces the scattergram and illustrates how the prediction would be made.

The predicted score on Y—which is symbolized as Y9 to distinguish predictions of Y from actual Y scores—is found by first locating the relevant score on X(X 5 6 in this case) and then drawing a straight line from that point to the regression line. From the regression line, another straight line parallel to the X axis is drawn across to the Y axis. The predicted Y score (Y′) is found at the point where the line crosses the Y axis. In our example, we would predict that, in a dual-wage-earner family with six children, the husband would devote about five hours per week to housework.

The Regression Line. Of course, this prediction technique is crude, and the value of Y' can change, depending on how accurately the freehand regression line is drawn. One way to eliminate this source of error would be to find the straight line that most accurately summarizes the pattern of the observation points and so best describes the relationship between the two variables. Is there such a "best-fitting" straight line? If there is, how is it defined?

Recall that our criterion for the freehand regression line was that it touch all the dots or come as close to doing so as possible. Also, recall that the dots above each value of X can be thought of as conditional distributions of Y, the dependent variable. Within each conditional distribution of Y, the mean is the point around which the variation of the scores is at a minimum. In Chapter 3, we noted that the mean of any distribution of scores is the point around which the variation of the scores, as measured by squared deviations, is minimized:

$$\sum (X_i - \overline{X})^2 = \text{minimum}$$

Thus, if the regression line is drawn so that it touches each **conditional mean of Y,** it would be the straight line that comes as close as possible to all the scores.

Conditional means are found by summing all Y values for each value of X and then dividing by the number of cases. For example, four families had one child ($X = 1$), and the husbands of these four families devoted 1, 2, 3, and 5 hours per week to housework. Thus, for $X = 1$, $Y = 1$, 2, 3, and 5, and the conditional mean of Y for $X = 1$ is 2.75 (11/4 = 2.75). Husbands in families with one child worked an average of 2.75 hours per week doing housekeeping chores. Conditional means of Y are computed in the same way for each value of X displayed in Table 15.2 and plotted in Figure 15.5.

Let us quickly remind ourselves of the reason for these calculations. We are seeking the single best-fitting regression line for summarizing the relationship between X and Y, and we have seen that a line drawn through the conditional means of Y will minimize the spread of the observation points. It will come as close to all the scores as possible and will therefore be the single best-fitting regression line.

Now, a line drawn through the points on Figure 15.5 (the conditional means of Y) will be the best-fitting line we are seeking, but you can see from the scattergram that the line will not be straight. In fact, only rarely (when there is a perfect relationship between X and Y) will conditional means fall in a perfectly straight line. Since we still must meet the condition of linearity, let us revise our criterion and define the regression line as the unique straight line that touches all conditional means of Y or comes as close to doing so as possible. Formula 15.1 defines the "least-squares" regression line, or the single straight regression line that best fits the pattern of the data points.

FORMULA 15.1

$$Y = a + bX$$

where Y = score on the dependent variable
a = the Y intercept, or the point where the regression line crosses the Y axis
b = the slope of the regression line, or the amount of change produced in Y by a unit change in X
X = score on the independent variable

The formula introduces two new concepts. First, the ***Y* intercept (*a*)** is the point at which the regression line crosses the vertical, or Y, axis. Second, the ***slope* (*b*)** of the least-squares regression line is the amount of change pro-

FIGURE 15.5 CONDITIONAL MEANS OF *Y*

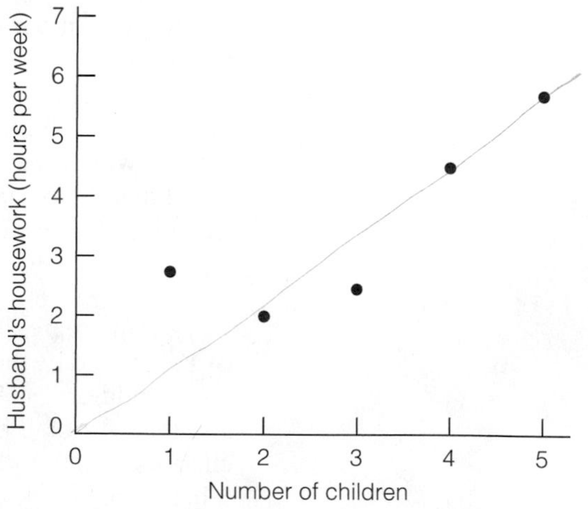

TABLE 15.2 CONDITIONAL MEANS OF *Y* (husband's housework) FOR VARIOUS VALUES OF *X* (number of children)

Number of Children (X)	Husband's Housework (Y)	Conditional Mean of Y
1	1,2,3,5	2.75
2	3,1	2.00
3	5,0	2.50
4	6,3	4.50
5	7,4	5.50

duced in the dependent variable (Y) by a unit change in the independent variable (X). Think of the slope of the regression line as a measure of the effect of the X variable on the Y variable. If the variables have a strong association, then changes in the value of X will be accompanied by substantial changes in the value of Y, and the slope (b) will have a high value. The weaker the effect of X on Y (the weaker the association between the variables), the lower the value of the slope (b). If the two variables are unrelated, the least-squares regression line would be parallel to the X axis, and b would be 0.00 (the line would have no slope).

With the least-squares formula (Formula 15.1), we can predict values of *Y* in a much less arbitrary and impressionistic way than through mere eyeballing. This will be so, remember, because the least-squares regression line as defined by Formula 15.1 is the single straight line that best fits the data, because it comes as close as possible to all of the conditional means of *Y*. Before seeing how predictions of *Y* can be made, however, we must first calculate *a* and *b*. *(For practice in using the regression line to predict scores on Y from scores on X, see problems 15.1 to 15.3 and 15.5.)*

15.4 COMPUTING *a* AND *b*

In this section, we cover how to compute and interpret the coefficients in the equation for the regression line: the slope (*b*) and the *Y* intercept (*a*). Since the value of *b* is needed to compute *a*, we begin with the computation of the slope.

Computing the Slope (*b*). The formula for the slope is

FORMULA 15.2

$$b = \frac{\Sigma(X - \overline{X})(Y - \overline{Y})}{\Sigma(X - \overline{X})^2}$$

The numerator of this formula is called the *covariation* of *X* and *Y*. It is a measure of how *X* and *Y* vary together, and its value will reflect both the direction and the strength of the relationship. The denominator is simply the sum of the squared deviations around the mean of *X*.

The calculations necessary for computing the slope should be organized into a computational table, as in Table15.3, which has a column for each of the quantities needed to solve the formula. The data are from the dual-wage-earner family sample (see Table 15.1).

In Table 15.3, the first column lists the original *X* scores for each case and the second column shows the deviations of these scores around their mean. The third and fourth columns repeat this information for the *Y* scores and the deviations of the *Y* scores. Column 5 shows the covariation of the *X* and *Y* scores. The entries in this column are found by multiplying the deviation of the *X* score (column 2) by the deviation of the *Y* score (column 4) for each case. Finally, the entries in column 6 are found by squaring the value in column 2 for each case.

Table 15.3 gives us all the quantities we need to solve Formula 15.2. Substitute the total of column 5 in Table 15.3 in the numerator and the total of column 6 in the denominator:

$$b = \frac{\Sigma(X - \overline{X})(Y - \overline{Y})}{\Sigma(X - \overline{X})^2}$$

$$b = \frac{18.33}{26.67}$$

$$b = 0.69$$

TABLE 15.3 COMPUTATION OF THE SLOPE (*b*)

1 X	2 $X - \bar{X}$	3 Y	4 $Y - \bar{Y}$	5 $(X - \bar{X})(Y - \bar{Y})$	6 $(X - \bar{X})^2$
1	−1.67	1	−2.33	3.89	2.79
1	−1.67	2	−1.33	2.22	2.79
1	−1.67	3	−0.33	0.55	2.79
1	−1.67	5	1.67	−2.79	2.79
2	−0.67	3	−0.33	0.22	0.45
2	−0.67	1	−2.33	1.56	0.45
3	0.33	5	1.67	0.55	0.11
3	0.33	0	−3.33	−1.10	0.11
4	1.33	6	2.67	3.55	1.77
4	1.33	3	−0.33	−0.44	1.77
5	2.33	7	3.67	8.55	5.43
5	2.33	4	0.67	1.56	5.43
32	−0.04	40	0.04	18.33	26.67

$$\bar{X} = \frac{32}{12} = 2.67$$

$$\bar{Y} = \frac{40}{12} = 3.33$$

A slope of 0.69 indicates that, for each unit change in X, there is an increase of 0.69 units in Y. For our example, the addition of each child (an increase of one unit in X) results in an increase of 0.69 hours of housework being done by the husband (an increase of 0.69 units—or hours—in Y).

Computing the *Y* intercept (*a*). Once the slope has been calculated, finding the intercept (a) is relatively easy. We computed the means of X and Y while calculating the slope, and we enter these figures into Formula 15.3:

FORMULA 15.3

$$a = \bar{Y} - b\bar{X}$$

For our sample problem, the value of a would be

$$a = \bar{Y} - b\bar{X}$$
$$a = 3.33 - (0.69)(2.67)$$
$$a = 3.33 - 1.84$$
$$a = 1.49$$

Thus, the least-squares regression line will cross the Y axis at the point where Y equals 1.49.

Now that we have values for the slope and the Y intercept, we can state the full least-squares regression line for our sample data:

$$Y = a + bX$$
$$Y = 1.49 + (0.69)X$$

Predicting Scores on *Y* with the Least-Squares Regression Line. The regression formula can be used to estimate, or predict, scores on Y for any value of X. In Section 15.3, we used the freehand regression line to predict a score on Y (husband's housework) for a family with six children ($X = 6$). Our prediction

ONE STEP AT A TIME **Computing the Slope (b)**

To compute the slope (b), solve Formula 15.2:

$$b = \frac{\Sigma(X - \overline{X})(Y - \overline{Y})}{\Sigma(X - \overline{X})^2}$$

Step 1: Set up a computing table like Table 15.3 to help organize the computations. List the scores of the cases on the independent variable (X) in column 1.

Step 2: Compute the mean of $X(\overline{X})$ by dividing the total of column 1 (ΣX) by the number of cases (N).

Step 3: Subtract the mean of $X(\overline{X})$ from each X score and list the results in column 2.

Step 4: Find the sum of column 2. This value must be zero (except for rounding error). If this sum is not zero, you have made a mistake in computations.

Step 5: List the score of each case on Y in column 3. Compute the mean of $Y(\overline{Y})$ by dividing the total of column 3 (ΣY) by the number of cases (N).

Step 6: Subtract the mean of $Y(\overline{Y})$ from each Y score and list the results in column 4.

Step 7: Find the sum of column 4. This value must be zero (except for rounding error). If this sum is not zero, you have made a mistake in computations.

Step 8: For each case, multiply the value in column 2 by the value in column 4. Place the result in column 5. Find the sum of this column.

Step 9: Square each value in column 2 and place the result in column 6. Find the sum of this column.

Step 10: Divide the sum of column 5 by the sum of column 6. The result is the slope.

ONE STEP AT A TIME **Computing the Y Intercept (a)**

To compute the Y intercept (a), solve Formula 15.3:

$$a = \overline{Y} - b\overline{X}$$

Step 1: The values for the mean of X and Y were computed while finding b.

Step 2: Multiply the slope (b) by the mean of $X(\overline{X})$.

Step 3: Subtract the value you found in step 2 from the mean of $Y(\overline{Y})$. This value is a, or the Y intercept.

was that, in families of six children, husbands would contribute about five hours per week to housekeeping chores. By using the least-squares regression line, we can see how close our impressionistic, eyeball prediction was.

$$Y = a + bX$$
$$Y = 1.49 + (0.69)(6)$$
$$Y = 1.49 + 4.14$$
$$Y = 5.63$$

Based on the least-squares regression line, we would predict that in a dual-wage-earner family with six children, husbands would devote 5.63 hours a week to housework. What would our prediction of husband's housework be for a family of seven children ($X = 7$)?

Note that our predictions of Y scores are basically "educated guesses." We will be unlikely to predict values of Y exactly except in the (relatively rare) case

ONE STEP AT A TIME **Using the Regression Line to Predict Scores on *Y***

Step 1: Choose a value for *X*. Multiply this value by the value of the slope (*b*).

Step 2: Add the value you found in step 1 to the value of *a*, the *Y* intercept. The resulting value is the predicted score on *Y*.

where the bivariate relationship is perfect and perfectly linear. Note also, however, that the accuracy of our predictions will increase as relationships become stronger. This is because the dots are more clustered around the least-squares regression line in stronger relationships. *(The slope and Y intercept may be computed for any problem at the end of this chapter, but see problems 15.1 to 15.5 in particular. These problems have smaller data sets and will provide good practice until you are comfortable with these calculations.)*

15.5 THE CORRELATION COEFFICIENT (PEARSON'S *r*)

I pointed out in Section 15.4 that the slope of the least-squares regression line (*b*) is a measure of the effect of *X* on *Y*. Since the slope is the amount of change produced in *Y* by a unit change in *X*, *b* will increase in value as the relationship increases in strength. However, *b* does not vary between zero and 1 and is therefore awkward to use as a measure of association. Instead, researchers rely heavily (almost exclusively) on a statistic called **Pearson's *r*,** or the correlation coefficient, to measure association between interval-ratio variables. Like the ordinal measures of association discussed in Chapter 14, Pearson's *r* varies from

ONE STEP AT A TIME **Computing Pearson's *r***

Step 1: Add a column to the computing table you used to compute the slope (*b*). Square the value of $(Y - \overline{Y})$ and record the result in this column (column 7).

Step 2: Find the sum of column 7.

Step 3: Find the value of Pearson's *r* by solving Formula 15.4:

$$r = \frac{\Sigma(X - \overline{X})(Y - \overline{Y})}{\sqrt{[\Sigma(X - \overline{X})^2][\Sigma(Y - \overline{Y})^2]}}$$

Step 4: Multiply the sum of column 6, or $\Sigma(X - \overline{X})^2$, by the sum of column 7, or $\Sigma(Y - \overline{Y})^2$.

Step 5: Take the square root of the value you found in step 4.

Step 6: Divide the quantity you found in step 5 into the sum of column 5, or the sum of the cross products $(X - \overline{X})(Y - \overline{Y})$. The result is Pearson's *r*.

Step 7: To interpret the strength of Pearson's *r*, you can either

a. Use Table 14.2 to describe strength in general terms.

b. Square the value of *r* and multiply by 100. This value represents the percentage of variation in *Y* that is explained by *X*.

Step 8: To interpret the direction of the relationship, look at the sign of *r*. If *r* has a plus sign (or if there is no sign), the relationship is positive and the variables change in the same direction. If *r* has a minus sign, the relationship is negative and the variables change in opposite directions.

TABLE 15.4 COMPUTATION OF PEARSON'S *r*

1 X	2 $X - \bar{X}$	3 Y	4 $Y - \bar{Y}$	5 $(X - \bar{X})(Y - \bar{Y})$	6 $(X - \bar{X})^2$	7 $(Y - \bar{Y})^2$
1	−1.67	1	−2.33	3.89	2.79	5.43
1	−1.67	2	−1.33	2.22	2.79	1.77
1	−1.67	3	−0.33	0.55	2.79	0.11
1	−1.67	5	1.67	−2.79	2.79	2.79
2	−0.67	3	−0.33	0.22	0.45	0.11
2	−0.67	1	−2.33	1.56	0.45	5.43
3	0.33	5	1.67	0.55	0.11	2.79
3	0.33	0	−3.33	−1.10	0.11	11.09
4	1.33	6	2.67	3.55	1.77	7.13
4	1.33	3	−0.33	−0.44	1.77	0.11
5	2.33	7	3.67	8.55	5.43	13.47
5	2.33	4	0.67	1.56	5.43	0.45
32	−0.04	40	0.04	18.33	26.67	50.67

0.00 to ±1.00, with 0.00 indicating no association and +1.00 and −1.00 indicating perfect positive and perfect negative relationships, respectively. The formula for Pearson's *r* is

FORMULA 15.4

$$r = \frac{\Sigma(X - \bar{X})(Y - \bar{Y})}{\sqrt{[\Sigma(X - \bar{X})^2][\Sigma(Y - \bar{Y})^2]}}$$

Note that the numerator of this formula is the covariation of X and Y, as was the case with Formula 15.2. A computing table like Table 15.3, with a column added for the sum of $(Y - \bar{Y})^2$, is strongly recommended as a way of organizing the quantities needed to solve this equation (see Table 15.4).

For our sample problem involving dual-wage-earner families, the quantities displayed in Table 15.4 can be substituted directly into Formula 15.4:

$$r = \frac{\Sigma(X - \bar{X})(Y - \bar{Y})}{\sqrt{[\Sigma(X - \bar{X})^2][\Sigma(Y - \bar{Y})^2]}}$$

$$r = \frac{18.33}{\sqrt{(26.67)(50.67)}}$$

$$r = \frac{18.33}{\sqrt{1351.37}}$$

$$r = \frac{18.33}{36.76}$$

$$r = 0.50$$

An r value of 0.50 indicates a moderately strong, positive linear relationship between the variables. As the number of children in the family increases, the hourly contribution of husbands to housekeeping duties also increases. *(Every problem at the end of this chapter requires the computation of Pearson's r. It is probably a good idea to practice with smaller data sets and easier computations first—see problem 15.1 in particular.)*

15.6 INTERPRETING THE CORRELATION COEFFICIENT: r^2

Pearson's r is an index of the strength of the linear relationship between two variables. While a value of 0.00 indicates no linear relationship and a value of ± 1.00 indicates a perfect linear relationship, values between these extremes have no direct interpretation. We can, of course, describe relationships in terms of how closely they approach the extremes (for example, coefficients approaching 0.00 can be described as "weak" and those approaching ± 1.00 as "strong"), but this description is somewhat subjective. Also, we can use the guidelines stated in Table 14.2 for gamma to attach descriptive words to the specific values of Pearson's r. In other words, values between 0.00 and 0.30 would be described as weak, values between 0.30 and 0.60 would be moderate, and values greater than 0.60 would be strong. Remember, of course, that these labels are arbitrary guidelines and will not be appropriate or useful in all possible research situations.

Fortunately, we can develop a less arbitrary, more direct interpretation of r by calculating an additional statistic called the **coefficient of determination.** This statistic, which is simply the square of Pearson's r (r^2), can be interpreted with a logic akin to proportional reduction in error (PRE). As you recall, the logic of PRE measures of association is to predict the value of the dependent variable under two different conditions. First, Y is predicted while ignoring the information supplied by X; second, the independent variable is taken into account. With r^2, both the method of prediction and the construction of the final statistic are somewhat different and require the introduction of some new concepts.

When working with variables measured at the interval-ratio level, the predictions of the Y scores under the first condition (while ignoring X) will be the mean of the Y. Given no information on X, this prediction strategy will be optimal because we know that the mean of any distribution is closer to all the scores than any other point in the distribution. I remind you of the principle of minimized variation introduced in Chapter 3 and expressed as

$$\sum(Y - \overline{Y})^2 = \text{minimum}$$

The scores of any variable vary less around the mean than around any other point. If we predict the mean of Y for every case, we will make fewer errors of prediction than if we predict any other value for Y.

Of course, we will still make many errors in predicting Y even if we faithfully follow this strategy. The amount of error is represented in Figure 15.6, which displays the relationship between number of children and husband's housework with the mean of Y noted. The vertical lines from the actual scores to the predicted score represent the amount of error we would make when predicting Y while ignoring X.

We can define the extent of our prediction error under the first condition (while ignoring X) by subtracting the mean of Y from each actual Y score and squaring and summing these deviations. The resultant figure, which can be noted as $\Sigma(Y - \overline{Y})^2$, is called the **total variation** in Y. We now have a visual representation (Figure 15.6) and a method for calculating the error we incur by predicting Y without knowledge of X. As we shall see, we do not actually need to calculate the total variation to find the value of the coefficient of determination, r^2.

Our next step will be to determine the extent to which knowledge of X improves our ability to predict Y. If the two variables have a linear relationship, then predicting scores on Y from the least-squares regression equation will

FIGURE 15.6 PREDICTING *Y* WITHOUT *X* (dual-career families)

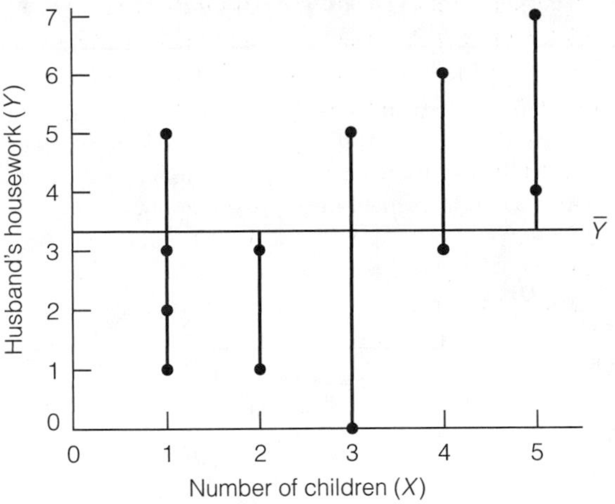

FIGURE 15.7 PREDICTING *Y* WITH *X* (dual-career families)

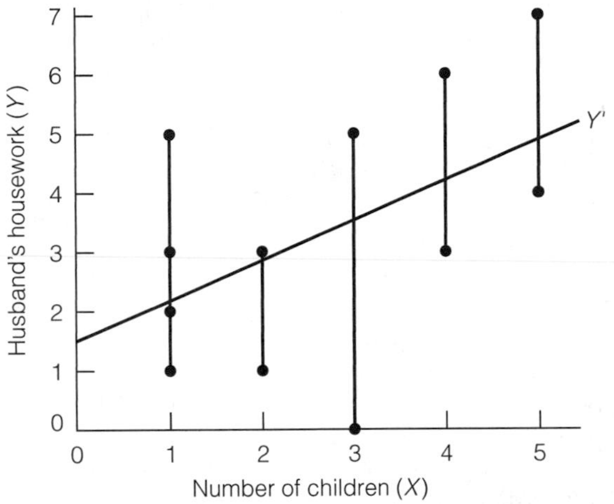

incorporate knowledge of *X* and reduce our errors of prediction. So, under the second condition, our predicted *Y* score for each value of *X* will be

$$Y' = a + bX$$

Figure 15.7 displays the data from the dual-career families with the regression line, as determined by the preceding formula, drawn in. The vertical lines from each data point to the regression line represent the amount of error in predicting *Y* that remains even after *X* has been taken into account.

As was the case under the first condition, we can precisely define the reduction in error that results from taking *X* into account. Specifically, two different sums can be found and then compared with the total variation of *Y* to construct a statistic that will indicate the improvement in prediction.

Application 15.1

For five cities, information has been collected on number of civil disturbances (riots, strikes, and so forth) over the past year and on unemployment rate. Are these variables associated? The data are presented in the following table. Columns have been added for all sums necessary for the computation of the slope (b) and Pearson's r.

City	Unemployment Rate (X)	$X - \bar{X}$	Civil Disturbances (Y)	$Y - \bar{Y}$	$(X - \bar{X})(Y - \bar{Y})$	$(X - \bar{X})^2$	$(Y - \bar{Y})^2$
A	22	6.8	25	14.4	97.92	46.24	207.36
B	20	4.8	13	2.4	11.52	23.04	5.76
C	10	−5.2	10	−0.6	3.12	27.04	0.36
D	15	−0.2	5	−5.6	1.12	0.04	31.36
E	9	−6.2	0	−10.6	65.72	38.44	112.36
	76	0.0	53	0.0	179.40	134.80	357.20

$$X = 76/5 = 15.2$$
$$Y = 53/5 = 10.6$$

The slope (b) is

$$b = \frac{\Sigma(X - \bar{X})(Y - \bar{Y})}{\Sigma(X - \bar{X})^2}$$

$$b = \frac{179.4}{134.8}$$

$$b = 1.33$$

A slope of 1.33 means that for every unit change in X (for every increase of 1 in the unemployment rate), there was a change of 1.33 units in Y (the number of civil disturbances increased by 1.33). The Y intercept (a) is

$$a = \bar{Y} - b\bar{X}$$

$$a = \frac{53}{5} - (1.33)\left(\frac{76}{5}\right)$$

$$a = 10.6 - (1.33)(15.20)$$

$$a = 10.6 - 20.2$$

$$a = -9.6$$

The least-squares regression equation is

$$Y = a + bX = -9.6 + (1.33)X$$

The correlation coefficient is

$$r = \frac{\Sigma(X - \bar{X})(Y - \bar{Y})}{\sqrt{[\Sigma(X - \bar{X})^2][\Sigma(Y - \bar{Y})^2]}}$$

$$r = \frac{179.40}{\sqrt{(134.80)(357.20)}}$$

$$r = \frac{179.40}{\sqrt{48150.56}}$$

$$r = \frac{179.40}{219.43}$$

$$r = 0.82$$

These variables have a strong, positive association. The number of civil disturbances increases as the unemployment rate increases. The coefficient of determination, r^2, is $(0.82)^2$, or 0.67. This indicates that 67% of the variance in civil disturbances is explained by the unemployment rate.

The first sum, called the **explained variation,** represents the improvement in our ability to predict Y when taking X into account. This sum is found by subtracting $\bar{Y}$ (our predicted Y score without X) from the score predicted by the regression equation (Y', or the Y score predicted with knowledge of X) for each case and then squaring and summing these differences. These operations can be summarized as $\Sigma(Y' - \bar{Y})^2$, and the resultant figure could then be compared with the total variation in Y to ascertain the extent to which our knowledge of X improves our ability to predict Y. Specifically, it can be shown mathematically that

FORMULA 15.5

$$r^2 = \frac{\Sigma(Y' - \bar{Y})}{\Sigma(Y - \bar{Y})} = \frac{\text{Explained variation}}{\text{Total variation}}$$

Thus, the coefficient of determination, or r^2, is the proportion of the total variation in Y attributable to, or explained by, X. Like other PRE measures, r^2 indicates precisely the extent to which X helps us predict, understand, or explain Y.

Earlier, we referred to the improvement in predicting Y with X as the explained variation. The use of this term suggests that some of the variation in Y will be "unexplained," or not attributable to the influence of X. In fact, the vertical lines in Figure 15.7 represent the **unexplained variation,** or the difference between our best prediction of Y with X and the actual scores. The unexplained variation is thus the scattering of the actual scores around the regression line and can be found by subtracting the predicted Y scores from the actual Y scores for each case and then squaring and summing the differences. These operations can be summarized as $\Sigma(Y - Y')^2$, and the resultant sum would measure the amount of error in predicting Y that remains even after X has been taken into account. The proportion of the total variation in Y unexplained by X can be found by subtracting the value of r^2 from 1.00. Unexplained variation is usually attributed to the influence of some combination of other variables, measurement error, and random chance.

As you may have recognized by this time, the explained and unexplained variations bear a reciprocal relationship with one another. As one of these sums increases in value, the other decreases. Furthermore, the stronger the linear relationship between X and Y, the greater the value of the explained variation and the lower the unexplained variation. In the case of a perfect relationship ($r = \pm 1.00$), the unexplained variation would be 0 and r^2 would be 1.00. This would indicate that X explains, or accounts for, all the variation in Y and that we could predict Y from X without error. On the other hand, when X and Y are not linearly related ($r = 0.00$), the explained variation would be 0 and r^2 would be 0.00. In such a case, we would conclude that X explains none of the variation in Y and does not improve our ability to predict Y.

Relationships intermediate between these two extremes can be interpreted in terms of how much X increases our ability to predict, or explain, Y. For the dual-career families, we calculated an r of 0.50. Squaring this value yields a coefficient of determination of 0.25 ($r^2 = 0.25$), which indicates that number of children (X) explains 25% of the total variation in husband's housework (Y). When predicting the number of hours per week that husbands in such families would devote to housework, we will make 25% fewer errors by basing the predictions on number of children and predicting from the regression line, as opposed to ignoring this variable and predicting the mean of Y for every case. Also, 75% of the variation in Y is unexplained by X and presumably due to some combination of the influence of other variables, measurement error, and random chance. *(For practice in the interpretation of r^2, see any of the problems at the end of this chapter.)*

15.7 THE CORRELATION MATRIX

Social science research projects usually include many variables, and the data analysis phase of a project often begins with the examination of a **correlation matrix:** a table that shows the relationships between all possible pairs of variables. The correlation matrix gives a quick, easy-to-read overview of the interrelationships in the data set and may suggest strategies or "leads" for further analysis. These tables are commonly included in the professional research literature, and it will be useful to have some experience reading them.

An example of a correlation matrix, using cross-national data, is presented in Table 15.5. The matrix uses variable names as rows and columns, and the cells in the table show the bivariate correlation (usually a Pearson's *r*) for each combination of variables. Note that the row headings duplicate the column headings. To read the table, begin with GDP per capita, the variable in the far left-hand column (column 1) and top row (row 1). Read down column 1 or across row 1 to see the correlations of this variable with all other variables, including the correlation of GDP per capita with itself (1.00) in the top cell. To see the relationships between other variables, move from column to column or row to row.

Note that the diagonal from upper left to lower right of the matrix presents the correlation of each variable with itself. Values along this diagonal will always be exactly 1.00 and, since this information is not useful, it could easily be deleted from the table. Also note that the cells below and to the left of the diagonal are redundant with the cells above and to the right of the diagonal. For example, look at the second cell down (row 2) in column 1. This cell displays the correlation between GDP per capita and inequality, as does the cell in the top row (row 1) of column 2. In other words, the cells below and to the left of the diagonal are mirror images of the cells above and to the right of the diagonal. Commonly, research articles in the professional literature will delete the redundant cells in order to make the table more readable.

What does this matrix tell us? Starting at the upper left of the table (column 1), we can see that GDP per capita has a moderate negative relationship with

TABLE 15.5 A CORRELATION MATRIX SHOWING INTERRELATIONSHIPS FOR FIVE VARIABLES ACROSS 161 NATIONS

	(1)	(2)	(3)	(4)	(5)
	GDP per capita	Inequality	Unemployment Rate	Literacy	Voter Turnout
(1) GDP per capita	1.00	−0.43	−0.34	0.46	0.28
(2) Inequality	−0.43	1.00	0.33	−0.15	−0.36
(3) Unemployment Rate	−0.34	0.33	1.00	−0.48	−0.28
(4) Literacy	0.46	−0.15	−0.48	1.00	0.40
(5) Voter Turnout	0.28	−0.36	−0.28	0.40	1.00

VARIABLES:

(1) *GDP per capita*: Gross domestic product (the total value of all goods and services) divided by population size. This variable is an indicator of the level of affluence and prosperity in the society. Higher scores mean greater prosperity.
(2) *Inequality*: An index of income inequality. Higher scores mean greater inequality.
(3) *Unemployment Rate*: The annual rate of joblessness.
(4) *Literacy Rate*: Number of people over 15 able to read and write per 1000 population.
(5) *Voter Turnout*: Percentage of eligible voters who participated in the most recent election.

Application 15.2

Are nations that have more educated populations more likely to engage in discussions about politics? Random samples from 10 nations have been asked how often they discuss politics with friends. Information has also been gathered on the average years of school completed for people over 25 in each nation. How are these variables related? The data are presented in the following table. The "X" variable is average years of schooling completed for the population 25 years of age and older. The "Y" variable is the percentage of respondents who say they discuss politics with friends "frequently." Columns have been added for all necessary sums.

Nation	X	$X - \overline{X}$	Y	$Y - \overline{Y}$	$(X - \overline{X})(Y - \overline{Y})$	$(X - \overline{X})^2$	$(Y - \overline{Y})^2$
China	6	−2.7	24	9.3	−25.11	7.29	86.49
Argentina	9	0.3	18	3.3	0.99	0.09	10.89
United States	12	3.3	17	2.3	7.59	10.89	5.29
Japan	10	1.3	7	−7.7	−10.01	1.69	59.29
Mexico	7	−1.7	13	−1.7	2.89	2.89	2.89
India	5	−3.7	16	1.3	−4.81	13.69	1.69
South Africa	6	−2.7	11	−3.7	9.99	7.29	13.69
Finland	10	1.3	7	−7.7	−10.01	1.69	59.29
Canada	12	3.3	12	−2.7	−8.91	10.89	7.29
Germany	10	1.3	22	7.3	9.49	1.69	53.29
Totals	87	0.0	147	0.0	−27.90	58.10	300.10

$$\overline{X} = \frac{87}{10} = 8.7$$

$$\overline{Y} = \frac{147}{10} = 14.7$$

*Data are from the World Values Survey and www.nationmaster.com.

The slope (b) is

$$b = \frac{\Sigma(X - \overline{X})(Y - \overline{Y})}{\Sigma(X - \overline{X})^2}$$

$$b = \frac{-27.90}{58.10}$$

$$b = -0.48$$

A slope of −0.48 means that for every increase in years of education (a unit change in X), there is a decrease of 0.48 points in the percentage of people who frequently discuss politics with their friends.

The Y intercept (a) is

$$a = \overline{Y} - b\overline{X}$$

$$a = 14.7 - (-0.48)(8.7)$$

$$a = 14.7 - (-4.18)$$

$$a = 18.88$$

The least-squares regression equation is

$$Y = a + bX = 18.87 + (-0.48)X$$

The correlation coefficient is

$$r = \frac{\Sigma(X - \overline{X})(Y - \overline{Y})}{\sqrt{[\Sigma(X - \overline{X})^2][\Sigma(Y - \overline{Y})^2]}}$$

$$r = \frac{-27.90}{\sqrt{(58.10)(300.10)}}$$

$$r = \frac{-27.90}{\sqrt{17435.81}}$$

$$r = \frac{-27.90}{132.05}$$

$$r = -0.21$$

For these 10 nations, education and frequency of discussing politics have a weak, negative relationship. Frequency of discussing politics decreases as education increases. The coefficient of determination, r^2, is $(0.21)^2$, or 0.04. This indicates that 4% of the variance in frequency of discussing politics is explained by education for this sample of 10 nations.

inequality and unemployment rate, which means that more affluent nations tend to have less inequality and lower rates of joblessness. GDP per capita also has a moderate positive relationship with literacy (more affluent nations have higher levels of literacy) and a weak-to-moderate positive relationship with voter turnout (more affluent nations tend to have higher levels of participation in the electoral process).

To assess the other relationships in the data set, move from column to column and row to row, one variable at a time. For each subsequent variable, there will be one fewer cell of new information. For example, consider inequality, the variable in column 2 and row 2. We have already noted its moderate negative relationship with GDP per capita, and, of course, we can ignore the correlation of the variable with itself. This leaves only three new relationships, which can be read by moving down column 2 or across row 2. Inequality has a positive moderate relationship with unemployment (the greater the inequality, the greater the unemployment), a weak negative relationship with literacy (nations with more inequality tend to have lower literacy rates), and a moderate negative relationship with voter turnout (the greater the inequality, the lower the turnout).

For unemployment, the variable in column 3, there are only two new relationships: a moderate negative correlation with literacy (the higher the unemployment, the lower the literacy) and a weak-to-moderate negative relationship with voter turnout (the higher the unemployment rate, the lower the turnout). For voter turnout, the variable in column 5, there is only one new relationship. Voter turnout has a moderate positive relationship with literacy (turnout increases as literacy goes up).

In closing, we should note that the cells in a correlation matrix will often include other information in addition to the bivariate correlations. It is common, for example, to include the number of cases on which the correlation is based and, if relevant, an indication of the statistical significance of the relationship.

15.8 CORRELATION, REGRESSION, LEVEL OF MEASUREMENT, AND DUMMY VARIABLES

Correlation and regression are very powerful and useful techniques, so much so that they are often used to analyze relationships between variables that are not interval-ratio in level of measurement. This practice is generally not a problem for "continuous" ordinal-level variables that have a broad range of possible scores, even though the variables may lack true zero points and equal distances from score to score. We considered an example of these types of variables when we discussed Spearman's rho in Chapter 14.

Researchers also use correlation and regression when working with "collapsed" ordinal variables. These are variables that have a limited number of scores (usually between two and five), such as survey items that ask respondents about their support for capital punishment or gay marriage (see Chapter 14). As was the case with continuous ordinal variables, this violation of level of measurement is not a particular problem as long as results are treated with a suitable amount of caution.

While researchers have a good deal of leeway in including ordinal-level variables for correlation and regression, this flexibility does not extend to nominal-level variables. Computing correlation or regression coefficients for variables such as marital status and religious denomination simply does not make sense. Why? Remember that the "scores" of nominal-level variables are not numbers and have no mathematical quality. We might represent a Protestant with a score of "2" and a Catholic with a score of "1," but the former score is not "twice as much" as the latter. The scores of nominal-level variables are labels, not numbers. Because these variables are nonmathematical, it makes no sense to compute a slope or to discuss positive or negative relationships.

This is an unfortunate situation. Many of the variables that are most important in everyday social life—gender, marital status, race, and ethnicity—are nominal in level of measurement and cannot be included in a regression equation or a correlational analysis, two of the most powerful and sophisticated tools available for social science research.

Fortunately, researchers have developed a way to solve this problem and include nominal-level variables by creating **dummy variables**. Dummy variables can be any level of measurement, including nominal, and have exactly two categories, one coded as 0 and the other coded as 1. Treated this way, nominal-level variables, such as gender (for example, with males coded as 0 and females coded as 1), race (with whites coded as 0 and blacks as 1), and religious denomination (Catholics coded as 1 and non-Catholics coded as 0) are commonly included in regression equations.

To illustrate, imagine that we were concerned with the relationship between race and education as measured by number of years of schooling completed. If we coded whites as 0 and blacks as 1, we could compute a slope and Y intercept, write a regression equation using race as an independent variable, and examine the correlation between the two variables. Suppose we measured the education of a sample of black (coded as 1) and white (coded as 0) Americans and found the following regression equation:

$$Y = a + bX$$
$$Y = 12.0 + (-0.5)X$$

Education is the dependent (or Y) variable and race (X) is the independent variable. The regression line crosses the vertical axis of the scattergram at the point where $Y = 12.0$. The value for the slope ($b = -0.5$) indicates a negative relationship: As race "increases" (or moves toward the higher score associated with being black), education tends to decrease. In other words, the black respondents in this sample averaged fewer years of schooling than the white respondents. Note that the *sign* of the slope (b) would have been positive had we reversed the coding scheme and labeled whites as 1 and blacks as 0, but the *value* of b would have stayed exactly the same. The coding scheme for dummy variables is arbitrary, and, as with ordinal-level variables, the researcher needs to be clear about what the values of a dummy variable indicate.

We can also use Pearson's r to assess the strength and direction of relationships with dummy variables. If we found an r of -0.23 between race and education, we would conclude that there was a weak-to-moderate, "negative" relationship between these variables for this sample. Consistent with the sign of the slope, we could also say that education decreased as race "increased," or moved from white to black. Also, using the coefficient of determination, we can say that race explains, or accounts for, 5% ($r^2 = 0.23^2 = .05$) of the variance in education. *(For experience in working with dummy variables, see problem 15.9b.)*

15.9 TESTING PEARSON'S *r* FOR SIGNIFICANCE

When the relationship measured by Pearson's r is based on data from a random sample, it will usually be necessary to test r for its statistical significance. That is, we will need to know if a relationship between the variables can be assumed to exist in the population from which the sample was drawn. To illustrate this test, the r of 0.50 from the dual-wage-earner family sample will be used. As was the case when testing gamma and Spearman's rho, the null hypothesis states that there is no linear association between the two variables in the population from which the sample was drawn. The population parameter is symbolized as ρ (rho), and the appropriate sampling distribution is the t distribution.

To conduct this test, we need to make a number of assumptions in step 1. Most should be quite familiar, but several are new. First, we must assume that both variables are normal in distribution **(bivariate normal distributions).** Second, we must assume that the relationship between the two variables is roughly linear in form.

The third assumption involves a new concept: **homoscedasticity.** Basically, a homoscedastistic relationship is one where the variance of the Y scores is uniform for all values of X. That is, if the Y scores are evenly spread above and below the regression line for the entire length of the line, the relationship is homoscedastistic.

A visual inspection of the scattergram will usually be sufficient to appraise the extent to which the relationship conforms to the assumptions of linearity and homoscedasticity. As a rule of thumb, if the data points fall in a roughly symmetrical, cigar-shaped pattern, whose shape can be approximated with a straight line, then it is appropriate to proceed with this test of significance. Any significant evidence of nonlinearity or marked departures from homoscedasticity may indicate the need for an alternative measure of association and thus a different test of significance.

Step 1. Making Assumptions and Meeting Test Requirements.

Model: Random sampling
Level of measurement is interval-ratio
Bivariate normal distributions
Linear relationship
Homoscedasticity
Sampling distribution is normal

Step 2. Stating the Null Hypothesis.

$$H_0: \rho = 0$$
$$(H_1: \rho \neq 0)$$

Step 3. Selecting the Sampling Distribution and Establishing the Critical Region. With the null hypothesis of "no relationship" in the population, the sampling distribution of all possible sample r's is approximated by the t distribution. Degrees of freedom are equal to $(N - 2)$.

$$\text{Sampling distribution} = t \text{ distribution}$$
$$\text{Alpha} = 0.05$$
$$\text{Degrees of freedom} = N - 2 = 10$$
$$t(\text{critical}) = -2.28$$

Step 4. Computing the Test Statistic. The equation for computing the test statistic is given in Formula 15.6.

FORMULA 15.6

$$t(\text{obtained}) = r\sqrt{\frac{N-2}{1-r^2}}$$

Substituting the values into the formula, we would have:

$$t(\text{obtained}) = r\sqrt{\frac{N-2}{1-r^2}}$$

$$t(\text{obtained}) = (0.50)\sqrt{\frac{12-2}{1-(0.50)^2}}$$

$$t(\text{obtained}) = (0.50)\sqrt{\frac{10}{0.75}}$$

$$t(\text{obtained}) = (0.50)(3.65)$$

$$t(\text{obtained}) = 1.83$$

Step 5. Making a Decision and Interpreting the Results of the Test. Since the test statistic does not fall into the critical region as marked by t(critical), we fail to reject the null hypothesis. Even though the variables are substantially related in the sample, we do not have sufficient evidence to conclude that the variables are also related in the population. The test indicates that the sample value of $r = 0.50$ could have occurred by chance alone if the null hypothesis is true and the variables are unrelated in the population. *(For practice in conducting and interpreting tests of significance with Pearson's r, see problems 15.1, 15.2, 15.4, 15.6, 15.8, and 15.9.)*

15.10 INTERPRETING STATISTICS: THE CORRELATES OF CRIME

What causes crime? Sociologists have been researching this question since the discipline was first founded and have conducted an enormous amount of research and theory construction. While we cannot contribute to this voluminous body of work in a text devoted to statistical analysis, we can investigate some of the relationships and correlations that are of continuing interest to criminologists.

One prominent school of criminological thought argues that crime is related to poverty. A central proposition of this approach might be phrased as: "Crime rates will be highest among the most disadvantaged and impoverished groups, those with the highest rates of unemployment and the lowest levels of 'economic viability' (job skills, levels of education) in the legitimate economy."[2] In this installment of Interpreting Statistics, we will take a look at some empirical relationships between crime and poverty using the 50 states as our sample. We'll measure crime with the homicide rate (the number of homicides per 100,000 population). The homicide rate is one of the more trustworthy measures of crime. The rates for many other types of crimes, such as rape and assault, are gross underestimates because victims of these crimes are less likely to report the

[2]For example, see Currie, E., and Skolnock, J. 1997. *America's Problems: Social Issues and Social Problems.* New York: Longman, p.347.

incident to the police and, thus, the incident is not included in the official counts of criminal activity. Homicides are much more likely to come to the attention of the authorities, if for no other reason than the dead body that must be dealt with. For the independent variable, poverty, we will use the percent of the population of the state that is living below the poverty line.

The basic descriptive statistics for the variables are reported in Table 15.6 for the year 2004, the latest year for which full information is available. In this year, homicide rates ranged from a high of 12.7 murders per 100,000 population in Louisiana to a low of 1.4 murders per 100,000 population in Maine, New Hampshire, and North Dakota. Poverty rates ranged from a high of 21.3% in Mississippi to a low of 7.6% in Connecticut.

The first step in assessing an association between interval-ratio-level variables is to produce a scatterplot. Figure 15.8 plots the homicide rate on the vertical, or *Y*, axis and the poverty rate along the horizontal, or *X*, axis. What can we say about this relationship? The regression line is not horizontal, so there is a relationship between these variables. The dots (states) are fairly well scattered around the regression line, and we can see immediately that this will be, at best,

TABLE 15.6 BASIC DESCRIPTIVE STATISTICS, 2004

Variable	Mean	Standard Deviation	Range	Number of Cases
Homicide rate (2004)	4.64	2.39	11.3	50
Poverty rate (2004)	12.70	3.24	14.0	50

FIGURE 15.8 HOMICIDE RATE BY PERCENT POOR (50 states)

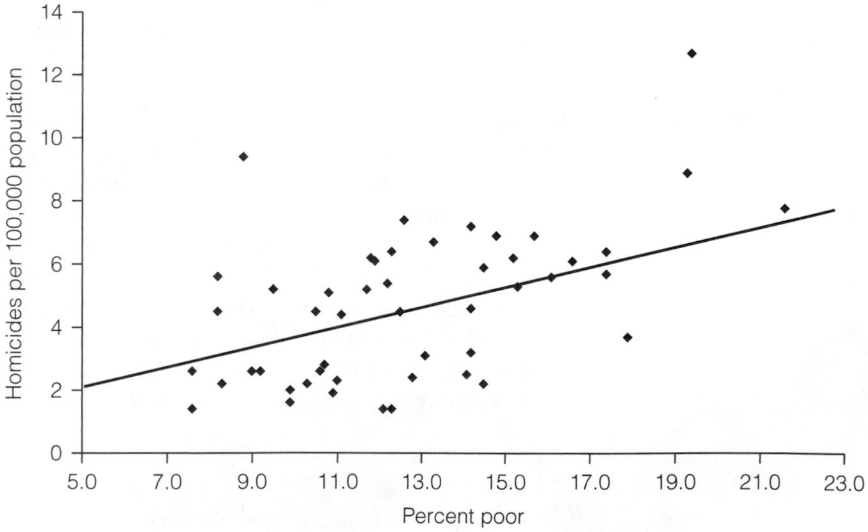

a moderately strong relationship. The regression line slopes up from left to right, so this is a positive relationship: States that have higher poverty rates tend to have higher homicide rates (that is, homicide rates tend to increase as poverty rates increase).

The second step in assessing this relationship is to specify the regression line. We'll skip the mechanics of computation here and simply report the values for the regression coefficients (a and b).

$$Y = a + bX$$
$$Y = -0.48 + (0.40)X$$

The regression line will cross the Y axis at the point where $Y = -0.48$, and every unit increase in X (percent poor) produces an increase of 0.40 in the homicide rate.

Next, we will calculate and interpret Pearson's r. Again, we'll skip the details of the computation and report that $r = 0.54$, which reinforces the impression that there is a moderate-to-strong relationship between homicide and poverty. The coefficient of determination (r^2) is .29, which means that poverty rate, by itself, explains 29% of the variation in the homicide rate.

In sum, the linear regression equation, the correlation coefficient, and the coefficient of determination suggest that there is a moderately strong relationship between inequality and crime. The amount of unexplained variation (71%) suggests that many other variables besides poverty have an important influence on the homicide rate.

We should also note two other limitations of this simple test. First, correlation is not the same thing as causation. Just because two variables are correlated does not mean that they have a causal relationship. This moderate-to-strong, positive relationship might be taken as support for the proposition that higher rates of poverty lead to higher rates of violence, but the mere existence of a relationship—even a strong relationship in the direction predicted by the theory—does not prove that one variable causes the other.

Second, consider a related point: The crime of homicide is by definition an individual act of behavior, but the data we used in this test were collected from states. Just because there is an association between the variables at the macro (state) level does not necessarily mean that the variables are related in the same way at the micro (individual) level.[3] All we know from our analysis is that states with higher rates of poverty tend to have higher rates of homicide. Our theory would lead us to assume that it is the victims of inequality (poor people) who are the offenders, but this conclusion is not proven by this analysis. In fact, it may be the wealthier residents of the poorer states who are committing the murders. We would need much more information—on both the micro and macro levels—before we could come to any final conclusions with respect to the theory that poverty and crime are related.

[3]This problem is called the *ecological fallacy*.

SUMMARY

This summary is based on the example used throughout the chapter.

1. We began with a question: Is the number of children in dual-wage-earner families related to the number of hours per week husbands devote to housework? We presented the observations in a scattergram (Figure 15.1), and our visual impression was that the variables were associated in a positive direction. The pattern formed by the observation points in the scattergram could be approximated with a straight line; thus, the relationship was roughly linear.
2. Values of Y can be predicted with the freehand regression line, but predictions are more accurate if the least-squares regression line is used. The least-squares regression line is the line that best fits the data by minimizing the variation in Y. Using the formula that defines the least-squares regression line ($Y = a + bX$), we found a slope (b) of 0.69, which indicates that each additional child (a unit change in X) is accompanied by an increase of 0.69 hours of housework per week for the husbands. We also predicted, based on this formula, that in a dual-wage-earner family with six children ($X = 6$), husbands would contribute 5.63 hours of housework a week ($Y' = 5.63$ for $X = 6$).
3. Pearson's r is a statistic that measures the overall linear association between X and Y. Our impression from the scattergram of a substantial positive relationship was confirmed by the computed r of 0.50. We also saw that this relationship yields an r^2 of .25, which indicates that 25% of the total variation in Y (husband's housework) is accounted for, or explained, by X (number of children).
4. Assuming that the 12 families represented a random sample, we tested the Pearson's r for its statistical significance and found that, at the .05 level, we could not assume that these two variables were also related in the population.
5. We acquired a great deal of information about this bivariate relationship. We know the strength and direction of the relationship and have also identified the regression line that best summarizes the effect of X on Y. We know the amount of change we can expect in Y for a unit change in X. In short, we have a greater volume of more precise information about this association between interval-ratio variables than we ever did about associations between ordinal or nominal variables. This is possible, of course, because the data generated by interval-ratio measurement are more precise and flexible than those produced by ordinal or nominal measurement techniques.

SUMMARY OF FORMULAS

Least-squares regression line:

15.1 $$Y = a + bX$$

Slope (b):

15.2 $$b = \frac{\Sigma(X - \overline{X})(Y - \overline{Y})}{\Sigma(X - \overline{X})^2}$$

Y intercept (a):

15.3 $$a = \overline{Y} - b\overline{X}$$

Pearson's r:

15.4. $$r = \frac{\Sigma(X - \overline{X})(Y - \overline{Y})}{\sqrt{[\Sigma(X - \overline{X})^2][\Sigma(Y - \overline{Y})^2]}}$$

Coefficient of determination:

15.5 $$r^2 = \frac{\Sigma(Y' - \overline{Y})}{\Sigma(Y - \overline{Y})}$$

15.6 $$t(\text{obtained}) = r\sqrt{\frac{N - 2}{1 - r^2}}$$

GLOSSARY

Bivariate normal distributions. The model assumption in the test of significance for Pearson's r that both variables are normally distributed.

Coefficient of determination (r^2). The proportion of all variation in Y that is explained by X. Found by squaring the value of Pearson's r.

Conditional means of Y. The mean of all scores on Y for each value of X.

Dummy variable. A nominal-level variable dichotomized so that it can be used in regression analysis. A dummy variable has two scores, one coded as 0 and the other as 1.

Explained variation. The proportion of all variation in Y that is attributed to the effect of X. Equal to $\Sigma(Y' - \overline{Y})^2$.

Homoscedasticity. The model assumption in the test of significance for Pearson's r that the variance of the Y scores is uniform across all values of X.

Linear relationship. A relationship between two variables in which the observation points (dots) in the scattergram can be approximated with a straight line.

Pearson's r (r). A measure of association for variables that have been measured at the interval-ratio level; ρ (Greek letter rho) is the symbol for the *population* value of Pearson's r.

Regression line. The single, best-fitting straight line that summarizes the relationship between two variables. Regression lines are fitted to the data points by the least-squares criterion, whereby the line touches all conditional means of Y or comes as close to doing so as possible.

Scattergram. Graphic display device that depicts the relationship between two variables.

Slope (b). The amount of change in one variable per unit change in the other; b is the symbol for the slope of a regression line.

Total variation. The spread of the Y scores around the mean of Y. Equal to $\Sigma(Y - \bar{Y})^2$.

Unexplained variation. The proportion of the total variation in Y that is not accounted for by X. Equal to $\Sigma(Y - Y')^2$.

Y intercept (a). The point where the regression line crosses the Y axis.

Y′. Symbol for predicted score on Y.

PROBLEMS

15.1 PS Why does voter turnout vary from election to election? For municipal elections in five different cities, information has been gathered on the percent of registered voters who actually voted, unemployment rate, average years of education for the city, and the percentage of all political ads that used "negative campaigning" (personal attacks, negative portrayals of the opponent's record, etc.). For each relationship:

a. Draw a scattergram and a freehand regression line.

b. Compute the slope (b) and find the Y intercept (a). *(HINT: Remember to compute b before computing a. A computing table such as Table 15.3 is highly recommended.)*

c. State the least-squares regression line and predict the voter turnout for a city in which the unemployment rate was 12, a city in which the average years of schooling was 11, and an election in which 90% of the ads were negative.

d. Compute r and r^2. *(HINT: A computing table such as Table 15.3 is highly recommended. If you constructed one for computing b, you already have most of the quantities you will need to solve for r.)*

e. Assume these cities are a random sample and conduct a test of significance for each relationship.

f. Describe the strength and direction of the relationships in a sentence or two. Which (if any) relationships were significant? Which factor had the strongest effect on turnout?

TURNOUT AND UNEMPLOYMENT

City	Turnout	Unemployment Rate
A	55	5
B	60	8
C	65	9
D	68	9
E	70	10

TURNOUT AND LEVEL OF EDUCATION

City	Turnout	Average Years of School
A	55	11.9
B	60	12.1
C	65	12.7
D	68	12.8
E	70	13.0

TURNOUT AND NEGATIVE CAMPAIGNING

City	Turnout	% of Negative Ads
A	55	60
B	60	63
C	65	55
D	68	53
E	70	48

15.2 SOC Occupational prestige scores for a sample of fathers and their oldest son and oldest daughter are shown in the table on page 386.

Family	Father's Prestige	Son's Prestige	Daughter's Prestige
A	80	85	82
B	78	80	77
C	75	70	68
D	70	75	77
E	69	72	60
F	66	60	52
G	64	48	48
H	52	55	57

Analyze the relationship between father's and son's prestige and the relationship between father's and daughter's prestige. For each relationship:

a. Draw a scattergram and a freehand regression line.

b. Compute the slope (b) and find the Y intercept (a).

c. State the least-squares regression line. What prestige score would you predict for a son whose father had a prestige score of 72? What prestige score would you predict for a daughter whose father had a prestige score of 72?

d. Compute r and r^2.

e. Assume these families are a random sample and conduct a test of significance for both relationships.

f. Describe the strength and direction of the relationships in a sentence or two. Does the occupational prestige of the father have an impact on his children? Does it have the same impact for daughters as it does for sons?

15.3 GER The residents of a housing development for senior citizens have completed a survey in which they indicated how physically active they are and how many visitors they receive each week. Are these two variables related for the 10 cases reported here? Draw a scattergram and compute r and r^2. Find the least-squares regression line. What would be the predicted number of visitors for a person whose level of activity was a 5? How about a person who scored 18 on level of activity?

Case	Level of Activity	Number of Visitors
A	10	14
B	11	12
C	12	10
D	10	9
E	15	8
F	9	7
G	7	10
H	3	15
I	10	12
J	9	2

15.4 PS The following variables were collected for a random sample of 10 precincts during the last national election. Draw scattergrams and compute r and r^2 for each combination of variables and test the correlations for their significance. Write a paragraph interpreting the relationship between these variables.

Precinct	Percent Democrat	Percent Minority	Voter Turnout
A	50	10	56
B	45	12	55
C	56	8	52
D	78	15	60
E	13	5	89
F	85	20	25
G	62	18	64
H	33	9	88
I	25	0	42
J	49	9	36

15.5 SOC/CJ The table on page 387 presents the scores of 10 states on each of six variables: three measures of criminal activity and three measures of population structure. Crime rates are number of incidents per 100,000 population as of 2004.

For each combination of crime rate and population characteristic:

a. Draw a scattergram and a freehand regression line.

b. Compute the slope (b) and find the Y intercept (a).

c. State the least-squares regression line. What homicide rate would you predict for a state with a growth rate of -1? What robbery rate would you predict for a state with a population density of 250? What auto theft rate would you predict for a state in which 50% of the population lived in urban areas?

d. Compute r and r^2.

e. Assume these states are a random sample and conduct a test of significance for both relationships.

f. Describe the strength and direction of each of these relationships in a sentence or two.

15.6 Data on three variables have been collected for 15 nations. The variables are fertility rate (average number of children born to each woman), education for females (expressed as a percentage of all students at the secondary level who are female), and maternal mortality (death rate for mothers per 100,000 live births).

Table for Problem 15.5

	Crime Rates			Population		
State	Homicide	Robbery	Car Theft	Growth[1]	Density[2]	Urban[3]
Maine	1	22	99	4	43	70
New York	5	174	213	2	408	88
Ohio	5	153	357	1	280	77
Iowa	2	38	183	1	53	61
Virginia	5	93	233	7	191	73
Kentucky	6	38	183	3	105	56
Texas	6	159	418	10	87	82
Arizona	7	134	963	16	52	88
Washington	3	95	696	7	95	82
California	7	172	703	7	232	94

[1]Percentage change in population from 2000 to 2005.
[2]Population per square mile of land area, 2005.
[3]Percent of population living in urban areas, 2000.
Source: United States Bureau of the Census, *Statistical Abstracts of the United States: 2007*. Washington, DC, 2007.

Table for Problem 15.6

Nation	Fertility	Education of Females	Maternal Mortality
Niger	7.6	38	590
Cambodia	3.4	36	440
Guatemala	3.8	47	190
Ghana	4.0	45	210
Bolivia	2.9	48	390
Kenya	4.9	48	590
Dominican Republic	2.8	55	230
Mexico	2.4	51	55
Vietnam	1.9	47	95
Turkey	1.9	38	130
United States	2.1	49	8
China	1.7	45	55
United Kingdom	1.7	53	7
Japan	1.4	49	8
Italy	1.3	48	7

Data are from www.nationmaster.com.

a. Compute r and r^2 for each combination of variables.

b. Summarize these relationships in terms of strength and direction.

15.7 The basketball coach at a small local college believes that his team plays better and scores more points in front of larger crowds. The number of points scored and attendance for all home games last season are reported here. Do these data support the coach's argument?

Game	Points Scored	Attendance
1	54	378
2	57	350
3	59	320
4	80	478
5	82	451
6	75	250
7	73	489
8	53	451
9	67	410
10	78	215
11	67	113
12	56	250
13	85	450
14	101	489
15	99	472

(*continued next column*)

(*continued*)

15.8 The following table presents the scores of 15 states on three variables. Compute r and r^2 for each combination of variables. Assume that these 15 states are a random sample of all states, and test the correlations for their significance. Write a paragraph interpreting the relationship among these three variables.

State	Per Capita Expenditures on Education, 2004	Percent High School Graduates, 2005*	Rank in Per Capita Income, 2005
Arkansas	1158	81	48
Colorado	1599	89	7
Connecticut	2142	90	1
Florida	1286	87	23
Illinois	2334	87	14
Kansas	1418	91	25
Louisiana	1367	80	42
Maryland	1627	87	4
Michigan	1880	89	24
Mississippi	1178	80	49
Nebraska	1373	90	20
New Hampshire	1632	92	6
North Carolina	1233	84	37
Pennsylvania	1617	86	18
Wyoming	1896	90	12

*Based on percentage of population age 25 and older.
Source: United States Bureau of the Census, *Statistical Abstracts of the United States: 2007*. Washington, DC, 2007.

15.9 Fifteen individuals were randomly selected from the respondents to a public opinion survey and their scores are reproduced here.

a. Calculate Pearson's *r* for age (*X*) and four dependent variables (*Y*'s): prestige, number of children, support for abortion, and hours of TV watching.

b. Calculate Pearson's *r* for gender (*X*) and four dependent variables (*Y*'s): support for abortion, prestige, hours of TV watching, and number of children.

c. Test the four relationships in either a or b for significance.

KEY TO VARIABLES:

"Number of Children," "Age," and "Hours of TV" are actual values. "Occupational Prestige" is a continuous ordinal scale that indicates the amount of respect or esteem associated with occupation. The higher the score, the greater the prestige. "Support for legal abortion" is a collapsed ordinal scale that measures respondent's strength of agreement or disagreement with the statement "A women should be able to get a legal abortion for any reason," where 1 = strongly agree, 2 = agree, 3 = neither agree nor disagree, 4 = disagree, and 5 = strongly disagree. "Gender" is a nominal-level variable coded as a dummy variable, with 0 = female and 1 = male.

Occupational Prestige	Number of Children	Age	Support for legal abortion	Hours of TV per Day	Gender
32	3	34	3	1	1
50	0	41	1	3	1
17	0	52	5	2	1
69	3	67	1	5	0
17	0	40	1	5	0
52	0	22	2	3	1
32	3	31	3	4	0
50	0	23	4	4	0
19	9	64	1	6	1
37	4	55	4	2	0
14	3	66	5	5	1
51	0	22	2	0	0
45	0	19	3	7	0
44	0	21	4	1	1
46	4	58	2	0	1

SPSS for Windows

Using *SPSS for Windows* to Produce Pearson's *r*

SPSS DEMONSTRATION 15.1 What Are the Correlates of Occupational Prestige?

To what extent is a person's social class, as measured by the prestige of her or his occupation, a result of personal efforts and talents (or lack thereof)? How does the social class of our parents shape our ability to rise in the class structure? We can investigate these issues by analyzing the relationships between three variables: *prestg80*, or the prestige of the respondent's occupation; *educ*, the respondent's years of schooling; and *papres80*, the prestige of the respondent's father's occupation.

If a person's social class position reflects her or his own preparation for the job market, there should be a strong positive correlation between *educ* and *prestg80*. On the other hand, if it's *who* you know and not *what* you know—if class position is greatly affected by the social class of one's family of origin—then there should be a strong positive relationship between *papres80* and *prestg80*. By comparing the strength of these bivariate correlations, we may be able to make some judgment about the relative importance of these factors in determining occupational prestige.

To request Pearson's *r*, click **Analyze, Correlate,** and **Bivariate.** The **Bivariate Correlations** window appears, with the variable list on the left. Find *educ*, *papres80*, and *prestg80* in the list and click the arrow to move them into the **Variables** box. If you wish, you can get descriptive statistics for each variable by clicking the **Options** button and requesting means and standard deviations. Unless you request otherwise, the program will conduct two-tailed tests of significance on the correlations. Note that Spearman's rho (see Chapter 14) is also an option. Click **OK,** and your output should look like this:

Correlations

		HIGHEST YEAR OF SCHOOL COMPLETED	FATHERS OCCUPATIONAL PRESTIGE SCORE (1980)	RS OCCUPATIONAL PRESTIGE SCORE (1980)
HIGHEST YEAR OF SCHOOL COMPLETED	Pearson Correlation	1	.338(**)	.504(**)
	Sig. (2-tailed)		.000	.000
	N	1424	758	1346
FATHERS OCCUPATIONAL PRESTIGE SCORE (1980)	Pearson Correlation	.338(**)	1	.297(**)
	Sig. (2-tailed)	.000		.000
	N	758	758	719
RS OCCUPATIONAL PRESTIGE SCORE (1980)	Pearson Correlation	.504(**)	.297(**)	1
	Sig. (2-tailed)	.000	.000	
	N	1346	719	1347

**Correlation is significant at the 0.01 level (2-tailed).

The output is in the form of a correlation matrix showing the relationships between all variables, including the correlation of a variable with itself. For each possible relationship, Pearson's *r* is reported in the top row, the results of the test of significance in the second row, and sample size in the third row (*N* varies because not everyone answered every question).

For our questions, we need to focus on the correlations between *prestg80, educ*, and *papres80.* The relationships are statistically significant ($p < .000$) and positive. The relationship between *prestg80* and *educ* is stronger (0.504) than the relationship between *prestg80* and *papres80* (0.297). This suggests that preparation for the job market (*educ*) is more important than the relative advantage or disadvantage conferred by one's family of origin (*papres80*). Success may come more easily to people from an advantaged family background (note that the correlation between father's occupational prestige and respondent's education is 0.338), but preparation (education) matters more. To further disentangle these relationships would require more sophisticated statistics; fortunately, some will become available in Chapter 17.

SPSS DEMONSTRATION 15.2 Are the Correlates of Occupational Prestige Affected by Gender?

Are the pathways to success and higher prestige open to all equally? We can begin to address this question by analyzing separately the relationships reported in Demonstration 15.1 for men and women. If gender has no effect on a person's opportunities for success, *prestg80, papres80*, and *educ* should be related in the same way for both men and women.

To observe the effect of sex, we will split the GSS sample into two subfiles. Men and women will then be processed separately, and SPSS will produce one correlation matrix for men and another for women.

Click **Data** from the main menu and then click **Split File.** On the **Split File** window, click the button next to "organize output by groups." This will generate separate outputs for whatever groups we select. Choose *sex* from the variable list and click the arrow to move the variable name into the **Groups Based On** window. Click **OK,** and all procedures requested will be done separately for men and women. To restore the full sample, call up the **Split File** window again and click the **Reset** button. For now, click **Statistics, Correlate,** and **Bivariate** and rerun the correlation matrix for *prestg80, papres80*, and *educ*. The output will look like this:

FOR MALES:

Correlations(a)

		HIGHEST YEAR OF SCHOOL COMPLETED	FATHERS OCCUPATIONAL PRESTIGE SCORE (1980)	RS OCCUPATIONAL PRESTIGE SCORE (1980)
HIGHEST YEAR OF SCHOOL COMPLETED	Pearson Correlation	1	.397(**)	.517(**)
	Sig. (2-tailed)		.000	.000
	N	644	353	617
FATHERS OCCUPATIONAL PRESTIGE SCORE (1980)	Pearson Correlation	.397(**)	1	.314(**)
	Sig. (2-tailed)	.000		.000
	N	353	353	344
RS OCCUPATIONAL PRESTIGE SCORE (1980)	Pearson Correlation	.517(**)	.314(**)	1
	Sig. (2-tailed)	.000	.000	
	N	617	344	617

**Correlation is significant at the 0.01 level (2-tailed).
[a]RESPONDENTS SEX = MALE.

FOR FEMALES:

Correlations(a)

		HIGHEST YEAR OF SCHOOL COMPLETED	FATHERS OCCUPATIONAL PRESTIGE SCORE (1980)	RS OCCUPATIONAL PRESTIGE SCORE (1980)
HIGHEST YEAR OF SCHOOL COMPLETED	Pearson Correlation	1	.291(**)	.492(**)
	Sig. (2-tailed)		.000	.000
	N	780	405	729
FATHERS OCCUPATIONAL PRESTIGE SCORE (1980)	Pearson Correlation	.291(**)	1	.278(**)
	Sig. (2-tailed)	.000		.000
	N	405	405	375
RS OCCUPATIONAL PRESTIGE SCORE (1980)	Pearson Correlation	.492(**)	.278(**)	1
	Sig. (2-tailed)	.000	.000	
	N	729	375	730

**Correlation is significant at the 0.01 level (2-tailed).
[a]RESPONDENTS SEX = FEMALE.

The correlations for both groups are statistically significant and positive. The value and direction of the correlations we are interested in are essentially the same for both genders. For males, education has a stronger effect on current occupational prestige (0.517) than father's occupational prestige (0.314). Essentially the same point can be made for females: The correlation of occupational prestige with education (0.492) is stronger than the correlation with father's occupational prestige (0.278). The correlations for females are very similar to those for males and suggest that gender makes little difference for these relationships.

Exercises

15.1 Run the analysis in Demonstration 15.1 again, with *income06* rather than *prestg80* as the indicator of social class standing. Use the same independent variables (*papres80, educ*) and see if the patterns are similar to those identified in Demonstration 15.1. (Ignore the fact that *income06* is only ordinal in level of measurement.) Write up your conclusions.

15.2 Run the analysis in Demonstration 15.2 again, but substitute *income06* for *prestg80*. Divide the sample by sex as before. Is there a gender difference in the correlates of *income06* as there was with *prestg80?*

15.3 Switch topics entirely and see if you can explain the variation in television viewing habits. What are the correlates of *tvhours?* Choose three or four possible independent variables (perhaps *age* and *educ*). Write up your results.

PART III CUMULATIVE EXERCISES

1. For each situation, compute and interpret the appropriate measure of association. Also, compute and interpret percentages for bivariate tables. Describe relationships in terms of the strength and pattern or direction.

a. For 10 cities, data have been gathered on total crime rate (major felonies per 100,000 population) and the percentage of people who are new immigrants (arrived in the United States within the past five years). Are the variables related?

City	Total Crime Rate	Percent Immigrants
A	1500	10
B	1200	18
C	2000	9
D	1700	11
E	1600	15
F	1000	20
G	1700	9
H	1300	22
I	900	10
J	700	15

b. There is some evidence that people's involvement in their communities (membership in voluntary organizations, participation in local politics, and so forth) has been declining, and television has been cited as the cause. Do the following data support the idea that TV is responsible for the decline?

Hours of Community Service:	Television Viewing			
	Low	Moderate	High	Totals
Low	5	10	18	33
Moderate	10	12	10	32
High	15	8	7	30
Totals	30	30	35	95

c. A national magazine has rated the states in terms of "quality of life" (a scale that includes health care, availability of leisure facilities, unemployment rate, and a number of other variables) and the quality of the state system of higher education. Both scales range from 1 (low) to 20 (high). Is there a correlation between these two variables for the 10 states listed?

State	Quality of Life	Quality of Higher Education
A	10	10
B	12	13
C	15	18

(*continued next page*)

(continued)

State	Quality of Life	Quality of Higher Education
D	18	20
E	10	15
F	9	11
G	11	12
H	8	6
I	13	9
J	6	8

d. Racial intermarriages are increasing in the United States, and the number of mixed-race people is growing. The following table shows the relationship between racial mixture of parents and the racial category, if any, with which a sample of people of mixed race identifies. Is there a relationship between these variables?

Do You Consider Yourself to Be:	Racial Mixture of Parents			Totals
	Black/White	White/Asian	Black/Asian	
Black	2	0	3	5
White	8	4	0	12
Asian	0	3	3	6
None of the above	10	8	4	22
Totals	20	15	10	45

2. A number of research questions are stated here. Each can be answered by at least one of the techniques presented in Chapters 12 through 15. For each research situation, compute the most appropriate measure of association and write a sentence or two in response to the question. The questions are presented in random order. In selecting a measure of association, you need to consider the number of possible values and the level of measurement of the variables.

The research questions refer to the database shown at the end of the exercise, which is taken from the 2006 General Social Survey (GSS). The actual questions asked and the complete response codes for the GSS are presented in Appendix G. Abbreviated codes are listed here. Some variables have been recoded for this exercise.

a. Are scores on "frequency of sex during the last year" associated with income or age? Compute a measure of association, assuming that the variables are interval-ratio in level of measurement.

b. Is fear associated with sex? Is it associated with marital status?

c. Is support for spanking associated with church attendance? Is it associated with marital status?

Survey Items:

1. Marital status of respondent (MARITAL):

1.. Married

2. Not married (includes widowed, divorced, etc.)

2. How often do you attend church? (ATTEND) See Appendix G for original codes:
   0. Never
   1. Rarely
   2. Often
3. Age of respondent (AGE) (Values are actual numbers.).
4. Respondent's total family income (INCOME98) See Appendix G for codes.
5. Sex:
   1. Male
   2. Female
6. Support for spanking (SPANKING):
   1. Favor
   2. Oppose
7. Fear of walking alone at night (FEAR):
   1. Yes
   2. No
8. Frequency of sex during past year. See Appendix G for codes.

Case	Marital Status	Church Attendance	Age	Income	Gender	Spanking	Fear	Frequency of Sex
1	2	0	22	12	2	1	2	5
2	1	0	52	13	1	1	2	2
3	1	2	44	16	1	2	1	3
4	1	0	56	10	1	1	2	1
5	1	2	61	8	2	2	2	0
6	1	0	28	19	2	1	2	5
7	2	1	59	9	1	2	1	0
8	1	1	69	11	1	2	2	5
9	2	0	23	4	1	1	2	6
10	2	2	31	20	2	1	1	6
11	1	2	67	21	2	2	1	1
12	1	1	46	9	1	1	1	2
13	2	0	19	10	2	1	1	5
14	1	2	34	11	1	2	2	2
15	2	0	29	18	2	1	1	4
16	1	1	31	16	1	1	1	5
17	2	1	88	6	2	2	2	1
18	1	0	24	15	2	1	1	4
19	2	2	69	11	1	1	1	2
20	1	2	60	14	1	2	2	3
21	1	1	29	12	2	1	2	4
22	1	2	43	13	2	2	1	2
23	1	0	35	20	1	1	1	4
24	2	2	38	9	2	2	2	4
25	2	0	83	19	1	2	1	0

Part IV Multivariate Techniques

The two chapters in this part introduce multivariate analytical techniques or statistics that allow us to analyze the relationships among more than two variables at a time. These statistics are extremely useful for probing possible causal relationships between variables and are commonly reported in the literature. In particular, Chapter 17 introduces regression analysis, which is the basis for many of the most popular and powerful statistical techniques used in social science research today.

Chapter 16 covers multivariate analysis for nominal- and ordinal-level variables that have been organized into table format. The chapter discusses the logic and procedures for analyzing the effect of a third (or control) variable on the relationship between independent and dependent variables. The possible outcomes of controlling for a third variable are introduced and analyzed one at a time, and the discussion is summarized in Table 16.5.

Chapter 17 introduces partial and multiple correlation and regression for variables at the interval-ratio level of measurement. Partial correlation is analogous to controlling for a third variable with bivariate tables, as presented in Chapter 16. I make several references to the concepts presented in Chapter 16 and to Table 16.5. However, it is not necessary to cover Chapter 16 in order to understand the material presented in Chapter 17.

Multiple regression and correlation are some of the most powerful and useful tools available to social science researchers. The mathematics underlying these techniques can become very complicated, and the chapter focuses on the simplest possible applications.

16 Elaborating Bivariate Tables

LEARNING OBJECTIVES

By the end of this chapter, you will be able to

1. Explain the purpose of multivariate analysis in terms of observing the effect of a control variable.
2. Construct and interpret partial tables.
3. Compute and interpret partial measures of association.
4. Recognize and interpret direct, spurious or intervening, and interactive relationships.
5. Compute and interpret partial gamma.
6. Cite and explain the limitations of elaborating bivariate tables.

16.1 INTRODUCTION

Few questions can be answered by a statistical analysis of only two variables; therefore, the typical research project will include many variables. In Chapters 12–15, we saw how various statistical techniques can be applied to bivariate relationships. In this chapter and Chapter 17, we see how some of these techniques can be extended to probe the relationships among three or more variables. This chapter presents some multivariate techniques appropriate for variables that have been measured at the nominal or ordinal level and organized into table format. Chapter 17 presents some techniques that can be used when the variables have been measured at the interval-ratio level.

Before considering the techniques themselves, we should consider why they are important and what they might be able to tell us. There are two general reasons for utilizing multivariate techniques. First, and most fundamental, is the goal of simply gathering additional information about a specific bivariate relationship by observing how that relationship is affected (if at all) by the presence of a third variable (or a fourth or a fifth). Multivariate techniques will increase the amount of information we have on the basic bivariate relationship and (we hope) enhance our understanding of that relationship.

A second and very much related rationale for multivariate statistics involves the issue of causation. While multivariate statistical techniques cannot prove the existence of causal connections between variables, they can provide valuable evidence in support of causal arguments and are very important tools for testing and revising theory.

16.2 CONTROLLING FOR A THIRD VARIABLE

For variables arrayed in tables, multivariate analysis proceeds by observing the effects of other variables on the bivariate relationship. That is, we observe the relationship between the independent (X) and dependent (Y) variables after a third variable (which we will call Z) has been controlled. We do this by reconstructing the relationship between X and Y for each value or score of Z. If the

TABLE 16.1 SATISFACTION WITH COLLEGE BY NUMBER OF MEMBERSHIPS IN STUDENT ORGANIZATIONS

	Memberships (X)		
Satisfaction (Y)	None	At Least One	Totals
Low	57 (*54.3%*)	56 (*33.9%*)	113
High	48 (*45.7%*)	109 (*66.1%*)	157
Totals	105 (*100.0%*)	165 (*100.0%*)	270
	Gamma = 0.40		

control variable has an effect, the relationship between X and Y will change under the various conditions of Z.[1]

The Bivariate Relationship. To illustrate, suppose that a researcher wishes to analyze the relationship between how well an individual is integrated into a group or organization and that individual's level of satisfaction with the group. The researcher has decided to focus on college students and their level of satisfaction with the college as a whole. The necessary information on satisfaction (Y) is gathered from a sample of 270 students, and all are asked to list the student clubs or organizations to which they belong. Integration (X) is measured by dividing the students into two categories: The first includes students who are not members of any organizations and the second includes students who are members of at least one organization. The researcher suspects that membership and satisfaction are positively related and that members will report higher levels of satisfaction than nonmembers. The relationship between these two variables is displayed in Table 16.1.

Inspection of the table suggests that these two variables are associated. The conditional distributions of satisfaction (Y) change across the two conditions of membership (X), and membership is associated with high satisfaction, whereas nonmembership is associated with low satisfaction. The existence and direction of the relationship are confirmed by the computation of a gamma of +0.40 for this table.[2] In short, Table 16.1 strongly suggests the existence of a relationship between integration and morale, and these results can be taken as evidence for a causal or direct relationship between the two variables. The causal relationship is summarized symbolically in Figure 16.1, where the arrow represents the effect of X on Y.

FIGURE 16.1 A DIRECT RELATIONSHIP BETWEEN TWO VARIABLES

$X \longrightarrow Y$

Partial Tables. The researcher recognizes, of course, that membership is not the sole cause of satisfaction (if it were, the gamma would be +1.00). Other variables may alter this relationship, and the researcher will need to consider the effects of these third variables (Z's) in a systematic and logical way. By doing so, the re-

[1] For the sake of brevity, we focus on the simplest application of multivariate statistical analysis, the case where the relationship between an independent (X) and dependent (Y) variable is analyzed in the presence of a single control variable (Z). Once the basic techniques are grasped, they can be readily expanded to situations involving more than one control variable.
[2] We will not display the computation of gamma here. See Chapter 14 for a review of this measure of association.

searcher will accumulate further information and detail about the bivariate relationship and can further probe the possible causal connection between *X* and *Y*.

For example, perhaps both satisfaction (*Y*) and membership (*X*) are affected by grades (*Z*) and students with higher grade-point averages are more satisfied with the university and are also more likely to participate actively in the life of the campus by joining organizations. How can the effects of this third variable on the bivariate relationship be investigated?

Consider Table 16.1 again. This table provides information about the distribution of 270 cases on two variables. By analyzing the table, we can see that 165 students are members, that 48 students are both nonmembers and highly satisfied, that the majority of the students are highly satisfied, and so forth. What the bivariate table cannot show us, of course, is the distribution of these students on GPA. For all we know at this point, the 109 highly satisfied members could all have high GPA's, low GPA's, or any combination of scores on this third variable. GPA is "free to vary" in this table, since the distribution of the cases on this variable is not accounted for.

We control for the effect (if any) of third variables by fixing their distributions so that they are no longer free to vary. We do this by sorting all cases in the sample according to their score on the third variable (*Z*) and then observing the relationship between *X* and *Y* for each value (or score) of *Z*. In the example at hand, we construct separate tables displaying the relationship between membership and satisfaction for each category of GPA. Tables that display the relationship between *X* and *Y* for each value of *Z* are called **partial tables.** In Table 16.2, the student sample has been subdivided into two groups based on GPA (*Z*): 135 students had "low" GPAs and the same number had "high" GPAs. The two partial tables in Table 16.2 show the relationship between membership (*X*) and satisfaction (*Y*) for each value of the control variable.

This type of multivariate analysis is called **elaboration** because the partial tables present the original bivariate relationship in a more detailed or elaborate

TABLE 16.2 SATISFACTION BY MEMBERSHIP, CONTROLLING FOR GPA

A. High GPA

	Memberships (*X*)		
Satisfaction (*Y*)	None	At Least One	Totals
Low	29 (*54.7%*)	28 (*34.2%*)	57
High	24 (*45.3%*)	54 (*65.9%*)	78
Totals	53 (*100.0%*)	82 (*100.0%*)	135

Gamma = 0.40

B. Low GPA

	Memberships (*X*)		
Satisfaction (*Y*)	None	At Least One	Totals
Low	28 (*53.9%*)	28 (*33.7%*)	56
High	24 (*46.2%*)	55 (*66.3%*)	79
Totals	52 (*100.0%*)	83 (*100.0%*)	135

Gamma = 0.39

form. The cell frequencies in the partial tables are subdivisions of the cell frequencies reported in Table 16.1. For example, if the frequencies of the cells in the partial tables are added together, the original frequencies of Table 16.1 will be reproduced.[3]

Also note how this method of controlling for other variables can be extended. In our example, the control variable (grades) had two categories, and, thus, there were two partial tables, one for each value of the control variable. If the control variable had had more than two categories, we would have had more partial tables. By the same token, we can control for more than one variable at a time by sorting the cases on all scores of all control variables and producing partial tables for each combination of scores on the control variables. Thus, if we had controlled for both GPA and gender, we would have had four partial tables to consider. There would have been one partial table for males with low GPA's, a second for males with high GPA's, and two partial tables for females with high and low GPA's.

Summary. For nominal and ordinal variables, we begin multivariate analysis by constructing partial tables, that is, tables that display the relationship between *X* and *Y* for each value of *Z*. The next step is to trace the effect of *Z* by comparing the partial tables with each other and with the original bivariate table. By analyzing the partial tables, we can observe the effects (if any) of the control variable on the original bivariate relationship.

16.3 INTERPRETING PARTIAL TABLES

The cell frequencies in the partial tables can follow a variety of forms, but we will concentrate on three basic patterns, determined by comparing the partial tables with each other and with the original bivariate table:

1. **Direct relationships:** The relationship between *X* and *Y* is the same in all partial tables and in the bivariate table.
2. **Spurious relationships or intervening relationships:** The relationship between *X* and *Y* is the same in all partial tables but much weaker than in the bivariate table.
3. **Interaction:** Each partial table and the bivariate table all show different relationships between *X* and *Y*.

Each pattern has different implications for the relationships among the variables and for the subsequent course of the statistical analysis. I next describe each pattern in detail and then summarize the discussion in Table 16.5.

Direct Relationships. This pattern is often called **replication,** because the partial tables reproduce (or replicate) the bivariate table; the cell frequencies are the same in the partial tables and the bivariate table. Measures of association

[3] The total of the cell frequencies in the partial tables will always equal the corresponding cell frequencies in the bivariate table except when, as often happens in "real life" research situations, the researcher is missing scores on the third variable for some cases. These cases must be deleted from the analysis; as a consequence, the partial tables will have fewer cases than the bivariate table.

calculated on the partial tables have the same value as the measure of association calculated for the bivariate table.

This outcome indicates that the control variable (*Z*) has no effect on the relationship between *X* and *Y*. Table 16.2 provides an example of this outcome. In this table, the relationship between membership (*X*) and satisfaction (*Y*) was investigated with GPA (*Z*) controlled. Table 16.2A shows the relationship for high-GPA students, and Table 16.2B shows the relationship for low-GPA students. The partial tables show the same conditional distributions of *Y*: For both high- and low-GPA students, about 45% of the nonmembers are highly satisfied, versus about 66% of the members. This same pattern was observed in the bivariate table (Table 16.1). Thus, the conditional distributions of *Y* are the same in each partial table as they were in the bivariate table.

The pattern of the cell frequencies will be easier to identify if we calculate measures of association. Working from the cell frequencies presented in Table 16.2, we see that the gamma for high-GPA students is 0.40 and that the gamma for low-GPA students is 0.39. The bivariate gamma (from Table 16.1) is 0.40. The fact that the partial and bivariate gammas are essentially the same value reinforces our conclusion that the relationship between *X* and *Y* is the same in the partial tables and the bivariate table.

This pattern of outcomes indicates that the control variable has no important impact on the bivariate relationship (if it did, the pattern in the partial tables would be different from that in the bivariate table) and may be ignored in any further analysis. In terms of the original research problem, the researcher may conclude that students who are members are more likely to express high satisfaction with the university regardless of their GPA. In other words, GPA has no effect on the relationship between membership and satisfaction. Low-GPA students who are members of at least one organization are just as likely to report high satisfaction as high-GPA students who are members of at least one organization. *(For practice in dealing with direct relationships, see Problems 16.1 and 16.2.)*

Spurious or Intervening Relationships. In this pattern, the relationship between *X* and *Y* is much weaker in the partial tables than in the bivariate table but the same across all partials. Measures of association for the partial tables are much lower in value (perhaps even dropping to 0.00) than the measure computed for the bivariate table. This outcome is consistent with two different causal relationships among the three variables. The first is called a spurious relationship or **explanation;** in this situation, *Z* is conceptualized as being antecedent to both *X* and *Y* (that is, *Z* is thought to occur earlier in time than the other two variables). In this pattern, *Z* causes both *X* and *Y* and the original bivariate relationship is said to be spurious. The apparent bivariate relationship between *X* and *Y* is explained by, or due to, the effect of *Z*. Once *Z* is controlled, the association between *X* and *Y* disappears and the value of the measures of association for the partial tables is dramatically lower than the measure of association for the bivariate table.

To illustrate, suppose that the researcher in our example had also controlled for class standing by dividing the sample into upperclasses (seniors and juniors) and underclasses (sophomores and freshmen). The reasoning of the researcher might be that dissatisfied students would have transferred or dropped out

Application 16.1

Seventy-eight juvenile males in a sample have been classified as high or low on a scale that measures involvement in delinquency. Also, each subject has been classified, using school records, as either a good or a poor student. The following table displays a strong relationship between these two variables for this sample ($G = -0.69$).

Delinquency	Academic Record: Poor	Academic Record: Good	Totals
Low	13 *(27.1%)*	20 *(66.7%)*	33 *(42.3%)*
High	35 *(72.9%)*	10 *(33.3%)*	45 *(57.7%)*
Totals	48 *(100.0%)*	30 *(100.0%)*	78 *(100.0%)*

Gamma = −0.69

Judging by the column percentages and the gamma, it is clear that juvenile males with poor academic records are especially prone to delinquency. Is this relationship between delinquency and academic record affected by whether the subject resides in an urban or nonurban area?

A. Urban areas:

Delinquency	Academic Record: Poor	Academic Record: Good	Totals
Low	10 *(27.8%)*	3 *(30.0%)*	13 *(28.3%)*
High	26 *(72.2%)*	7 *(70.0%)*	33 *(71.7%)*
Totals	36 *(100.0%)*	10 *(100.0%)*	46 *(100.0%)*

Gamma = −0.05

B. Nonurban areas:

Delinquency	Academic Record: Poor	Academic Record: Good	Totals
Low	3 *(25.0%)*	17 *(85.0%)*	20 *(62.5%)*
High	9 *(75.0%)*	3 *(15.0%)*	12 *(37.5%)*
Totals	12 *(100.0%)*	20 *(100.0%)*	32 *(100.0%)*

Gamma = −0.89

For urban juvenile males, the relationship between academic record and delinquency disappears. The gamma for this table is −0.05, and the column percentages are very similar. For this group, delinquency is not associated with experience in school.

For nonurban males, on the other hand, there is a very strong relationship between the two variables. Gamma is −0.89, and there is a dramatic difference in the column percentages. Poor students living in nonurban areas are especially prone to delinquency.

Comparing the partial tables with each other and with the bivariate table reveals an interactive relationship. Although urban juvenile males are more delinquent than nonurban males (71.7% of the urban males were highly delinquent, as compared to 37.5% of the nonurban males), their delinquency is not associated with academic record. For nonurban males, academic record is very strongly associated with delinquency. Urban males are more delinquent than nonurban males but not because of their experience in school. Nonurban males who are also poor students are especially likely to become involved in delinquency.

FIGURE 16.2 A SPURIOUS RELATIONSHIP

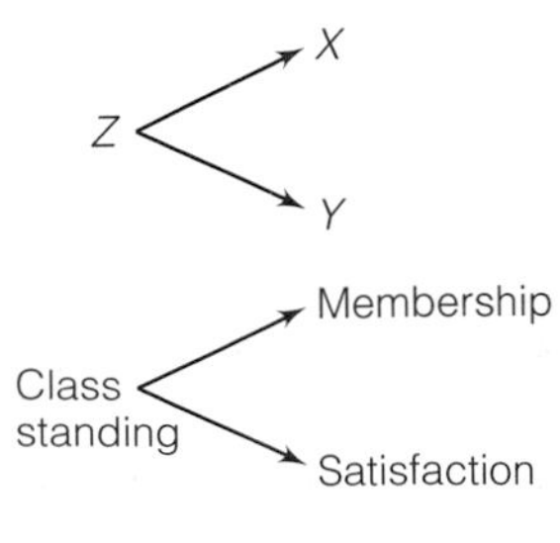

before they achieve upperclass standing and upperclass students are more satisfied because of simple self-selection. Also, students who have been on campus longer will be more likely to locate an organization of sufficient appeal or interest to join. This might especially be the case for organizations based on major field (such as the Accounting Club), which underclass students are less likely to join, or honorary organizations, for which underclass students are unlikely to qualify.

These thoughts about the possible relationships among these variables are expressed in diagram form in Figure 16.2. The absence of an arrow from membership (X) to satisfaction (Y) indicates that these variables are not truly

TABLE 16.3 SATISFACTION BY MEMBERSHIP, CONTROLLING FOR CLASS STANDING

A. Upperclass Students

	Memberships (X)		
Satisfaction (Y)	None	At Least One	Totals
Low	8 (*25.0%*)	32 (*24.8%*)	40
High	24 (*75.0%*)	97 (*75.2%*)	121
Totals	32 (*100.0%*)	129 (*100.0%*)	161

Gamma = 0.01

B. Underclass Students

	Memberships (X)		
Satisfaction (Y)	None	At Least One	Totals
Low	49 (*67.1%*)	24 (*66.7%*)	73
High	24 (*32.9%*)	12 (*33.3%*)	36
Totals	73 (*100.0%*)	36 (*100.0%*)	109

Gamma = 0.01

associated with each other but rather are both caused by class standing (Z). If this causal diagram is a correct description of the relationship between these three variables (if the association between X and Y is spurious), then the association between membership and satisfaction should disappear once class standing has been controlled. That is, even though the bivariate gamma was 0.40 (Table 16.1), the gammas computed on the partial tables will approach 0. Table 16.3 displays the partial tables generated by controlling for class standing.

The partial tables indicate that, once class standing is controlled, membership is no longer related to satisfaction. Regardless of their number of memberships, upperclass students are more likely to be high on satisfaction and underclass students are more likely to be low on satisfaction. In the partial tables, the distributions of Y no longer vary by the conditions of X, and the gammas computed on the partial tables are virtually 0. These results indicate that X and Y have no direct relationship: Their apparent relationship is spurious. Class standing (Z) is a key factor in accounting for satisfaction, and the analysis must be reoriented with class standing as an independent variable.

This outcome (partial measures much weaker than the original measure but equal to each other) is consistent with a second conception of the causal links among the variables. In addition to a causal scheme wherein Z is antecedent to both X and Y, Z might intervene between the two variables. This pattern, also called **interpretation,** is illustrated in Figure 16.3, where X is causally linked to Z, which is in turn linked to Y. This pattern indicates that, although X and Y are related, they are associated primarily through the control variable Z. This pattern of outcomes does not allow the researcher to distinguish between spurious relationships (Figure 16.2) and intervening relationships (Figure 16.3). The differentiation between these two types of causal patterns may be made on temporal or theoretical grounds but not on statistical grounds. *(For practice in dealing with spurious or intervening relationships see problem 16.6b.)*

FIGURE 16.3 AN INTERVENING RELATIONSHIP

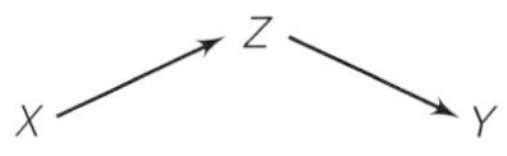

FIGURE 16.4 AN INTERACTIVE RELATIONSHIP

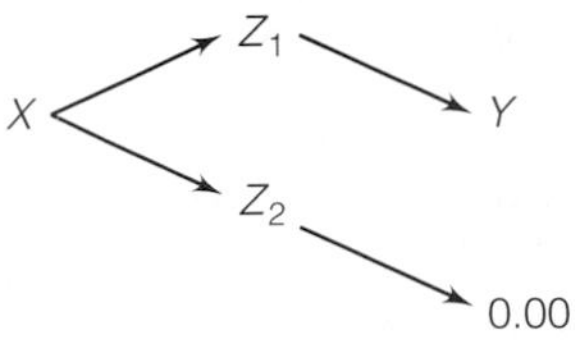

Interaction. In this pattern, also called **specification,** the relationship between *X* and *Y* changes markedly for the different values of the control variable. The partial tables differ from each other and from the bivariate table. Interaction can be manifested in various ways in the partial tables. One possible pattern, for example, is for one partial table to display a stronger relationship between *X* and *Y* than that displayed in the bivariate table, while the relationship between *X* and *Y* drops to 0 in a second partial table. Symbolically, this outcome could be represented as in Figure 16.4, which would indicate that *X* and the first category of Z (Z_1) have strong effects on *Y*, but, for the second category of Z (Z_2), there is no association between *X* and *Y*.

Interaction might be found, for example, in a situation in which *all* employees of a corporation were required to attend a program (*X*) designed to reduce antiblack prejudice (*Y*). Such a program would be likely to have stronger effects on white employees (Z_1) than on African American employees (Z_2). That is, the program would have an effect on prejudice only for certain types of subjects. The partial tables for white employees (Z_1) might show a strong relationship between program attendance and reduction in prejudice, whereas the partial tables for African American employees showed no relationship. African American employees, being unprejudiced against themselves in the first place, would be less affected by the program.

Interaction can take other forms. For example, the relationship between *X* and *Y* can vary not only in strength but also in direction between the partial tables. This causal relationship is symbolically represented in Figure 16.5, which indicates a situation where *X* and *Y* are positively related for the first category of Z (Z_1) and negatively related for the second category of Z (Z_2).

FIGURE 16.5 AN INTERACTIVE RELATIONSHIP

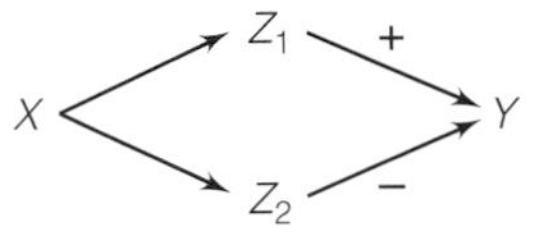

The researcher investigating the relationship between club membership and satisfaction with college life establishes a final control for race and divides the sample into white and black students. The partial tables are displayed in Table 16.4 and show an interactive relationship. The relationship between membership

TABLE 16.4 SATISFACTION BY MEMBERSHIP, CONTROLLING FOR RACE

A. White Students

	Memberships (*X*)		
Satisfaction (*Y*)	None	At Least One	Totals
Low	40 (*50.0%*)	20 (*16.7%*)	60
High	40 (*50.0%*)	100 (*83.3%*)	140
Totals	80 (*100.0%*)	120 (*100.0%*)	200

Gamma = 0.67

B. Black Students

	Memberships (*X*)		
Satisfaction (*Y*)	None	At Least One	Totals
Low	17 (*68.0%*)	36 (*80.0%*)	53
High	8 (*32.0%*)	9 (*20.0%*)	17
Totals	25 (*100.0%*)	45 (*100.0%*)	70

Gamma = −0.31

Application 16.2

Cohabitation has become increasingly popular in U.S. society over the past several decades. What are the sources of support for this living arrangement? One obvious hypothesis would be that attitude towards cohabitation is related to religiosity and is less supported by people who attend church more frequently.

A recent public opinion survey asked respondents if they agree or disagreed with the statement "It's a good idea for a couple who intend to get married to live together first." The following table shows the relationship between church attendance and response to this item.

	Church Attendance			
Opinion on Cohabitation	Never (1)	Yearly or Monthly (2)	Daily or Weekly (3)	Totals
(1) Agree	141 *(63.8%)*	347 *(54.7%)*	60 *(20.5%)*	548 *(47.8%)*
(2) Neutral	47 *(21.3%)*	130 *(20.6%)*	49 *(16.8%)*	226 *(19.7%)*
(3) Disagree	33 *(14.9%)*	157 *(24.8%)*	183 *(62.7%)*	373 *(32.5%)*
Totals	221 *(100.0%)*	634 *(100.0%)*	292 *(100.0%)*	1147 *(100.0%)*

$\chi^2 = 179.74\ (\alpha < .001)$ $\quad G = 0.51$

The chi square test indicates that the relationship is statistically significant, and the gamma of 0.51 indicates a moderate-to-strong positive relationship. Given the way the responses are coded (higher scores indicate greater *dis*approval), the positive gamma means that, consistent with the hypothesis stated earlier, disapproval increases as church attendance increases. Disapproval is highest among frequent attendees (62.7%) and lowest among nonchurchgoers (14.9%).

Is this relationship between opinion and church attendance affected by the gender of the respondent? The gammas for the partial tables are fairly close in

A. Males

	Church Attendance			
Opinion on Cohabitation	Never (1)	Yearly or Monthly (2)	Daily or Weekly (3)	Totals
(1) Agree	81 *(68.6%)*	149 *(52.5%)*	21 *(25.6%)*	251 *(51.9%)*
(2) Neutral	17 *(14.43%)*	62 *(21.8%)*	13 *(15.9%)*	92 *(19.0%)*
(3) Disagree	20 *(16.9%)*	73 *(25.7%)*	48 *(58.5%)*	141 *(29.1%)*
Totals	118 *(100.0%)*	284 *(100.0%)*	82 *(100.0%)*	484 *(100.0%)*

$G = 0.44$

(continued next page)

and satisfaction is different for white students (Z_1) and for black students (Z_2), and each partial table is different from the bivariate table. For white students, the relationship is positive and stronger than in the bivariate table. White students who are also members are much more likely to express high overall satisfaction with the university, as indicated by both the percentage distribution (83% of the white students who are members are highly satisfied) and the measure of association (gamma = +0.67 for white students).

Application 16.2: *(continued)*

B. Females

Opinion on Cohabitation	Church Attendance: Never (1)	Yearly or Monthly (2)	Daily or Weekly (3)	Totals
(1) Agree	60 *(58.3%)*	198 *(56.6%)*	39 *(18.6%)*	251 *(44.8%)*
(2) Neutral	30 *(29.1%)*	68 *(19.4%)*	36 *(17.1%)*	134 *(20.2%)*
(3) Disagree	13 *(12.6%)*	84 *(24.0%)*	135 *(64.3%)*	141 *(35.0%)*
Totals	103 *(100.0%)*	350 *(100.0%)*	210 *(100.0%)*	663 *(100.0%)*

$G = 0.54$

value to the bivariate gamma (0.51), which would indicate a direct relationship between church attendance and attitude, but the gammas show a weaker relationship for males and a slightly stronger relationship for females, a pattern that is more consistent with an interactive relationship between the three variables. In other words, these results are somewhat ambiguous and do not precisely match the pattern we would expect in either a direct relationship or an interactive relationship (see Table 16.5). Such messy results are extremely common in real-life research situations and they call for some caution when making conclusions.

What can we say? Church attendance has a weaker effect on support for cohabitation for males than for females. Males are generally more supportive of cohabitation (compare the top rows—"agree"—of the two partial tables) and non-churchgoing males are especially approving of cohabitation (68.6% agreed). Although females are less supportive, the higher value for gamma means that church attendance makes a bigger difference for them. Another way to visualize this pattern is to note that the maximum difference in column percentages (see Chapter 12) is greater for females (52% vs. 43% for males). Also, females who attend church frequently are especially disapproving of cohabitation (64.3% disagreed).

Looking at the overall patterns in the partial and bivariate tables, we would probably conclude that this relationship is closer to a direct than to an interactive relationship. The difference between the gammas for the partial tables is certainly worthy of note but is not dramatic enough to call this a case of interaction. We would probably eliminate gender from further analysis (see Table 16.5) and select another control variable (e.g., age or education) to test the relationship between church attendance and support for cohabitation further.

For black students, the relationship between membership and satisfaction is very different, nearly the reverse of the pattern shown by white students. The great majority (80%) of the black students who are members report low satisfaction, and the gamma for this partial table is negative (−0.31). Thus, satisfaction increases with membership for white students but decreases with membership for black students. Based on these results, one might conclude that the social meanings and implications of joining student clubs and organizations vary by race and that the function of joining has different effects for black students. It may be, for example, that black students join different kinds of organizations than do white students. If the black students belonged primarily to a black student association that had an antagonistic relationship with the university, then belonging to such an organization could increase dissatisfaction with the university as it

TABLE 16.5 A SUMMARY OF THE POSSIBLE RESULTS OF CONTROLLING FOR THIRD VARIABLES

Partial Tables (compared with bivariate table) Show	Pattern	Implications for Further Analysis	Likely Next Step in Statistical Analysis	Theoretical Implications
Same relationship between *X* and *Y*	Direct relationship, replication	Disregard *Z*	Analyze another control	Theory that *X* causes *Y* is supported
Weaker relationship between *X* and *Y*	Spurious relationship	Incorporate *Z*	Focus on relationship between *Z* and *Y*	Theory that *X* causes *Y* is not supported
	Intervening relationship	Incorporate *Z*	Focus on relationships between *X*, *Y*, and *Z*	Theory that *X* causes *Y* is partially supported but must be revised to take *Z* into account
Mixed	Interaction	Incorporate *Z*	Analyze subgroups (categories of *Z*) separately	Theory that *X* causes *Y* is partially supported but must be revised to take *Z* into account

increased awareness of racial problems. *(For practice in dealing with interaction, see problem 16.4.)*

Conclusion. In closing this section, let me stress that, for the sake of clarity, I have presented examples that have been unusually "clean." In an actual research project, controlling for third variables will probably have results that are considerably more ambiguous and open to interpretation than the examples presented here. In the case of spurious relationships, for example, the measures of association computed for the partial tables will probably not actually drop to zero, even though they may be dramatically lower than the bivariate measure. (The pattern where the partial measures are roughly equivalent and much lower than the bivariate measure but not zero is sometimes called *attenuation.*) It is probably best to consider the foregoing examples as ideal types against which any given empirical result can be compared.

Table 16.5 summarizes this discussion by outlining guidelines for decision making for each of the three outcomes discussed in this section. Since your own results will probably be more ambiguous than the ideal types presented here, you should regard this table as a set of suggestions and not as a substitute for your own creativity and sensitivity to the problem under consideration.

16.4 PARTIAL GAMMA (G_p)

When the results of controlling for a third variable indicate a direct, spurious, or intervening relationship, it is often useful to compute an additional measure that indicates the overall strength of the association between *X* and *Y* after the effects of the control variable (*Z*) have been removed. This statistic, called partial gamma (G_p), is somewhat easier to compare with the bivariate gamma than are the gammas computed on the partial tables separately. G_p is computed across all partial tables by Formula 16.1:

FORMULA 16.1

$$G_p = \frac{\Sigma N_s - \Sigma N_d}{\Sigma N_s + \Sigma N_d}$$

where N_s = the number of pairs of cases ranked the same across all partial tables
N_d = the number of pairs of cases ranked differently across all partial tables

In words, ΣN_s is the total of all N_s's from the partial tables, and ΣN_d is the total of all N_d's from all partial tables. (See Chapter 14 to review the computation of N_s and N_d.)

To illustrate the computation of G_p, let us return to Table 16.2, which showed the relationship between satisfaction and membership while controlling for GPA. The gammas computed on the partial tables were essentially equal to each other and to the bivariate gamma. Our conclusion was that the control variable had no effect on the relationship, and this conclusion can be confirmed by also computing partial gamma.

From Table 16.2A High GPA	From Table 16.2B Low GPA
$N_s = (29)(54) = 1566$	$N_s = (28)(55) = 1540$
$N_d = (28)(24) = 672$	$N_d = (28)(24) = 672$

$$\Sigma N_s = 1566 + 1540 = 3106$$
$$\Sigma N_d = 672 + 672 = 1344$$

$$G_p = \frac{\Sigma N_s - \Sigma N_d}{\Sigma N_s + \Sigma N_d}$$

$$G_p = \frac{3106 - 1344}{3106 + 1344}$$

$$G_p = \frac{1762}{4450}$$

$$G_p = 0.40$$

The partial gamma measures the strength of the association between *X* and *Y* once the effects of *Z* have been removed. In this instance, the partial gamma is the same value as the bivariate gamma ($G_p = G = 0.40$) and indicates that GPA has no effect on the relationship between satisfaction and membership.

When class standing was controlled (Table 16.3), clear evidence of a spurious relationship was found, since the gammas computed on the partial tables dropped almost to zero. Let us see what the value of partial gamma (G_p) would be for this second control:

From Table 16.3A Upperclassmen	From Table 16.3B Underclassmen
$N_s = (8)(97) = 776$	$N_s = (49)(12) = 588$
$N_d = (32)(24) = 768$	$N_d = (24)(24) = 576$

$$\Sigma N_s = 776 + 588 = 1364$$
$$\Sigma N_d = 768 + 576 = 1344$$

$$G_p = \frac{\Sigma N_s - \Sigma N_d}{\Sigma N_s + \Sigma N_d}$$

$$G_p = \frac{1364 - 1344}{1364 + 1344}$$

$$G_p = \frac{20}{2708}$$

$$G_p = 0.01$$

ONE STEP AT A TIME **Computing Partial Gamma (G_p)**

Step 1: Compute N_s for all partial tables. Add up all values for N_s from all partial tables to find ΣN_s.

Step 2: Compute N_d for all partial tables. Add up all values for N_d from all partial tables to find ΣN_d.

Step 3: Find partial gamma (G_p) by solving Formula 16.1:

$$G_p = \frac{\Sigma N_s - \Sigma N_d}{\Sigma N_s + \Sigma N_d}$$

a. Subtract ΣN_d from ΣN_s.
b. Add ΣN_d to ΣN_s.
c. Divide the quantity you found in step a by the quantity you found in step b. The result is G_p.

Once the effects of the control variable are removed, there is no relationship between X and Y. The very low value of G_p confirms our previous conclusion that the bivariate relationship between membership and satisfaction is spurious and actually due to the effects of class standing.

In a sense, G_p tells us no more about the relationships than we can see for ourselves from a careful analysis of the percentage distributions of Y in the partial tables or by a comparison of the measures of association computed on the partial tables. The advantage of G_p is that it tells us, in a single number (that is, in a compact and convenient way), the precise effects of Z on the relationship between X and Y. While G_p is no substitute for the analysis of the partial tables per se, it is a convenient way of stating our results and conclusions when working with direct or spurious relationships.

Although G_p can be calculated in cases of interactive relationships (see Table 16.4), it is rather difficult to interpret in these instances. If substantial interaction is found in the partial tables, this indicates that the control variable has a profound effect on the bivariate relationship. Thus, we should not attempt to separate the effects of Z from the bivariate relationship, and, since G_p involves exactly this kind of separation, it should not be computed.

16.5 WHERE DO CONTROL VARIABLES COME FROM?

In one sense, this question is easy to answer: Control variables come mainly from theory. Social research proceeds in many different ways and is begun in response to a variety of problems. However, virtually all research projects are guided by a more-or-less-explicit theory or by some question about the relationship between two or more variables. The ultimate goal of social research is to develop defensible generalizations that improve our understanding of the variables under consideration and link back to theory at some level.

Thus, research projects are anchored in theory; and the concepts of interest, which will later be operationalized as variables, are first identified and their interrelationships first probed at the theoretical level. The social world is exceedingly complex, and any attempt to explain it with a simple bivariate relationship will fail. Theories are, almost by definition, multivariate, even when they focus on a particular bivariate relationship. Thus, to the extent that a research project is anchored in theory, the theory itself will suggest the control variables that need to be incorporated into the analysis. In the example used throughout this chapter, I tried to suggest that any researcher attempting to probe the relationship between involvement in an organization and satisfaction would, in the course of thinking over the possibilities, identify a number of additional variables that needed to be explicitly incorporated into the analysis.

Of course, textbook descriptions of the research process are oversimplified and suggest that research flows smoothly from conceptualization to operationalization to quantification to generalization. In reality, research is full of surprises, unexpected outcomes, and unanticipated results. Research in "real life" is more loosely structured and requires more imagination and creativity than textbooks can fully convey. My point is that the control variables that might be appropriate to incorporate in the data-analysis phase will be suggested or implied in the theoretical backdrop of the research project along with the researcher's imagination and sensitivity to the problem being addressed.

These considerations have taken us well beyond the narrow realm of statistics and back to the planning stages of the research project. At this early time the researcher must make decisions about which variables to measure during the data-gathering phase and, thus, which variables might be incorporated as potential controls. Careful thinking and an extended consideration of possible outcomes at the planning stage will pay significant dividends during the data-analysis phase. Ideally, all relevant control variables will be incorporated and readily available for statistical analysis.

Thus, control variables come from the theory underlying the research project and from creative and imaginative thinking and planning during the early phases of the project. Nonetheless, it is not unheard of for a researcher to realize during data analysis that the control variable now so obviously relevant was never measured during data gathering and is thus unavailable for statistical analysis.

16.6 THE LIMITATIONS OF ELABORATING BIVARIATE TABLES

The basic limitation of this technique involves sample size. Elaboration is a relatively inefficient technique for multivariate analysis because it requires that the researcher divide the sample into a series of partial tables. If the control variable has more than two or three possible values or if we attempt to control for more than one variable at a time, many partial tables will be produced. The greater the number of partial tables, the more likely we are to run out of cases to fill all of the cells. Empty or small cells, in turn, can create serious problems in terms of generalizability and confidence in our findings.

To illustrate, the example used throughout this chapter began with two dichotomized variables and a four-cell table (Table 16.1). Each of the control variables was also dichotomized, and we never confronted more than two partial tables with four cells each, or eight cells for each control variable. If we had used a control variable with three values, we would have had 12 cells to fill up, and if we had attempted to control for two dichotomized variables, we would have had 16 cells in four different partial tables. Clearly, as control variables become more elaborate and/or as the process of controlling becomes more complex, the phenomenon of empty or small cells will increasingly become a problem.

Two potential solutions to this dilemma immediately suggest themselves. The easy solution is to reduce the number of cells in the partial tables by collapsing categories within variables. If all variables are dichotomized, for example, the number of cells will be kept to a minimum. The best solution is to work with only very large samples so that there will be plenty of cases to fill up the cells. Unfortunately, the easy solution will often violate common sense (collapsed categories are more likely to group dissimilar elements together); and the best solution is not always feasible (mundane matters of time and money rear their ugly heads).

A third solution to the problem of empty cells requires the (sometimes risky) assumption that the variables of interest are measured at the interval-ratio level. At that level the techniques of partial and multiple correlation and regression, to be introduced in Chapter 17, are available. These multivariate techniques are more efficient than elaboration because they utilize all cases simultaneously and do not require that the sample be divided among the various partial tables.

16.7 INTERPRETING STATISTICS: ANALYZING SOCIAL INVOLVEMENT

In Chapter 14, we analyzed the sources of social involvement in the United States. In this section, we analyze how the bivariate relationship behaves after controlling for a third variable. As you recall from Chapter 14, social involvement was measured with a composite variable that ranked people as high, moderate, or low in their involvement across a variety of different types of social activities. We analyzed the relationship between this variable and three different independent variables: marital status, work status, and amount of TV viewing. To keep the number of tables manageable, we will concentrate on the bivariate relationship between TV viewing and engagement while controlling for gender and race. The bivariate relationship is reproduced here as Table 16.6. Since both variables are ordinal in level of measurement, we will focus on gamma as our measure of the strength and direction of the relationship.

The original theory investigated in Chapter 14 was that high levels of TV viewing decrease involvement in social life. The gamma of −0.11 indicates a weak, negative association between the variables. There is a relationship between the variables and the relationship is in the direction predicted by the theory. Also, the pattern of column percentages is largely what would be expected if the theory is true (e.g., people who don't watch much TV are more engaged).

TABLE 16.6 INVOLVEMENT BY LEVEL OF TV WATCHING

	TV Watching per Day			
Involvement	Low (0–1 hours)	Moderate (2–3 hours)	High (4 or more hours)	Totals
Low	160 (*32.1%*)	283 (*31.0%*)	241 (*41.8%*)	684 (*34.4%*)
Moderate	128 (*25.7%*)	285 (*31.2%*)	137 (*23.7%*)	550 (*27.7%*)
High	211 (*42.3%*)	345 (*37.8%*)	199 (*34.5%*)	755 (*38.0%*)
Totals	499 (*100.0%*)	913 (*100.0%*)	577 (*100.0%*)	1,989 (*100%*)
		Gamma = −0.11		

TABLE 16.7 INVOLVEMENT BY LEVEL OF TV WATCHING, CONTROLLING FOR GENDER

A. Males

Involvement	TV Watching per Day: Low (0–1 hours)	Moderate (2–3 hours)	High (4 or more hours)	Totals
Low	70 (*30.4%*)	121 (*30.8%*)	101 (*40.9%*)	292 (*33.6%*)
Moderate	60 (*26.1%*)	109 (*27.7%*)	54 (*21.9%*)	223 (*25.6%*)
High	100 (*43.5%*)	163 (*41.5%*)	92 (*37.2%*)	355 (*40.8%*)
Totals	230 (*100.0%*)	393 (*100.0%*)	247 (*100.0%*)	870 (*100.0%*)

Gamma = −0.10

B. Females

Involvement	TV Watching per Day: Low (0–1 hours)	Moderate (2–3 hours)	High (4 or more hours)	Totals
Low	90 (*33.5%*)	162 (*31.2%*)	140 (*42.4%*)	392 (*35.0%*)
Moderate	68 (*25.3%*)	176 (*33.8%*)	83 (*25.2%*)	327 (*29.2%*)
High	111 (*41.3%*)	182 (*35.0%*)	107 (*32.4%*)	400 (*35.7%*)
Totals	269 (*100.1%*)	520 (*100.0%*)	330 (*100.0%*)	1,119 (*99.9%*)

Gamma = −0.11

Even though the relationship is weak, this pattern is consistent with the idea that heavy TV viewing leads to lower levels of social activity.[4]

Will this relationship retain its strength and direction after controlling for other variables? Table 16.7 presents the partial tables and gammas after controlling for sex, and Table 16.8 does the same for race as a control variable. We will analyze the partial tables separately and then come to some conclusions about the original bivariate relationship.

For both male respondents (part A of the table) and female (part B), partial gamma is the same strength and direction as the bivariate gamma. Using Table 16.5 as a guide, we see that these partial tables, compared to the bivariate table, show a direct relationship. The fact that gender has no effect reinforces the original theory (even though relationships are weak).

As an additional step in the analysis, we can compute G_p for Table 16.7. We will not show the calculations here but simply report that $G_p = -0.11$, the same value as the bivariate gamma.

Table 16.8 presents the results of controlling for race. We use only two values (black and white) of this control variable because the other categories (Asian Americans, Native Americans, Hispanic Americans) had very few cases. Remember that sample size is a major limitation of this form of statistical analysis.

[4]The pattern is also consistent with the reverse causal argument: Lower levels of participation lead to higher levels of TV viewing.

TABLE 16.8 INVOLVEMENT BY LEVEL OF TV WATCHING, CONTROLLING FOR RACE

A. Whites

	TV Watching per Day			
Involvement	Low (0–1 hours)	Moderate (2–3 hours)	High (4 or more hours)	Totals
Low	110 (*28.1%*)	191 (*27.5%*)	151 (*42.4%*)	452 (*31.3%*)
Moderate	112 (*28.6%*)	233 (*33.6%*)	99 (*27.8%*)	444 (*30.8%*)
High	170 (*43.4%*)	270 (*38.9%*)	106 (*29.8%*)	546 (*37.9%*)
Totals	392 (*100.0%*)	694 (*100.0%*)	356 (*100.0%*)	1,442 (*100.0%*)
		Gamma = −0.16		

B. Blacks

	TV Watching per Day			
Involvement	Low (0–1 hours)	Moderate (2–3 hours)	High (4 or more hours)	Totals
Low	16 (*42.1%*)	37 (*35.6%*)	51 (*36.4%*)	104 (*36.9%*)
Moderate	7 (*18.4%*)	31 (*29.8%*)	25 (*17.9%*)	63 (*22.3%*)
High	15 (*39.5%*)	36 (*34.6%*)	64 (*45.7%*)	115 (*40.8%*)
Totals	38 (*100.0%*)	104 (*100.0%*)	140 (*100.0%*)	282 (*100.0%*)
		Gamma = 0.09		

The relationship for white respondents is almost exactly the same as the relationship in the bivariate table, both in terms of the pattern of column percentages and in terms of gamma. For black respondents, however, the relationship is quite different. The gamma is about the same strength but is positive rather than negative. This means that involvement in social life tends to increase as TV watching increases, exactly the opposite of whites. Needless to say, this pattern is contrary to the theory and, at the very least, suggests that the effects of TV on social life are not the same for every group in the population.

The control for race reveals an interactive relationship between these three variables (see Table 16.5). Any further analysis will have to incorporate racial (and maybe ethnic?) group as an additional variable and attempt to develop an explanation for the patterns observed in Table 16.8.

In conclusion, we can say that controlling for gender generally supports the original theory, but not so with the control for race. Our likely next step would be to analyze the effect of additional control variables (e.g., education and social class) in order to develop more information about the bivariate relationship and more evidence for (or against) the idea that TV viewing and involvement in social life are causally related.

SUMMARY

1. Most research questions require the analysis of the interrelationship among many variables, even when the researcher is primarily concerned with a specific bivariate relationship. Multivariate statistical techniques provide the researcher with a set of tools by which additional information can be gathered about the variables of interest and by which causal interrelationships can be probed.

2. When variables have been organized in bivariate tables, multivariate analysis proceeds by controlling for a third variable. Partial tables are constructed and compared with each other and with the original bivariate table. Comparisons are made easier if appropriate measures of association are computed for all tables.
3. A direct relationship exists between the independent (X) and dependent (Y) variables if, after controlling for the third variable (Z), the relationship between X and Y is the same across all partial tables and the same as in the bivariate table. This pattern suggests a causal relationship between X and Y.
4. If the relationship between X and Y is the same across all partial tables but much weaker than in the bivariate table, the relationship is either spurious (Z causes both X and Y) or intervening (X causes Z, which causes Y). Either pattern suggests that Z must be explicitly incorporated into the analysis.
5. Interaction exists if the relationship between X and Y varies across the partial tables and between each partial and the bivariate table. This pattern suggests that no simple or direct causal relationship exists between X and Y and that Z must be explicitly incorporated into the causal scheme.
6. Partial gamma (G_p) is a useful summary statistic that measures the strength of the association between X and Y after the effects of the control variable (Z) have been removed. Partial gamma should not be computed when an analysis of the partial tables shows substantial interaction.
7. Potential control variables must be identified before the data-gathering phase of the research project. The theoretical backdrop of the research project, along with creative thinking and some imagination, will suggest the variables that should be controlled for and measured.
8. Controlling for third variables by constructing partial tables is inefficient, in that the cases must be spread out across many cells. If the variables have many categories and/or the researcher attempts to control for more than one variable simultaneously, "empty cells" may become a problem. It may be possible to deal with this problem by either collapsing categories or gathering very large samples. If interval-ratio level of measurement can be assumed, the multivariate techniques presented in the next chapter will be preferred, since they do not require the partitioning of the sample.

SUMMARY OF FORMULAS

Partial gamma 16.1 $$G_p = \frac{\Sigma N_s - \Sigma N_d}{\Sigma N_s + \Sigma N_d}$$

GLOSSARY

Direct relationship. A multivariate relationship in which the control variable has no effect on the bivariate relationship.

Elaboration. The basic multivariate technique for analyzing variables arrayed in tables. Partial tables are constructed to observe the bivariate relationship in a more detailed or elaborated format.

Explanation. See **spurious relationship.**

Interaction. A multivariate relationship wherein a bivariate relationship changes across the categories of the control variable.

Interpretation. See **intervening relationship.**

Intervening relationship. A multivariate relationship wherein a bivariate relationship becomes substantially weaker after a third variable is controlled for. The independent and dependent variables are linked primarily through the control variable.

Partial gamma (G_p). A statistic that indicates the strength of the association between two variables after the effects of a third variable have been removed.

Partial tables. Tables produced when controlling for a third variable.

Replication. See **direct relationship.**

Specification. See **interaction.**

Spurious relationship. A multivariate relationship in which a bivariate relationship becomes substantially weaker after a third variable is controlled for. The independent and dependent variables are not causally linked. Rather, both are caused by the control variable.

***Z*.** Symbol for any control variable.

PROBLEMS

16.1 SOC Problem 14.1 concerned the assimilation of a small sample of immigrants. One table showed the relationship between length of residence in the United States and facility with the English language:

English Facility	Length of Residence: Less than Five Years (Low)	More than Five Years (High)	Totals
Low	20	10	30
High	5	15	20
Totals	25	25	50

a. If necessary, find the column percentages and compute gamma for this table. Describe the bivariate relationship in terms of strength and direction.

Is the relationship between residence and language affected by the gender of the immigrant? Here are the partial tables showing the bivariate relationship for males and females:

A. Males:

English Facility	Length of Residence: Less than Five Years (Low)	More than Five Years (High)	Totals
Low	10	5	15
High	2	8	10
Totals	12	13	25

B. Females:

English Facility	Length of Residence: Less than Five Years (Low)	More than Five Years (High)	Totals
Low	10	5	15
High	3	7	10
Totals	13	12	25

b. Find the column percentages and compute gamma for each of the partial tables. Compare these percentages and gammas with each other and with the bivariate table. Does controlling for gender change the bivariate relationship?

Is the relationship between residence and language affected by the origin of the immigrant? Here are the partial tables showing the bivariate relationship for Asian and Hispanic immigrants:

A. Asian immigrants:

English Facility	Length of Residence: Less than Five Years (Low)	More than Five Years (High)	Totals
Low	8	4	12
High	2	5	7
Totals	10	9	19

B. Hispanic immigrants:

English Facility	Length of Residence: Less than Five Years (Low)	More than Five Years (High)	Totals
Low	12	6	18
High	3	10	13
Totals	15	16	31

c. Find the column percentages and compute gamma for each of the partial tables. Compare these percentages and gammas with each other and with the bivariate table. Does controlling for the origin of the immigrants change the bivariate relationship?

16.2 SOC Data on suicide rates, age structure, and unemployment rates have been gathered for 100 census tracts. Suicide and unemployment rates have been dichotomized at the median so that each tract could be rated as high or low. Age structure is measured in terms of the percentage of the population age 65 and older. This variable has also been dichotomized, and tracts have been rated as high or low. The following tables display the bivariate relationship between suicide rate and age structure and the same relationship controlling for unemployment.

Suicide Rate	Population 65 and Older (%): Low	High	Totals
Low	45	20	65
High	10	25	35
Totals	55	45	100

a. Calculate percentages and gamma for the bivariate table. Describe the bivariate relationship in terms of strength and direction.

SUICIDE RATE BY AGE, CONTROLLING FOR UNEMPLOYMENT:

A. High unemployment:

Suicide Rate	Population 65 and Older (%)		
	Low	High	Totals
Low	23	10	33
High	5	12	17
Totals	28	22	50

B. Low unemployment:

Suicide Rate	Population 65 and Older (%)		
	Low	High	Totals
Low	22	10	32
High	5	13	18
Totals	27	23	50

b. Calculate percentages and gamma for each partial table. Compare the partial tables with each other and with the bivariate table. Compute partial gamma (G_p).

c. Summarize the results of the control. Does unemployment rate have any effect on the relationship between age and suicide rate? Describe the effect of the control variable in terms of the pattern of percentages, the value of the gammas, and the possible causal relationships among these variables.

16.3 SW Do long-term patients in mental health facilities become more withdrawn and reclusive over time? A sample of 608 institutionalized patients was rated by a standard "reality orientation scale." Is there a relationship with length of institutionalization? Does gender have any effect on the relationship?

Reality orientation by length of institutionalization:

Reality Orientation	Length of Institutionalization		
	Less Than 5 Years	More Than 5 Years	Totals
Low	200	213	413
High	117	78	195
Totals	317	291	608

REALITY ORIENTATION BY LENGTH OF INSTITUTIONALIZATION, CONTROLLING FOR GENDER:

A. Females:

Reality Orientation	Length of Institutionalization		
	Less Than 5 Years	More Than 5 Years	Totals
Low	95	120	215
High	60	37	97
Totals	155	157	312

B. Males:

Reality Orientation	Length of Institutionalization		
	Less Than 5 Years	More Than 5 Years	Totals
Low	105	93	198
High	57	41	98
Totals	162	134	296

16.4 SOC Is there a relationship between attitudes on sexuality and age? Are older people more conservative with respect to questions of sexual morality? A national sample of 925 respondents has been questioned about attitudes on premarital sex. Responses have been collapsed into two categories: those who believe that premarital sex is "always wrong" and those who believe it is not wrong under certain conditions ("sometimes wrong").

These responses have been cross-tabulated by age, and the results are reported here: Attitude toward premarital sex by age:

Premarital Sex Is:	Age		
	Younger Than 35	35 and Older	Totals
Always wrong	90	235	325
Sometimes wrong	420	180	600
Totals	510	415	925

a. Calculate percentages and gamma for the table. Is there a relationship between these two variables? Describe the bivariate relationship in terms of its strength and direction. Next, the bivariate relationship is reproduced after controlling for the sex of the respondent.

Does gender have any effect on the relationship?

ATTITUDE TOWARD PREMARITAL SEX BY AGE, CONTROLLING FOR GENDER:

A. Males:

Premarital Sex Is:	Age: Younger Than 35	Age: 35 and Older	Totals
Always wrong	70	55	125
Sometimes wrong	190	80	270
Totals	260	135	395

B. Females:

Premarital Sex Is:	Age: Younger Than 35	Age: 35 and Older	Totals
Always wrong	20	180	200
Sometimes wrong	230	100	330
Totals	250	280	530

b. Summarize the results of this control. Does gender have any effect on the relationship between age and attitude? If so, describe the effect of the control variable in terms of the pattern of percentages, the value of the gammas, and the possible causal interrelationships among these variables.

16.5 SOC A job-training center is trying to justify its existence to its funding agency. To this end, data on four variables have been collected for each of the 403 trainees served over the past three years: (1) whether or not the trainee completed the program, (2) whether or not the trainee got and held a job for at least a year after training, (3) the sex of the trainee, and (4) the race of the trainee. Is employment related to completion of the program? Is the relationship between completion and employment affected by race or sex?

Employment by training:

Held Job for at Least One Year?	Training Completed? Yes	Training Completed? No	Totals
Yes	145	60	205
No	72	126	198
Totals	217	186	403

EMPLOYMENT BY TRAINING, CONTROLLING FOR RACE:

A. Whites:

Held Job for at Least One Year?	Training Completed? Yes	Training Completed? No	Totals
Yes	85	33	118
No	38	47	85
Totals	123	80	203

B. Blacks:

Held Job for at Least One Year?	Training Completed? Yes	Training Completed? No	Totals
Yes	60	27	87
No	34	79	113
Totals	94	106	200

EMPLOYMENT BY TRAINING, CONTROLLING FOR SEX:

C. Males:

Held Job for at Least One Year?	Training Completed? Yes	Training Completed? No	Totals
Yes	73	30	103
No	36	62	98
Totals	109	92	201

D. Females:

Held Job for at Least One Year?	Training Completed? Yes	Training Completed? No	Totals
Yes	72	30	102
No	36	64	100
Totals	108	94	202

16.6 SOC What are the social sources of support for the environmental movement? A recent survey gathered information on level of concern for such issues as global warming and acid rain. Is concern for the environment related to level of education? What effects do the control variables have? Write a paragraph summarizing your conclusions.

Concern for the environment by level of education:

	Level of Education		
Concern	Low	High	Totals
Low	27	35	62
High	22	48	70
Totals	49	83	132

CONCERN FOR THE ENVIRONMENT BY LEVEL OF EDUCATION, CONTROLLING FOR GENDER:

A. Males:

	Level of Education		
Concern	Low	High	Totals
Low	14	17	31
High	11	22	33
Totals	25	39	64

B. Females:

	Level of Education		
Concern	Low	High	Totals
Low	13	18	31
High	11	26	37
Totals	24	44	68

CONCERN FOR THE ENVIRONMENT BY LEVEL OF EDUCATION, CONTROLLING FOR "LEVEL OF TRUST IN THE NATION'S LEADERSHIP":

C. Low levels of trust:

	Level of Education		
Concern	Low	High	Totals
Low	6	22	28
High	10	40	50
Totals	16	62	78

D. High levels of trust:

	Level of Education		
Concern	Low	High	Totals
Low	21	13	34
High	12	8	20
Totals	33	21	54

CONCERN FOR THE ENVIRONMENT BY LEVEL OF EDUCATION, CONTROLLING FOR RACE:

E. Whites:

	Level of Education		
Concern	Low	High	Totals
Low	19	11	30
High	18	44	62
Totals	37	55	92

F. Blacks:

	Level of Education		
Concern	Low	High	Totals
Low	8	24	32
High	4	4	8
Totals	12	28	40

16.7 In problem 14.15, we investigated the relationships between income and five dependent variables. Let's return to two of those relationships and see if sex has any effect on the bivariate relationships. The bivariate and partial tables are presented here along with the bivariate gammas that were computed in the previous exercise. Compute percentages and gammas for each of the partial tables and state your conclusions. Does controlling for sex have any effect?

a. Support for abortion by income (from problem 14.15a):

	Income			
Right to an Abortion?	(1) Low	(2) Moderate	(3) High	Totals
Yes (1)	220	218	226	664
No (2)	366	299	250	915
Totals	586	517	476	1579

Gamma = −0.14

(HINT: Note that the higher score on the variable that measures support for abortion is associated with "No." The negative sign of gamma means that as score on income increases, score on the dependent variable decreases. In other words, people with higher incomes were more supportive of the legal right to abortion (more likely to say "Yes," or have a score of 1). This is actually a positive relation-

ship: As income increases, support for abortion increases.

SUPPORT FOR THE LEGAL RIGHT TO ABORTION BY INCOME, CONTROLLING FOR SEX:

A. Males:

Right to an Abortion?	Income (1) Low	(2) Moderate	(3) High	Totals
Yes (1)	89	96	100	285
No (2)	130	141	122	393
Totals	219	237	222	678

B. Females:

Right to an Abortion?	Income (1) Low	(2) Moderate	(3) High	Totals
Yes (1)	131	122	126	379
No (2)	236	158	128	522
Totals	367	280	254	901

b. Support for the right to commit suicide for people with an incurable disease by income (from problem 14.15c):

Right to Suicide?	Income (1) Low	(2) Moderate	(3) High	Totals
Approve (1)	343	341	338	1022
Oppose (2)	227	194	147	568
Totals	570	535	485	1590

Gamma = −0.22

HINT: Be careful in interpreting direction for this table.

APPROVAL OF SUICIDE FOR PEOPLE WITH AN INCURABLE DISEASE BY INCOME, CONTROLLING FOR SEX:

A. Males:

Right to Suicide?	Income (1) Low	(2) Moderate	(3) High	Totals
Approve (1)	140	165	170	475
Oppose (2)	82	80	61	223
Totals	222	245	231	698

B. Females:

Right to Suicide?	Income (1) Low	(2) Moderate	(3) High	Totals
Approve (1)	203	176	168	547
Oppose (2)	145	114	86	345
Totals	348	290	254	892

16.8 In this problem, we return to the analysis of the relationship between age and attitude towards premarital sex. This time, we will observe the relationship in two different nations and also control for gender. The bivariate relationships for Canada and the United States are presented along with the bivariate gamma. For both nations, there is a weak-to-moderate positive relationship between age and opinion about sexual freedom. Given the way the dependent variable is scored (higher scores mean *less* approval), we can say that support for sexual freedom decreases with age. Are the bivariate relationships affected by gender? For each partial table, find gamma, and write a paragraph of analysis comparing the partial tables with each other and with the bivariate table.

a. Support for sexual freedom by age (Canada)

"People should enjoy sexual freedom"	Age 18–34	35–54	55+	Totals
(1) Agree	378	174	66	618
(2) Neither agree nor disagree	626	710	586	1922
(3) Disagree	163	101	74	338
Totals	1167	985	726	2878

Gamma = 0.24

b. Support for sexual freedom by age (United States)

"People should enjoy sexual freedom"	Age 18–34	35–54	55+	Totals
(1) Agree	583	288	147	1018
(2) Neither agree nor disagree	877	982	1061	2920
(3) Disagree	113	72	53	238
Totals	1573	1342	1261	4176

Gamma = 0.32

c. Support for sexual freedom by age, controlling for gender (Canada)

A. Males:

"People should enjoy sexual freedom"	Age 18–34	35–54	55+	Totals
(1) Agree	230	101	39	370
(2) Neither agree nor disagree	273	339	282	894
(3) Disagree	82	37	38	157
Totals	585	477	359	1421

B. Females:

"People should enjoy sexual freedom"	Age 18–34	35–54	55+	Totals
(1) Agree	148	73	27	248
(2) Neither agree nor disagree	252	371	314	937
(3) Disagree	81	63	36	180
Totals	481	507	377	1365

d. Support for sexual freedom by age, controlling for gender (United States)

A. Males:

"People should enjoy sexual freedom"	Age 18–34	35–54	55+	Totals
(1) Agree	302	169	87	558
(2) Neither agree nor disagree	393	451	470	1314
(3) Disagree	49	57	29	135
Totals	744	677	586	2007

B. Females:

"People should enjoy sexual freedom"	Age 18–34	35–54	55+	Totals
(1) Agree	277	116	57	450
(2) Neither agree nor disagree	483	525	583	1591
(3) Disagree	65	31	24	120
Totals	825	672	664	2161

SPSS for Windows

Using *SPSS for Windows* to Elaborate Bivariate Tables

SPSS DEMONSTRATION 16.1 Analyzing the Effects of Age and Sex on Sexual Attitudes

Since we have been concerned with bivariate tables in this chapter, it will come as no surprise to find that we will once again make use of the **Crosstabs** procedure. To control for a third variable with **Crosstabs,** add the name of the control variable to the box at the bottom of the **Crosstabs** dialog window. In other words, on the **Crosstabs** window, place the name of your dependent variable(s) in the top (row) box, the name of your independent variable(s) in the middle (column) box, and the name of your control variable(s) in the box at the bottom of the screen.

To illustrate, in Demonstration 12.3 and again in Demonstration 14.1, we looked at the relationships between *premarsx* and recoded *age*, or *ager* (see Demonstrations 12.3 and 14.1 for the recoding scheme), and found weak-to-moderate, negative relationships (gamma = −0.14). That is, approval of premarital sex decreased as age increased. Now let's see if gender has any effect on these bivariate relationships. Are older women more opposed to premarital sex than older men?

If necessary, recode *age* as in Demonstration 12.3. On the **Crosstabs** dialog window, *premarsx* should be in the top (row) box, recoded *age* in the middle (column) box, and *sex* in the box at the bottom. Make sure you request gamma and column percentages. The output will consist of two partial tables, one for males and one for females, and gammas computed for each of these partial tables.

The bivariate relationships and statistics were displayed in Demonstration 14.1. To conserve space, we will concentrate on gamma and not reproduce the partial tables here. Gamma is −0.14 for males and −0.15 for females. If you compare the percentages in the partial tables, you will see that males are generally less opposed to premarital sex than females. Specifically, compare the top row of the two partial tables, and you'll see that males are less likely to say that premarital sex is "always wrong," regardless of age group. Overall, however, this is a direct relationship: Opposition to premarital sex increases with age regardless of sex (although males are slightly more permissive in every age group).

SPSS DEMONSTRATION 16.2 Are Prestigious Jobs More Satisfying? Another Look

In Demonstration 14.3, we looked at the relationship between recoded *satjob* and *prestg80.* We found a bivariate gamma of −0.31, indicating that people with higher-prestige jobs had higher levels of satisfaction with their jobs. While not exactly surprising, we might be able to learn a little more about the relationship with some additional controls. In this demonstration, the bivariate relationship is reexamined with *sex* and *degree* as control variables. Does the relationship between satisfaction and prestige vary by sex? Does the educational level of the respondent affect the relationship between job satisfaction and prestige?

The variable *degree* has a total of five values—too many to permit the variable to be used as a control. To simplify the variable, I grouped all respondents with any level of college education into a single category. Here is the recoding scheme I used:

0 = 0 (less than high school)
1 = 1 (high school)
2 thru 4 = 2 (at least some college)

I called the recoded version of this variable *rdeg.* If necessary, see Demonstration 14.3 for the recode schemes for *satjob* (*rsat*) and *prestg80* (*rprest*).

Run the **Crosstabs** procedure and move *rsat* into the **Row Variable** box, *rprest* into the **Column Variable** box, and *rdeg* and *sex* into the bottom box. Don't forget gamma and column percents. To conserve space, the partial tables are not reproduced here. Instead, we concentrate on the gammas. (NOTE: Recall from Demonstration 14.3 that *rsat* is scored so that higher scores indicate lower satisfaction.) The results of the test are presented in two summary tables.

	Gamma
For the bivariate table (*rsat by rpres*)	−0.31
Controlling for education (*rdeg*)	
Less than high school	0.18
High school	−0.20
At least some college	−0.37

There is an interactive relationship between these variables. Satisfaction increases with prestige for people with high school degrees and those with at least some college but *decreases* (note the positive gamma) for people with less than a high school degree. Looking at the partial tables, note that the distribution of job satisfaction is generally the same for people in lower-prestige jobs at all levels of education: 43% to 45% are less satisfied and 55% to 59% are more satisfied. The difference occurs among those with higher-prestige jobs. People with the highest level of education (*rdeg* = 2) in higher-prestige jobs are especially satisfied (64%). People at the middle level of education (*rdeg* = 1) in higher-prestige jobs are also more satisfied (52%), but the relationship is weaker.

For people with the lowest level of education (*rdeg* = 0), the pattern reverses itself. Those in higher-prestige jobs are more likely to report *lower* levels of job satisfaction (67%). We can only speculate about the reasons for this relationship, but perhaps they feel out of place or some form of status strain as they bring lower levels of preparation (education) to the more prestigious jobs.

The control for gender shows some interaction, but this is, overall, a direct relationship. For both males and females, job satisfaction increases with the prestige of the job, although the effect is a little stronger for males than for females. The gamma is almost exactly the same value for the two groups, and it would be justified to discard *sex* from further analysis.

	Gamma
For the bivariate table (*rsat by rpres*)	−0.31
Controlling for sex	
Males	−0.37
Females	−0.25

Exercises

16.1 Use Demonstration 16.1 as a guide and analyze the relationship of recoded *age* on *cappun, grass,* and *marhomo,* using *sex* as the control variable. Summarize the results in a paragraph or two. Be sure to characterize the relationships as direct, spurious, or interactive.

16.2 Using Demonstration 16.2 as a guide, analyze the relationships between income (*income06*) and recoded job satisfaction, controlling for *sex* and recoded *degree.* Recode *income06* at the median before conducting the analysis. Summarize the results in a paragraph or two. Be sure to characterize the relationships as direct, spurious, or interactive.

17

Partial Correlation and Multiple Regression and Correlation

LEARNING OBJECTIVES

By the end of this chapter, you will be able to

1. Compute and interpret partial correlation coefficients.
2. Find and interpret the least-squares multiple regression equation with partial slopes.
3. Calculate and interpret the multiple correlation coefficient (R^2).
4. Explain the limitations of partial and multiple regression analysis.

17.1 INTRODUCTION

As mentioned at the beginning of Chapter 16, social science research is, by nature, multivariate and involves the simultaneous analysis of scores of variables. Some of the most powerful and widely used statistical tools for multivariate analysis are introduced in this chapter. We cover techniques that are used to analyze causal relationships and to make predictions, both crucial endeavors in any science.

These techniques are based on Pearson's *r* (see Chapter 15) and are most appropriately used with high-quality, precisely measured interval-ratio variables. As we have noted on many occasions, such data are relatively rare in social science research, and the techniques presented in this chapter are commonly used on variables measured at the ordinal level and with nominal-level variables in the form of dummy variables (see Section 15.8).

The techniques presented in this chapter are generally more flexible than those presented in Chapter 16. They produce more information and provide a wider variety of ways of disentangling the interrelationships among the variables.

We first consider partial correlation analysis, a technique analogous to controlling for a third variable by constructing partial tables (see Chapter 16). The second technique involves multiple regression and correlation and allows the researcher to assess the effects, separately and in combination, of more than one independent variable on the dependent variable.

Throughout this chapter, we focus our attention on research situations involving three variables. This is the least complex application of these techniques, but extensions to situations involving four or more variables are relatively straightforward. To deal efficiently with the computations required by the more complex applications, I refer you to any of the computerized statistical packages (such as SPSS) probably available at your local computer center.

17.2 PARTIAL CORRELATION

In Chapter 15, we used Pearson's *r* to measure the strength and direction of bivariate relationships. To provide an example, we looked at the relationship between husband's contribution to housework (the dependent, or *Y*, variable) and number of children (the independent, or *X*, variable) for a sample of 12 families. We found a positive relationship of moderate strength ($r = +0.50$) and concluded that husbands tend to make a larger contribution to housework as the number of children increases.

You might wonder, as researchers commonly do, if this relationship holds true for *all* types of families? For example, might husbands in strongly religious families respond differently than those in less religious families? Would politically conservative husbands behave differently than husbands who are politically liberal? How about more educated husbands? Would they respond differently than less educated husbands? Or perhaps husbands in some ethnic or racial groups would have different responses than husbands in other groups. We can address these kinds of issues by means of a technique called **partial correlation,** in which we observe how the bivariate relationship changes when a third variable, such as religiosity, education, or ethnicity, is introduced. Third variables are often referred to as *Z* variables or **control variables.**

Partial correlation proceeds by first computing Pearson's *r* for the bivariate (or zero-order) relationship and then computing the partial (or first-order) correlation coefficient. If the partial correlation coefficient differs from the zero-order correlation coefficient, we conclude that the third variable has an effect on the bivariate relationship. If, for example, well-educated husbands respond differently to an additional child than less well-educated husbands, the partial correlation coefficient will differ in strength (and perhaps in direction) from the bivariate correlation coefficient.

Before considering matters of computation, we'll consider the possible relationships between the partial and bivariate correlation coefficients and what they might mean. These relationships were introduced in Chapter 16 and summarized in Table 16.5. I repeat the information here for the convenience of those readers who did not cover Chapter 16 and to place these patterns in the context of partial correlation.

Types of Relationships.

Direct Relationship. One possible outcome is that the partial correlation coefficient is essentially the same value as the bivariate coefficient. Imagine, for example, that after we controlled for husband's education, we found a partial correlation coefficient of +0.49, compared to the zero-order Pearson's r of +0.50. This would mean that the third variable (husband's education) has no effect on the relationship between number of children and husband's hours of housework. In other words, regardless of their education, husbands respond in a similar way to additional children. This outcome is consistent with the conclusion that there is a **direct** or casual relationship (see Figure 17.1) between *X* and *Y* and that the third variable (*Z*) is irrelevant to the investigation. In this case, the researcher would discard that particular *Z* variable from further consideration but might well run additional tests with other likely control variables (e.g., the researcher might control for the religion or ethnicity of the family).

FIGURE 17.1 A DIRECT RELATIONSHIP BETWEEN *X* AND *Y*

$$X \longrightarrow Y$$

Spurious and Intervening Relationships. A second possible outcome occurs when the partial correlation coefficient is much weaker than the bivariate correlation, perhaps even dropping to zero. This outcome is consistent with two different relationships between the variables. The first is called a **spurious** relationship: The control variable (*Z*) is a cause of both the independent (*X*) and the dependent (*Y*) variable (see Figure 17.2). This outcome would mean that *X* and *Y* are not actually related. They appear to be related only because both are dependent on a common cause (*Z*). Once *Z* is taken into account, the apparent relationship between *X* and *Y* disappears.

FIGURE 17.2 A SPURIOUS RELATIONSHIP BETWEEN *X* AND *Y*

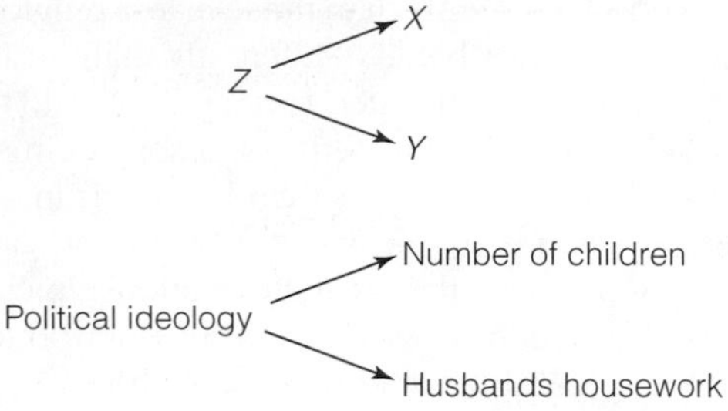

What would a spurious relationship look like? Imagine that we controlled for the political ideology of the parents in our sample and found that the partial correlation coefficient was much weaker than the bivariate Pearson's *r*. This would indicate that the number of children does not actually change the husband's contribution to housework (that is, the relationship between *X* and *Y* is not direct). Rather, political ideology is the mutual cause of both of the other variables: Perhaps more conservative families are more likely to follow traditional gender role patterns (in which husbands contribute less to housework) *and* have more children.

This pattern (partial correlation much weaker than the bivariate correlation) is also consistent with an **intervening** relationship between the variables (see Figure 17.3). In this situation, *X* and *Y* are not linked directly but are casually connected through the *Z* variable. Again, once *Z* is controlled, the apparent relationship between *X* and *Y* disappears.

FIGURE 17.3 AN INTERVENING RELATIONSHIP BETWEEN *X* AND *Y*

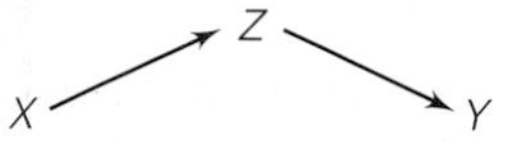

How can we tell the difference between spurious and intervening relationships? This distinction cannot be made on statistical grounds: Spurious and intervening relationships look exactly the same in terms of statistics. The researcher may be able to distinguish between these two relationships in terms of the time order of the variables (i.e., which came first) or on theoretical grounds, but not on statistical grounds.

Interaction. A final possible relationship between variables should be mentioned, even though it *cannot* be detected by partial correlation analysis. This relationship, called interaction, occurs when the relationship between *X* and *Y* changes markedly under the various values of *Z*. For example, if we controlled for social class and found that husbands in middle-class families increased their contribution to housework as the number of children increased while husbands in working-class families did just the reverse, we would conclude that there was interaction between these three variables. In other words, there would be a positive relationship between *X* and *Y* for one category of *Z* and a negative relationship for the other category, as illustrated in Figure 17.4

FIGURE 17.4 AN INTERACTIVE RELATIONSHIP BETWEEN *X*, *Y*, AND *Z*

Interactive relationships are explored in Chapter 16. We will not examine them further here, however, since they cannot be detected by partial correlation.

Computing and Interpreting the Partial Correlation Coefficient.

Terminology and Formula. The formula for partial correlation requires some new terminology. We will be dealing with more than one bivariate relationship and need to differentiate between them with subscripts. Thus, the symbol r_{yx} will refer to the correlation coefficient between variable *Y* and variable *X*; r_{yz} will re-

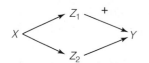

ONE STEP AT A TIME **Computing and Interpreting the Partial Correlation Coefficient**

Computation

Step 1: Compute Pearson's *r* for all pairs of variables. Be clear about which variable is independent (*X*), which is dependent (*Y*), and which is the control (*Z*).

Step 2: Find the partial correlation coefficient by solving Formula 17.1:

$$r_{yx.z} = \frac{r_{yx} - (r_{yz})(r_{xz})}{\sqrt{1 - r_{yz}^2}\sqrt{1 - r_{xz}^2}}$$

a. Multiply r_{yz} by r_{xz}.

b. Subtract the value you found in step a from r_{yx}. This value is the numerator of Formula 17.1.

c. Square the value of r_{yz}.

d. Subtract the quantity you found in step c from 1.

e. Take the square root of the quantity you found in step d.

f. Square the value of r_{xz}.

g. Subtract the quantity you found in step f from 1.

h. Take the square root of the quantity you found in step g.

i. Multiply the quantity you found in step h by the value you found in step e. This value is the denominator of Formula 17.1.

j. Divide the quantity you found in step b by the value you found in step i. This value is the partial correlation coefficient.

Interpretation

Step 3: Compare the value of the partial correlation coefficient ($r_{yx.z}$) with the value of the zero-order correlation (r_{yx}). From the following scenarios, choose the one that comes closest to describing the relationship between the two values:

a. The partial correlation coefficient is roughly the same value as the zero-order or bivariate correlation. A good rule of thumb for "roughly the same" is a difference of less than 0.10. This outcome is evidence that the control variable (*Z*) has no effect and that the relationship between *X* and *Y* is direct.

b. The partial correlation coefficient is much less (say, more than 0.10 less) than the bivariate correlation. This is evidence that the control variable (*Z*) changes the relationship between *X* and *Y*. The relationship between *X* and *Y* is either spurious (*Z* causes both *X* and *Y*) or intervening (*X* and *Y* are linked by *Z*).

c. Be aware that *X*, *Y*, and *Z* may have an interactive relationship in which the relationship between *X* and *Y* changes for each category of *Z*. Partial correlation analysis cannot detect interactive relationships.

fer to the correlation coefficient between *Y* and *Z*; and r_{xz} will refer to the correlation coefficient between *X* and *Z*. Recall that correlation coefficients calculated for bivariate relationships are often referred to as **zero-order correlations.**

Partial correlation coefficients, or first-order partials, are symbolized as $r_{yx.z}$. The variable to the right of the dot is the control variable. Thus, $r_{yx.z}$ refers to the partial correlation coefficient that measures the relationship between variables *X* and *Y* while controlling for variable *Z*. The formula for the first-order partial is

FORMULA 17.1

$$r_{yx.z} = \frac{r_{yx} - (r_{yz})(r_{xz})}{\sqrt{1 - r_{yz}^2}\sqrt{1 - r_{xz}^2}}$$

Note that you must first calculate the zero-order coefficients between all possible pairs of variables (variables *X* and *Y*, *X* and *Z*, and *Y* and *X*) before solving this formula.

Computation. To illustrate the computation of a first-order partial, we will return to the relationship between number of children (*X*) and husband's

TABLE 17.1 SCORES ON THREE VARIABLES FOR 12 DUAL-WAGE-EARNER FAMILIES AND ZERO-ORDER CORRELATIONS

Family	Husband's Housework (*Y*)	Number of Children (*X*)	Husband's Years of Education (*Z*)
A	1	1	12
B	2	1	14
C	3	1	16
D	5	1	16
E	3	2	18
F	1	2	16
G	5	3	12
H	0	3	12
I	6	4	10
J	3	4	12
K	7	5	10
L	4	5	16

contribution to housework (Y) for 12 dual-career families. The zero-order r between these two variables ($r_{yx} = 0.50$) indicated a moderate, positive relationship (as number of children increased, husbands tended to contribute more to housework). Suppose the researcher wished to investigate the possible effects of husband's education on the bivariate relationship. The original data (from Table 15.1) and the scores of the 12 families on the new variable are presented in Table 17.1.

The zero-order correlations, as presented in the correlation matrix in Table 17.2, indicate that the husband's contribution to housework is positively related to number of children ($r_{yx} = 0.50$), that better-educated husbands tend to do less housework ($r_{yz} = -0.30$), and that families with better-educated husbands have fewer children ($r_{xz} = -0.47$).

Is the relationship between husband's housework (X) and number of children (Y) affected by husband's years of education? Substituting the zero-order correlations into Formula 17.1, we would have

$$r_{yx.z} = \frac{r_{yx} - (r_{yz})(r_{xz})}{\sqrt{1 - r_{yz}^2}\sqrt{1 - r_{xz}^2}}$$

$$r_{yx.z} = \frac{(0.50) - (-0.30)(-0.47)}{\sqrt{1 - (-0.30)^2}\sqrt{1 - (0.47)^2}}$$

$$r_{yx.z} = \frac{(0.50) - (0.14)}{\sqrt{1 - 0.09}\sqrt{1 - 0.22}}$$

$$r_{yx.z} = \frac{0.36}{\sqrt{0.91}\sqrt{0.78}}$$

$$r_{yx.z} = \frac{0.36}{(0.95)(0.88)}$$

$$r_{yx.z} = \frac{0.36}{0.84}$$

$$r_{yx.z} = 0.43$$

TABLE 17.2 ZERO-ORDER CORRELATIONS

	Husband's Housework (*Y*)	Number of Children (*X*)	Husband's Years of Education (*Z*)
Husband's Housework (*Y*)	1.00	0.50	−0.30
Number of Children (*X*)		1.00	−0.47
Husband's Years of Education (*Z*)			1.00

Interpretation. The first-order partial ($r_{yx.z} = 0.43$), which measures the strength of the relationship between husband's housework (*Y*) and number of children (*X*) while controlling for husband's education (*Z*), is lower in value than the zero-order coefficient ($r_{yx} = 0.50$), but the difference in the two values is not great. This result suggests a direct relationship between variables *X* and *Y*. That is, when controlling for husband's education, the statistical relationship between husband's housework and number of children is essentially unchanged. Regardless of education, husband's hours of housework increase with the number of children.

Our next step in statistical analysis would probably be to select another control variable. The more the bivariate relationship retains its strength across a series of controls for third variables (*Z*'s), the stronger the evidence for a direct relationship between *X* and *Y*.

In closing, I should mention an additional possible outcome, in which the partial correlation coefficient is greater in value than the zero-order coefficient ($r_{yx.z} > r_{yx}$). This outcome would be consistent with a causal model in which the variable taken as independent and the control variable each had a separate effect on the dependent variable and were uncorrelated with each other. This relationship is depicted in Figure 17.5. The absence of an arrow between *X* and *Z* indicates that they have no mutual relationship.

FIGURE 17.5 A POSSIBLE CAUSAL RELATIONSHIP AMONG THREE VARIABLES

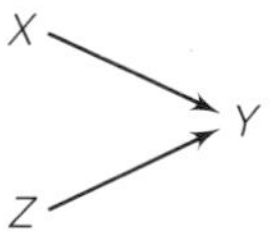

This pattern means that both *X* and *Z* should be treated as independent variables, and the next step in the statistical analysis would probably involve multiple correlation and regression. As we shall see in Sections 17.3 and 17.4, these techniques enable the researcher to isolate the separate effects of several independent variables on the dependent variable and thus to make judgments about which independent variable has the strongest effect on the dependent. *(For practice in computing and interpreting partial correlation coefficients, see problems 17.1 to 17.3.)*

17.3 MULTIPLE REGRESSION: PREDICTING THE DEPENDENT VARIABLE

In Chapter 15, the least-squares regression line was introduced as a way of describing the overall linear relationship between two interval-ratio variables and of predicting scores on *Y* from scores on *X*. This line was the best-fitting line to summarize the bivariate relationship and was defined by this formula:

FORMULA 17.2

$$Y = a + bX$$

where a = the *Y* intercept
b = the slope

The least-squares regression line can be modified to include (theoretically) any number of independent variables. This technique is called **multiple regression.**

For ease of explication, we will confine our attention to the case involving two independent variables. The least-squares multiple regression equation is

FORMULA 17.3

$$Y = a + b_1X_1 + b_2X_2$$

where b_1 = the partial slope of the linear relationship between the first independent variable and Y

b_2 = the partial slope of the linear relationship between the second independent variable and Y

Some new notation and some new concepts are introduced in this formula. First, while the dependent variable is still symbolized as Y, the independent variables are differentiated by subscripts. Thus, X_1 identifies the first independent variable and X_2 the second. The symbol for the slope (b) is also subscripted to identify the independent variable with which it is associated.

Partial Slopes. A major difference between the multiple and bivariate regression equations concerns the slopes (b's). In the case of multiple regression, the b's are called **partial slopes,** and they show the amount of change in Y for a unit change in the independent while controlling for the effects of the other independent variables in the equation. The partial slopes are thus analogous to partial correlation coefficients and represent the direct effect of the associated independent variable on Y.

ONE STEP AT A TIME Computing and Interpreting Partial Slopes

NOTE: These procedures apply when there are two independent variables and one dependent variable. For more complex situations, use a computerized statistical package such as SPSS to do the calculations.

Computation

Step 1: Compute the partial slope associated with the first independent variable by using Formula 17.4:

$$b_1 = \left(\frac{s_y}{s_1}\right)\left(\frac{r_{y1} - r_{y2}r_{12}}{1 - r_{12}^2}\right)$$

a. Divide s_y by s_1.

b. Multiply r_{y2} by r_{12}.

c. Subtract the value you computed in step b from r_{y1}.

d. Square the value of r_{12}.

e. Subtract the value you computed in step d (r_{12}^2) from 1.

f. Divide the value you computed in step c by the value you computed in step e.

g. Multiply the value you computed in step f by the value you computed in step a. This value is the partial slope associated with the first independent variable.

Step 2: Compute the partial slope associated with the second independent variable by using Formula 17.5:

$$b_2 = \left(\frac{s_y}{s_2}\right)\left(\frac{r_{y2} - r_{y1}r_{12}}{1 - r_{12}^2}\right)$$

a. Divide s_y by s_2.

b. Multiply r_{y1} by r_{12}.

c. Subtract the value you computed in step b from r_{y2}.

d. Square the value of r_{12}.

e. Subtract the value you computed in step d from 1.

f. Divide the value you computed in step c by the value you computed in step e.

g. Multiply the value you computed in step f by the value you computed in step a. This value is the partial slope associated with the second independent variable.

Interpretation

Step 3: The value of a partial slope is the increase in the value of Y for a unit increase in the value of the associated independent variable while controlling for the effects of the other independent variable.

Computing Partial Slopes. The partial slopes for the independent variables are determined by Formulas 17.4 and 17.5:[1]

FORMULA 17.4

$$b_1 = \left(\frac{s_y}{s_1}\right)\left(\frac{r_{y1} - r_{y2}r_{12}}{1 - r_{12}^2}\right)$$

FORMULA 17.5

$$b_2 = \left(\frac{s_y}{s_2}\right)\left(\frac{r_{y2} - r_{y1}r_{12}}{1 - r_{12}^2}\right)$$

where b_1 = the partial slope of X_1 on Y
b_2 = the partial slope of X_2 on Y
s_y = the standard deviation of Y
s_1 = the standard deviation of the first independent variable (X_1)
s_2 = the standard deviation of the second independent variable (X_2)
r_{y1} = the bivariate correlation between Y and X_1
r_{y2} = the bivariate correlation between Y and X_2
r_{12} = the bivariate correlation between X_1 and X_2

To illustrate the computation of the partial slopes, we will assess the combined effects of number of children (X_1) and SES (X_2) on husband's contribution to housework. All the relevant information can be calculated from Table 17.1 and is reproduced here:

Husband's Housework	Number of Children	Husband's Education
$\bar{Y} = 3.3$	$\bar{X}_1 = 2.7$	$\bar{X}_2 = 13.7$
$s_y = 2.1$	$s_1 = 1.5$	$s_2 = 2.6$

Zero-order correlations

$r_{y1} = 0.50$
$r_{y2} = -0.30$
$r_{12} = -0.47$

The partial slope for the first independent variable (X_1) is

$$b_1 = \left(\frac{s_y}{s_1}\right)\left(\frac{r_{y1} - r_{y2}r_{12}}{1 - r_{12}^2}\right)$$

$$b_1 = \left(\frac{2.1}{1.5}\right)\left(\frac{0.50 - (-0.30)(-0.47)}{1 - (-0.47)^2}\right)$$

$$b_1 = (1.4)\left(\frac{0.50 - 0.14}{1 - 0.22}\right)$$

$$b_1 = (1.4)\left(\frac{0.36}{0.78}\right)$$

$$b_1 = (1.4)(0.46)$$

$$b_1 = 0.65$$

[1] Partial slopes can be computed from zero-order slopes, but Formulas 17.4 and 17.5 are somewhat easier to use.

ONE STEP AT A TIME **Computing the Y Intercept**

Find the Y intercept by using Formula 17.6:

$$a = \overline{Y} - b_1\overline{X}_1 - b_2\overline{X}_2$$

Step 1: Multiply the mean of X_2 by b_2.

Step 2: Multiply the mean of X_1 by b_1.

Step 3: Subtract the quantity you found in step 1 from the quantity you found in step 2.

Step 4: Subtract the quantity you found in step 3 from the mean of Y. The result is the value of a, the Y intercept.

ONE STEP AT A TIME **Using the Multiple Regression Line to Predict Scores on Y**

Step 1: Choose a value for X_1. Multiply this value by the value of b_1.

Step 2: Choose a value for X_2. Multiply this value by the value of b_2.

Step 3: Add the values you found in steps 1 and 2 to the value of a, the Y intercept. The resulting value is the predicted score on Y.

For the second independent variable, SES or X_2, the partial slope is

$$b_2 = \left(\frac{s_y}{s_2}\right)\left(\frac{r_{y2} - r_{y1}r_{12}}{1 - r_{12}^2}\right)$$
$$b_2 = \left(\frac{2.1}{2.6}\right)\left(\frac{-0.30 - (+.50)(-0.47)}{1 - (-0.47)^2}\right)$$
$$b_2 = (0.81)\left(\frac{-0.30 - (-0.24)}{1 - 0.22}\right)$$
$$b_2 = (0.81)\left(\frac{-0.30 + 0.24}{0.78}\right)$$
$$b_2 = (0.81)\left(\frac{-0.06}{0.78}\right)$$
$$b_2 = (0.81)(-0.08)$$
$$b_2 = -0.07$$

Finding the *Y* Intercept. Now that partial slopes have been determined for both independent variables, the Y intercept (a) can be found. Note that a is calculated from the mean of the dependent variable (symbolized as $\overline{Y}$) and the means of the two independent variables ($\overline{X}_1$ and $\overline{X}_2$).

FORMULA 17.6
$$a = \overline{Y} - b_1\overline{X}_1 - b_2\overline{X}_2$$

Substituting the proper values for the example problem at hand, we would have

$$a = \overline{Y} - b_1\overline{X}_1 - b_2\overline{X}_2$$
$$a = 3.3 - (0.65)(2.7) - (-0.07)(13.7)$$
$$a = 3.3 - (1.8) - (-1.0)$$
$$a = 3.3 - 1.8 + 1.0$$
$$a = 2.5$$

The Least-Squares Multiple Regression Line and Predicting Y'. For our example problem, the full least-squares multiple regression equation would be

$$Y = a + b_1X_1 + b_2X_2$$
$$Y = 2.5 + (0.65)X_1 + (-0.07)X_2$$

As was the case with the bivariate regression line, this formula can be used to predict scores on the dependent variable from scores on the independent variables. For example, what would be our best prediction of husband's housework (Y') for a family of four children ($X_1 = 4$) where the husband had completed 11 years of schooling ($X_2 = 11$)? Substituting these values into the least-squares formula, we would have

$$Y' = 2.5 + (0.65)(4) + (-0.07)(11)$$
$$Y' = 2.5 + 2.6 - 0.08$$
$$Y' = 4.3$$

Our prediction would be that this husband would contribute 4.3 hours per week to housework. This prediction is, of course, a kind of "educated guess," which is unlikely to be perfectly accurate. However, we will make fewer errors of prediction using the least-squares line (and, thus, incorporating information from the independent variables) than we would using any other method of prediction (assuming, of course, that there is a linear association between the independent and the dependent variables). *(For practice in predicting Y scores and in computing slopes and the Y intercept, see problems 17.1 to 17.6.)*

17.4 MULTIPLE REGRESSION: ASSESSING THE EFFECTS OF THE INDEPENDENT VARIABLES

The least-squares multiple regression equation (Formula 17.3) is used to isolate the separate effects of the independents and to predict scores on the dependent variable. However, in many situations, using this formula to determine the relative importance of the various independent variables will be awkward—especially when the independent variables differ in terms of units of measurement (e.g., number of children vs. years of education). When the units of measurement differ, a comparison of the partial slopes will not necessarily tell us which independent variable has the strongest effect and is thus the most important. Comparing the partial slopes of variables that differ in units of measurement is a little like comparing apples and oranges.

We can make it easier to compare the effects of the independent variables by converting all variables in the equation to a common scale and thereby eliminating variations in the values of the partial slopes that are solely a function of differences in units of measurement. We can, for example, standardize all distributions by changing the scores of all variables to Z scores. Each distribution of scores would then have a mean of 0 and a standard deviation of 1 (see Chapter 5), and comparisons between the independent variables would be much more meaningful.

Computing the Standardized Regression Coefficients.

Beta-Weights. To standardize the variables to the normal curve, we could actually convert all scores into the equivalent Z scores and then recompute the slopes and the Y intercept. This would require a good deal of work; fortunately, a shortcut is available for computing the slopes of the standardized scores directly. These **standardized partial slopes,** called **beta-weights,** are symbolized b^*.

The beta-weights show the amount of change in the standardized scores of *Y* for a one-unit change in the standardized scores of each independent variable while controlling for the effects of all other independent variables.

Formulas and Computation for Beta-Weights. When we have two independent variables, the beta-weight for each is found by using Formulas 17.7 and 17.8:

FORMULA 17.7

$$b_1^* = b_1\left(\frac{s_1}{s_y}\right)$$

FORMULA 17.8

$$b_2^* = b_2\left(\frac{s_2}{s_y}\right)$$

We can now compute the beta-weights for our sample problem to see which of the two independent variables has the stronger effect on the dependent. For the first independent variable, number of children (X_1):

$$b_1^* = b_1\left(\frac{s_1}{s_y}\right)$$

$$b_1^* = (0.65)\left(\frac{1.5}{2.1}\right)$$

$$b_1^* = (0.65)(0.71)$$

$$b_1^* = 0.46$$

For the second independent variable, SES (X_2):

$$b_2^* = b_2\left(\frac{s_2}{s_y}\right)$$

$$b_2^* = (-0.07)\left(\frac{2.6}{2.1}\right)$$

$$b_2^* = (-0.07)(1.24)$$

$$b_2^* = -0.09$$

Comparing the value of the beta-weights, we see that number of children has a stronger effect than SES on husband's housework. Furthermore, the net effect (after controlling for the effect of SES) of the first independent variable is positive while the net effect of the second independent variable is negative.

The Standardized Least-Squares Regression Line. Using standardized scores, the least-squares regression equation can be written as

FORMULA 17.9

$$Z_y = a_z + b_1^* Z_1 + b_2^* Z_2$$

where *Z* indicates that all scores have been standardized to the normal curve

The standardized regression equation can be further simplified by dropping the term for the *Y* intercept, since this term will always be zero when scores have been standardized. This value is the point where the regression line crosses the *Y* axis and is equal to the mean of *Y* when all independent variables

Application 17.1

Five recently divorced men have been asked to rate subjectively the success of their adjustment to single life, on a scale ranging from 5 (very successful adjustment) to 1 (very poor adjustment). Is adjustment related to the length of time married? Is adjustment related to socioeconomic status as measured by yearly income?

Case	Adjustment (Y)	Years Married (X_1)	Income (dollars) (X_2)
A	5	5	30,000
B	4	7	45,000
C	4	10	25,000
D	3	2	27,000
E	1	15	17,000
A	5	5	30,000
	$\bar{Y} = 3.4$	$\bar{X}_1 = 7.8$	$\bar{X}_2 = 28{,}800.00$
	$s = 1.4$	$s = 4.5$	$s = 9{,}173.88$

The zero-order correlations among these three variables are

	Adjustment (Y)	Years Married (X_1)	Income (X_2)
Adjustment (Y)	1.00	−0.62	−0.62
Years married (X_1)		1.00	−.49
Income (X_2)			1.00

These results suggest a strong but opposite relationship between each independent variable and adjustment. Adjustment decreases as years married increases, and it increases as income increases.

To find the multiple regression equation, we must find the partial slopes.

For years married (X_1):

$$b_1 = \left(\frac{s_y}{s_1}\right)\left(\frac{r_{y1} - r_{y2}r_{12}}{1 - r_{12}^2}\right)$$
$$b_1 = \left(\frac{1.4}{4.5}\right)\left(\frac{(-0.62) - (0.62)(-0.49)}{1 - (-0.49)^2}\right)$$
$$b_1 = (0.31)\left(\frac{(-0.62) - (-0.30)}{1 - 0.24}\right)$$
$$b_1 = (0.31)\left(\frac{-0.32}{0.76}\right)$$
$$b_1 = (0.31)(-0.42)$$
$$b_1 = -0.13$$

For income (X_2):

$$b_2 = \left(\frac{s_y}{s_2}\right)\left(\frac{r_{y2} - r_{y1}r_{12}}{1 - r_{12}^2}\right)$$
$$b_2 = \left(\frac{1.4}{9173.88}\right)\left(\frac{(0.62) - (0.62)(-0.49)}{1 - (-0.49)^2}\right)$$
$$b_2 = (0.00015)\left(\frac{(0.62) - (0.30)}{1 - 0.24}\right)$$
$$b_2 = (0.00015)\left(\frac{0.32}{0.76}\right)$$
$$b_2 = (0.00015)(0.42)$$
$$b_2 = 0.000063$$

The Y intercept would be

$$a = \bar{Y} - b_1\bar{X}_1 - b_2\bar{X}_2$$
$$a = 3.4 - (-0.13)(7.8) - (0.000063)(28{,}800)$$
$$a = 3.4 - (-1.01) - (1.81)$$
$$a = 3.4 + 1.01 - 1.81$$
$$a = 2.60$$

The multiple regression equation is

$$Y = a + b_1X_1 + b_2X_2$$
$$Y = 2.60 + (-0.13)X_1 + (0.000063)X_2$$

What adjustment score could we predict for a male who had been married 30 years ($X_1 = 30$) and had an income of \$50,000 ($X_2 = 50{,}000$)?

$$Y' = 2.60 + (-0.13)(30) + (0.000063)(50{,}000)$$
$$Y' = 2.60 + (-3.90) + (3.15)$$
$$Y' = 1.85$$

To assess which of the two independents has the stronger effect on adjustment, the standardized partial slopes must be computed.

For years married (X_1):

$$b_1^* = b_1\left(\frac{s_1}{s_y}\right)$$
$$b_1^* = (-0.13)\left(\frac{4.5}{1.4}\right)$$
$$b_1^* = -0.42$$

(*continued next page*)

Application 17.1: *(continued)*

For income (X_2):

$$b_2^* = b_2\left(\frac{s_2}{s_y}\right)$$

$$b_2^* = (0.000063)\left(\frac{9173.88}{1.4}\right)$$

$$b_2^* = 0.41$$

The standardized regression equation is

$$Z_y = b_1^* Z_1 + b_2^* Z_2$$

$$Z_y = (-0.42)Z_1 + (0.41)Z_2$$

and the independent variables have nearly equal but opposite effects on adjustment. To assess the combined effects of the two independents on adjustment, the coefficient of multiple determination must be computed:

$$R^2 = r_{y1}^2 + r_{y2.1}^2(1 - r_{y1}^2)$$

$$R^2 = (-0.62)^2 + (0.46)^2(1 - (-0.62)^2)$$

$$R^2 = 0.38 + (0.21)(1 - 0.38)$$

$$R^2 = 0.38 + (0.21)(0.62)$$

$$R^2 = 0.38 + 0.13$$

$$R^2 = 0.51$$

The first independent variable, years married, explains 38% of the variation in adjustment by itself. To this quantity, income explains an additional 13% of the variation in adjustment. Taken together, the two independent variables explain a total of 51% of the variation in adjustment.

equal 0. This relationship can be seen by substituting 0 for all independent variables in Formula 17.6:

$$a = \overline{Y} - b_1\overline{X}_1 - b_2\overline{X}_2$$

$$a = \overline{Y} - b_1(0) - b_2(0)$$

$$a = \overline{Y}$$

Since the mean of any standardized distribution of scores is zero, the mean of the standardized Y scores will be zero and the Y intercept will also be zero ($a = \overline{Y} = 0$). Thus, Formula 17.9 simplifies to

FORMULA 17.10

$$Z_y = b_1^* Z_1 + b_2^* Z_2$$

The standardized regression equation, with beta-weights noted, would be

$$Z_y = (0.46)Z_1 + (-0.09)Z_2$$

and it is immediately obvious that the first independent variable has a much stronger direct effect on Y than does the second independent variable.

Summary. Multiple regression analysis permits the researcher to summarize the linear relationship among two or more independents and a dependent variable. The unstandardized regression equation (Formula 17.2) permits values of Y to be predicted from the independent variables in the original units of the variables. The standardized regression equation (Formula 17.10) allows the researcher to assess easily the relative importance of the various independent variables by comparing the beta-weights. *(For practice in computing and interpreting beta-weights, see any of the problems at the end of this chapter. It is probably a good idea to start with problem 17.1 since it has the smallest data set and the least complex computations.)*

Application 17.2

The following table presents information on three variables for a small sample of 10 nations. The dependent variable is the percentage of respondents who said they are "very happy" on a survey administered to random samples from each nation. The independent variables measure health and physical well-being (life expectancy, or the number of years the average citizen can expect to live at birth) and income inequality (the amount of total income that goes to the richest 20% of the population). Our expectation is that happiness will have a positive correlation with life expectancy (the greater the health, the happier the population) and a negative relationship with inequality (the greater the inequality, the greater the discontent and the lower the level of happiness). In this analysis, we focus on R^2 and the beta-weights only.

The scores of the nations, along with descriptive statistics, are as follows:

Nation	Percent "Very Happy" (Y)	Life Expectancy (X_1)	Income Inequality (X_2)
Brazil	22	69	64
Belgium	37	79	37
Canada	32	80	39
China	25	73	39
Dominican Republic	32	71	53
Ghana	26	58	47
India	23	65	46
Japan	23	81	36
Mexico	31	75	57
Ukraine	5	70	38
Mean =	24.7	72.8	45.6
Standard deviation =	9.4	6.8	9.6

The zero-order correlations for these variables are given in the following correlation matrix:

	Happy	Life Expectancy	Inequality
Happiness	1.00	0.32	0.12
Life Expectancy		1.00	−0.39
GDP per capita			1.00

Consistent with our expectations, there is a positive, weak-to-moderate relationship between life expectancy and happiness. Unexpectedly, however, the relationship between inequality and happiness is positive, although weak in strength. The relationship between the two independent variables is moderate and negative, indicating that nations with more income inequality have lower life expectancy.

The combined effect of life expectancy and inequality on happiness is found by computing R^2:

$$R^2 = r_{y1}^2 + r_{y2.1}^2(1 - r_{y1}^2)$$
$$R^2 = (0.32)^2 + (0.27)^2(1 - 0.26^2)$$
$$R^2 = 0.10 + (0.07)(0.93)$$
$$R^2 = 0.10 + 0.07$$
$$R^2 = 0.17$$

By itself, life expectancy explains 10% of the variance in happiness. To this, income inequality adds another 7%, for a total of 17%. This leaves about 83% of the variance unexplained, a sizeable proportion but not unusually large in social science research.

To assess the separate effects of the two independent variables, the beta-weights must be calculated. We need values for the unstandardized partial slopes to compute beta-weights, and we will simply report the values as 0.53 for X_1 (life expectancy) and 0.26 for X_2 (income inequality).

For the first independent variable (life expectancy):

$$b_1^* = b_1\left(\frac{s_1}{s_y}\right)$$
$$b_1^* = (0.53)\left(\frac{7.17}{8.77}\right)$$
$$b_1^* = (0.53)(0.82)$$
$$b_1^* = 0.43$$

For the second independent variable (income inequality):

$$b_2^* = b_2\left(\frac{s_2}{s_y}\right)$$
$$b_2^* = (0.26)\left(\frac{9.64}{8.77}\right)$$
$$b_2^* = (0.26)(1.10)$$
$$b_2^* = 0.28$$

Recall that the beta-weights show the effect of each independent variable on the dependent variable while controlling for the other independent variables in the equation. In this case, life expectancy has the stronger effect and the relationship is positive. The effect of income inequality is also positive.

In summary, for these nations, level of happiness has a moderate positive relationship with life expectancy and a weaker positive relationship with income inequality. Taken together, the independent variables explain 17% of the variation in happiness.

ONE STEP AT A TIME **Computing and Interpreting Beta-Weights (b^*)**

Computation

NOTE: These procedures apply when there are two independent variables and one dependent variable. For more complex situations, use a computerized statistical package such as SPSS to do the calculations.

Step 1: Compute the beta-weight associated with the first independent variable by using Formula 17.7:

$$b_1^* = b_1\left(\frac{s_1}{s_y}\right)$$

a. Divide s_1 by s_y.

b. Multiply the value you found in step a by the partial slope of the first independent variable (b_1). This value is the beta-weight associated with the first independent variable.

Step 2: Compute the beta-weight associated with the second independent variable by using Formula 17.8:

$$b_2^* = b_2\left(\frac{s_2}{s_y}\right)$$

a. Divide s_2 by s_y.

b. Multiply the value you found in step a by the partial slope of the second independent variable. (b_2). This value is the beta-weight associated with the second independent variable.

Interpretation

Step 3: A beta-weight (or standardized partial slope) shows the increase in the value of *Y* for a unit increase in the value of the associated independent variable while controlling for the effects of the other independent variable after all variables have been standardized (or transformed to *Z* scores).

17.5 MULTIPLE CORRELATION

We use the multiple regression equations to disentangle the separate direct effects of each independent variable on the dependent variable. Using **multiple correlation** techniques, we can also ascertain the combined effects of all independent variables on the dependent variable. We do so by computing the **multiple correlation coefficient (*R*)** and the **coefficient of multiple determination $(R)^2$.** The value of the latter statistic represents the proportion of the variance in *Y* that is explained by all the independent variables combined.

In terms of zero-order correlation, we have seen that number of children (X_1) explains 25% of the variance in $Y(r_{y1}^2 = (.50)^2 = .25 \times 100 = 25\%)$ by itself and that husband's education explains 9% of the variance in $Y(r_{y2}^2 = (-.30)^2 = .09 \times 100 = 9\%)$. The two zero-order correlations cannot simply be added together to ascertain their combined effect on *Y*, because the two independent variables are also correlated with each other, and, therefore, they will "overlap" in their effects on *Y* and explain some of the same variance. This overlap is eliminated in Formula 17.11:

FORMULA 17.11

$$R^2 = r_{y1}^2 + r_{y2.1}^2(1 - r_{y1}^2)$$

where R^2 = the multiple correlation coefficient
r_{y1}^2 = the zero-order correlation between *Y* and X_1, the quantity squared
$r_{y2.1}^2$ = the partial correlation of *Y* and X_2, while controlling for X_1, the quantity squared

The first term in this formula (r_{y1}^2) is the coefficient of determination for the relationship between *Y* and X_1. It represents the amount of variation in *Y* explained by X_1 by itself. To this quantity we add the amount of the variation re-

ONE STEP AT A TIME

Computing and Interpreting the Multiple Correlation Coefficient (R^2)

Computation

NOTE: These procedures apply when there are two independent variables and one dependent variable. For more complex situations, use a computerized statistical package such as SPSS to do the calculations.

Step 1: Compute the multiple correlation coefficient by using Formula 17.11:

$$R^2 = r_{y1}^2 + r_{y2.1}^2(1 - r_{y1}^2)$$

a. Find the value of the partial correlation coefficient for $r_{y2.1}$.

b. Square the value you found in step a.

c. Square the value of r_{y1}.

d. Subtract the value you found in step c from 1.

e. Multiply the value you found in step d by the value you found in step b.

f. Add the value you found in step e to the value you found in step c. The result is the multiple correlation coefficient.

Interpretation

Step 2: The multiple correlation coefficient is the total amount of the variation in Y explained by all independent variables combined.

maining in Y (given by $1 - r_{y1}^2$) that can be explained by X_2 after the effect of X_1 is controlled ($r_{y2.1}^2$). Basically, Formula 17.11 allows X_1 to explain as much of Y as it can and then adds in the effect of X_2 after X_1 is controlled (thus eliminating the "overlap" in the variance of Y that X_1 and X_2 have in common).

Computing and Interpreting R and R^2. To observe the combined effects of number of children (X_1) and husband's education (X_2) on husband's housework (Y), we need two quantities. The correlation between X_1 and Y ($r_{y1} = .50$) has already been found. Before we can solve Formula 17.11, we must first calculate the partial correlation of Y and X_2 while controlling for X_1 ($r_{y2.1}$):

$$r_{y2.1} = \frac{r_{y2} - (r_{y1})(r_{12})}{\sqrt{1 - r_{y1}^2}\sqrt{1 - r_{12}^2}}$$

$$r_{y2.1} = \frac{(-0.30) - (0.50)(-0.47)}{\sqrt{1 - (0.50)^2}\sqrt{1 - (-0.47)^2}}$$

$$r_{y2.1} = \frac{(-0.30) - (-0.24)}{\sqrt{0.75}\sqrt{0.78}}$$

$$r_{y2.1} = \frac{-0.06}{0.77}$$

$$r_{y2.1} = -0.08$$

Formula 17.11 can now be solved for our sample problem:

$$R^2 = r_{y1}^2 + r_{y2.1}^2(1 - r^2_{y1})$$

$$R^2 = (0.50)^2 + (-0.08)^2(1 - 0.50^2)$$

$$R^2 = 0.25 + (0.006)(1 - 0.25)$$

$$R^2 = 0.25 + 0.005$$

$$R^2 = 0.255$$

The first independent variable (X_1), number of children, explains 25% of the variance in Y by itself. To this total, the second independent (X_2), husband's education, adds only a half a percent, for a total explained variance of 25.5%. In combination, the two independent variables explain a total of 25.5% of the variation in the dependent variable. *(For practice in computing and interpreting R and R^2, see any of the problems at the end of this chapter. It is probably a good idea to start with problem 17.1 since it has the smallest data set and the least complex computations.)*

17.6 INTERPRETING STATISTICS: ANOTHER LOOK AT THE CORRELATES OF CRIME

In Chapter 15, we assessed the association between poverty and crime. We found a moderate-to-strong, positive relationship ($r = 0.54$) between a measure of poverty and the homicide rate for the 50 states, an indication that there may be an important relationship between these two variables. In this installment of Interpreting Statistics, we return to this relationship and add several independent variables to the analysis.[2]

The second independent variable is a measure of age: the percentage of the population less than 18. Research on street crimes and serious felonies like homicide commonly find moderate-to-strong relationships with age. Rates of violent crime are highest for people in their teens and twenties but decline dramatically as age rises. Thus, we can expect a substantial positive relationship between this measure of age and the homicide rate (the higher the proportion of younger people in a population, the higher the homicide rate).

The third independent variable comes from a body of research that argues that there is a subculture of violence, a stronger tendency to use force and aggression, in the South. In other words, the norms and values of the Southern regional subculture sanction and endorse the everyday use of force more strongly than in other regions of the nation. Thus, we should expect to find higher rates of homicide in Southern states.

Although there are compelling reasons for including region in this analysis, we must first confront an important problem. Region (South, North, Midwest, etc.) is a nominal-level variable, but regression analysis requires that all variables be interval-ratio in level of measurement. We can resolve this problem by treating region as a dummy variable, as described in Section 15.8. As you recall, dummy variables have exactly two categories, one coded as a score of 0 and the other as a score of 1. Treated this way, nominal-level variables such as gender and race are commonly included as independent variables in regression equations. In this case, we will create a variable that is coded so that Southern states are scored as "1" and non-Southern states as "0." Coded this way, region should have a positive relationship with homicide (since "South" is coded as 1 and all other regions as 0).

Before beginning the multivariate analysis, we should examine the correlation coefficients between all possible pairs of variables. Table 17.3 reports the zero-order correlations.

---

[2]This analysis considers the effects of three independent variables, one more than were included in previous examples in this chapter. As you will see, the addition of an independent variable will not complicate interpretation and analysis unduly. However, the mathematics underlying multiple regression with three independent variables are complex and should be attempted only with the aid of a computerized statistics package such as SPSS. We will not show the underlying computations in this analysis.

TABLE 17.3 ZERO-ORDER CORRELATIONS

	Homicide	Poverty	Age	Region
Homicide	1.00	0.54	0.41	0.46
Poverty		1.00	0.10	0.61
Age			1.00	0.03
Region				1.00

As we saw in Chapter 15, there is a moderate-to-strong, positive relationship between poverty and homicide: States with higher rates of poverty have higher rates of homicide ($r = 0.54$). Turning to the other variables, there are, as expected, moderate positive relationships between age and homicide and between region and homicide. This indicates that Southern states and states with a higher percentage of young people have higher homicide rates. Southern states tend to have higher poverty rates ($r = 0.61$), and, finally, the relationship between age and poverty is weak ($r = 0.10$) and there appears to be no relationship between age and region ($r = 0.03$).

We will begin the multivariate analysis by asking if the original relation between homicide and poverty persists after controlling for the effects of age and region, controlling for these variables one at a time. Recall that a partial correlation coefficient indicates the strength and direction of a bivariate relation after controlling for the effects of a third variable. Table 17.4 reports the results of controlling for age and then region.

The partial correlation using age as the control variable is almost exactly equal in value to the zero-order correlation. Thus, we can conclude that age has no effect, an outcome that would increase our confidence that there is a direct, causal relationship between poverty and the homicide rate.

However, the partial correlation when controlling for region (0.37) is noticeably weaker than the zero-order correlation (0.54). On one hand, this outcome weakens the argument that there is a direct causal relationship between poverty and violence: Region has a substantial effect on the bivariate relationship and should be taken into account in any further analysis. On the other hand, the bivariate relationship between homicide and poverty is positive and moderate in strength even after controlling for region. Thus, we would probably conclude that there is a relationship between poverty and homicide rate, even though the relationship is affected by region.

Next, we can assess the relative importance of the three independent variables on homicide. Table 17.5 presents the essential information—beta-weights

TABLE 17.4 PARTIAL CORRELATION COEFFICIENTS FOR THE RELATIONSHIP BETWEEN HOMICIDE AND POVERTY

Zero-order Correlation Between Homicide and Poverty	Partial Correlation After Controlling for	
	Age	Region
0.54	0.55	0.37

TABLE 17.5. REGRESSION ANALYSIS WITH HOMICIDE RATE AS THE DEPENDENT VARIABLE

Variable	Beta-Weight
Poverty	.37
Age	.37
Region	.22
	$R^2 = 0.45$

and R^2—that will help us sort out the relative importance and total impact of these variables on the homicide rate.

By itself, poverty explained 29% of the variance in homicide rates among the states ($r^2 = (0.54)^2 = 0.29$). To this, age and region add another 16%, for a total of 45% ($R^2 = 0.45$). These three variables, together, account for almost half of the variation in homicide rates from state to state. This is quite a substantial percentage, even though more than half of the variation remains unaccounted for or unexplained.

We might be able to raise the value of R^2 by adding more independent variables to the equation. However, it is common to find that additional independent variables explain smaller and smaller proportions of the remaining variance and have a diminishing effect on R^2. This phenomenon can be caused by many factors, including the fact that the independent variables will usually be correlated with each other (e.g., see the relationship between region and poverty in Table 17.3) and will overlap in their effect on the dependent variable.

The beta-weights in Table 17.5 show the effect of each variable while controlling for all other variables in the equation (as opposed to controlling for other variables one at a time as we did in Table 17.4). The direction of all three independent variables is positive. Poverty and age have equal effects on homicide rate, and region has a substantially weaker effect when the other variables are controlled.

In conclusion, this multivariate analysis indicates that poverty, age, and region all have important, positive effects on homicide rate. The beta-weights (and the partial correlation coefficients) show that poverty and age are equal in their effects, with region having a somewhat weaker effect. Taken together, the three variables account for 45% of the variance in homicide rate from state to state.

17.7 THE LIMITATIONS OF MULTIPLE REGRESSION AND CORRELATION

Multiple regression and correlation are very powerful tools for analyzing the interrelationships among three or more variables. The techniques presented in this chapter permit the researcher to predict scores on one variable from two or more other variables, to distinguish between independent variables in terms of the importance of their direct effects on a dependent, and to ascertain the total effect of a set of independent variables on a dependent variable. In terms of the flexibility of the techniques and the volume of information they can supply, multiple regression and correlation represent some of the most powerful statistical techniques available to social science researchers.

Powerful tools are not cheap. They demand high-quality data, and measurement at the interval-ratio level is difficult to accomplish at this stage in the development of the social sciences. Furthermore, these techniques assume that the interrelationships among the variables follow a particular form. First, they assume that each independent variable has a linear relationship with the dependent variable. How well a given set of variables meets this assumption can be quickly checked with scattergrams.

Second, the techniques presented in this chapter assume that there is no interaction among the variables in the equation. If there is interaction among the variables, it will not be possible to estimate or predict the dependent variable accurately by simply adding the effects of the independents. There are techniques for handling interaction among the variables in the set, but these techniques are beyond the scope of this text.

READING STATISTICS 12: Regression and Correlation

Research projects that analyze the interrelationships among many variables are particularly likely to employ regression and correlation as central statistical techniques. The results of these projects will typically be presented in summary tables that report the multiple correlations, the slopes, and, if applicable, the significance of the results. The zero-order correlations are often presented in the form of a matrix that displays the value of Pearson's *r* for every possible bivariate relationship in the data set. An example of such a matrix can be found in Section 17.6.

Usually, tables that summarize the multivariate analysis will report R^2 and the slope for each independent variable in the regression equation. An example of this kind of summary table would look like this:

Independent Variables	Multiple R^2	Beta-weights
X_1	.17	.47
X_2	.23	.32
X_3	.27	.16
⋮	⋮	⋮

This table reports that the first independent variable, X_1, has the strongest direct relationship with the dependent variable and explains 17% of the variance in the dependent variable by itself ($R^2 = 0.17$). The second independent variable, X_2, adds 6% to the explained variance ($R^2 = 0.23$ after X_2 is entered into the equation). The third independent variable, X_3, adds 4% to the explained variance of the dependent ($R^2 = 0.27$ after X_3 is entered into the equation).

Statistics in the Professional Literature

Researchers have documented a sharp racial difference in support for the death penalty in American society. In public opinion surveys, for example, capital punishment is supported by large majorities of white respondents but by only a minority of blacks. What accounts for this difference? One commonly researched possibility is that support among whites is associated with antiblack racism and that opposition among blacks is associated with the perception that the sanction—indeed the entire criminal justice system—is biased against blacks. The former possibility is investigated by sociologists James Unnever and Francis Cullen. Using a nationally representative sample of over 1000 respondents, they attempt to ascertain the extent to which white support for the death penalty reflects a perception of blacks as threatening, criminally dangerous, and unintelligent.

The researchers constructed several different regression models with support for capital punishment as the dependent variable. They begin their analysis by reporting the bivariate relationship between race and support, summarized as Model 1 in Table 1.

TABLE 1 MULTIPLE REGRESSION RESULTS WITH SUPPORT FOR CAPITAL PUNISHMENT AS THE DEPENDENT VARIABLE ($N = 1117$)†

	Model 1		Model 4	
	Unstandardized Coefficient	Beta-weight	Unstandardized Coefficient	Beta-weight
Race (black = 1)	−0.60	−0.18***	−0.36	−0.11***
Gender (male = 1)			−0.04	−0.01
Education			−0.01	−0.03
Religiosity			−0.05	−0.17***
Political Conservatism			0.04	0.07**
Income			0.02	0.04
Racism			0.05	0.11***
R^2	0.03***		0.15***	

*$p < .05$; **$p < .01$; ***$p < .001$.
†Adapted from Table 1 in the original article.

(*continued next page*)

READING STATISTICS 12: *(continued)*

Note that race is a dummy variable (see Section 15.8), with blacks coded as 1 and whites as 0. Thus, the negative values of the regression coefficients indicate that, as expected, whites are more supportive. In other words, support decreases as race "increases," or moves towards the higher scores associated with black respondents. Note also that the relationship is statistically significant at the .001 level.

Unnever and Cullen introduce almost a score of independent variables in the equation, including many measures of social and political traits that previous research has shown are correlated with support for capital punishment. For example, support has been shown to be greater among males, Southerners, political conservatives, and the more religious. The great power of regression analysis is that the researchers can examine the effect of race on support with all of these other factors controlled. Model 4 in Table 1 summarizes some of the final results of the researchers.

Model 4 shows that support for the death penalty is significantly associated with their measure of white racism. Indeed the beta-weight shows that white racism is one of the strongest predictors of support for the death penalty. Note that the beta-weight associated with race loses about a third of its strength (that is, it declines from −0.18 to −0.11) after the measure of racism is added to the equation. This indicates that, with all other variables controlled, white racism by itself accounts for about 33% of the racial divide in support for capital punishment.

Unnever, James, and Francis Cullen. 2007. "The Racial Divide in Support for the Death Penalty: Does Racism Matter?" *Social Forces* 85: 1281–1301.

Third, the techniques of multiple regression and correlation assume that the independent variables are uncorrelated with each other. Strictly speaking, this condition means that the zero-order correlation among all pairs of independents should be zero; but, practically, we act as if this assumption has been met if the intercorrelations among the independents are low.

To the extent that these assumptions are violated, the regression coefficients (especially partial and standardized slopes) and the coefficient of multiple determination (R^2) become less and less trustworthy and the techniques less and less useful. If the assumptions of the model cannot be met, the alternative might be to turn to the multivariate techniques described in Chapter 16. Unfortunately, those techniques, in general, supply a lower volume of less precise information about the interrelationships among the variables.

Finally, we should note that we have covered only the simplest applications of partial correlation and multiple regression and correlation. In terms of logic and interpretation, the extensions to situations involving more variables are relatively straightforward. However, the computations for these situations are extremely complex. If you are faced with a situation involving more than three variables, turn to one of the computerized statistical packages commonly available on college campuses (e.g., SPSS or SAS). These programs require minimal computer literacy and can handle complex calculations in, literally, the blink of an eye. Efficient use of these packages will enable you to avoid drudgery and will free you to do what social scientists everywhere enjoy doing most: pondering the meaning of your results and, by extension, the nature of social life.

SUMMARY

1. Partial correlation involves controlling for third variables in a manner analogous to that introduced in the previous chapter. Partial correlations permit the detection of direct and spurious or intervening relationships between X and Y.
2. Multiple regression includes statistical techniques by which predictions of the dependent variable from more than one independent variable can be made (by partial slopes and the multiple regression equation) and by which we can disentangle the relative importance of the independent variables (by standardized partial slopes).
3. The multiple correlation coefficient (R^2) summarizes the combined effects of all independent variables on the dependent variable in terms of the proportion of the total variation in Y that is explained by all of the independent variables.
4. Partial correlation and multiple regression and correlation are some of the most powerful tools available to the researcher and demand high-quality measurement and relationships among the variables that are linear and noninteractive. Further, correlations among the independent variables must be low (preferably zero). Although the price is high, these techniques pay considerable dividends in the volume of precise and detailed information they generate about the interrelationships among the variables.

SUMMARY OF FORMULAS

Partial correlation coefficient:

17.1 $$r_{yx.z} = \frac{r_{yx} - (r_{yz})(r_{xz})}{\sqrt{1 - r_{yz}^2}\sqrt{1 - r_{xz}^2}}$$

Least-squares regression line (bivariate):

17.2 $$Y = a + bX$$

Least-squares multiple regression line:

17.3 $$Y = a + b_1X_1 + b_2X_2$$

Partial slope for X_1:

17.4 $$b_1 = \left(\frac{s_y}{s_1}\right)\left(\frac{r_{y1} - r_{y2}r_{12}}{1 - r_{12}^2}\right)$$

Partial slope for X_2:

17.5 $$b_2 = \left(\frac{s_y}{s_2}\right)\left(\frac{r_{y2} - r_{y1}r_{12}}{1 - r_{12}^2}\right)$$

Y intercept:

17.6 $$a = \bar{Y} - b_1\bar{X}_1 - b_2\bar{X}_2$$

Standardized partial slope (beta-weight) for X_1:

17.7 $$b_1^* = b_1\left(\frac{s_1}{s_y}\right)$$

Standardized partial slope (beta-weight) for X_2:

17.8 $$b_2^* = b_2\left(\frac{s_2}{s_y}\right)$$

Standardized least-squares regression line:

17.9 $$Z_y = a_z + b_1^*Z_1 + b_2^*Z_2$$

Standardized least-squares regression line (simplified):

17.10 $$Z_y = b_1^*Z_1 + b_2^*Z_2$$

Coefficient of multiple determination:

17.11 $$R^2 = r_{y1}^2 + r_{y2.1}^2(1 - r_{y1}^2)$$

GLOSSARY

Beta-weights (b^*). Standardized partial slopes.

Coefficient of multiple determination (R^2). A statistic that equals the total variation explained in the dependent variable by all independent variables combined.

Multiple correlation. A multivariate technique for examining the combined effects of more than one independent variable on a dependent variable.

Multiple correlation coefficient (R). A statistic that indicates the strength of the correlation between a dependent variable and two or more independent variables.

Multiple regression. A multivariate technique that breaks down the separate effects of the independent variables on the dependent variable; used to make predictions of the dependent variable.

Partial correlation. A multivariate technique for examining a bivariate relationship while controlling for other variables.

Partial correlation coefficient. A statistic that shows the relationship between two variables while controlling for other variables; $r_{yx.z}$ is the symbol for the partial correlation coefficient when controlling for one variable.

Partial slopes. In a multiple regression equation, the slope of the relationship between a particular independent variable and the dependent variable while controlling for all other independent variables in the equation.

Standardized partial slopes (beta-weights). The slope of the relationship between a particular independent variable and the dependent variable when all scores have been normalized.

Zero-order correlations. Correlation coefficients for bivariate relationships.

PROBLEMS

17.1 PS In problem 15.1, data regarding voter turnout in five cities was presented. For the sake of convenience, the data for three of the variables are presented again here along with descriptive statistics and zero-order correlations.

City	Turnout	Unemployment Rate	% Negative Ads
A	55	5	60
B	60	8	63
C	65	9	55
D	68	9	53
E	70	10	48
Mean =	63.6	8.2	55.8
s =	5.5	1.7	5.3

	Unemployment Rate	Negative Ads
Turnout	0.95	−0.87
Unemployment Rate		−0.70

a. Compute the partial correlation coefficient for the relationship between turnout (Y) and unemployment (X) while controlling for the effect of negative advertising (Z). What effect does this control variable have on the bivariate relationship? Is the relationship between turnout and unemployment direct? *(HINT: Use Formula 17.1 and see Section 17.2.)*

b. Compute the partial correlation coefficient for the relationship between turnout (Y) and negative advertising (X) while controlling for the effect of unemployment (Z). What effect does this have on the bivariate relationship? Is the relationship between turnout and negative advertising direct? *(HINT: Use Formula 17.1 and see Section 17.2. You will need this partial correlation to compute the multiple correlation coefficient.)*

c. Find the unstandardized multiple regression equation with unemployment (X_1) and negative ads (X_2) as the independent variables. What turnout would be expected in a city in which the unemployment rate was 10% and 75% of the campaign ads were negative? *(HINT: Use Formulas 17.4 and 17.5 to compute the partial slopes and then use Formula 17.6 to find a, the Y intercept. The regression line is stated in Formula 17.3. Substitute 10 for X_1 and 75 for X_2 to compute predicted Y.)*

d. Compute beta-weights for each independent variable. Which has the stronger impact on turnout? *(HINT: Use Formulas 17.7 and 17.8 to calculate the beta-weights.)*

e. Compute the multiple correlation coefficient (R) and the coefficient of multiple determination (R^2). How much of the variance in voter turnout is explained by the two independent variables? *(HINT: Use Formula 17.11. You calculated $r^2_{y2.1}$ in part b of this problem.)*

f. Write a paragraph summarizing your conclusions about the relationships among these three variables.

17.2 SOC A scale measuring support for increases in the national defense budget has been administered to a sample. The respondents have also been asked to indicate how many years of school they have completed and how many years, if any, they served in the military. Take "support" as the dependent variable.

Case	Support	Years of School	Years of Service
A	20	12	2
B	15	12	4
C	20	16	20
D	10	10	10
E	10	16	20
F	5	8	0
G	8	14	2
H	20	12	20
I	10	10	4
J	20	16	0

a. Compute the partial correlation coefficient for the relationship between support (Y) and years of school (X) while controlling for the effect of years of service (Z). What effect does this have on the bivariate relationship? Is the relationship between support and years of school direct?

b. Compute the partial correlation coefficient for the relationship between support (Y) and years of service (X) while controlling for the effect of years of school (Z). What effect does this have on the bivariate relationship? Is the relationship between support and years of service direct? *(HINT: You will need this partial correlation to compute the multiple correlation coefficient.)*

c. Find the unstandardized multiple regression equation with school (X_1) and service (X_2) as the independent variables. What level of support would be expected in a person with 13 years of school and 15 years of service?

d. Compute beta-weights for each independent variable. Which has the stronger impact on turnout?

e. Compute the multiple correlation coefficient (R) and the coefficient of multiple determination (R^2). How much of the variance in support is explained by the two independent variables? *(HINT: You calculated $r^2_{y2.1}$ in part b of this problem.)*

f. Write a paragraph summarizing your conclusions about the relationships among these three variables.

17.3 SOC Data on civil strife (number of incidents), unemployment, and urbanization have been gathered for 10 nations. Take civil strife as the dependent variable. Compute the zero-order correlations among all three variables.

Number of Incidents of Civil Strife	Unemployment Rate	Percentage of Population Living in Urban Areas
0	5.3	60
1	1.0	65
5	2.7	55
7	2.8	68
10	3.0	69
23	2.5	70
25	6.0	45
26	5.2	40
30	7.8	75
53	9.2	80

a. Compute the partial correlation coefficient for the relationship between strife (Y) and unemployment (X) while controlling for the effect of urbanization (Z). What effect does this have on the bivariate relationship? Is the relationship between strife and unemployment direct?

b. Compute the partial correlation coefficient for the relationship between strife (Y) and urbanization (X) while controlling for the effect of unemployment (Z). What effect does this have on the bivariate relationship? Is the relationship between strife and urbanization direct? *(HINT: You will need this partial correlation to compute the multiple correlation coefficient.)*

c. Find the unstandardized multiple regression equation with unemployment (X_1) and urbanization (X_2) as the independent variables. What level of strife would be expected in a nation in which the unemployment rate was 10% and 90% of the population lived in urban areas?

d. Compute beta-weights for each independent variable. Which has the stronger impact on turnout?

e. Compute the multiple correlation coefficient (R) and the coefficient of multiple determination (R^2). How much of the variance in strife is explained by the two independent variables?

f. Write a paragraph summarizing your conclusions about the relationships among these three variables.

17.4 SOC/CJ In problem 15.5, crime and population data were presented for each of 10 states. The data are reproduced here.

	Crime Rates			Population*		
State	Homicide	Robbery	Car Theft	Growth	Density	Urban
Maine	1	22	99	4	43	70
New York	5	174	213	2	408	88
Ohio	5	153	357	1	280	77
Iowa	2	38	183	1	53	61
Virginia	5	93	233	7	191	73
Kentucky	6	38	183	3	105	56
Texas	6	159	418	10	87	82
Arizona	7	134	963	16	52	88
Washington	3	95	696	7	95	82
California	7	172	703	7	232	94

**Growth*: percentage change in population from 2000 to 2005; *density*: population per square mile of land area, 2005; *urban*: percent of population living in urban areas, 2000.
Source: United States Bureau of the Census, *Statistical Abstracts of the United States: 2007*. Washington, DC, 2007.

Take the three crime variables as the dependent variables (one at a time) and do the following:

a. Find the multiple regression equations (unstandardized) with growth and urbanization as independent variables.

b. Make a prediction for each crime variable for a state with a 5% growth rate and a population that is 90% urbanized.

c. Compute beta-weights for each independent variable in each equation and compare their relative effect on each dependent.

d. Compute R and R^2 for each crime variable, using two of the population variables as independent variables.

e. Write a paragraph summarizing your findings.

17.5 PS Problem 15.4 presented data on 10 precincts. The information is reproduced here.

Precinct	Percent Democrat	Percent Minority	Voter Turnout
A	50	10	56
B	45	12	55
C	56	8	52
D	78	15	60
E	13	5	89
F	85	20	25
G	62	18	64
H	33	9	88
I	25	0	42
J	49	9	36

Take voter turnout as the dependent variable and do the following:

a. Find the multiple regression equations (unstandardized).

b. What turnout would you expect for a precinct in which 0% of the voters were Democrats and 5% were minorities?

c. Compute beta-weights for each independent variable and compare their relative effect on turnout. Which was the more important factor?

d. Compute R and R^2.

e. Write a paragraph summarizing your findings.

17.6 SW Twelve families have been referred to a counselor, and she has rated each of them on a cohesiveness scale. Also, she has information on family income and number of children currently living at home. Take family cohesion as the dependent variable.

Family	Cohesion Score	Income	Number of Children
A	10	30,000	5
B	10	70,000	4
C	9	35,000	4
D	5	25,000	0
E	1	55,000	3
F	7	40,000	0
G	2	60,000	2
H	5	30,000	3
I	8	50,000	5
J	3	25,000	4
K	2	45,000	3
L	4	50,000	0

a. Find the multiple regression equations (unstandardized).

b. What level of cohesion would be expected in a family with an income of $20,000 and 6 children?

c. Compute beta-weights for each independent variable and compare their relative effect on cohesion. Which was the more important factor?

d. Compute R and R^2.

e. Write a paragraph summarizing your findings.

17.7 Problem 15.8 presented per capita expenditures on education for 15 states, along with rank on income per capita and the percentage of the population that has graduated from high school. The data are reproduced here.

State	Per Capita Expenditures on Education, 2004	Percent High School Graduates, 2005*	Rank in Per Capita Income, 2005
Arkansas	1158	81	48
Colorado	1599	89	7
Connecticut	2142	90	1
Florida	1286	87	23
Illinois	2334	87	14
Kansas	1418	91	25
Louisiana	1367	80	42
Maryland	1627	87	4
Michigan	1880	89	24
Mississippi	1178	80	49
Nebraska	1373	90	20
New Hampshire	1632	92	6
North Carolina	1233	84	37
Pennsylvania	1617	86	18
Wyoming	1896	90	12

*Based on percentage of population age 25 and older.
Source: United States Bureau of the Census, *Statistical Abstracts of the United States: 2007*. Washington, DC, 2007.

Take educational expenditures as the dependent variable.

a. Compute beta-weights for each independent variable and compare their relative effect on expenditures. Which was the more important factor?

b. Compute R and R^2.

c. Write a paragraph summarizing your findings.

17.8 SOC The scores on four variables for 20 individuals are reported here: hours of TV (average number of hours of TV viewing each day), occupational prestige (higher scores indicate greater prestige), number of children, and age. Take TV viewing as the dependent variable and select two of the remaining variables as independent variables.

Hours of TV	Occupational Prestige	Number of Children	Age
4	50	2	43
3	36	3	58
3	36	1	34
4	50	2	42

(*continued next column*)

(*continued*)

Hours of TV	Occupational Prestige	Number of Children	Age
2	45	2	27
3	50	5	60
4	50	0	28
7	40	3	55
1	57	2	46
3	33	2	65
1	46	3	56
3	31	1	29
1	19	2	41
0	52	0	50
2	48	1	62
4	36	1	24
3	48	0	25
1	62	1	87
5	50	0	45
1	27	3	62

a. Compute beta-weights for each of the independent variables you selected and compare their relative effect on the hours of television watching. Which was the more important factor?

b. Compute R and R^2.

c. Write a paragraph summarizing your findings.

SPSS for Windows

Using *SPSS for Windows* for Regression Analysis

SPSS DEMONSTRATION 17.1 What Are the Correlates of Occupational Prestige? Another Look

In Demonstration 15.1, we used the **Correlate** procedure to calculate zero-order correlation coefficients between *prestg80, papres80*, and *educ*. In this section, we use a significantly more complex and flexible procedure called **Regression** to analyze the effects of *papres80* and *educ* on *prestg80*. The **Regression** procedure permits the user to control many aspects of the regression formula, and it can produce a much greater volume of output than the **Correlate** procedure. Among other things, **Regression** displays the slope (*b*) and the *Y* intercept (*a*), so we can use this procedure to find least-squares regression lines. This demonstration represents a very sparing use of the power of this command and an extremely economical use of all the options available. I urge you to explore some of the variations and capabilities of this powerful data-analysis procedure.

With the 2006 GSS loaded, click **Analyze, Regression,** and **Linear,** and the **Linear Regression** window will appear. Move *prestg80* into the **Dependent** box and *educ* and *papres80* into the **Independent(s)** box. If you wish, you can click the **Statistics** button and then click **Descriptives** to get zero-order correlations, means, and standard deviations for the variables. Click **Continue** and **OK,** and the following output will appear (descriptive information about the variables and the zero-order correlations are omitted here to conserve space).

Model Summary

Model	R	R Square	Adjusted R Square	Std. Error of the Estimate
1	.540(a)	.292	.290	12.035

[a]Predictors: (Constant), HIGHEST YEAR OF SCHOOL COMPLETED, FATHERS OCCUPATIONAL PRESTIGE SCORE (1980)

ANOVA(b)

Model		Sum of Squares	df	Mean Square	F	Sig.
1	Regression	42766.270	2	21383.135	147.641	.000(a)
	Residual	103699.970	716	144.832		
	Total	146466.239	718			

[a]Predictors: (Constant), HIGHEST YEAR OF SCHOOL COMPLETED, FATHERS OCCUPATIONAL PRESTIAGE STORE (1980)
[b]Dependent Variable: RS OCCUPATIONAL PRESTIGE SCORE (1980)

Coefficients(a)

Model		Unstandardized Coefficients		Standardized Coefficients	t	Sig.
		B	Std. Error	Beta		
1	(Constant)	10.677	2.105		5.073	.000
	FATHERS OCCUPATIONAL PRESTIGE SCORE (1980)	.147	.037	.132	3.959	.000
	HIGHEST YEAR OF SCHOOL COMPLETED	2.061	.144	.480	14.358	.000

[a]Dependent Variable: RS OCCUPATIONAL PRESTIGE SCORE (1980)

The Model Summary block reports the multiple *R* (.540) and *R* square (.292). The ANOVA output block shows the significance of the relationship (Sig. = .000). So far, we know that the independent variables explain about 29% of the variance in *prestg80* and that this result is statistically significant. In the last output block, we see the slopes (*B*) of the independent variables on *prestg80*, the standardized partial slopes (Beta), and the *Y* intercept (reported as a constant of 10.677). From this information, we can build a regression equation to predict scores on *prestg80*. The beta for *educ* (.480) is greater than the beta for *papres80* (.132), so education is the more important independent variable.

What does all this mean? At least for this sample, a person's occupational prestige is more affected by his or her education than by the social class of his or her family of origin.

SPSS DEMONSTRATION 17.2 What Are the Correlates of Sexual Activity?

The 2006 GSS includes an item (*sexfreq*) that asks respondents about the frequency of their sexual activity over the past year. What causes variations in sexual activity? Some obvious correlates include gender and age, since it is commonly supposed that men are more sexually active than women and younger people more active than older. What are the effects of social class? Does the level of sexual activity—like so many other aspects of the social world—vary by a person's economic status?

Gender is a nominal-level variable, so it will be included in the analysis as a dummy variable. Click **Analyze, Regression,** and **Linear,** and name *sexfreq* as the dependent variable and *age, income06*, and *sex* as the independent variables. Request descriptive statistics, if you wish, by clicking the **Statistics** button and making the appropriate selection. The output will look like this:

Model Summary

Model	R	R Square	Adjusted R Square	Std. Error of the Estimate
1	.499(a)	.249	.246	1.755

[a]Predictors: (Constant), RESPONDENTS SEX, AGE OF RESPONDENT, TOTAL FAMILY INCOME

ANOVA(b)

Model		Sum of Squares	df	Mean Square	F	Sig.
1	Regression	670.713	3	223.571	72.595	.000(a)
	Residual	2020.281	656	3.080		
	Total	2690.994	659			

[a]Predictors: (Constant), RESPONDENTS SEX, AGE OF RESPONDENT, TOTAL FAMILY INCOME
[b]Dependent Variable: FREQUENCY OF SEX DURING LAST YEAR

Coefficients(a)

Model		Unstandardized Coefficients		Standardized Coefficients	t	Sig.
		B	Std. Error	Beta		
1	(Constant)	4.601	.367		12.524	.000
	AGE OF RESPONDENT	−.057	.004	−.474	−13.998	.000
	TOTAL FAMILY INCOME	.058	.012	.159	4.657	.000
	RESPONDENTS SEX	−.104	.138	−.026	−.750	.453

[a]Dependent Variable: FREQUENCY OF SEX DURING LAST YEAR

The R and R^2 indicate that the independent variables account for about 25% of the variation in frequency of sexual activity. The slopes (B) indicate that the dependent variable increases with income (.058) and decreases with age (−.057). The slope for gender is also negative (−.104). Since a male is coded as 1 and a female as 2, this indicates that the frequency of sex is greater for men. In other words, the frequency of sex increases as gender "decreases"—its rate is higher for people with the "lower" score on gender. The beta-weights (Beta) suggest that age has the greatest influence on frequency of sexual activity, followed by income and gender. The rate of sexual activity increases as affluence increases, decreases as age increases, and is more associated with males than females.

Exercises

17.1 Conduct the analysis in Demonstration 17.1 again, with *income06* as the dependent variable. Compare your conclusions with those that you made in Demonstration 15.1. Write a paragraph summarizing the relationships.

17.2 Conduct the analysis in Demonstration 17.2 again, with *partnrs5* (number of different sexual partners over the past five years) as the dependent variable. Compare and contrast with the analysis of *sexfreq*.

17.3 Conduct a regression analysis of *tvhours, childs*, or *attend.* Choose two or three potential independent variables and follow the instructions in Demonstrations 17.1 and 17.2. Ideally, your independent variables should be interval-ratio in level of measurement, but ordinal variables with a broad range of scores and nominal variables with only two scores will work as well.

PART IV CUMULATIVE EXERCISES

1. Two research questions that can be answered by one of the techniques presented in Chapters 16 and 17 are stated here. For each research situation, choose either the elaboration technique or regression analysis. The level of measurement of the variables should have a strong influence on your decision.

   a. For a sample of college graduates, did having a job during the school year interfere with academic success? For 20 students, data have been gathered on college GPA, the average number of hours each student worked each week, and College Board scores (a measure of preparedness for college-level work). Take GPA as the dependent variable.

   b. Only about half of this sample graduated within four years (1 = yes, 2 = no). Was their progress affected by their level of social activity (1 = low, 2 = high)? Is the relationship between these variables the same for both males (1) and females (2)?

GPA	Hours Worked per Week	Average College Board Scores	Graduated in Four Years	Social Activity	Sex
3.14	13	550	1	1	1
2.00	20	375	1	2	2
2.11	22	450	1	1	2
3.00	10	575	1	2	1
3.75	0	600	1	1	1
3.11	21	650	1	1	2
3.22	7	605	1	2	1
2.75	25	630	1	1	2
2.50	30	680	1	1	1
2.10	32	610	1	1	2
2.45	20	580	2	2	2
2.01	40	590	2	1	1
3.90	0	675	2	2	2
3.45	0	650	2	2	1
2.30	25	550	2	2	1
2.20	18	470	2	1	2
2.60	25	600	2	2	1
3.10	15	525	2	2	1
2.60	27	480	2	1	2
2.20	20	500	2	2	1

2. A research project has been conducted on the audiences of Christian religious television programs. Several research questions have been developed with regard to the following variables:

How many hours per week do you watch religious programs on television? (aactual hours)
What is your age? (years)

(*continued next page*)

(continued)

How many years of formal schooling have you completed? (years)

Have you ever donated money to a religious television program?

1. Yes
2. No

What is your religious preference?

1. Protestant
2. Catholic

What is your sex?

1. Female
2. Male

a. The amount of time devoted to religious television will increase with age and decrease with education, but age will have the strongest effect.

b. Protestants will be more likely to donate money. Protestant females will be especially likely to donate money.

Do the data support these hypotheses?

Hours	Age	Education	Ever Donate?	Denomination	Sex
14	52	10	1	1	1
10	45	12	1	1	2
8	18	12	1	1	1
5	45	12	1	1	1
10	57	10	1	1	2
14	65	8	2	1	1
3	23	12	1	1	1
12	47	11	2	1	1
2	30	14	2	1	2
1	20	16	2	1	2
21	60	16	1	2	1
15	55	12	1	2	2
12	47	9	2	2	1
8	32	14	1	2	2
15	50	12	1	2	1
10	45	10	2	2	2
20	72	16	1	2	1
12	40	16	2	2	2
14	42	14	2	2	2
10	38	12	1	2	1

Appendix A

Area Under the Normal Curve

Column (a) lists *Z* scores from 0.00 to 4.00. Only positive scores are displayed, but, since the normal curve is symmetrical, the areas for negative scores will be exactly the same as areas for positive scores. Column (b) lists the proportion of the total area between the *Z* score and the mean. Figure A.1 displays areas of this type. Column (c) lists the proportion of the area beyond the *Z* score, and Figure A.2 displays this type of area.

FIGURE A.1 AREA BETWEEN MEAN AND *Z*

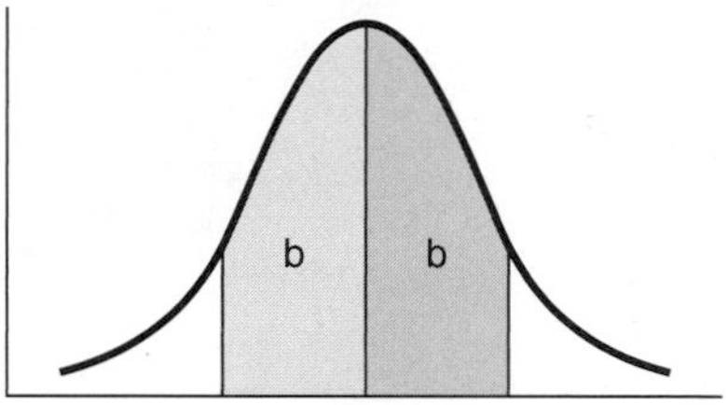

FIGURE A.2 AREA BEYOND *Z*

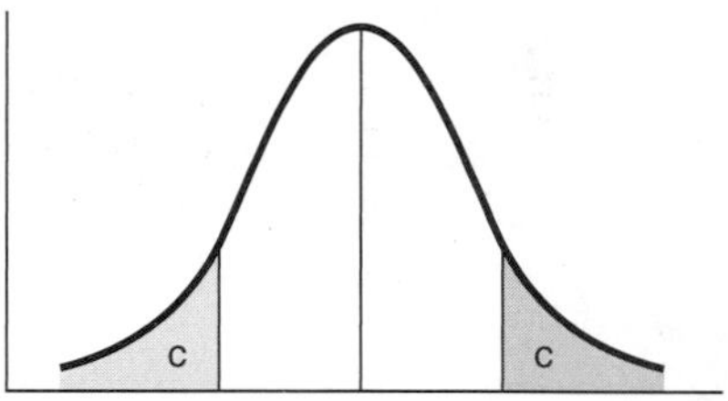

(a) *Z*	(b) Area Between Mean and *Z*	(c) Area Beyond *Z*
0.00	0.0000	0.5000
0.01	0.0040	0.4960
0.02	0.0080	0.4920
0.03	0.0120	0.4880
0.04	0.0160	0.4840
0.05	0.0199	0.4801
0.06	0.0239	0.4761
0.07	0.0279	0.4721
0.08	0.0319	0.4681
0.09	0.0359	0.4641
0.10	0.0398	0.4602
0.11	0.0438	0.4562
0.12	0.0478	0.4522
0.13	0.0517	0.4483
0.14	0.0557	0.4443
0.15	0.0596	0.4404
0.16	0.0636	0.4364
0.17	0.0675	0.4325
0.18	0.0714	0.4286
0.19	0.0753	0.4247
0.20	0.0793	0.4207

(a) *Z*	(b) Area Between Mean and *Z*	(c) Area Beyond *Z*
0.21	0.0832	0.4168
0.22	0.0871	0.4129
0.23	0.0910	0.4090
0.24	0.0948	0.4052
0.25	0.0987	0.4013
0.26	0.1026	0.3974
0.27	0.1064	0.3936
0.28	0.1103	0.3897
0.29	0.1141	0.3859
0.30	0.1179	0.3821
0.31	0.1217	0.3783
0.32	0.1255	0.3745
0.33	0.1293	0.3707
0.34	0.1331	0.3669
0.35	0.1368	0.3632
0.36	0.1406	0.3594
0.37	0.1443	0.3557
0.38	0.1480	0.3520
0.39	0.1517	0.3483
0.40	0.1554	0.3446

(a) Z	(b) Area Between Mean and Z	(c) Area Beyond Z
0.41	0.1591	0.3409
0.42	0.1628	0.3372
0.43	0.1664	0.3336
0.44	0.1700	0.3300
0.45	0.1736	0.3264
0.46	0.1772	0.3228
0.47	0.1808	0.3192
0.48	0.1844	0.3156
0.49	0.1879	0.3121
0.50	0.1915	0.3085
0.51	0.1950	0.3050
0.52	0.1985	0.3015
0.53	0.2019	0.2981
0.54	0.2054	0.2946
0.55	0.2088	0.2912
0.56	0.2123	0.2877
0.57	0.2157	0.2843
0.58	0.2190	0.2810
0.59	0.2224	0.2776
0.60	0.2257	0.2743
0.61	0.2291	0.2709
0.62	0.2324	0.2676
0.63	0.2357	0.2643
0.64	0.2389	0.2611
0.65	0.2422	0.2578
0.66	0.2454	0.2546
0.67	0.2486	0.2514
0.68	0.2517	0.2483
0.69	0.2549	0.2451
0.70	0.2580	0.2420
0.71	0.2611	0.2389
0.72	0.2642	0.2358
0.73	0.2673	0.2327
0.74	0.2703	0.2297
0.75	0.2734	0.2266
0.76	0.2764	0.2236
0.77	0.2794	0.2206
0.78	0.2823	0.2177
0.79	0.2852	0.2148
0.80	0.2881	0.2119
0.81	0.2910	0.2090
0.82	0.2939	0.2061
0.83	0.2967	0.2033
0.84	0.2995	0.2005
0.85	0.3023	0.1977
0.86	0.3051	0.1949
0.87	0.3078	0.1922
0.88	0.3106	0.1894
0.89	0.3133	0.1867
0.90	0.3159	0.1841
0.91	0.3186	0.1814
0.92	0.3212	0.1788
0.93	0.3238	0.1762
0.94	0.3264	0.1736
0.95	0.3289	0.1711

(a) Z	(b) Area Between Mean and Z	(c) Area Beyond Z
0.96	0.3315	0.1685
0.97	0.3340	0.1660
0.98	0.3365	0.1635
0.99	0.3389	0.1611
1.00	0.3413	0.1587
1.01	0.3438	0.1562
1.02	0.3461	0.1539
1.03	0.3485	0.1515
1.04	0.3508	0.1492
1.05	0.3531	0.1469
1.06	0.3554	0.1446
1.07	0.3577	0.1423
1.08	0.3599	0.1401
1.09	0.3621	0.1379
1.10	0.3643	0.1357
1.11	0.3665	0.1335
1.12	0.3686	0.1314
1.13	0.3708	0.1292
1.14	0.3729	0.1271
1.15	0.3749	0.1251
1.16	0.3770	0.1230
1.17	0.3790	0.1210
1.18	0.3810	0.1190
1.19	0.3830	0.1170
1.20	0.3849	0.1151
1.21	0.3869	0.1131
1.22	0.3888	0.1112
1.23	0.3907	0.1093
1.24	0.3925	0.1075
1.25	0.3944	0.1056
1.26	0.3962	0.1038
1.27	0.3980	0.1020
1.28	0.3997	0.1003
1.29	0.4015	0.0985
1.30	0.4032	0.0968
1.31	0.4049	0.0951
1.32	0.4066	0.0934
1.33	0.4082	0.0918
1.34	0.4099	0.0901
1.35	0.4115	0.0885
1.36	0.4131	0.0869
1.37	0.4147	0.0853
1.38	0.4162	0.0838
1.39	0.4177	0.0823
1.40	0.4192	0.0808
1.41	0.4207	0.0793
1.42	0.4222	0.0778
1.43	0.4236	0.0764
1.44	0.4251	0.0749
1.45	0.4265	0.0735
1.46	0.4279	0.0721
1.47	0.4292	0.0708
1.48	0.4306	0.0694
1.49	0.4319	0.0681
1.50	0.4332	0.0668

(a) Z	(b) Area Between Mean and Z	(c) Area Beyond Z	(a) Z	(b) Area Between Mean and Z	(c) Area Beyond Z
1.51	0.4345	0.0655	2.06	0.4803	0.0197
1.52	0.4357	0.0643	2.07	0.4808	0.0192
1.53	0.4370	0.0630	2.08	0.4812	0.0188
1.54	0.4382	0.0618	2.09	0.4817	0.0183
1.55	0.4394	0.0606	2.10	0.4821	0.0179
1.56	0.4406	0.0594	2.11	0.4826	0.0174
1.57	0.4418	0.0582	2.12	0.4830	0.0170
1.58	0.4429	0.0571	2.13	0.4834	0.0166
1.59	0.4441	0.0559	2.14	0.4838	0.0162
1.60	0.4452	0.0548	2.15	0.4842	0.0158
1.61	0.4463	0.0537	2.16	0.4846	0.0154
1.62	0.4474	0.0526	2.17	0.4850	0.0150
1.63	0.4484	0.0516	2.18	0.4854	0.0146
1.64	0.4495	0.0505	2.19	0.4857	0.0143
1.65	0.4505	0.0495	2.20	0.4861	0.0139
1.66	0.4515	0.0485	2.21	0.4864	0.0136
1.67	0.4525	0.0475	2.22	0.4868	0.0132
1.68	0.4535	0.0465	2.23	0.4871	0.0129
1.69	0.4545	0.0455	2.24	0.4875	0.0125
1.70	0.4554	0.0446	2.25	0.4878	0.0122
1.71	0.4564	0.0436	2.26	0.4881	0.0119
1.72	0.4573	0.0427	2.27	0.4884	0.0116
1.73	0.4582	0.0418	2.28	0.4887	0.0113
1.74	0.4591	0.0409	2.29	0.4890	0.0110
1.75	0.4599	0.0401	2.30	0.4893	0.0107
1.76	0.4608	0.0392	2.31	0.4896	0.0104
1.77	0.4616	0.0384	2.32	0.4898	0.0102
1.78	0.4625	0.0375	2.33	0.4901	0.0099
1.79	0.4633	0.0367	2.34	0.4904	0.0096
1.80	0.4641	0.0359	2.35	0.4906	0.0094
1.81	0.4649	0.0351	2.36	0.4909	0.0091
1.82	0.4656	0.0344	2.37	0.4911	0.0089
1.83	0.4664	0.0336	2.38	0.4913	0.0087
1.84	0.4671	0.0329	2.39	0.4916	0.0084
1.85	0.4678	0.0322	2.40	0.4918	0.0082
1.86	0.4686	0.0314	2.41	0.4920	0.0080
1.87	0.4693	0.0307	2.42	0.4922	0.0078
1.88	0.4699	0.0301	2.43	0.4925	0.0075
1.89	0.4706	0.0294	2.44	0.4927	0.0073
1.90	0.4713	0.0287	2.45	0.4929	0.0071
1.91	0.4719	0.0281	2.46	0.4931	0.0069
1.92	0.4726	0.0274	2.47	0.4932	0.0068
1.93	0.4732	0.0268	2.48	0.4934	0.0066
1.94	0.4738	0.0262	2.49	0.4936	0.0064
1.95	0.4744	0.0256	2.50	0.4938	0.0062
1.96	0.4750	0.0250	2.51	0.4940	0.0060
1.97	0.4756	0.0244	2.52	0.4941	0.0059
1.98	0.4761	0.0239	2.53	0.4943	0.0057
1.99	0.4767	0.0233	2.54	0.4945	0.0055
2.00	0.4772	0.0228	2.55	0.4946	0.0054
2.01	0.4778	0.0222	2.56	0.4948	0.0052
2.02	0.4783	0.0217	2.57	0.4949	0.0051
2.03	0.4788	0.0212	2.58	0.4951	0.0049
2.04	0.4793	0.0207	2.59	0.4952	0.0048
2.05	0.4798	0.0202	2.60	0.4953	0.0047

(a) Z	(b) Area Between Mean and Z	(c) Area Beyond Z
2.61	0.4955	0.0045
2.62	0.4956	0.0044
2.63	0.4957	0.0043
2.64	0.4959	0.0041
2.65	0.4960	0.0040
2.66	0.4961	0.0039
2.67	0.4962	0.0038
2.68	0.4963	0.0037
2.69	0.4964	0.0036
2.70	0.4965	0.0035
2.71	0.4966	0.0034
2.72	0.4967	0.0033
2.73	0.4968	0.0032
2.74	0.4969	0.0031
2.75	0.4970	0.0030
2.76	0.4971	0.0029
2.77	0.4972	0.0028
2.78	0.4973	0.0027
2.79	0.4974	0.0026
2.80	0.4974	0.0026
2.81	0.4975	0.0025
2.82	0.4976	0.0024
2.83	0.4977	0.0023
2.84	0.4977	0.0023
2.85	0.4978	0.0022
2.86	0.4979	0.0021
2.87	0.4979	0.0021
2.88	0.4980	0.0020
2.89	0.4981	0.0019
2.90	0.4981	0.0019
2.91	0.4982	0.0018
2.92	0.4982	0.0018
2.93	0.4983	0.0017
2.94	0.4984	0.0016
2.95	0.4984	0.0016
2.96	0.4985	0.0015
2.97	0.4985	0.0015
2.98	0.4986	0.0014
2.99	0.4986	0.0014
3.00	0.4986	0.0014
3.01	0.4987	0.0013
3.02	0.4987	0.0013
3.03	0.4988	0.0012
3.04	0.4988	0.0012
3.05	0.4989	0.0011
3.06	0.4989	0.0011
3.07	0.4989	0.0011
3.08	0.4990	0.0010
3.09	0.4990	0.0010
3.10	0.4990	0.0010
3.11	0.4991	0.0009
3.12	0.4991	0.0009
3.13	0.4991	0.0009
3.14	0.4992	0.0008
3.15	0.4992	0.0008
3.16	0.4992	0.0008
3.17	0.4992	0.0008
3.18	0.4993	0.0007
3.19	0.4993	0.0007
3.20	0.4993	0.0007
3.21	0.4993	0.0007
3.22	0.4994	0.0006
3.23	0.4994	0.0006
3.24	0.4994	0.0006
3.25	0.4994	0.0006
3.26	0.4994	0.0006
3.27	0.4995	0.0005
3.28	0.4995	0.0005
3.29	0.4995	0.0005
3.30	0.4995	0.0005
3.31	0.4995	0.0005
3.32	0.4995	0.0005
3.33	0.4996	0.0004
3.34	0.4996	0.0004
3.35	0.4996	0.0004
3.36	0.4996	0.0004
3.37	0.4996	0.0004
3.38	0.4996	0.0004
3.39	0.4997	0.0003
3.40	0.4997	0.0003
3.41	0.4997	0.0003
3.42	0.4997	0.0003
3.43	0.4997	0.0003
3.44	0.4997	0.0003
3.45	0.4997	0.0003
3.46	0.4997	0.0003
3.47	0.4997	0.0003
3.48	0.4997	0.0003
3.49	0.4998	0.0002
3.50	0.4998	0.0002
3.60	0.4998	0.0002
3.70	0.4999	0.0001
3.80	0.4999	0.0001
3.90	0.4999	<0.0001
4.00	0.4999	<0.0001

Appendix B

Distribution of *t*

Degrees of Freedom (df)	Level of Significance for One-tailed Test					
	.10	.05	.025	.01	.005	.0005
	Level of Significance for Two-tailed Test					
	.20	.10	.05	.02	.01	.001
1	3.078	6.314	12.706	31.821	63.657	636.619
2	1.886	2.920	4.303	6.965	9.925	31.598
3	1.638	2.353	3.182	4.541	5.841	12.941
4	1.533	2.132	2.776	3.747	4.604	8.610
5	1.476	2.015	2.571	3.365	4.032	6.859
6	1.440	1.943	2.447	3.143	3.707	5.959
7	1.415	1.895	2.365	2.998	3.499	5.405
8	1.397	1.860	2.306	2.896	3.355	5.041
9	1.383	1.833	2.262	2.821	3.250	4.781
10	1.372	1.812	2.228	2.764	3.169	4.587
11	1.363	1.796	2.201	2.718	3.106	4.437
12	1.356	1.782	2.179	2.681	3.055	4.318
13	1.350	1.771	2.160	2.650	3.012	4.221
14	1.345	1.761	2.145	2.624	2.977	4.140
15	1.341	1.753	2.131	2.602	2.947	4.073
16	1.337	1.746	2.120	2.583	2.921	4.015
17	1.333	1.740	2.110	2.567	2.898	3.965
18	1.330	1.734	2.101	2.552	2.878	3.922
19	1.328	1.729	2.093	2.539	2.861	3.883
20	1.325	1.725	2.086	2.528	2.845	3.850
21	1.323	1.721	2.080	2.518	2.831	3.819
22	1.321	1.717	2.074	2.508	2.819	3.792
23	1.319	1.714	2.069	2.500	2.807	3.767
24	1.318	1.711	2.064	2.492	2.797	3.745
25	1.316	1.708	2.060	2.485	2.787	3.725
26	1.315	1.706	2.056	2.479	2.779	3.707
27	1.314	1.703	2.052	2.473	2.771	3.690
28	1.313	1.701	2.048	2.467	2.763	3.674
29	1.311	1.699	2.045	2.462	2.756	3.659
30	1.310	1.697	2.042	2.457	2.750	3.646
40	1.303	1.684	2.021	2.423	2.704	3.551
60	1.296	1.671	2.000	2.390	2.660	3.460
120	1.289	1.658	1.980	2.358	2.617	3.373
∞	1.282	1.645	1.960	2.326	2.576	3.291

Source: Table III of Fisher & Yates: *Statistical Tables for Biological, Agricultural and Medical Research,* published by Longman Group Ltd., London (1974), 6th edition (previously published by Oliver & Boyd Ltd., Edinburgh).

Appendix C

Distribution of Chi Square

df	.99	.98	.95	.90	.80	.70	.50	.30	.20	.10	.05	.02	.01	.001
1	$.0^{3}157$	$.0^{3}628$	.00393	.0158	.0642	.148	.455	1.074	1.642	2.706	3.841	5.412	6.635	10.827
2	.0201	.0404	.103	.211	.446	.713	1.386	2.408	3.219	4.605	5.991	7.824	9.210	13.815
3	.115	.185	.352	.584	1.005	1.424	2.366	3.665	4.642	6.251	7.815	9.837	11.341	16.268
4	.297	.429	.711	1.064	1.649	2.195	3.357	4.878	5.989	7.779	9.488	11.668	13.277	18.465
5	.554	.752	1.145	1.610	2.343	3.000	4.351	6.064	7.289	9.236	11.070	13.388	15.086	20.517
6	.872	1.134	1.635	2.204	3.070	3.828	5.348	7.231	8.558	10.645	12.592	15.033	16.812	22.457
7	1.239	1.564	2.167	2.833	3.822	4.671	6.346	8.383	9.803	12.017	14.067	16.622	18.475	24.322
8	1.646	2.032	2.733	3.490	4.594	5.527	7.344	9.524	11.030	13.362	15.507	18.168	20.090	26.125
9	2.088	2.532	3.325	4.168	5.380	6.393	8.343	10.656	12.242	14.684	16.919	19.679	21.666	27.877
10	2.558	3.059	3.940	4.865	6.179	7.267	9.342	11.781	13.442	15.987	18.307	21.161	23.209	29.588
11	3.053	3.609	4.575	5.578	6.989	8.148	10.341	12.899	14.631	17.275	19.675	22.618	24.725	31.264
12	3.571	4.178	5.226	6.304	7.807	9.034	11.340	14.011	15.812	18.549	21.026	24.054	26.217	32.909
13	4.107	4.765	5.892	7.042	8.634	9.926	12.340	15.119	16.985	19.812	22.362	25.472	27.688	34.528
14	4.660	5.368	6.571	7.790	9.467	10.821	13.339	16.222	18.151	21.064	23.685	26.873	29.141	36.123
15	5.229	5.985	7.261	8.547	10.307	11.721	14.339	17.322	19.311	22.307	24.996	28.259	30.578	37.697
16	5.812	6.614	7.962	9.312	11.152	12.624	15.338	18.418	20.465	23.542	26.296	29.633	32.000	39.252
17	6.408	7.255	8.672	10.085	12.002	13.531	16.338	19.511	21.615	24.769	27.587	30.995	33.409	40.790
18	7.015	7.906	9.390	10.865	12.857	14.440	17.338	20.601	22.760	25.989	28.869	32.346	34.805	42.312
19	7.633	8.567	10.117	11.651	13.716	15.352	18.338	21.689	23.900	27.204	30.144	33.687	36.191	43.820
20	8.260	9.237	10.851	12.443	14.578	16.266	19.337	22.775	25.038	28.412	31.410	35.020	37.566	45.315
21	8.897	9.915	11.591	13.240	15.445	17.182	20.337	23.858	26.171	29.615	32.671	36.343	38.932	46.797
22	9.542	10.600	12.338	14.041	16.314	18.101	21.337	24.939	27.301	30.813	33.924	37.659	40.289	48.268
23	10.196	11.293	13.091	14.848	17.187	19.021	22.337	26.018	28.429	32.007	35.172	38.968	41.638	49.728
24	10.856	11.992	13.848	15.659	18.062	19.943	23.337	27.096	29.553	33.196	36.415	40.270	42.980	51.179
25	11.524	12.697	14.611	16.473	18.940	20.867	24.337	28.172	30.675	34.382	37.652	41.566	44.314	52.620
26	12.198	13.409	15.379	17.292	19.820	21.792	25.336	29.246	31.795	35.563	38.885	42.856	45.642	54.052
27	12.879	14.125	16.151	18.114	20.703	22.719	26.336	30.319	32.912	36.741	40.113	44.140	46.963	55.476
28	13.565	14.847	16.928	18.939	21.588	23.647	27.336	31.391	34.027	37.916	41.337	45.419	48.278	56.893
29	14.256	15.574	17.708	19.768	22.475	24.577	28.336	32.461	35.139	39.087	42.557	46.693	49.588	58.302
30	14.953	16.306	18.493	20.599	23.364	25.508	29.336	33.530	36.250	40.256	43.773	47.962	50.892	59.703

Source: Table IV of Fisher & Yates: *Statistical Tables for Biological, Agricultural and Medical Research,* published by Longman Group Ltd., London (1974), 6th edition (previously published by Oliver & Boyd Ltd., Edinburgh). Reprinted by permission of Addison Wesley Longman Ltd.

Appendix D

Distribution of *F*

$p = .05$

n_1 / n_2	1	2	3	4	5	6	8	12	24	∞
1	161.4	199.5	215.7	224.6	230.2	234.0	238.9	243.9	249.0	254.3
2	18.51	19.00	19.16	19.25	19.30	19.33	19.37	19.41	19.45	19.50
3	10.13	9.55	9.28	9.12	9.01	8.94	8.84	8.74	8.64	8.53
4	7.71	6.94	6.59	6.39	6.26	6.16	6.04	5.91	5.77	5.63
5	6.61	5.79	5.41	5.19	5.05	4.95	4.82	4.68	4.53	4.36
6	5.99	5.14	4.76	4.53	4.39	4.28	4.15	4.00	3.84	3.67
7	5.59	4.74	4.35	4.12	3.97	3.87	3.73	3.57	3.41	3.23
8	5.32	4.46	4.07	3.84	3.69	3.58	3.44	3.28	3.12	2.93
9	5.12	4.26	3.86	3.63	3.48	3.37	3.23	3.07	2.90	2.71
10	4.96	4.10	3.71	3.48	3.33	3.22	3.07	2.91	2.74	2.54
11	4.84	3.98	3.59	3.36	3.20	3.09	2.95	2.79	2.61	2.40
12	4.75	3.88	3.49	3.26	3.11	3.00	2.85	2.69	2.50	2.30
13	4.67	3.80	3.41	3.18	3.02	2.92	2.77	2.60	2.42	2.21
14	4.60	3.74	3.34	3.11	2.96	2.85	2.70	2.53	2.35	2.13
15	4.54	3.68	3.29	3.06	2.90	2.79	2.64	2.48	2.29	2.07
16	4.49	3.63	3.24	3.01	2.85	2.74	2.59	2.42	2.24	2.01
17	4.45	3.59	3.20	2.96	2.81	2.70	2.55	2.38	2.19	1.96
18	4.41	3.55	3.16	2.93	2.77	2.66	2.51	2.34	2.15	1.92
19	4.38	3.52	3.13	2.90	2.74	2.63	2.48	2.31	2.11	1.88
20	4.35	3.49	3.10	2.87	2.71	2.60	2.45	2.28	2.08	1.84
21	4.32	3.47	3.07	2.84	2.68	2.57	2.42	2.25	2.05	1.81
22	4.30	3.44	3.05	2.82	2.66	2.55	2.40	2.23	2.03	1.78
23	4.28	3.42	3.03	2.80	2.64	2.53	2.38	2.20	2.00	1.76
24	4.26	3.40	3.01	2.78	2.62	2.51	2.36	2.18	1.98	1.73
25	4.24	3.38	2.99	2.76	2.60	2.49	2.34	2.16	1.96	1.71
26	4.22	3.37	2.98	2.74	2.59	2.47	2.32	2.15	1.95	1.69
27	4.21	3.35	2.96	2.73	2.57	2.46	2.30	2.13	1.93	1.67
28	4.20	3.34	2.95	2.71	2.56	2.44	2.29	2.12	1.91	1.65
29	4.18	3.33	2.93	2.70	2.54	2.43	2.28	2.10	1.90	1.64
30	4.17	3.32	2.92	2.69	2.53	2.42	2.27	2.09	1.89	1.62
40	4.08	3.23	2.84	2.61	2.45	2.34	2.18	2.00	1.79	1.51
60	4.00	3.15	2.76	2.52	2.37	2.25	2.10	1.92	1.70	1.39
120	3.92	3.07	2.68	2.45	2.29	2.17	2.02	1.83	1.61	1.25
∞	3.84	2.99	2.60	2.37	2.21	2.09	1.94	1.75	1.52	1.00

Values of n_1 and n_2 represent the degrees of freedom associated with the between and within estimates of variance, respectively.

Source: Table V of Fisher and Yates: *Statistical Tables for Biological, Agricultural and Medical Research,* published by Longman Group Ltd., London (1974), 6th edition (previously published by Oliver and Boyd Ltd., Edinburgh). Reprinted by permission of Addison Wesley Longman Ltd.

$p = .01$

n_1 / n_2	1	2	3	4	5	6	8	12	24	∞
1	4052	4999	5403	5625	5764	5859	5981	6106	6234	6366
2	98.49	99.01	99.17	99.25	99.30	99.33	99.36	99.42	99.46	99.50
3	34.12	30.81	29.46	28.71	28.24	27.91	27.49	27.05	26.60	26.12
4	21.20	18.00	16.69	15.98	15.52	15.21	14.80	14.37	13.93	13.46
5	16.26	13.27	12.06	11.39	10.97	10.67	10.27	9.89	9.47	9.02
6	13.74	10.92	9.78	9.15	8.75	8.47	8.10	7.72	7.31	6.88
7	12.25	9.55	8.45	7.85	7.46	7.19	6.84	6.47	6.07	5.65
8	11.26	8.65	7.59	7.01	6.63	6.37	6.03	5.67	5.28	4.86
9	10.56	8.02	6.99	6.42	6.06	5.80	5.47	5.11	4.73	4.31
10	10.04	7.56	6.55	5.99	5.64	5.39	5.06	4.71	4.33	3.91
11	9.65	7.20	6.22	5.67	5.32	5.07	4.74	4.40	4.02	3.60
12	9.33	6.93	5.95	5.41	5.06	4.82	4.50	4.16	3.78	3.36
13	9.07	6.70	5.74	5.20	4.86	4.62	4.30	3.96	3.59	3.16
14	8.86	6.51	5.56	5.03	4.69	4.46	4.14	3.80	3.43	3.00
15	8.68	6.36	5.42	4.89	4.56	4.32	4.00	3.67	3.29	2.87
16	8.53	6.23	5.29	4.77	4.44	4.20	3.89	3.55	3.18	2.75
17	8.40	6.11	5.18	4.67	4.34	4.10	3.79	3.45	3.08	2.65
18	8.28	6.01	5.09	4.58	4.25	4.01	3.71	3.37	3.00	2.57
19	8.18	5.93	5.01	4.50	4.17	3.94	3.63	3.30	2.92	2.49
20	8.10	5.85	4.94	4.43	4.10	3.87	3.56	3.23	2.86	2.42
21	8.02	5.78	4.87	4.37	4.04	3.81	3.51	3.17	2.80	2.36
22	7.94	5.72	4.82	4.31	3.99	3.76	3.45	3.12	2.75	2.31
23	7.88	5.66	4.76	4.26	3.94	3.71	3.41	3.07	2.70	2.26
24	7.82	5.61	4.72	4.22	3.90	3.67	3.36	3.03	2.66	2.21
25	7.77	5.57	4.68	4.18	3.86	3.63	3.32	2.99	2.62	2.17
26	7.72	5.53	4.64	4.14	3.82	3.59	3.29	2.96	2.58	2.13
27	7.68	5.49	4.60	4.11	3.78	3.56	3.26	2.93	2.55	2.10
28	7.64	5.45	4.57	4.07	3.75	3.53	3.23	2.90	2.52	2.06
29	7.60	5.42	4.54	4.04	3.73	3.50	3.20	2.87	2.49	2.03
30	7.56	5.39	4.51	4.02	3.70	3.47	3.17	2.84	2.47	2.01
40	7.31	5.18	4.31	3.83	3.51	3.29	2.99	2.66	2.29	1.80
60	7.08	4.98	4.13	3.65	3.34	3.12	2.82	2.50	2.12	1.60
120	6.85	4.79	3.95	3.48	3.17	2.96	2.66	2.34	1.95	1.38
∞	6.64	4.60	3.78	3.32	3.02	2.80	2.51	2.18	1.79	1.00

Values of n_1 and n_2 represent the degrees of freedom associated with the between and within estimates of variance, respectively.

Appendix E

Using Statistics: Ideas for Research Projects

This appendix presents outlines for four research projects, each of which requires you to use SPSS to analyze the 2006 General Social Survey, the database that has been used throughout this text. The research projects should be completed at various intervals during the course, and each project permits a great deal of choice on the part of the student. The first project stresses description and should be done after completing Chapters 2–4. The second involves estimation and should be completed in conjunction with Chapter 7. The third project uses inferential statistics and should be done after completing Part II, and the fourth project combines inferential statistics with measures of association (with an option for multivariate analysis) and should be done after Part III (or IV).

PROJECT 1—DESCRIPTIVE STATISTICS

1. Select five variables from the 2006 General Social Survey *(NOTE: Your instructor may specify a different number of variables)* and use the **Frequencies** command to get frequency distributions and summary statistics for each variable. Click the **Statistics** button on the **Frequencies** command window and request the mean, median, mode, standard deviation, and range. See Demonstrations 3.1, 3.2, and 4.1 for guidelines and examples. Make a note of all relevant information when it appears on screen or make a hard copy. See Appendix G for a list of variables available in the 2006 GSS.

2. For each variable, get bar or line charts to summarize the overall shape of the distribution of the variable. See Demonstration 2.3 for guidelines and examples.

3. Inspect the frequency distributions and graphs and choose appropriate measures of central tendency and, for ordinal- and interval-ratio-level variables, dispersion. Also, for interval-ratio and ordinal variables with many scores, check for skew both by using the line chart and by comparing the mean and median (see Sections 3.6 and 3.8). Write a sentence or two of description for each variable, being careful to include a description of the overall shape of the distribution (see Chapter 2), the central tendency (Chapter 3), and the dispersion (Chapter 4). For nominal- and ordinal-level variables, be sure to explain any arbitrary numerical codes. For example, on the variable *class* in the 2006 GSS (see Appendix G), a 1 is coded as "lower class," a 2 indicates "working class," and so forth. This is an ordinal-level variable, so you might choose to report the median as a measure of central tendency. If the median score on *class* were 2.45, for example, you might place that value in context by reporting that "the median is 2.45, about halfway between 'working class' and 'middle class.'"

4. Given here are examples of *minimal* summary sentences, using fictitious data:

For a nominal-level variable (e.g., gender), report the mode and some detail about the overall distribution. For example: "Most respondents were married (57.5%), but divorced (17.4%) and single (21.3%) individuals were also common."

For an ordinal-level variable (e.g., occupational prestige), use the median (and perhaps the mean and mode) and the range. For example: "The median prestige score was 44.3, and the range extended from 34 to 87. The most common score was 42 and the average score was 40.8."

For an interval-ratio level variable (e.g., age), use the mean (and perhaps the median or mode) and the standard deviation (and perhaps the range). For example: "Average age for this sample was 42.3. Respondents ranged from 18 to 94 years of age with a standard deviation of 15.37."

PROJECT 2—ESTIMATION

In this exercise, you will use the 2006 GSS sample to estimate the characteristics of the U.S. population. You will use SPSS to generate the sample statistics and then use either Formula 7.2 or Formula 7.3 to find the confidence interval and state each interval in words.

A. Estimating Means

1. There are relatively few interval-ratio variables in the 2006 GSS, and for this part of the project you may use ordinal variables that have *at least* three categories or scores. Choose a total of three variables that fit this description *other than* the variables you used in Exercise 7.1. *(NOTE: Your instructor may specify a different number of variables).*

2. Use the **Descriptives** command to get means, standard deviations, and sample size (N), and use this information to construct 95% confidence intervals for each of your variables. Make a note of the mean, standard deviation, and sample size or keep a hard copy. Use Formula 7.2 to compute the confidence intervals. Repeat this procedure for the remaining variables.

3. For each variable, write a summary sentence reporting the variable, the interval itself, the confidence level, and the sample size. Write in plain English, as if you were reporting results in a newspaper. Most importantly, you should make it clear that you are estimating characteristics of the population of the entire United States. For example, a summary sentence might look like this: "Based on a random sample of 1231, I estimate at the 95% level that U.S. drivers average between 64.46 and 68.22 miles per hour when driving on interstate highways."

B. Estimating Proportions

4. Choose three variables that are nominal or ordinal *other than* the variables you used in Exercise 7.2. *(NOTE: Your instructor may specify a different number of variables.)*

5. Use the **Frequencies** command to get the percentage of the sample in the various categories of each variable. Change the percentages (remember to use the "valid percents" column) to proportions and construct confidence intervals for one category of each variable (e.g., the % female for *sex*) using Formula 7.3.

6. For each variable, write a summary sentence reporting the variable, the interval, the confidence level, and the sample size. Write in plain English, as if you were reporting results in a newspaper. Remember to make it clear that you are estimating a characteristic of the United States population.

7. For any one of the intervals you constructed in either Part A or Part B, identify each of the following concepts and terms and briefly explain their role in estimation: sample, population, statistic, parameter, EPSEM, representative, confidence level.

PROJECT 3—SIGNIFICANCE TESTING

A. Two-Sample *t* Test (Chapter 9)

1. Choose two different dependent variables from the interval-ratio or ordinal variables that have three or more scores. Choose independent variables that might logically be a cause of your dependent variables. Remember that, for a *t* test, independent variables can have *only* two categories, but you can still use independent variables with more than two categories by (a) using the **Grouping Variable** box to specify the exact categories (e.g., select scores of 1 and 5 on *marital* to compare married with never-married respondents), or (b) collapsing the scores of variables with more than two categories by using the **Recode** command. Independent variables can be any level of measurement, and you may use the same independent variable for both tests.

2. Click **Analyze, Compare Means,** and then **Independent Samples T Test.** Name your dependent variable(s) in the **Test Variable** window and your independent variable in the **Grouping Variable** window. You will also need to specify the scores used to define the groups on the independent variable. See SPSS Demonstration 9.1 for examples. Make a note of the test results (group means, obtained *t* score, significance, sample size) or keep a hard copy. Repeat the procedure for the second dependent variable.

3. Write up the results of the test. See Reading Statistics 6 for some ideas about how to report results. At a minimum, your report should clearly identify the independent and dependent variables, the sample statistics, the value of the test statistic (step 4), the results of the test (step 5), and the alpha level you used.

B. Analysis of Variance (Chapter 10)

1. Choose two different dependent variables from the interval-ratio or ordinal variables that have three or more scores. Choose independent variables that might logically be a cause of your dependent variables and that have between three and five categories. You may use the same independent variables for both tests.

2. Click **Analyze, Compare Means,** and then **One-way Anova.** The **One-way Anova** window will appear. Find your dependent variable in the variable list on the left and click the arrow to move the variable name into the **Dependent List** box. Note that you can request more than one dependent variable at a time. Next, find the name of your independent variable and move it to the **Factor** box. Click **Options** and then click the box next to **Descriptive** in the **Statistics** box to request means and standard deviations. Click **Continue** and **OK.** Make a note of the test results or keep a hard copy. Repeat, if necessary, for your second dependent variable.

3. Write up the results of the test. At a minimum, your report should clearly identify the independent and dependent variables, the sample statistics (category means), the value of the test statistic (step 4), the results of the test (step 5), the degrees of freedom, and the alpha level you used.

C. Chi Square (Chapter 11)

1. Choose two different dependent variables of any level of measurement that have five or fewer (preferably two to three) scores. For each dependent variable, choose an independent variable that might logically be a cause. Independent variables can be any level of measurement as long as they have five or fewer (preferably two to three) categories. Output will be easier to analyze if you use variables with few categories. You may use the same independent variable for both tests.

2. Click **Analyze, Descriptive Statistics,** and then **Crosstabs.** The **Crosstabs** dialog box will appear. Highlight your first dependent variable and move it into the **Rows** box. Next, highlight your independent variable and move it into the **Columns** box. Click the **Statistics** button at the bottom of the window and click the box next to chi square. Click **Continue** and **OK.** Make a note of the results or a get a hard copy. Repeat for your second dependent variable.

3. Write up the results of the test. At a minimum, your report should clearly identify the independent and dependent variables, the value of the test statistic (step 4), the results of the test (step 5), the degrees of freedom, and the alpha level you used. It is almost always desirable to report the column percentages as well (see Section 11.5)

PROJECT 4—ANALYZING THE STRENGTH AND SIGNIFICANCE OF RELATIONSHIPS

A. Using Bivariate Tables

1. From the 2006 GSS data set, select either
   a. One dependent variable and three independent variables (possible causes)

   or

   b. One independent variable and three possible dependent variables (possible effects).

   Variables can be from any level of measurement but must have only a few (two to five) categories or scores. Develop research questions or hypotheses about the relationships between variables. Make sure the causal links you suggest are sensible and logical.

2. Use the **Crosstabs** procedure to generate bivariate tables. See any of the Demonstrations at the end of Chapter 12, 13, or 14 for examples. Click **Analyze, Descriptive Statistics,** and **Crosstabs** and place your dependent variable(s) in the rows and independent variable(s) in the columns. On the **Crosstabs** dialog box, click the **Statistics** button and choose chi square, phi or V, and gamma for every table you request. On the **Crosstabs** dialog box, click the **Cells** button and get column percentages for every table you request. Make a note of results as they appear on the screen or get hard copies.

3. Write a report that presents and analyzes these relationships. Be clear about which variables are dependent and which are independent. For each combination of variables, report the test of significance and measure of association. In addition, for each relationship, report and discuss column percentages, pattern or direction of the relationship, and strength of the relationship.

4. *OPTIONAL MULTIVARIATE ANALYSIS*: Pick one of the bivariate relationships you produced in step 2 and find a logical control variable for this relationship. Run **Crosstabs** for the bivariate relationship again while controlling for the third variable. Compare the partial tables with each other and with the bivariate table. Is the original bivariate relationship direct? Is there evidence of a spurious or intervening relationship? Do the variables have an interactive relationship? Write up the results of this analysis and include them in your summary paper for this project.

B. Using Interval-Ratio Variables

1. From the 2006 GSS, select either
   a. One dependent variable and three independent variables (possible causes)

   or

   b. One independent variable and three dependent variables (possible effects).

   Variables should be interval-ratio in level of measurement, but you may use ordinal-level variables as long as they have more than three scores. Develop research questions or hypotheses about the relationships between variables. Make sure the causal links you suggest are sensible and logical.

2. Use the **Regression** and **Scatterplot** (click Graphs and then Scatter) procedures to analyze the bivariate relationships. Make a note of results (including r, r^2, slope, beta-weights, and a) as they appear on the screen or get hard copies.

3. Write a report that presents and analyzes these relationships. Be clear about which variables are dependent and which are independent. For each combination of variables, report the significance of the relationship (if relevant) and the strength and direction of the relationship. Include r, r^2, and the beta-weights in your report.

4. *OPTIONAL MULTIVARIATE ANALYSIS*: Pick one of the bivariate relationships you produced in step 2 and find another logical independent variable. Run **Regression** again with both independent variables and analyze the results. How much improvement is there in the explained variance after the second independent variable is included? Write up the results of this analysis and include them in your summary paper for this project.

Appendix F An Introduction to *SPSS for Windows*

Computers have affected virtually every aspect of human society and, as you would expect, their impact on the conduct of social research has been profound. Researchers routinely use computers to organize data and compute statistics—activities that humans often find dull, tedious, and difficult but that computers accomplish with accuracy and ease. This division of labor allows social scientists to spend more time on analysis and interpretation—activities that humans typically enjoy but that are beyond the power of computers (so far, at least).

These days, the skills needed to use computers successfully are quite accessible, even for people with little or no experience. This appendix will prepare you to use a statistics program called *SPSS for Windows* (SPSS stands for Statistical Package for the Social Sciences). If you have used a mouse to "point and click" and run a computer program, you are ready to learn how to use this program. Even if you are completely unfamiliar with computers, you will find this program accessible. After you finish this appendix, you will be ready to do the exercises found at the end of most chapters of this text.

A word of caution before we begin: This appendix is intended only as an *introduction* to SPSS. It will give you an overview of the program and enough information so that you can complete the assignments in the text. It is unlikely, however, that this appendix will answer all your questions or provide solutions to all the problems you might encounter. So, this is a good place to tell you that SPSS has an extensive and easy-to-use "help" facility that will provide assistance as you request it. You should familiarize yourself with this feature and use it as needed. To get help, simply click on the **Help** command on the toolbar across the top of the screen.

SPSS is a **statistical package** (or **statpak**), that is, a set of computer programs that work with data and compute statistics as requested by the user (you). Once you have entered the data for a particular group of observations, you can easily and quickly produce an abundance of statistical information without doing any computations or writing any computer programs yourself.

Why bother to learn this technology? The truth is that the laborsaving capacity of computers is sometimes exaggerated, and there are research situations in which they are unnecessary. If you are working with a small number of observations or need only a few, uncomplicated statistics, then statistical packages are probably not going to be helpful. However, as the number of cases increases and as your requirements for statistics become more sophisticated, computers and statpaks will become more and more useful.

An example should make this point clearer. Suppose you have gathered a sample of 150 respondents and the *only* thing you want to know about these people is their average age. To compute an average, as you know, you add the scores and divide by the number of cases. How long do you think it would take you to add 150 two-digit numbers (ages) with a hand calculator? If you entered the scores at the rate of one per second—60 scores a minute—it would take about 3 or 4 minutes to enter the ages and get the average. Even if you worked slowly and carefully and did the addition a second and third time to check your math, you could probably complete all calculations in less than 15 or 20 minutes. If this were all the information you needed, computers and statpaks would not save you any time.

Such a simple research project is not very realistic, however. Typically, researchers deal with not one but scores or even hundreds of variables, and samples have hundreds or thousands of cases. While you could add 150 numbers in perhaps 3 or 4 minutes, how long would it take to add the scores for 1500 cases? What are the chances of adding 1500 numbers without making significant errors of arithmetic? The more complex the research situation, the more valuable and useful statpaks become. SPSS can produce statistical information in a few keystrokes or clicks of the mouse that might take you minutes, hours, or even days to produce with a hand calculator.

Clearly, this is technology worth mastering by any social researcher. With SPSS, you can avoid the drudgery of mere computation, spend more time on analysis and interpretation, and conduct research projects with very large data sets. Mastery of this technology might be very handy indeed in your senior-level courses, in a wide variety of jobs, or in graduate school.

F.1 GETTING STARTED—DATABASES AND COMPUTER FILES

Before statistics can be calculated, SPSS must have some data to process. A **database** is an organized collection of related information, such as the responses to a survey. For purposes of computer analysis, a database is organized into a **file:** a collection of information that is stored under the same name in the memory of the computer, on a disk or flash drive, or on some other medium. Words as well as numbers can be saved in files. If you've ever used a word processing program to type a letter or term paper, you probably saved your work in a file so that you could update or make corrections at a later time. Data can be stored in files indefinitely. Since it can take months to conduct a thorough data analysis, the ability to save a database is another advantage of using computers.

For the SPSS exercises in this text, we will use a database that contains some of the results of the General Social Survey (GSS) for 2006. This database contains the responses of a sample of adult Americans to questions about a wide variety of social issues. The GSS has been conducted regularly since 1972 and has been the basis for hundreds of research projects by professional social researchers. It is a rich source of information about public opinion in the United States and includes data on everything from attitudes about abortion to opinions on assisted suicide.

The GSS is especially valuable because the respondents are chosen so that the sample as a whole is representative of the entire U.S. population. A representative sample reproduces, in miniature form, the characteristics of the population from which it was taken (see Chapters 6 and 7). So when you analyze the 2006 General Social Survey database, you are in effect analyzing U.S. society as of 2006. The data are real, and the relationships you will analyze reflect some of the most important and sensitive issues in American life.

The complete General Social Survey for 2006 includes hundreds of items of information (age, sex, opinion about such social issues as capital punishment, and so forth) for about 4000 respondents. Some of you will be using a student version of *SPSS for Windows*, which is limited in the number of cases and variables it can process. To accommodate those limits, I have reduced the database to about 50 items of information and fewer than 1500 respondents.

The GSS data file is summarized in Appendix G. Please turn to this appendix and familiarize yourself with it. Note that the variables are listed alphabetically by their variable names. In SPSS, the names of variables must be no more than eight characters long. In many cases, the resultant need for brevity is not a problem, and variable names (e.g., *age*) are easy to figure out. In other cases, the eight-character limit necessitates extreme abbreviation, and some variable names (like *abany* or *fefam*) are not so obvious. Appendix G also shows the wording of the item that generated the variable. For example, the *abany* variable consists of responses to a question about legal abortion: Should it be possible for a woman to have an abortion for "any reason"? Note that the variable name is formed from the question: Should an *ab*ortion be possible for *any* reason? The *fefam* variable consists of responses to the statement "It is much better for everyone involved if the man is the achiever outside the home and the woman takes care of the home and family." Appendix G is an example of a code book for the database since it lists all the codes (or scores) for the survey items along with their meanings.

Notice that some of the possible responses to *abany* and *fefam* and other variables in Appendix G are labeled NAP ("Not applicable"), NA, or DK. The first of these responses means that the item in question was not given to a respondent. The full GSS is very long, and, to keep the time frame for completing the survey reasonable, not all respondents are asked every question. NA stands for "No Answer" and means that the respondent was asked the question but refused to answer. DK stands for "Don't Know," which means that the respondent did not have the requested information. All three of these scores are *Missing Values*, and, as "noninformation," they should be eliminated from statistical analysis. Missing values are common on surveys, and, as long as they are not too numerous, they are not a particular problem.

It's important that you understand the difference between a statpak (SPSS) and a database (the GSS) and what we are ultimately after here. A database consists of information. A statpak processes the information in the database and produces statistics. Our goal is to apply the statpak to the database to produce output (for example, statistics and graphs) that we can analyze and use to answer questions. The process might be diagrammed as in Figure F.1.

Statpaks like SPSS are general research tools that can be used to analyze databases of all sorts; they are not limited to the 2006 GSS. In the same way, the 2006 GSS could be analyzed with statpaks other than those used in this text. Other widely used statpaks include Microcase, SAS, and Stata—each of which may be available on your campus.

FIGURE F.1 THE DATA ANALYSIS PROCESS

DataBase	→	**Statpak**	→	**Output**	→	**Analysis**
(raw information)		(computer programs)		(statistics and graphs)		(interpretation)

F.2 STARTING *SPSS FOR WINDOWS* AND LOADING THE 2006 GSS

If you are using the complete, professional version of *SPSS for Windows,* you will probably be working in a computer lab, and you can begin running the program immediately. If you are using the student version of the program on your personal computer, the first thing you need to do is install the software. Follow the instructions that came with the program and return to this appendix when the installation is complete.

To start *SPSS for Windows,* find the icon (or picture) on the screen of your monitor that has an "SPSS" label attached to it. Use the computer mouse to move the arrow on the monitor screen over this icon and then double-click the left button on the mouse. This will start up the SPSS program.

After a few seconds the *SPSS for Windows* screen will appear and ask, at the top of the screen, "What would you like to do?" Of the choices listed, the button next to "Open an existing file" will be checked (or preselected), and there will probably be a number of data sets listed in the window at the bottom of the screen. Find the 2006 General Social Survey data set, probably labeled *GSS06.sav* or something similar. If the data set is on a disk or flash drive, you will need to specify the correct drive. Check with your instructor to make sure you know where to find the 2006 GSS.

Once you've located the data set, click on the name of the file with the left-hand button on the mouse, and SPSS will load the data. The next screen you will see is the SPSS Data Editor screen.

Note that there is a list of commands across the very top of the screen. These commands begin with **File** at the far left and end with **Help** at the far right. This is the main menu bar for SPSS. When you click any of these words, a **menu** of commands and choices will drop down. Basically, you tell SPSS what to do by clicking on your desired choices from these menus. Sometimes submenus will appear, and you will need to specify your choices further.

SPSS provides the user with a variety of options for displaying information about the data file and output on the screen. I recommend that you tell the program to display lists of variables by name (e.g., *age, abany*) rather than by labels (e.g., AGE OF RESPONDENT, ABORTION IF WOMAN WANTS FOR ANY REASON). Lists displayed this way will be easier to read and compare to Appendix G. To do this, click **Edit** on the main menu bar and then click **Options** from the drop-down submenu. A dialog box labeled "Options" will appear with a series of "tabs" along the top. The "General" options should be displayed; if not, click on this tab. On the "General" screen, find the box labeled "Variable Lists" and, if they are not already selected, click "Display names" and "alphabetical" and then click **OK.** If you make changes, a message may appear on the screen that tells you that changes will take effect the next time a data file is opened.

In this section, you learned how to start up *SPSS for Windows,* load a data file, and set some of the display options for this program. These procedures are summarized in Table F.1.

TABLE F.1 SUMMARY OF COMMANDS

To start *SPSS for Windows*	Click the SPSS icon on the screen of the computer monitor
To open a data file	Double-click on the data file name
To set display options for lists of variables	Click **Edit** from the main menu bar, then click **Options.** On the "General" tab make sure that "Display names" and "alphabetical" are selected and then click **OK.**

F.3 WORKING WITH DATABASES

Note that in the SPSS Data Editor Window the data are organized into a two-dimensional grid, with columns running up and down (vertically) and rows running across (horizontally). Each column is a variable or item of information from the survey. The names of the variables are listed at the tops of the columns. Remember that you can find the meaning of these variable names in the GSS 2006 code book in Appendix G.

Another way to decipher the meaning of variable names is to click **Utilities** on the menu bar and then click **Variables.** The Variables window opens. This window has two parts. On the left is a list of all variables in the database, arranged in alphabetical order and with the first variable highlighted. On the right is the Variable Information window, with information about the highlighted variable. The first variable is listed as *abany*. The Variable Information window displays a fragment of the question that was actually asked during the survey ("ABORTION IF WOMAN WANTS FOR ANY REASON") and shows the possible scores on this variable (a score of 1 = yes and a score of 2 = no) along with some other information.

The same information can be displayed for any variable in the data set. For example, find the variable *marital* in the list. You can do this by using the arrow keys on your keyboard or the slider bar on the right of the variable list window. You can also move through the list by typing the first letter of the variable name you are interested in. For example, type "m" and you will be moved to the first variable name in the list that begins with that letter. Now you can see that the variable measures marital status and that a score of "1" indicates that the respondent was married, and so forth. What do *prestg80* and *marhomo* measure? Close this window by clicking the **Close** button at the bottom of the window.

Examine the window displaying the 2006 GSS a little more. Each row of the window (reading across, or from left to right) contains the scores of a particular respondent on all the variables in the database. Note that the upper-left-hand cell is highlighted (outlined in a darker border than the other cells). This cell contains the score of respondent 1 on the first variable. The second row contains the scores of respondent 2, and so forth. You can move around in this window with the arrow keys on your keyboard. The highlight moves in the direction of the arrow, one cell at a time.

In this section, you learned to read information in the data display window and to decipher the meaning of variable names and scores. These commands are summarized in Table F.2. We are now prepared to perform some statistical operations with the 1998 GSS database.

TABLE F.2 SUMMARY OF COMMANDS

To move around in the Data Editor window	1. Click the cell you want to highlight or 2. Use the arrow keys on your keyboard or 3. Move the slider buttons or 4. Click the arrows on the right-hand and bottom margins.
To get information about a variable	1. From the menu bar, click Utilities and then click Variables. Scroll through the list of variable names until you highlight the name of the variable in which you are interested. Variable information will appear in the window on the right. 2. See Appendix G.

F.4 PUTTING SPSS TO WORK: PRODUCING STATISTICS

At this point, the database on the screen is just a mass of numbers with little meaning for you. That's okay because you will not actually have to read any information from this screen. Virtually all of the statistical operations you will conduct will begin by clicking the **Analyze** command from the menu bar, selecting a procedure and statistics, and then naming the variable or variables you would like to process.

To illustrate, let's have *SPSS for Windows* produce a frequency distribution for the variable *sex*. Frequency distributions are tables that display the number of times each score of a variable occurred in the sample (see Chapter 2). So when we complete this procedure, we will know the number of males and females in the 2006 GSS sample.

With the 2006 GSS loaded, begin by clicking the **Analyze** command on the menu bar. From the menu that drops down, click **Descriptive Statistics** and then **Frequencies.** The Frequencies window appears, with the variables listed in alphabetical order in the box on the left. The first variable (*abany*) will be highlighted. Use the slider button or the arrow keys on the right-hand margin of this box to scroll through the variable list until you highlight the variable *sex*, or type "s" to move to the approximate location.

Once the variable you want to process has been highlighted, click the arrow button in the middle of the screen to move the variable name to the box on the right-hand side of the screen. SPSS will produce frequency distributions for all variables listed in this box, but for now we will confine our attention to *sex*. Click the **OK** button in the upper-right-hand corner of the Frequencies window and, in seconds, a frequency distribution will be produced.

SPSS sends all tables and statistics to the Output window, or SPSS viewer. This window is now "closest" to you, and the Data Editor window is "behind" the Output window. If you want to return to the Data Editor, click on any part of it if it is visible, and it will move to the "front" and the Output window will be "behind" it. To display the Data Editor window if it is not visible, minimize the Output window by clicking the "-" box in the upper-right-hand corner.

Frequencies The output from SPSS, slightly modified, is reproduced as Table F.3. What can we tell from this table? The score labels (male and female) are printed at the left, with the number of cases (frequency) in each category of the variable one column to the right. As you can see, there are 644 males and 782 females in the sample. The next two columns give information about percentages and the last column to the right displays cumulative percentages. We will defer a discussion of this last column until a later exercise.

One of the percentage columns is labeled Percent and the other is labeled Valid Percent. The difference between these two columns lies in the handling

TABLE F.3 AN EXAMPLE OF SPSS OUTPUT (respondents' sex)

		Frequency	Percent	Valid Percent	Cumulative Percent
Valid	MALE	644	45.2	45.2	45.2
	FEMALE	782	54.8	54.8	100.0
	Total	1426	100.0	100.0	

of missing values. The Percent column is based on all cases, including people who did not respond to the item (NA) and people who said they did not have the requested information (DK). The Valid Percent column excludes all missing scores. Since we will almost always want to ignore missing scores, we will pay attention only to the Valid Percent column. Note that for sex, there are no missing scores (gender was determined by the interviewer), and the two columns are identical.

F.5 PRINTING AND SAVING OUTPUT

Once you've gone to the trouble of producing statistics, a table, or a graph, you will probably want to keep a permanent record. There are two ways to do this. First, you can print a copy of the contents of the Output Window to take with you. To do this, click on **File** and then click **Print** from the **File** menu. Alternatively, find the icon of a printer (third from the left) in the row of icons just below the menu bar and click on it.

The other way to create a permanent record of SPSS output is to save the Output window to the computer's memory or to a diskette. To do this, click **Save** from the **File** menu. The Save dialog box opens. Give the output a name (some abbreviation such as "freqsex" might do) and, if necessary, specify the name of the drive in which your diskette is located. Click **OK,** and the table will be permanently saved.

F.6 ENDING YOUR *SPSS FOR WINDOWS* SESSION

Once you have saved or printed your work, you may end your SPSS session. Click on **File** from the menu bar and then click **Exit.** If you haven't already done so, you will be asked if you want to save the contents of the Output window. You may save the frequency distribution at this point if you wish. Otherwise, click **NO.** The program will close, and you will be returned to the screen from which you began.

Appendix G

Code Book for the General Social Survey, 2006

The General Social Survey (GSS) is a public opinion poll that has been regularly conducted by the National Opinion Research Council. A version of the 2006 GSS is available at the web site for this text and is used for all end-of-chapter exercises. Our version of the 2006 GSS includes about 50 variables for a randomly selected subsample of about 1500 of the original respondents. This code book lists each item in the data set. The variable names are those used in the data files. The questions have been reproduced exactly as they were asked (with a few exceptions to conserve space), and the numbers beside each response are the scores recorded in the data file.

The data set includes variables that measure demographic or background characteristics of the respondents, including sex, age, race, religion, and several indicators of socioeconomic status. Also included are items that measure opinion on such current and controversial topics as abortion, capital punishment, and homosexuality.

Most variables in the data set have codes for "missing data." These codes are italicized in the listings below for easy identification. The codes refer to various situations in which the respondent does not or cannot answer the question and are excluded from all statistical operations. The codes are: NAP ("not applicable," that is, the respondent was not asked the question), DK ("Don't know," that is, the respondent didn't have the requested information), and NA ("No answer," that is, the respondent refused to answer).

Please tell me if you think it should be possible for a woman to get a legal abortion if . . .

abany — She wants it for any reason.

1. Yes
2. No

0. NAP, 8. DK, 9. NA

abrape — The pregnancy is the result of rape

(Same scoring as abany)

affrmact — Some people say that because of past discrimination, blacks should be given preference in hiring and promotion. Others say that such preference is wrong because it discriminates against whites. Are you for or against preferential hiring and promotion of blacks?

1. Strongly supports preferences
2. Supports preferences
3. Opposes preferences
4. Strongly opposes preferences

0. NAP, 8. DK, 9. NA

age	Age of respondent 18–89. Actual age in years *99 NA*
attend	How often do you attend religious services? 0. Never 1. Less than once per year 2. Once or twice a year 3. Several times per year 4. About once a month 5. 2–3 times a month 6. Nearly every week 7. Every week 8. Several times a week *9. DK or NA*
cappun	Do you favor or oppose the death penalty for persons convicted of murder? 1. Favor 2. Oppose *0. NAP, 8. DK, 9. NA*
childs	How many children have you ever had? Please count all that were born alive at any time (including any from a previous marriage). 0–7. Actual number 8. Eight or more *9. NA*
chldidel	What do you think is the ideal number of children for a family to have? 0–6. Actual values 7. Seven or more *−1. NAP, 8. As many as want, 9. DK, NA*
class	Subjective class identification 1. Lower class 2. Working class 3. Middle class 4. Upper class *0. NAP, 8. DK, 9. NA*
degree	Respondent's highest degree 0. Less than high school 1. High school 2. Assoc./Junior college 3. Bachelor's 4. Graduate *7. NAP, 8. DK, 9. NA*
educ	Highest year of school completed 0–20. Actual number of years *97. NAP, 98. DK, 99. NA*
fear	Is there any area right around here—that is, within a mile—where you would be afraid to walk alone at night? 1. Yes 2. No *0. NAP, 8. DK, 9. NA*

fefam — It is much better for everyone involved if the man is the achiever outside the home and the woman takes care of the home and family.

1. Strongly agree
2. Agree
3. Disagree
4. Strongly disagree

0. NAP, 8. DK, 9. NA

fepresch — A preschool child is likely to suffer if his or her mother works.

1. Strongly agree
2. Agree
3. Disagree
4. Strongly disagree

0. NAP, 8. DK, 9. NA

grass — Do you think the use of marijuana should be made legal or not?

1. Should
2. Should not

0. NAP, 8. DK, 9. NA

gunlaw — Would you favor or oppose a law which would require a person to obtain a police permit before he or she could buy a gun?

1. Favor
2. Oppose

0. NAP, 8. DK, 9. NA

happy — Taken all together, how would you say things are these days—would you say that you are very happy, pretty happy, or not too happy?

1. Very happy
2. Pretty happy
3. Not too happy

0. NAP, 8. DK, 9. NA

helppoor — Some people think that the [federal] government ... should do everything possible to improve the standard of living of all poor Americans; other people think it is not the government's responsibility, and that each person should take care of himself. Where would you put yourself on this scale?

1. Government action
2.
3. Agree with both
4.
5. People should help selves

0. NAP, 8. DK, 9. NA

hrs1 — How many hours did you work last week?

1–89. Actual hours

−1. NAP, 98. DK, 99. NA

income06 — Respondent's total family income from all sources

1. Less than 1,000	2. 1,000 to 2,999
3. 3,000 to 3,999	4. 4,000 to 4,999

5. 5,000 to 5,999
6. 6,000 to 6,999
7. 7,000 to 7,999
8. 8,000 to 9,999
9. 10,000 to 12,499
10. 12,500 to 14,999
11. 15,000 to 17,499
12. 17,500 to 19,999
13. 20,000 to 22,499
14. 22,500 to 24,999
15. 25,000 to 29,999
16. 30,000 to 34,999
17. 35,000 to 39,999
18. 40,000 to 49,999
19. 50,000 to 59,999
20. 60,000 to 74,999
21. 75,000 to 89,999
22. 90,000 to 109,999
23. 110,000 to 129,999
24. 130,000 to 149,999
25. 150,000 or more
98. DK, 99. NA

letdie1 — When a person has a disease that cannot be cured, do you think doctors should be allowed by law to end the patient's life by some painless means if the patient and his family request it?

1. Yes
2. No

0. NAP, 8. DK, 9. NA

letin1 — Do you think the number of immigrants to America nowadays should be

1. Increased a lot
2. Increased a little
3. Remain the same as it is
4. Reduced a little
5. Reduced a lot

0. NAP, 8. DK, 9. NA

marblk — What about a close relative marrying a black person? Would you be

1. Strongly in favor
2. In favor
3. Neither favor nor oppose
4. Oppose
5. Strongly oppose

0. NAP, 8. DK, 9. NA

marhomo — Homosexual couples should have the right to marry one another.

1. Strongly agree
2. Agree
3. Neither agree or disagree
4. Disagree
5. Strongly disagree

0. NAP, 8. DK, 9. NA

marital — Are you currently married, widowed, divorced, separated, or have you never been married?

1. Married
2. Widowed
3. Divorced
4. Separated
5. Never married

9. NA

news — How often do you read the newspaper?

1. Every day
2. A few times a week
3. Once a week
4. Less than once a week
5. Never

0. NAP, 8. DK, 9. NA

obey — How important is it for a child to learn obedience to prepare him or her for life?

1. Most important
2. 2nd most important
3. 3rd most important
4. 4th most important
5. Least important

0. NAP, 8. DK, 9. NA

paeduc — Father's highest year of school completed

0–20. Actual number of years

97. NAP, 98. DK, 99. NA

papres80 — Prestige of father's occupation

17–86. Actual score

0. NAP, DK, NA

partnrs5 — How many sex partners have you had over the past five years?

0. No partners
1. 1 partner
2. 2 partners
3. 3 partners
4. 4 partners
5. 5–10 partners
6. 11–20 partners
7. 21–100 partners
8. More than 100 partners

9. 1 or more, don't know the number, 95 Several, 98 DK, 99 NA, −1. NAP

popleff3 — The average citizen has considerable influence on politics.

1. Strongly agree
2. Agree
3. Neither agree nor disagree
4. Disagree
5. Strongly disagree

0. NAP, 8. DK, 9. NA

polviews — I'm going to show you a seven-point scale on which the political views that people might hold are arranged from extremely liberal to extremely conservative. Where would you place yourself on this scale?

1. Extremely liberal
2. Liberal
3. Slightly liberal
4. Moderate
5. Slightly conservative
6. Conservative
7. Extremely conservative

0. NAP, 8. DK, 9. NA

premarsx — There's been a lot of discussion about the way morals and attitudes about sex are changing in this country. If a man and a woman have sex relations before marriage, do you think it is always wrong, almost always wrong, wrong only sometimes, or not wrong at all?

1. Always wrong
2. Almost always wrong
3. Wrong only sometimes
4. Not wrong at all

0. NAP, 8. DK, 9. NA

pres04 — In 2004, did you vote for Kerry (the Democratic candidate) or Bush (the Republican candidate)? (Includes only those who said they voted in this election.)

1. Kerry
2. Bush

3. Other, 6. No presidential vote, 0. NAP, 8. DK, 9. NA

prestg80 — Prestige of respondent's occupation

17–86. Actual score

0. NAP, DK, NA

racecen1 — Race of respondent

1. White
2. Black
3. American Indian or Alaska Native
4. Asian American or Pacific Islander
5. Hispanic

0. NAP, 98. DK

region — Region of interview

1. New England (ME, VT, NH, MA, CT, RI)
2. Mid-Atlantic (NY, NJ, PA)
3. East North Central (WI, IL, IN, MI, OH)
4. West North Central (MN, IA, MO, ND, SD, NE, KS)
5. South Atlantic (DE, MY, WV, VA, NC, SC, GA, FL, DC)
6. East South Central (KY, TN, AL, MS)
7. West South Central (AK, OK, LA, TX)
8. Mountain (MT, ID, WY, NV, UT, CO, AR, NM)
9. Pacific (WA, OR, CA, AL, HI)

relexper — Has there been a turning point in your life when you made a new and personal commitment to religion?

1. Yes
2. No

0. NAP, 8. DK, 9. NA

relig — What is your religious preference? Is it Protestant, Catholic, Jewish, some other religion, or no religion?

1. Protestant
2. Catholic
3. Jewish
4. None

5. Other
8. DK, 9. NA

satjob — All in all, how satisfied would you say you are with your job?
1. Very satisfied
2. Moderately satisfied
3. A little dissatisfied
4. Very dissatisfied
0. NAP, 8. DK, 9. NA

scresrch — Recently, there has been controversy over whether the government should provide any funds for scientific research that uses "stem cells" taken from human embryos. Would you say the government
1. Definitely should fund such research
2. Probably should fund such research
3. Probably should not fund such research
4. Definitely should not fund such research
0. NAP, 8. DK, 9. NA

sex — Respondent's gender
1. Male
2. Female

sexfreq — About how many times did you have sex during the last 12 months?
0. Not at all
1. Once or twice
2. About once a month
3. 2 or 3 times a month
4. About once a week
5. 2 or 3 times a week
6. More than 3 times a week
−1. NAP, 8. DK, 9. NA

size — Size of place, in thousands. Population figures from U.S. Census. Add three zeros to code for actual values.

spanking — Do you strongly agree, agree, disagree, or strongly disagree that it is sometimes necessary to discipline a child with a good, hard spanking?
1. Strongly agree
2. Agree
3. Disagree
4. Strongly disagree
0. NAP, 8. DK, 9. NA

tapphone — Suppose the government suspected that a terrorist act was about to happen. Do you think the authorities should have the right to tap people's telephone conversations?
1. Definitely should have the right
2. Probably should have the right
3. Probably should not have the right
4. Definitely should not have the right
0. NAP, 8. DK, 9. NA

trust	Generally speaking, would you say that most people can be trusted or that you can't be too careful in life? 1. Most people can be trusted 2. Can't be too careful 3. Depends *0. NAP, 8. DK, 9. NA*
tvhours	On the average day, about how many hours do you personally watch television? 00–22. Actual hours *−1. NAP, DK, NA*
wwwhr	Not counting e-mail, about how many hours per week do you use the web (or Internet)? 0–75. actual hours *−1. NAP, 998. DK, 999. NA*

Answers to Odd-Numbered End-of-Chapter Problems and Cumulative Exercises

In addition to answers, this section suggests some problem-solving strategies and provides some examples of how to interpret some statistics. You should try to solve and interpret the problems on your own before consulting this section.

In solving these problems, I let my calculator or computer do most of the work. I worked with whatever level of precision these devices permitted and didn't round off until the end or until I had to record an intermediate sum. I always rounded off to two places of accuracy (that is, to two places beyond the decimal point, or to 100ths). If you follow these same conventions, your answers will almost always match mine. However, there is no guarantee that our answers will always be exact matches. You should realize that small discrepancies might occur and that these differences almost always will be trivial. If the difference between your answer and mine doesn't seem trivial, you should double-check to make sure you haven't made an error or solve the problem again using a greater degree of precision.

Finally, please allow me a brief disclaimer about mathematical errors in this section. Let me assure you, first of all, that I know how important this section is for most students and that I worked hard to be certain that these answers are correct. Human fallibility being what it is, however, I know that I cannot make absolute guarantees. Should you find any errors, please let me know so that I can make corrections in the future.

Chapter 1

1.5
a. Nominal
b. Ordinal (the categories can be ranked in terms of degree of honesty with "Returned the wallet with money" the "most honest")
c. Ordinal
d. Interval-ratio ("years" has equal intervals and a true zero point)
e. Interval-ratio
f. Interval-ratio
g. Nominal (the various patterns are different from each other but cannot be ranked from high to low)
h. Interval-ratio
i. Ordinal
j. Number of accidents: interval-ratio; Severity of accident: ordinal

1.7

	Variable	Level of Measurement	Type	Application
a.	Opinion	Ordinal	Discrete	Inferential
b.	Grade	Interval-ratio	Continuous	Descriptive (two variables)
c.	Party	Nominal	Discrete	Inferential
	Sex	Nominal	Discrete	
	Opinion	Ordinal		
d.	Homicide rate	Interval-ratio	Continuous	Descriptive (two variables)
e.	Satisfaction	Ordinal	Discrete	Descriptive (one variable

Chapter 2

2.1 **a.** Complex A: (5/20) × 100 = 25.00%
Complex B: (10/20) × 100 = 50.00%
b. Complex A: 4:5 = 0.80
Complex B: 6:10 = 0.60
c. Complex A: (0/20) = 0.00
Complex B: (1/20) = 0.05
d. (6/(4 + 6)) = (6/10) = 60.00%
e. Complex A: 8:5 = 1.60
Complex B: 2:10 = 0.20

2.3 Bank robbery rate =
(23/211,732)(47/211,732) × 100,000 = 22.20
Homicide rate = (13/211,732) × 100,000 = 6.14
Auto theft rate = (23/211,732) × 100,000 = 10.86

2.5 For sex:

Sex	Frequency
Male	9
Female	6
Total	15

For age, we will follow the procedure established in Section 2.5 (see the "One Step at a Time" box). Set $k = 10$. $R = 77 - 23$, or 54, so we can round off interval size to 5 ($i = 5$). The first interval will be 20–24, to include the low score of 23, and the highest interval will be 75–79.

Age	Frequency
20–24	1
25–29	2
30–34	3
35–39	2
40–44	1
45–49	3
50–54	1
55–59	1
60–64	0
65–69	0
70–74	0
75–79	1
	15

2.9 Set $k = 10$. $R = 92 - 5$, or 87, so set i at 10.

Score	Frequency
0–9	3
10–19	7
20–29	6
30–39	0
40–49	2
50–59	2
60–69	3
70–79	0
80–89	0
90–99	2
	25

2.11 Answers should note and describe the rise in all crime rates except burglary up to a peak in the early 1990s, followed by a dramatic decline and then a leveling off in recent years.

2.13 Answers should note and describe the decrease in accidents in the later time period. The number of accidents was lower in almost every month, sometimes falling by half or more.

Chapter 3

3.1 "Region of birth" and "religion" are nominal-level variables, "support for legalization" and "opinion of food" are ordinal, and "expenses" and "number of movies" are interval-ratio. The mode, the most common score, is the only measure of central tendency available for nominal-level variables. For the two ordinal-level variables, *don't forget to array the scores from high to low* before locating the median. There are 10 freshmen (N is even), so the median for freshmen will be the score halfway between the scores of the two middle cases. There are 11 seniors (N is odd), so the median for seniors will be the score of the middle case. To find the mean for the interval-ratio variables, add the scores and divide by the number of cases.

Variable	Freshmen	Seniors
Region of birth	Mode = North	Mode = North
Legalization	Median = 3	Median = 5
Expenses	Mean = 48.50	Mean = 63.00
Movies	Mean = 5.80	Mean = 5.18
Food	Median = 6	Median = 4
Religion	Mode = Protestant	Mode = Protestant and None (4 cases each)

3.3

Variable	Level of Measurement	Measure of Central Tendency
Sex	Nominal	Mode = male
Social Class	Ordinal	Median = "medium" (the middle case is in this category)
Number of years in the party	I-R	Mean = 26.15
Education	Ordinal	Median = high school
Marital status	Nominal	Mode = married
Number of children	I-R	Mean = 2.39

3.5

Variable	Level of Measurement	Measure of Central Tendency
Marital status	Nominal	Mode = married
Race	Nominal	Mode = white
Age	I-R	Mean = 27.53
Attitude on abortion	Ordinal	Median = 7

3.7

Variable	Level of Measurement	Measure of Central Tendency
Sex	Nominal	There are 9 males and 6 females, so the mode is male
Support for gun control	Ordinal	Median = 1
Level of education	Ordinal	Median = 1
Age	I-R	Mean = 40.93

3.9 Attitude and opinion scales almost always generate ordinal-level data, so the appropriate measure of central tendency would be the median. The median is 9 for the students and 2 for the neighbors. Incidentally, the means are 7.80 for the students and 4.00 for the neighbors.

3.11 Mean = 40.25, median = 44.5. The lower value for the mean indicates a negative skew, or a few very *low* scores. For this small group of nations, the skew is caused by the score of Mexico (10), which is much lower than the scores of the other seven nations (which are grouped between 37 and 51).

3.13 To find the median, you must first rank the scores from high to low. Both groups have 25 cases (N = odd), so the median is the score of the 13th case. For freshman the median is 35, and for seniors the median is 30. The mean score for freshmen is 31.72. For seniors, the mean is 28.60.

3.15 The owners are using the mean while the players are citing the median. This distribution has a positive skew (the mean is greater than the median), as is typical for income data. Note the difference in wording in the two reports: The owners cite the "average" (mean) and the players cite the "typical player" (the score of the middle case or the median).

Chapter 4

4.1

Complex	IQV
A	0.89
B	0.99
C	0.71
D	0.74

Complex B has the highest IQV and is the most heterogeneous. Complex C is the least heterogeneous.

4.3 The high score is 50 and the low score is 10, so the range is 50 − 10, or 40. The standard deviation is 12.28.

4.5

Statistic	2000	2004
Mean	48,161.54	57,146.15
Median	47,300.00	54,100.00
Standard deviation	6,753.64	9,066.81
Range	23,400.00	33,700.00

In this time period, the mean and median increase. The distributions for both years show a positive skew (the mean is greater than the median). Note that the skew is greater for 2004. This is caused by the Northwest Territories, which has a much higher average income than the other provinces, especially for 2004 The standard deviation and the range also increase, a reflection of increasing skew.

4.7

Variable	Statistic	Males	Females
Labor force participation	Mean	77.60	58.40
	Standard deviation	2.73	6.73
% High school graduate	Mean	69.20	70.20
	Standard deviation	5.38	4.98
Mean income	Mean	33,896.60	29,462.40
	Standard deviation	4,443.16	4,597.93

Males and females are very similar in terms of educational level, but females are less involved in the labor force and, on the average, earn almost $4500 less than males per year. The females in these 10 states are much more variable in their labor force participation but are similar to males in dispersion on the other two variables. See Section 9.6 for more on gender disparities in income.

4.9 $s = 12.29$

4.11 $R = 50 - 6 = 44$, $s = 11.12$. The score for Los Angeles (50) is much higher than the other scores (the next highest score is 37), so removing this score would reduce the variation in the data set. The standard deviation of the scores without Los Angeles would be lower in value.

4.13

Division	Range	Standard Deviation
A	$R = 18$	$s = 5.32$
B	$R = 8$	$s = 2.37$
C	$R = 3$	$s = 1.03$
D	$R = 15$	$s = 7.88$

4.15 The means are virtually the same, so the experimental course did not raise the average grade. However, the standard deviation for the experimental course is much lower, indicating that student grades were less diverse in these sections.

Chapter 5

5.1

X_i	Z score	% Area Above	% Area Below
5	−1.67	95.25	4.75
6	−1.33	90.82	9.18
7	−1.00	84.13	15.87
8	−0.67	74.86	25.14
9	−0.33	62.93	37.07
11	0.33	37.07	62.93
12	0.67	25.14	74.86
14	1.33	9.18	90.82
15	1.67	4.75	95.25
16	2.00	2.28	97.72
18	2.67	0.38	99.62

5.3

	Z scores	Area
a.	0.10 & 1.10	32.45%
b.	0.60 & 1.10	13.86%
c.	0.60	27.43%
d.	0.90	18.41%
e.	0.60 & −0.40	38.11%
f.	0.10 & −0.40	19.52%
g.	0.10	53.98%
h.	0.30	61.79%
i.	0.60	72.57%
j.	1.10	86.43%

5.5

X_i	Z Score	Number of Students Above	Number of Students Below
60	−2.00	195	5
57	−2.50	199	1
55	−2.83	199	1
67	−0.83	159	41
70	−0.33	126	74
72	0.00	100	100
78	1.00	32	168
82	1.67	10	190
90	3.00	1	199
95	3.83	1	199

Note: Number of students (a discrete variable) has been rounded off to the nearest whole number.

5.7

	Z score	Area
a.	−2.20	1.39%
b.	1.80	96.41%
c.	−0.20 & 1.80	54.34%
d.	0.80 & 2.80	20.93%
e.	−1.20	88.49%
f.	0.80	21.19%

5.9

	Z score	Area
a.	−1.00 & 1.50	.7745
b.	0.25 & 1.50	.3345
c.	1.50	.0668
d.	0.25 & −2.25	.5865
e.	−1.00 & −2.25	.1465
f.	−1.00	.1587

5.11 Yes. The raw score of 110 translates into a Z score of +2.88. 99.80% of the area lies below this score, so this individual was in the top 1% on this test.

5.13 For the first event, the probability is .0919; for the second, the probability is .0655. The first event is more likely.

Part I Cumulative Exercises

1. The level of measurement is the most important criterion for selecting descriptive statistics. The following table presents all relevant statistics for each variable. Statistics that are not appropriate for a variable are noted with an "X."

Religion is nominal, so the only statistics available are the mode and the IQV. For the two ordinal-level variables ("strength" and "comfort"), the median and the range are the appropriate choices. However, for ordinal-level variables like "strength," which have a wide range of scores, it is common for social science researchers to report the standard deviation and mean. For interval-ratio variables like "pray" and "age," the mean and standard deviation are the preferred summary statistics.

	Religion	Strength	Pray	Com-fort	Age
	Level of Measurement				
	Nominal	Ordinal	I-R	Ordinal	I-R
Mode	Protestant				
Median	X	7		1	
Mean	X	6.07	1.40		41.53
IQV	0.86				
Range	X	9	9	4	49
Stnd. dev.	X	2.77	1.62		12.54

3. As always, the level of measurement is the primary guideline for choosing descriptive statistics; the most appropriate statistics are noted in the following table.

	Children	School	Race	Spanking	TV	Religion
	Level of Measurement					
	I-R	Ordinal	Nominal	Ordinal	I-R	Nominal
Mode			White			Prot.
Median		1.00		2		
Mean	2.44				3.08	
IQV			0.50			0.54
Range	9	4		3	10	
Stnd. dev.	2.06				2.04	

Chapter 7

7.1 **a.** 5.2 ± 0.11 **b.** 100 ± 0.71
c. 20 ± 0.40 **d.** 1020 ± 5.41
e. 7.3 ± 0.23 **f.** 33 ± 0.80

7.3

Confidence Level	Alpha	Area Beyond Z	Z score
95%	.05	.0250	±1.96
94%	.06	.0300	±1.88 or ±1.89
92%	.08	.0400	±1.75 or ±1.76
97%	.03	.0150	±2.17
98%	.02	.0100	±2.33
99.9%	.001	.0005	±2.58

7.5 **a.** 2.30 ± 0.04 **b.** 2.10 ± 0.01, 0.78 ± .07
c. 6.00 ± 0.37

7.7 **a.** 178.23 ± 1.97 The estimate is that students spent between $176.26 and $180.20 on books.
b. 1.5 ± 0.04 The estimate is that students visited the clinic between 1.46 and 1.54 times on the average.
c. 2.8 ± .13 **d.** 3.5 ± .19

7.9. 14 ± .07 The estimate is that between 7% and 21% of the population consists of unmarried couples living together.

7.11 **a.** $P_s = 823/1496 = 0.55$
Confidence interval: 0.55 ± 0.03
Between 52% and 58% of the population agrees with the statement.
b. $P_s = 650/1496 = 0.44$
Confidence interval: .44 ± 0.03
c. $P_s = 375/1496 = 0.25$
Confidence interval: 0.25 ± 0.03
d. $P_s = 1023/1496 = 0.68$
Confidence interval: 0.68 ± 0.03
e. $P_s = 800/1496 = 0.54$
Confidence interval: 0.54 ± 0.03

7.13

Alpha (α)	Confidence Level	Confidence Interval
0.10	90%	100 ± 0.74
0.05	95%	100 ± 0.88
0.01	99%	100 ± 1.16
0.001	99.9%	100 ± 1.47

7.15 The confidence interval is .51 ± .05. The estimate would be that between 46% and 56% of the population prefer candidate A. The population parameter (P_u) is equally likely to be anywhere in the interval (that is, it's just as likely to be 46% as it is to be 56%), so a winner cannot be predicted.

7.17 The confidence interval is 0.23 ± 0.08. At the 95% confidence level, the estimate would be that between 240 (15%) and 496 (31%) of the 1600 freshmen would be extremely interested. The estimated numbers are found by multiplying *N* (1600) by the upper (.31) and lower (.15) limits of the interval.

7.19 a. 43.87 ± 0.49
b. 2.86 ± 0.08
c. 1.81 ± 0.06
d. 0.29 ± 0.02 (About 27% of the population are Catholic.)
e. 0.18 ± 0.02 (About 24% of the population have never married.)
f. 0.52 ± 0.02 (About 52% of the electorate voted for Bush in 2004.)
g. 0.81 ± 0.02

Chapter 8

8.3 a. Z(obtained) = −41.00
b. Z(obtained) = 29.09

8.5 Z(obtained) = 6.04

8.7 a. Z(obtained) = −13.66
b. Z(obtained) = 25.50

8.9 t(obtained) = 4.50

8.11 Z(obtained) = 3.06

8.13 Z(obtained) = −1.48

8.15 a. Z(obtained) = −0.74
b. Z(obtained) = 2.19
c. Z(obtained) = −8.55
d. Z(obtained) = −18.07
e. Z(obtained) = 2.09
f. Z(obtained) = −53.33

8.17 t(obtained) = −1.14

Chapter 9

9.1 a. σ = 1.39, Z(obtained) = −2.53
b. σ = 1.60, Z(obtained) = 2.49

9.3 a. σ = 10.57, Z(obtained) = 1.70
b. σ = 11.28, Z(obtained) = −2.48

9.5 a. σ = 0.08 Z(obtained) = −11.25
b. σ = 0.12 Z(obtained) = −3.33
c. σ = 0.15 Z(obtained) = 20.00

9.7 These are small samples (combined *N*'s of less than 100), so be sure to use Formulas 9.5 and 9.6 in step 4.
a. σ = 0.12, t(obtained) = −1.33
b. σ = 0.13, t(obtained) = 14.85

9.9 a. (France) σ = 0.0095, Z(obtained) = −31.58
b. (Nigeria) σ = 0.0075, Z(obtained) = −146.67
c. (China) σ = 0.0065, Z(obtained) = 76.92
d. (Mexico) σ = 0.0107, Z(obtained) = −74.77
e. (Japan) σ = 0.0115, Z(obtained) = −43.48

The large values for the *Z* scores indicate that the differences are significant at very low alpha levels (i.e., they are extremely unlikely to have been caused by random chance alone). Note that women are significantly happier than men in every nation except China where men are significantly happier.

9.11 P_u = .45, σ_p = .06, Z(obtained) = 0.67

9.13 a. P_u = .46, σ_p = 0.06, Z(obtained) = 2.17
b. P_u = .80, σ_p = 0.07, Z(obtained) = 1.43
c. P_u = .72, σ_p = 0.08, Z(obtained) = 0.75

9.15 a. Z(obtained) = 1.50
b. Z(obtained) = −2.75
c. Z(obtained) = 4.00
d. σ = 0.43, Z(obtained) = 1.86
e. σ = 0.14, Z(obtained) = −5.71
f. σ = 0.08, Z(obtained) = −5.50

Chapter 10

10.1

Problem	Grand Mean	SST	SSB	SSW	*F* ratio
a.	12.17	231.67	173.17	58.5	13.32
b.	6.87	455.73	78.53	377.20	1.25
c.	31.65	8362.55	5053.35	3309.2	8.14

10.3

Problem	Grand Mean	SST	SSB	SSW	*F* ratio
a.	4.39	86.28	45.78	40.50	8.48
b.	16.44	332.44	65.44	267.00	1.84

For problem 10.3a, with alpha = 0.05 and df = 2, 15, the critical *F* ratio would be 3.68. We would reject the null hypothesis and conclude that decision making *does* vary significantly by type of relationship. By inspection of the group means, it seems that the "cohabitational" category accounts for most of the differences.

10.5

Grand Mean	SST	SSB	SSW	*F* Ratio
9.28	213.61	2.11	211.50	0.08

10.7

Grand Mean	SST	SSB	SSW	*F* Ratio
5.40	429.48	305.42	124.06	5.96

10.9

Nation	Grand Mean	SST	SSB	SSW	*F* ratio
Mexico	3.78	300.98	154.08	146.90	12.59
Canada	6.88	156.38	20.08	136.30	1.77
United States	5.13	286.38	135.28	151.10	10.74

At alpha = 0.05 and df = 3, 36, the critical *F* ratio is 2.92. There is a significant difference in support for suicide by class in Mexico and the United States but not in Canada. The category means for Mexico suggest that the upper class accounts for most of the differences. For the United States, there is more variation across the category means, and the working class seems to account for most of the differences. Going beyond the ANOVA test and comparing the grand means, we see that support is highest in Canada and lowest in Mexico.

Chapter 11

11.1 **a.** 1.11 **b.** 0.00 **c.** 1.52 **d.** 1.46

11.3 A computing table is highly recommended as a way of organizing the computations for chi square:

Computational Table for Problem 11.3

(1) f_o	(2) f_e	(3) $f_o - f_e$	(4) $(f_o - f_e)^2$	(5) $(f_o - f_e)^2/f_e$
6	5	1	1	.20
7	8	−1	1	.13
4	5	−1	1	.20
9	8	1	1	.13
N = 26	N = 26	0		χ^2(obtained) = 0.65

There is 1 degree of freedom in a 2 × 2 table. With alpha set at .05, the critical value for the chi square would be 3.841. The obtained chi square is 0.65, so we fail to reject the null hypothesis of independence between the variables. There is no statistically significant relationship between race and services received.

11.5 a.

Computational Table for Problem 11.5

(1) f_o	(2) f_e	(3) $f_o - f_e$	(4) $(f_o - f_e)^2$	(5) $(f_o - f_e)^2/f_e$
21	17.5	3.5	12.25	.70
29	32.5	−3.5	12.25	.38
14	17.5	−3.5	12.25	.70
36	32.5	3.5	12.25	.38
N = 100	N = 100.0	0		χ^2(obtained) = 2.15

With 1 degree of freedom and alpha set at .05, the critical region will begin at 3.841. The obtained chi square of 2.15 does not fall within this area, so the null hypothesis cannot be rejected. There is no statistically significant relationship between unionization and salary.

b. Column percentages:

	Status	
Salary	Union	Nonunion
High	60.00%	44.6%
Low	40.00%	55.4%
Totals	100.00%	100.0%

Although the relationship is not significant, unionized fire departments tend to have higher salary levels.

11.7 The obtained chi square is 5.12, which is significant (df = 1, alpha = .05). The column percentages show that more affluent communities have higher-quality schools.

11.9 The obtained chi square is 6.67, which is significant (df = 2, alpha = .05).The column percentages show that shorter marriages have higher satisfaction.

11.11 The obtained chi square is 12.59, which is significant (df = 4, alpha = .05). The column percentages show that proportionally more of the students living "off campus with roommates" are in the high-GPA category.

11.13 The obtained chi square is 19.34, which is significant (df = 3, alpha = .05). Legalization was not favored by a majority of any region, but the column percentages show that the West was most in favor.

11.15

Problem	Chi Square	Significant at α = 0.05?	Column Percentages
a.	25.19	Yes	The oldest age group was most opposed.
b.	1.80	No	The great majority (72%–75%) of all three age groups were in favor of capital punishment.
c.	5.23	Yes	The oldest age group was most likely to say yes. Although significant, the differences in column percentages are small.
d.	28.43	Yes	The youngest age group was most likely to support legalization.
e.	14.17	Yes	The oldest age group was least likely to support suicide.

11.17 The obtained chi square is 4.43. With df = 3 and alpha = .05, this is not a significant relationship.

Part II Cumulative Exercises

1. One of the challenges of empirical research is to make reasonable decisions about which statistical test to use in which situation. You can minimize the confusion and ambiguity by approaching the decision systematically. We'll use the first problem of this exercise to consider some ways in which reasonable decisions can be made. The situation calls for a test of hypotheses ("Is the difference significant?"), so our choice of procedures will be limited to Chapters 8–11. Next, determine the types of variables you are working with. Number of minutes is an interval-ratio-level variable, and the research question asks us to compare two groups or samples: males and females. Which test should we use? The techniques in Chapter 8 (one-sample tests) and Chapter 10 (tests involving more than two samples or categories) are not relevant. Chi square (Chapter 11) won't work unless we collapse the scores on Internet use into a few categories. This leaves Chapter 9. A test of sample means fits the situation, and we have a large sample (combined *N*s greater than 100), so it looks like we're going to wind up in Section 9.2.

   a. σ = .14, *Z*(obtained) = −35.51
   The difference in Internet minutes is significant. Men, on the average, use this technology more frequently.

   b. The table format is a sure tip-off that the chi square test is appropriate. The obtained chi square is 0.88, which is not significant at the .05 level. There is no statistically significant relationship between involvement and social class.

   c. "Number of partners" sounds like an interval-ratio-level variable, and education has three categories. Analysis of variance is an appropriate test for this situation. The *F* ratio is .13—not at all significant—so we must conclude that this dimension of sexuality does not vary by level of education.

Grand Mean	SST	SSB	SSW	*F* Ratio
4.04	262.96	3.24	259.71	0.13

   d. The problem asks for a characteristic of a population ("How many times do adult Americans move?") but gives information only for a random sample. The estimation procedures presented in Chapter 7 fit the situation, and, since the information is presented in the form of a mean, you should use Formula 7.2 to form the estimate. At an alpha level of .05, the confidence interval would be 3.5 ± 0.02.

   e. The research question focuses on the difference between a single sample and a population, so Chapter 8 is relevant. Since the population standard deviation is unknown and we have a large sample, Sections 8.1–8.5 and Formula 8.1 will be applicable. *Z*(obtained) is −2.50, which is significant at alpha = 0.05. The sample is significantly different from the population—rural school districts are different from the universe of all school districts in this state.

Chapter 12

12.1

	Authoritarianism	
Efficiency	Low	High
Low	37.04%	70.59%
High	62.96%	29.41%
Totals	100.00%	100.00%

The conditional distributions change, so there is a relationship between the variables. The change from column to column is quite large, and the maximum difference is (70.59 − 37.04) = 33.55. Using Table 12.5 as a guideline, we can say that this relationship is strong. From inspection of the percentages, we can see that efficiency decreases as authoritarianism increases—workers with dictatorial bosses are less productive (or, maybe, bosses become more dictatorial when workers are inefficient), so this relationship is negative in direction.

12.3

2004	2000 Election	
Election	Democrat	Republican
Democrat	87.31%	11.44%
Republican	12.69%	88.56%
Totals	100.00%	100.00%

The maximum difference for this table is (87.31 − 11.44), or 75.87. This is a very strong relationship. People are very consistent in their voting habits.

12.5 **a.**

Received	Race	
Services?	Black	White
Yes	60.00%	43.75%
No	40.00%	56.25%
Totals	100.00%	100.00%

The maximum difference is (60.00 − 43.75), or 16.25. This is a moderate relationship, and blacks were more likely to receive services.

b.

Party	Gender	
Preference	Male	Female
Democrats	40.00%	60.00%
Republicans	60.00%	40.00%
Totals	100.00%	100.00%

c.

	Status	
Salary	Union	Nonunion
High	60.00%	44.61%
Low	40.00%	55.39%
Totals	100.00%	100.00%

12.7

	Program	
Crime Rate	No	Yes
Low	25.44%	17.24%
Moderate	28.95%	31.03%
High	45.61%	51.72%
Totals	100.00%	100.00%

The maximum difference is (25.44 − 17.24), or 8.20. This is a weak relationship, but a higher percentage of the cities with the program have high crime rates.

12.9

Student Involvement by Extent of Press Coverage

	Coverage			
Involvement	None	Moderate	Extensive	Totals
None	3 (50.00%)	4 (36.36%)	0 (00.00%)	7 (28.00%)
Some	2 (33.33%)	4 (36.36%)	3 (37.50%)	9 (36.00%)
Extensive	1 (16.67%)	3 (27.27%)	5 (62.50%)	9 (36.00%)
Totals	6 (100.00%)	11 (99.99%)	8 (100.00%)	25 (100.00%)

The conditional distributions change, so there is a relationship between these two variables. The maximum difference is (50.00 − 00.00), or 50, so the relationship is strong. Campuses with extensive coverage had more extensive involvement and campuses with no coverage tended to have no involvement. This is a positive relationship: The greater the coverage, the more extensive the involvement.

12.11 a. Support for legal abortion by political ideology:

Supports Legal Abortion?	Political Ideology: Liberal	Moderate	Conservative
Yes	59.42	39.39	26.88
No	40.58	60.31	73.12
Totals	100.00	100.00	100.00

The maximum difference of 32.54 indicates a strong relationship. Liberals support legal abortion, conservatives are opposed, and moderates are intermediate.

b. Support for capital punishment by political ideology:

Supports Capital Punishment?	Political Ideology: Liberal	Moderate	Conservative
Yes	62.41%	76.41%	78.84%
No	37.59%	23.59%	21.16%
Totals	100.00%	100.00%	100.00%

c. Support for the right of people with an incurable disease to commit suicide by political ideology:

Supports the Right to suicide?	Political Ideology: Liberal	Moderate	Conservative
Yes	76.05%	63.24%	55.00%
No	23.95%	36.76%	45.00%
Totals	100.00%	100.00%	100.00%

d. Support for sex education in public schools by political ideology:

Supports Sex Education?	Political Ideology: Liberal	Moderate	Conservative
Yes	94.50%	90.08%	79.47%
No	5.50%	9.92%	20.53%
Totals	100.00%	100.00%	100.00%

e. Support for traditional gender roles by political ideology:

Supports Traditional Gender Roles?	Political Ideology: Liberal	Moderate	Conservative
Yes	11.50%	14.11%	18.24%
No	88.50%	85.89%	81.76%
Totals	100.00%	100.00%	100.00%

f. Support for legalizing marijuana:

Should Marijuana Be Legalized?	Political Ideology: Liberal	Moderate	Conservative
Yes	56.65%	47.27%	32.30%
No	43.35%	52.73%	67.70%
Totals	100.00%	100.00%	100.00%

Chapter 13

13.1 **a.** $\phi = 0.00, \lambda = 0.00$
b. $\phi = 0.09, \lambda = 0.00$
c. $\phi = 0.25, \lambda = 0.14$

13.3 $\phi = 0.17, \lambda = 0.00$

13.5 $\phi = 0.31, \lambda = 0.03$

13.7 $\phi = 0.00, \lambda = 0.00$

13.9 **a.** $\phi = 0.05, \lambda = 0.00$
b. $\phi = 0.49, \lambda = 0.42$
c. $\phi = 0.33, \lambda = 0.05$

Note that phi is greater than lambda for all three tables. This is a reflection of the much larger number of cases in the top row. Even though they differ in value, the two statistics rank the relationships consistently: Status has the strongest effect on attrition, followed by age and then race. Using phi, we can conclude that race is not an important correlate of attrition. Status and age have strong relationships with the dependent variable.

13.11 Cramer's $V = 0.05, \lambda = 0.00$

Chapter 14

14.1 **a.** $G = 0.71$ **b.** $G = 0.69$ **c.** $G = -0.88$

These relationships are strong. Facility in English and income increase with length of residence (+0.71). Use the percentages to help interpret the direction of a relationship. In the first table, 80% of the "newcomers" were "Low" in English facility, while 60% of the "old-timers" were "High". In this relationship, low scores on one variable are associated with low scores on the other, and scores increase together (as one increases, the other increases), so this is a positive relationship. In contrast, contact with the old country decreases with length of residence (−0.88). Most newcomers have higher levels of contact, and most old-timers have lower levels.

14.3 $G = -0.61$. This is a strong negative relationship. As authoritarianism increases, efficiency decreases.

14.5 $G = 0.27$. Be careful interpreting the direction of this relationship. The positive sign of gamma means that cases tend to fall along the diagonal from upper left to lower right. In this case, white-collar families are more associated with organized sports and blue-collar families with sandlot sports. Computing percentages will help you identify the direction of the relationship.

14.7 $G = 0.22$, $Z\text{(obtained)} = 0.92$

14.9 $G = -0.14$

14.11 $r_s = -0.46$, $t\text{(obtained)} = -1.55$

14.13 $r_s = 0.33$
For these nations, there is a moderate positive relationship between diversity and inequality. The greater the diversity, the greater the inequality.

14.15 **a.** $G = -0.14$ **b.** $G = -0.17$
c. $G = -0.14$ **d.** $G = -0.13$
e. $G = 0.39$

Income has weak negative relationships with the first four dependent variables. Be careful in interpreting direction for these tables, and remember that a negative gamma means that cases tend to be clustered along the diagonal from lower left to upper right. Computing percentages will clarify direction. For example, for the first table, low income is associated with opposition to abortion (62% of the people in this column said "No") and high income is associated with support (47% of the people in this column said "Yes" and this is the highest percentage of support across the three income groups).

Income has a moderate positive relationship with support for traditional gender roles. Note the way in which the dependent variable is coded: "agree" means support for traditional gender roles. A positive relationship means that cases tend to fall along the diagonal from upper left to lower right. In this case, agreement is greater for low income and declines as income increases. Is this truly a "positive" relationship? As always, percentages will help clarify the direction of the relationship.

Chapter 15

15.1 *(HINT: When finding the slope, remember that "Turnout" is the dependent, or Y, variable.)*

	For Turnout (Y) and		
	Unemployment	Education	Negative Campaigning
Slope (b)	3.00	12.67	−0.90
Y intercept (a)	39.00	−94.73	114.01
Reg. Eq.	$Y = 39 + 3X$	$Y = -94.73 + 12.67X$	$Y = 114.01 + (-0.90)X$
r	0.95	0.98	−0.87
r^2	0.90	0.97	0.76
t(obtained)	5.20	8.49	−3.08

15.3 *(HINT: When finding the slope, remember that "Number of visitors" is the dependent, or Y, variable.)*

Slope (*b*)	−0.56
Y intercept (*a*)	14.04
r	−0.37
r^2	0.14

15.5

Dependent Variable		Independent Variable		
		Density	Growth	Urbanization
Car theft	*a*	458.59	127.58	−838.55
	b	−0.35	47.80	16.13
	r	−0.15	0.78	0.68
	r^2	0.02	0.61	0.47
Robbery	*a*	58.74	86.00	−210.40
	b	0.32	3.76	4.13
	r	0.65	0.30	0.85
	r^2	0.42	0.09	0.73
Homicide	*a*	3.87	3.43	−1.40
	b	0.01	0.22	0.08
	r	0.31	0.50	0.47
	r^2	0.10	0.25	0.22

c. For a growth rate of −1, the predicted homicide rate would be 3.21. For a population density of 250, the predicted robbery rate would be 138.74. For a state with 50% urbanization, the predicted rate of auto theft would be −32.05. (Since negative crime rates are impossible, we would predict that the auto theft rates for this very low score on urbanization would approach zero.)

15.7 $b = 0.05$, $a = 53.18$, $r = 0.40$, $r^2 = 0.16$

15.9

		Prestige	Number of Children	Support for Abortion	Hours of TV per Day
Age	*r*	−0.30	0.67	−0.08	0.16
	r^2	0.09	0.45	0.01	0.03
	t	−1.13	3.26	−0.29	0.59
Sex	*r*	−0.28	0.19	0.11	−0.29
	r^2	0.08	0.04	0.01	0.08
	t	−1.01	0.70	0.40	−1.09

Part III Cumulative Exercises

1. a. Assuming that crime rates and "percent immigrant" are both measured at the interval-ratio level, Pearson's *r* would be the appropriate measure of association. This is a moderate negative relationship ($r = -0.47$) and percentage of immigrants explains 22% of the variation in crime rate for these 10 cities. As the percentage of immigrants increases, the crime rate decreases.

b. These variables are ordinal in level of measurement, so gamma would be the appropriate measure of association. Gamma is −0.43, indicating a moderate negative relationship. As TV viewing increases, involvement decreases.

c. The appropriate measure for these variables is Spearman's rho, which is 0.74. This is a strong positive relationship, and states with higher quality of life have superior systems of higher education.

d. Race is a nominal-level variable, and the table is larger than 2 × 2. Cramer's *V* is .40, indicating a strong relationship between the variables. The column percentages show that this sample tends to reject all of the singular racial categories in favor of "None of the above." A second, weaker trend is for people with one white parent to identify with that group. Because of the uneven row totals, lambda is zero even though there is an association between these variables.

Chapter 16

16.1

Bivariate gamma: 0.71

Partial Gammas		
Controlling for gender	Male	0.78
	Female	0.65
Controlling for origin	Asian	0.67
	Hispanic	0.74

The bivariate relationship is strong and positive. The longer the residence, the greater the facility in English. The bivariate relationship is not affected by the sex or origin of the immigrants. These results would be taken as strong evidence of a direct (causal) relationship between length of residence and facility in English.

16.3 Bivariate gamma: −0.23

Partial Gammas, Controlling for Gender	
Female	−0.34
Male	−0.10

This is an interactive relationship. Length of institutionalization has a greater effect on the reality orientation of females than of males.

16.5 Bivariate gamma: 0.62

Partial Gammas		
Controlling for race	Whites	0.52
	Blacks	0.68
	$G_p =$	0.60
Controlling for gender	Males	0.62
	Females	0.62
	$G_p =$	0.62

The bivariate relationship is strong and positive. Completion of the training program is closely associated with holding a job for at least one year. There is some interaction with race. The training has less impact for whites than for blacks. Blacks who did not complete the training were less likely than whites to have held a job for at least one year. Gender has no impact at all on the relationship. Overall, there is a direct relationship between completion of the training and holding a job for at least a year, though there is some interaction with race.

16.7 Support for the legal right to an abortion:

Bivariate Gamma: −0.14

Partial Gammas, Controlling for Gender	
Males	−0.06
Females	−0.19
$G_p =$	−0.15

Support for the right to suicide:

Bivariate gamma: −0.14

Partial Gammas, Controlling for Gender	
Males	−0.16
Females	−0.11
$G_p =$	−0.13

Chapter 17

17.1 **a.** For turnout (Y) and unemployment (X) while controlling for negative advertising (Z), $r_{yx.z} = 0.95$. The relationship between X and Y is not affected by the control variable Z.

b. For turnout (Y) and negative advertising (X) while controlling for unemployment (Z), $r_{yx.z} = -0.89$. The bivariate relationship is not affected by the control variable.

c. Turnout (Y) = 70.25 + (2.09)unemployment (X_1) + (−0.43)negative advertising (X_2). For unemployment (X_1) = 10 and negative advertising (X_2) = 75, turnout (Y) = 58.09.

d. For unemployment (X_1): $b_1^* = 0.66$. For negative advertising (X_2): $b_2^* = -0.41$. Unemployment has a stronger effect on turnout than negative advertising. Note that the independent variables' effect on turnout is in opposite directions.

e. $R^2 = 0.98$

17.3 **a.** For strife (Y) and unemployment (X), controlling for urbanization (Z), $r_{yx.z} = 0.79$.

b. For strife (Y) and urbanization (X), controlling for unemployment (Z), $r_{yx.z} = 0.20$.

c. Strife (Y) = −14.60 + (4.94)unemployment (X_1) + (0.16)urbanization (X_2). With unemployment = 10 and urbanization = 90, strife (Y') would be 49.19.

d. For unemployment (X_1): $b_1^* = 0.78$. For urbanization (X_2): $b_2^* = 0.13$.

e. $R^2 = 0.65$

17.5 **a.** Turnout (Y) = 83.80 + (−1.16)Democrat (X_1) + (2.89)minority (X_2).

b. For $X_1 = 0$ and $X_2 = 5$, $Y' = 98.25$

c. $Z_y = -1.27Z_1 + 0.84Z_2$

d. $R^2 = 0.51$

17.7 **a.** Z_y = (−0.17)HS grads (Z_1) + (−0.84) Rank (Z_2)

b. $R^2 = 0.50$

Part IV Cumulative Exercises

1. a. The choice of multivariate procedures will depend on the level of measurement and the number of possible scores for each variable. Regression analysis is appropriate for interval-ratio, continuous variables. In this situation, we have three variables: GPA, hours of work, and College Board scores. College Board scores are arguably only ordinal in level of measurement, but the scores have a wide range and seem continuous. GPA is the dependent variable, and the zero-order correlation with "hours worked" is -0.83, a strong and negative relationship, which indicates that having a job did interfere with academic success for this sample. The zero-order correlation between GPA and College Board scores is 0.55, a relationship that is consistent with the idea that College Board scores predict success in college.

The results of the regression analysis with "hours worked" (X_1) and College Board scores (X_2) as independents:

$$Y = 1.71 + (-0.04)X_1 + (0.003)X_2$$

$$Z_y = (-0.75)Z_1 + (0.43)Z_2$$

$$R^2 = 0.86$$

The beta-weights indicate that "hours worked" has a stronger direct effect on GPA than College Board scores. Even for students who were well prepared for college and had high College Board scores, having a job had a negative impact on GPA. The high value for R^2 means that only a small percentage of the variance in GPA (14%) is left unexplained by these two independent variables.

b. In this situation, we have three dichotomous variables, two ordinal level and one nominal. With the limited variation possible, the elaboration technique is appropriate to analyze these relationships. The bivariate table shows that there is a relationship between graduating and level of social activity for this sample: 70% of the students with low levels of social activity graduated in four years vs. only 30% of the students with high levels. Bivariate gamma is 0.69, indicating a strong relationship.

Controlling for sex reveals some interaction. The gamma for males (0.77) is stronger than the bivariate gamma, and the gamma for females (0.60) is weaker. In other words, while level of social activity has an effect on graduation rates, the effect is stronger for males than for females. This suggests that sex should be incorporated into the analysis, with a view to developing an understanding of why it would have different effects for males and females.

Glossary

Each entry includes a brief definition and notes the chapter that introduces the term.

Alpha (α). The probability of error, or the probability that a confidence interval does not contain the population value. Alpha levels are usually set at 0.10, 0.05, 0.01, or 0.001. Chapter 7

Alpha level (α). The proportion of area under the sampling distribution that contains unlikely sample outcomes, given that the null hypothesis is true. Also, the probability of Type I error. Chapter 8

Analysis of variance. A test of significance appropriate for situations in which we are concerned with the differences among more than two sample means. Chapter 10

ANOVA. See **Analysis of variance.** Chapter 10

Association. The relationship between two (or more) variables. Two variables are said to be associated if the distribution of one variable changes for the various categories or scores of the other variable. Chapter 12

Average deviation (AD). The average of the absolute deviations of the scores around the mean. Chapter 4

Bar chart. A graphic display device for discrete variables. Categories are represented by bars of equal width, the height of each corresponding to the number (or percentage) of cases in the category. Chapter 2

Beta-weights (b^*). Standardized partial slopes. Chapter 17

Bias. A criterion used to select sample statistics as estimators. A statistic is unbiased if the mean of its sampling distribution is equal to the population value of interest. Chapter 7

Bivariate normal distributions. The model assumption in the test of significance for Pearson's r that both variables are normally distributed. Chapter 15

Bivariate table. A table that displays the joint frequency distributions of two variables. Chapter 11

Cells. The cross-classification categories of the variables in a bivariate table. Chapter 11

Central Limit Theorem. A theorem that specifies the mean, standard deviation, and shape of the sampling distribution, given that the sample is large. Chapter 6

χ^2(critical). The score on the sampling distribution of all possible sample chi squares that marks the beginning of the critical region. Chapter 11

χ^2(obtained). The test statistic as computed from sample results. Chapter 11

Chi square test. A nonparametric test of hypothesis for variables that have been organized into a bivariate table. Chapter 11

Class intervals. The categories used in the frequency distributions for interval-ratio variables. Chapter 2

Cluster sample. A method of sampling by which geographical units are randomly selected and all cases within each selected unit are tested. Chapter 6

Coefficient of determination (r^2). The proportion of all variation in Y that is explained by X. Found by squaring the value of Pearson's r. Chapter 15

Coefficient of multiple determination (R^2). A statistic that equals the total variation explained in the dependent variable by all independent variables combined. Chapter 17

Column. The vertical dimension of a bivariate table. By convention, each column represents a score on the independent variable. Chapter 11

Column percentages. Percentages computed with each column of a bivariate table. Chapter 12

Conditional distribution of *Y*. The distribution of scores on the dependent variable for a specific score or category of the independent variable when the variables have been organized into table format. Chapter 12

Conditional means of *Y*. The mean of all scores on Y for each value of X. Chapter 15

Confidence interval. An estimate of a population value in which a range of values is specified. Chapter 7

Confidence level. A frequently used alternate way of expressing alpha, the probability that an interval estimate will not contain the population value. Confidence levels of 90%, 95%, 99%, and 99.9% correspond to alphas of 0.10, 0.05, 0.01, and 0.001, respectively. Chapter 7

Continuous variable. A variable with a unit of measurement that can be subdivided infinitely. Chapter 1

Cramer's *V*. A chi square–based measure of association. Appropriate for nominally measured variables that have been organized into a bivariate table of any number of rows and columns. Chapter 13

Critical region (region of rejection). The area under the sampling distribution that, in advance of the test itself, is defined as including unlikely sample outcomes, given that the null hypothesis is true. Chapter 8

Cumulative frequency. An optional column in a frequency distribution that displays the number of cases within an interval and all preceding intervals. Chapter 2

Cumulative percentage. An optional column in a frequency distribution that displays the percentage of cases within an interval and all preceding intervals. Chapter 2

Data. Any information collected as part of a research project and expressed as numbers. Chapter 1

Data reduction. Summarizing many scores with a few statistics. A major goal of descriptive statistics. Chapter 1

Deciles. The points that divide a distribution of scores into 10ths. Chapter 3

Dependent variable. A variable that is identified as an effect, result, or outcome variable. The dependent variable is thought to be caused by the independent variable. Chapters 1 and 12

Descriptive statistics. The branch of statistics concerned with (1) summarizing the distribution of a single variable or (2) measuring the relationship between two or more variables. Chapter 1

Deviations. The distances between the scores and the mean. Chapter 4

Direct relationship. A multivariate relationship in which the control variable has no effect on the bivariate relationship. Chapter 16

Discrete variable. A variable with a basic unit of measurement that cannot be subdivided. Chapter 1

Dispersion. The amount of variety, or heterogeneity, in a distribution of scores. Chapter 4

Dummy variable. A nominal-level variable dichotomized so that it can be used in regression analysis. A dummy variable has two scores, one coded as 0 and the other as 1. Chapter 15

E_1. For lambda, the number of errors of prediction made when predicting which category of the dependent variable cases will fall into while ignoring the independent variable. Chapter 13

E_2. For lambda, the number of errors of prediction made when predicting which category of the dependent variable cases will fall into while taking account of the independent variable. Chapter 13

Efficiency. The extent to which the sample outcomes are clustered around the mean of the sampling distribution. Chapter 7

Elaboration. The basic multivariate technique for analyzing variables arrayed in tables. Partial tables are constructed to observe the bivariate relationship in a more detailed or elaborated format. Chapter 16

EPSEM. The **E**qual **P**robability of **SE**lection **M**ethod for selecting samples. Every element or case in the population must have an equal probability of selection for the sample. Chapter 6

Expected frequency (f_e). The cell frequencies that would be expected in a bivariate table if the variables were independent. Chapter 11

Explained variation. The proportion of all variation in *Y* that is attributed to the effect of *X*. Equal to $\Sigma(Y' - \bar{Y})^2$. Chapter 15

Explanation. See **Spurious relationship.** Chapter 16

***F* ratio.** The test statistic computed in step 4 of the ANOVA test. Chapter 10

Five-step model. A step-by-step guideline for conducting tests of hypotheses. A framework that organizes decisions and computations for all tests of significance. Chapter 8

Frequency distribution. A table that displays the number of cases in each category of a variable. Chapter 2

Frequency polygon. A graphic display device for interval-ratio variables. Class intervals are represented by dots placed over the midpoints, the height of each corresponding to the number (or percentage) of cases in the interval. All dots are connected by straight lines. Same as a **line chart.** Chapter 2

Gamma (*G*). A measure of association appropriate for variables measured with "collapsed" ordinal scales that have been organized into table format; *G* is the symbol for any sample gamma, γ is the symbol for any population gamma. Chapter 14

Goodness-of-fit test. An additional use for chi square that tests the significance of the distribution of a single variable. Chapter 11

Histogram. A graphic display device for interval-ratio variables. Class intervals are represented by contiguous bars of equal width (equal to the class limits), the height of each corresponding to the number (or percentage) of cases in the interval. Chapter 2

Homoscedasticity. The model assumption in the test of significance for Pearson's *r* that the variance of the *Y* scores is uniform across all values of *X*. Chapter 15

Hypothesis. A statement about the relationship between variables that is derived from a theory. Hypotheses are more specific than theories, and all terms and concepts are fully defined. Chapter 1

Hypothesis testing. Statistical tests that estimate the probability of sample outcomes if assumptions about the population (the null hypothesis) are true. Chapter 8

Independence. The null hypothesis in the chi square test. Two variables are independent if, for all cases the classification of a case on one variable has no effect on the probability that the case will be classified in any particular category of the second variable. Chapter 11

Independence random samples. Random samples gathered in such a way that the selection of a particular case for one sample has no effect on the probability that any other particular case will be selected for the other samples. Chapter 9

Independent variable. A variable that is identified as a causal variable. The independent variable is thought to cause the dependent variable. Chapters 1 and 12

Index of qualitative variation (IQV). A measure of dispersion for variables that have been organized into frequency distributions. Chapter 4

Inferential statistics. The branch of statistics concerned with making generalizations from samples to populations. Chapter 1

Interaction. A multivariate relationship wherein a bivariate relationship changes across the categories of the control variable. Chapter 16

Interpretation. See **intervening relationship.** Chapter 16

Interquartile range (*Q*). The distance from the third quartile to the first quartile. Chapter 4

Intervening relationship. A multivariate relationship wherein a bivariate relationship becomes substantially weaker after a third variable is controlled for. The independent and dependent variables are linked primarily through the control variable. Chapter 16

Lambda (λ). A measure of association appropriate for nominally measured variables that have been organized into a bivariate table. Lambda is based on the logic of proportional reduction in error (PRE). Chapter 13

Level of measurement. The mathematical characteristic of a variable and the major criterion for selecting statistical techniques. Variables can be measured at any of three levels, each permitting certain mathematical operations and statistical techniques. The characteristics of the three levels are summarized in Table 1.2. Chapter 1

Line chart. See **Frequency polygon.** Chapter 2

Linear relationship. A relationship between two variables in which the observation points (dots) in the scattergram can be approximated with a straight line. Chapter 15

Marginals. The row and column subtotals in a bivariate table. Chapter 11

Maximum difference. A way to assess the strength of an association between variables that have been organized into a bivariate table. The maximum difference is the largest difference between column percentages for any row of the table. Chapter 12

Mean. The arithmetic average of the scores. $\bar{X}$ represents the mean of a sample, and μ, the mean of a population. Chapter 3

Mean square estimate. An estimate of the variance calculated by dividing the sum of squares within (SSW) or the sum of squares between (SSB) by the proper degrees of freedom. Chapter 10

Measures of association. Statistics that summarize the strength and direction of the relationship between variables. Chapters 1 and 12

Measures of central tendency. Statistics that summarize a distribution of scores by reporting the most typical or representative value of the distribution. Chapter 3

Measures of dispersion. Statistics that indicate the amount of variety, or heterogeneity, in a distribution of scores. Chapter 4

Median (Md). The point in a distribution of scores above and below which exactly half of the cases fall. Chapter 3

Midpoint. The point exactly halfway between the upper and lower limits of a class interval. Chapter 2

Mode. The most common value in a distribution or the largest category of a variable. Chapter 3

Mu(μ). The mean of a population. Chapter 6

μ_p (Mu-sub-p). The mean of a sampling distribution of sample proportions. Chapter 6

$\mu_{\bar{X}}$ (Mu-sub-X). The mean of a sampling distribution of sample means. Chapter 6

Multiple correlation. A multivariate technique for examining the combined effects of more than one independent variable on a dependent variable. Chapter 17

Multiple correlation coefficient (R). A statistic that indicates the strength of the correlation between a dependent variable and two or more independent variables. Chapter 17

Multiple regression. A multivariate technique that breaks down the separate effects of the independent variables on the dependent variable; used to make predictions of the dependent variable. Chapter 17

N_d. The number of pairs of cases ranked in different order on two variables. Chapter 14

N_s. The number of pairs of cases ranked in the same order on two variables. Chapter 14

Negative association. A bivariate relationship where the variables vary in opposite directions. As one variable increases, the other decreases, and high scores on one variable are associated with low scores on the other. Chapter 12

Nonparametric. A "distribution-free" test. These tests do not assume a normal sampling distribution. Chapter 11

Normal curve. A theoretical distribution of scores that is symmetrical, unimodal, and bell shaped. The standard normal curve always has a mean of 0 and a standard deviation of 1. Chapter 5

Normal curve table. Appendix A; a detailed description of the area between a Z score and the mean of any standardized normal distribution. Chapter 5

Null hypothesis (H_0). A statement of "no difference." In the context of single-sample tests of significance, the population from which the sample was drawn is assumed to have a certain characteristic or value. Chapter 8

Observed frequency (f_o). The cell frequencies actually observed in a bivariate table. Chapter 11

One-tailed test. A type of hypothesis test used when (1) the direction of the difference can be predicted or (2) concern focuses on outcomes in only one tail of the sampling distribution. Chapter 8

One-way analysis of variance. Applications of ANOVA in which the effect of a single independent variable on a dependent variable is observed. Chapter 10

P_s. (P-sub-s) Any sample proportion. Chapter 6

P_u. (P-sub-u) Any population proportion. Chapter 6

Parameter. A characteristic of a population. Chapter 6

Partial correlation. A multivariate technique for examining a bivariate relationship while controlling for other variables. Chapter 17

Partial correlation coefficient. A statistic that shows the relationship between two variables while controlling for other variables; $r_{yx.z}$ is the symbol for the partial correlation coefficient when controlling for one variable. Chapter 17

Partial gamma (G_p). A statistic that indicates the strength of the association between two variables after the effects of a third variable have been removed. Chapter 16

Partial slopes. In a multiple regression equation, the slope of the relationship between a particular independent variable and the dependent variable while controlling for all other independent variables in the equation. Chapter 17

Partial tables. Tables produced when controlling for a third variable. Chapter 16

Pearson's r (r). A measure of association for variables that have been measured at the interval-ratio level; ρ is the symbol for the *population* value of Pearson's r. Chapter 15

Percent change. A statistic that expresses the magnitude of change in a variable from time 1 to time 2. Chapter 2

Percentage. The number of cases in a category of a variable divided by the number of cases in all categories of the variable, the entire quantity multiplied by 100. Chapter 2

Percentile. A point below which a specific percentage of the cases fall. Chapter 3

Phi (ϕ). A chi square–based measure of association. Appropriate for nominally measured variables that have been organized into a 2×2 bivariate table. Chapter 13

Pie chart. A graphic display device especially for discrete variables with only a few categories. A circle (the pie) is divided into segments proportional in size to the percentage of cases in each category of the variable. Chapter 2

Point estimate. An estimate of a population value where a single value is specified. Chapter 7

Pooled estimate. An estimate of the standard deviation of the sampling distribution of the difference in sample means based on the standard deviations of both samples. Chapter 9

Population. The total collection of all cases in which the researcher is interested. Chapter 1

Positive association. A bivariate relationship where the variables vary in the same direction. As one variable increases, the other also increases, and high scores on one variable are associated with high scores on the other. Chapter 12

Post hoc test. A technique for determining which pairs of means are significantly different. Chapter 10

Proportion. The number of cases in one category of a variable divided by the number of cases in all categories of the variable. Chapter 2

Proportional reduction in error (PRE). The logic that underlies the definition and computation of lambda. The statistic compares the number of errors made when predicting the dependent variable while ignoring the independent variable (E_1) with the number of errors made while taking the independent variable into account (E_2). Chapter 13

Quartiles. The points that divide a distribution into quarters. Chapter 3

Range (*R*). The highest score minus the lowest score. Chapter 4

Rate. The number of actual occurrences of some phenomenon or trait divided by the number of possible occurrences per some unit of time. Chapter 2

Ratio. The number of cases in one category divided by the number of cases in some other category. Chapter 2

Real class limits. The class intervals of a frequency distribution when stated as continuous categories. Chapter 2

Regression line. The single, best-fitting straight line that summarizes the relationship between two variables. Regression lines are fitted to the data points by the least-squares criterion, whereby the line touches all conditional means of *Y* or comes as close to doing so as possible. Chapter 15

Replication. See **Direct relationship.** Chapter 16

Representative. The quality a sample is said to have if it reproduces the major characteristics of the population from which it was drawn. Chapter 6

Research. Any process of gathering information systematically and carefully to answer questions or test theories. Statistics are useful for research projects in which the information is represented in numerical form or as data. Chapter 1

Research hypothesis (H_1). A statement that contradicts the null hypothesis. In the context of single-sample tests of significance, the research hypothesis says that the population from which the sample was drawn does not have a certain characteristic or value. Chapter 8

Row. The horizontal dimension of a bivariate table, conventionally representing a score on the dependent variable. Chapter 11

Sample. A carefully chosen subset of a population. In inferential statistics, information is gathered from a sample and then generalized to a population. Chapter 1

Sampling distribution. The distribution of a statistic for all possible sample outcomes of a certain size. Under conditions specified in two theorems, the sampling distribution will be normal in shape, with a mean equal to the population value and a standard deviation equal to the population standard deviation divided by the square root of *N*. Chapter 6

Scattergram. Graphic display device that depicts the relationship between two variables. Chapter 15

Σ. "The summation of." Chapter 3

Significance testing. See Hypothesis testing. Chapter 8

Simple random sample. A method for choosing cases from a population by which every case and every combination of cases has an equal chance of being included. Chapter 6

Skew. The extent to which a distribution of scores has a few scores that are extremely high (positive skew) or extremely low (negative skew). Chapter 3

Slope (*b*). The amount of change in one variable per unit change in the other; *b* is the symbol for the slope of a regression line. Chapter 15

Spearman's rho (r_s). A measure of association appropriate for ordinally measured variables that are "continuous" in form; r_s is the symbol for any sample Spearman's rho; ρ_s is the symbol for any population Spearman's rho. Chapter 14

Specification. See **Interaction.** Chapter 16

Spurious relationship. A multivariate relationship in which a bivariate relationship becomes substantially weaker after a third variable is controlled for. The independent and dependent variables are not causally linked. Rather, both are caused by the control variable. Chapter 16

Standard deviation. The square root of the squared deviations of the scores around the mean, divided by *N*. The most important and useful descriptive measure of dispersion; *s* represents the standard deviation of a sample; Σ represents the standard deviation of a population. Chapter 4

σ_{p-p} (sigma-sub-p-minus-p). Symbol for the standard deviation of the sampling distribution of the differences in sample proportions. Chapter 9

$\sigma_{\bar{X}-\bar{X}}$ (sigma-sub-X-minus-sub-X). Symbol for the standard deviation of the sampling distribution of the differences in sample means. Chapter 9

Standard error of the mean. The standard deviation of a sampling distribution of sample means. Chapter 6

Standardized partial slopes (beta-weights). The slope of the relationship between a particular independent variable and the dependent variable when all scores have been normalized. Chapter 17

Stated class limits. The class intervals of a frequency distribution when stated as discrete categories. Chapter 2

Statistics. A set of mathematical techniques for organizing and analyzing data. Chapter 1

Stratified sample. A method of sampling by which cases are selected from sublists of the population. Chapter 6

Student's *t* distribution. A distribution used to find the critical region for tests of sample means when σ is unknown and sample size is small. Chapter 8

Sum of squares between (SSB). The sum of the squared deviations of the sample means from the overall mean, weighted by sample size. Chapter 10

Sum of squares within (SSW). The sum of the squared deviations of scores from the category means. Chapter 10

Systematic sampling. A method of sampling by which the first case from a list of the population is randomly selected. Thereafter, every *k*th case is selected. Chapter 6

***t*(critical).** The *t* score that marks the beginning of the critical region of a *t* distribution. Chapter 8

***t*(obtained).** The test statistic computed in step 4 of the five-step model. The sample outcome expressed as a *t* score. Chapter 8

Test statistic. The value computed in step 4 of the five-step model that converts the sample outcome into either a *t* score or a *Z* score. Chapter 8

Theory. A generalized explanation of the relationship between two or more variables. Chapter 1

Total sum of squares (SST). The sum of the squared deviations of the scores from the overall mean. Chapter 10

Total variation. The spread of the *Y* scores around the mean of *Y*. Equal to $\Sigma(Y - \bar{Y})^2$. Chapter 15

Two-tailed test. A type of hypothesis test used when (1) the direction of the difference cannot be predicted or (2) concern focuses on outcomes in both tails of the sampling distribution. Chapter 8

Type I error (alpha error). The probability of rejecting a null hypothesis that is, in fact, true. Chapter 8

Type II error (beta error). The probability of failing to reject a null hypothesis that is, in fact, false. Chapter 8

Unexplained variation. The proportion of the total variation in *Y* that is not accounted for by *X*. Equal to $\Sigma(Y - Y')^2$. Chapter 15

Variable. Any trait that can change values from case to case. Chapter 1

Variance. The squared deviations of the scores around the mean divided by *N*. A measure of dispersion used primarily in inferential statistics and also in correlation and regression techniques; s^2 represents the variance of a sample; σ^2 represents the variance of a population. Chapter 4

***X*.** Symbol used for any independent variable. Chapter 12

X_i ("*X* sub *i*"). Any score in a distribution. Chapter 3

***Y*.** Symbol used for any dependent variable. Chapter 12

***Y* intercept (*a*).** The point where the regression line crosses the *Y* axis. Chapter 15

Y'. Symbol for predicted score on *Y*. Chapter 15

***Z*.** Symbol for any control variable. Chapter 16

***Z* scores.** Standard scores; the way scores are expressed after they have been standardized to the theoretical normal curve. Chapter 5

***Z*(critical).** The *Z* score that marks the beginnings of the critical region on a *Z* distribution. Chapter 8

***Z*(obtained).** The test statistic computed in step 4 of the five-step model. The sample outcomes expressed as a *Z* score. Chapter 8

Zero-order correlations. Correlation coefficients for bivariate relationships. Chapter 17

Credits

Chapter 1, page 15: System suggested by Michael G. Bisciglia, Louisiana State University. **Chapter 3, page 82:** NationMaster.com. **Chapter 4, page 103:** McCarthy, Bill, Felmlee, Dine, and Hagan, John. 2004. "Girl Friends are Better: Gender, Friends, and Crime among School and Street Youth". Criminology:42:805-835. **Chapter 7, page 173:** Gender Differences in the Prevalence of Same-Sex Partnering: 1988–2002 by Amy Butler from SOCIAL FORCES, Volume 84, No. 1. Copyright (c) 2005 by the University of North Carolina Press. Used by permission of the publisher. www.uncpress.unc.edu. **Chapter 9, page 222:** "Residential Mobility and Adolescent Violence" by Dana Haynie and Scott South from SOCIAL FORCES, Volume 84, No. 1. Copyright (c) 2005 by the University of North Carolina Press. Used by permission of the publisher. www.uncpress.unc.edu. **Chapter 10, page 251:** Ruffolo, Mary, Sarri, Rosemary, and Goodkind, Sara. 2004. "Study of Delinquent, Diverted, and High-Risk Adolescent Girls: Implications for Mental Health Intervention." SOCIAL WORK RESEARCH: 28, pp. 237–245. Copyright 2004, National Association of Social Workers, Inc., Social Work Research. **Chapter 11, page 279:** Ruffolo, Mary, Sarri, Rosemary, and Goodkind, Sara. 2004. "Study of Delinquent, Diverted, and High-Risk Adolescent Girls: Implications for Mental Health Intervention." SOCIAL WORK RESEARCH: 28, pp. 237–245. Copyright 2004, National Association of Social Workers, Inc., Social Work Research. **Chapter 12, page 303:** Bullying and Victimization: Prevalence and Relationship to Gender, Grade Level, Ethnicity, Self-Esteem and Depression by Dorothy Seals and Jerry Young from ADOLESCENCE, Vol. 38, 2003, pp. 735–747. Reprinted by permission of Elsevier. **Chapter 12, page 305:** Cause and Effect Rule by Leonard. **Chapter 12, page 305:** Racial and Gender Report Card by Lapchick. **Chapter 14, page 351:** Militarist, Marxian and Non-Marxian Materialist Theories of Gender Inequality: A Cross-Cultural Test by Stephen Sanderson, D. Alex Heckert, and Joshua Dubrow from SOCIAL FORCES, Vol. 84, No. 1. Copyright (c) 2005 by the University of North Carolina Press. Used by permission of the publisher. www.uncpress.unc.edu. **Chapter 15, page 387:** NationMaster.com. **Chapter 17, page 441:** The Racial Divide in Support for the Dealth Penalty: Does White Racism Matter? by James Unnever and Francis Cullen from SOCIAL FORCES, Volume 84, No. 1. Copyright (c) 2005 by the University of North Carolina Press. Used by permission of the publisher. www.uncpress.unc.edu.

Index

Note: Page numbers followed by "n" refer to footnotes.

GLOSSARY OF SYMBOLS

The number in parentheses indicates the chapter in which the symbol is introduced.

a	Point at which the regression line crosses the *Y* axis (15)
ANOVA	The analysis of variance (10)
b	Slope of the regression line (15)
b_i	Partial slope of the linear relationship between the *i*th independent variable and the dependent variable (17)
b_i^*	Standardized partial slope of the linear relationship between the *i*th independent variable and the dependent variable (17)
df	Degrees of freedom (8)
f	Frequency (2)
F	The *F* ratio (10)
f_e	Expected frequency (11)
f_o	Observed frequency (11)
G	Gamma for a sample (14)
G_p	Partial gamma (16)
H_0	Null hypothesis (8)
H_1	Research or alternate hypothesis (8)
IQV	Index of qualitative variation (4)
Md	Median (3)
Mo	Mode (3)
N	Number of cases (2)
N_d	Number of pairs of cases ranked in different order on two variables (14)
N_s	Number of pairs of cases ranked in the same order on two variables (14)
%	Percentage (2)
P	Proportion (2)
P_s	A sample proportion (7)
P_u	A population proportion (7)
PRE	Proportional reduction in error (13)
Q	Interquartile range (4)
r	Pearson's correlation coefficient for a sample (15)
r^2	Coefficient of determination (15)
R	Range (4)
r_s	Spearman's rho for a sample (14)
$r_{xy.z}$	Partial correlation coefficient (17)
R^2	Multiple correlation coefficient (17)
s	Sample standard deviation (4)
SSB	The sum of squares between (10)
SST	The total sum of squares (10)
SSW	The sum of squares within (10)
s^2	Sample variance (4)
t	Student's *t* score (8)
V	Cramer's *V* (13)
X	Any independent variable (12)
$\overline{X}$	Mean of a sample (3)
X_i	Any score in a distribution (3)
Y	Any dependent variable (12)
Y'	A predicted score on *Y* (15)
Z scores	Standard scores (5)
Z	A control variable (16)